T0212400

Springer Biographies

The books published in the Springer Biographies tell of the life and work of scholars, innovators, and pioneers in all fields of learning and throughout the ages. Prominent scientists and philosophers will feature, but so too will lesser known personalities whose significant contributions deserve greater recognition and whose remarkable life stories will stir and motivate readers. Authored by historians and other academic writers, the volumes describe and analyse the main achievements of their subjects in manner accessible to nonspecialists, interweaving these with salient aspects of the protagonists' personal lives. Autobiographies and memoirs also fall into the scope of the series.

More information about this series at http://www.springer.com/series/13617

Deri Sheppard

Robert Le Rossignol

Engineer of the Haber Process

 Springer

Deri Sheppard
Cowbridge, Vale of Glamorgan, UK

ISSN 2365-0613   ISSN 2365-0621   (electronic)
Springer Biographies
ISBN 978-3-030-29716-9   ISBN 978-3-030-29714-5   (eBook)
https://doi.org/10.1007/978-3-030-29714-5

This Springer imprint is published by the registered company Springer Nature Switzerland AG
The registered company address is: Gewerbestrasse 11, 6330 Cham, Switzerland

Never be a spectator of unfairness … the grave will supply enough time for silence.
Christopher Hitchens, 'Letters to a Young Contrarian', April 2005.

For my grandchildren,
Benjamin Fionn and Eiriol Gwenllian.
Gyda chariad, Tad-cu.

Preface

In the abstract to a recent paper for the Royal Society in London,[1] this author described Le Rossignol's long life in the following, 'matter-of-fact', terms.

In March 1908, the BASF at Ludwigshafen provided financial support to Fritz Haber in his attempt to synthesise ammonia from the elements. The process that now famously bears his name was demonstrated to BASF in July 1909. However, its engineer was Haber's private assistant, Robert Le Rossignol, a young British chemist from the Channel Islands with whom Haber made a generous financial arrangement regarding subsequent royalties. Le Rossignol left Haber in August 1909 as BASF began the industrialisation of their process, and took a consultancy at the Osram works in Berlin. He was interned briefly during WWI before being released to resume his occupation. His position eventually led to His Majesty's Government formulating a national policy regarding released British internees in Germany. After the war, Le Rossignol spent his professional life at the GEC laboratories in the UK, first making fundamental contributions to the development of high-power radio transmitting valves, then later developing smaller valves used as mobile power sources in the airborne radars of WWII. Through his share of Haber's royalties, Le Rossignol became wealthy. In retirement, Le Rossignol and his wife gave their money away to charitable causes.

As befits an abstract, such a description only provides the outline of the paper, and although the body naturally continued to elaborate on many areas of Le

---

[1]Deri Sheppard, 'Robert Le Rossignol, 1884–1976; Engineer of the Haber Process', *Notes and Records*, **71**, 3, 263–296, (September 2017). Published online, **25 January 2017**. https://doi.org/10.1098/rsnr.2016.0019.

Rossignol's life, its focus—Le Rossignol as engineer—meant that many aspects of his humanity were left unexplored. This biography provides a first comprehensive contribution to the literature regarding Le Rossignol. His story is the 'missing link' in the history of the discovery of nitrogen 'fixation', a history which has undoubtedly been dominated by Fritz Haber. The omission has long been recognised by historians, but in writing this book, I have found it impossible to 'decant' Haber from Le Rossignol's life, the two men were partners and friends, and so aspects of their lives are often interwoven throughout.

Written in three parts, Part I covers the years from 1884 to 1909 and contains eight chapters. The first two describe the early lives of the two men, drawing parallels and contrasts relevant to the development of their professional and personal friendships—the latter lasting almost 30 years. Another chapter explains Germany's commitment to, and her dependence upon, the nitrate trade, whose perceived demise at the time provided the *reason d'être* for the creation of a 'fixation' technology. Yet another chapter describes the qualifications, experiences and interests Le Rossignol gathered before travelling to Karlsruhe. These in turn provide some insight as to why he was so well regarded by Haber, and so well placed to provide a solution to the problem of 'fixation'. The remaining four chapters are dedicated to the chemistry these men performed together, and the contracts Haber entered into with both BASF and Le Rossignol. With regard to the chemistry, I have re-visited their original papers of the period, and explained the work they performed together, revealing the boundaries between each man's contribution, so there can be no future misunderstanding of the 'division of work' these men undertook in partnership. The last chapter of the four explains the engineering Le Rossignol performed at Karlsruhe and his subsequent move to Osram in Berlin. Part I contains much archive material never published before. Some of the chapters are technical but they are supported by Appendices explaining thermodynamic theory if readers are so inclined to understand the latter rather than simply accept it.

Part II covers the years from 1909 to 1918 and contains a further six chapters. The first four of these describe Le Rossignol's move to Berlin and the nature of the work he did there. They continue with Haber's arrival at the Kaiser Wilhelm Institute in Berlin in 1911, the industrialisation of Le Rossignol's engineering by Carl Bosch, the 'patent wars' between BASF and rival firms upon which Haber's royalties depended, and the prelude to the First World War. These are relatively short chapters. The remaining chapters describe the wartime experiences of the two men, once again drawing attention to those aspects of Haber's life that were to impinge upon Le Rossignol. For example, Haber's membership of the 'Nitrate Commission' and his support of BASF's ammonia synthesis that maximised his royalties, his involvement in gas warfare, a 'toxic brand' that came to dominate his memory, the suicide of his wife Clara—a parallel tragedy being suffered later by Le Rossignol where chemistry again played a part—and the devastation of Europe laid firmly at the door of Germany's industrialisation of the 'Haber' process. The final chapter describes Le Rossignol's war, his internment and his release to Berlin, the work he did there during the war and the concerns of His Majesty's Government in the UK regarding this work. The chapter concludes by discussing Le Rossignol's financial arrangements with Haber, and the considerations of the Swedish Academy regarding the 1918 Nobel Prize for Chemistry.

Part III covers the years from 1919 until Le Rossignol's death in 1976. It contains five chapters. The first of these covers the years from 1919 to 1930, including Le Rossignol's return from Germany, his unexpected interest in airships, the creation of the GEC Laboratories, the migration of the technology of 'fixation' out of Germany to France, the UK and the USA, the 'engineered' award of an individual Nobel Prize to Haber, the furore which followed the award, Le Rossignol's involvement with the cooled anode transmitting (CAT) valve and his contributions to that technology over the decade. During this time, Le Rossignol met Haber on a number of occasions both in Germany and the UK, and the chapter also discusses what we know of their financial arrangement which ended in 1924. The next two chapters cover the years from 1930 to 1949. The deaths of Le Rossignol's sons Peter Walter and John Augustin are dealt with here, along with further developments of CAT valves, Le Rossignol's contribution(s) to the BBC Droitwich super-broadcasting station which opened in 1934 and his role at the GEC during the Second World War. The latter revealing how the British government was intimately bound to its scientific and industrial base during the war. The final two chapters deal with Le Rossignol's retirement with his wife Emily at Penn, Buckinghamshire, their remarkable benevolence, and the memories of family and others who can still recall the couple's time there.

Cowbridge, Vale of Glamorgan, UK   Deri Sheppard

Acknowledgements

In September 2005, my wife and I spent 2 weeks travelling the eastern seaboard of the USA acquainting ourselves with recently discovered family members whose forebears had emigrated from the UK at the turn of the last century. Having recently retired from my university lectureship in computer science, but being first-and-foremost a physical chemist—by training and inclination—I decided to use some of my 'down time' on the trip to 'gently' re-visit physical chemistry, not of course to keep up with modern developments, but to better understand the lives of some of those who had first defined the subject and attracted me to it.

One quiet evening with family in their elegant, traditional 'New England' home, I began reading Daniel Charles' biography of Fritz Haber.[2] For my generation, Fritz Haber was the 'stand out' physical chemist of his day. Many of us in the UK were of course familiar with the Haber process, the Haber-Bosch process and the Born-Haber cycle. Some were also familiar with Haber's contribution to electrochemistry. But few of us were familiar with the story of his life, he seemingly being a 'forgotten man' after his death in 1934.

On page 87 of Charles' book, I read the following passage:

First, an extraordinary scientist joined Haber's laboratory, a young Englishman named Robert Le Rossignol, with a gift for solving practical problems of engineering. Haber did not know it yet, but the task that lay before him would

[2]Daniel Charles, Between Genius and Genocide, Jonathan Cape, London, (2005). ISBN-0-224-06444-4.

require ingenious design of new experimental equipment. In this, Le Rossignol proved a master …

It was clear that Le Rossignol had a profound influence on the evolution of the 'Haber process', but I was amazed that, in my 'previous life', I had never heard of this fellow 'Brit'. Back home, an exhaustive search on the Internet revealed an obituary of Le Rossignol written in 1977 by friend and colleague Ralph Chirnside, but otherwise a dearth of biographical evidence. Robert *who*? was clearly a question that needed attention. So, intrigued by the lines from Charles' book, and guided by Chirnside's 'road map' to Le Rossignol's life, I embarked on a biographical Odyssey to find out more of the man who helped create the most influential technology of the twentieth century. My first acknowledgments are therefore to Daniel Charles and Ralph Chirnside. Had I never encountered the work of the former, Le Rossignol's life would still be an enigma. Had I never encountered the work of the latter, the biographical effort involved may well have been beyond me. For after only a short time into my Odyssey, I realised that Le Rossignol had left little of himself behind and that which *was* left was fragmented and widely distributed. Every new 'nugget' of information regarding this man had to be 'hunted down' down.

It follows, therefore, that this book would have been impossible for me to write if it were not for the Internet and its 'world wide web' of accessible digitised material. The various instantaneous communication systems of the Internet allowed this 'virgin' biographer to effortlessly contact academics, members of the public, libraries and archives, charities, newspapers, public record offices and other concerned organisations, worldwide. The efforts of thousands of workers in digitising billons of photographs, newspapers, scientific papers, textbooks, certificates of births, deaths and marriages and compiling online resources such as 'Wikipedia', built an unparalleled digital 'honey pot', made sensible by the 'search engines', whose software 'incisions' cut through the 'fog of big data'. My journey across the web introduced me to many people, far too many to thank on an individual basis, but where their efforts have resulted in an element included in the book, such as a photograph or a letter, I have recorded this in the legend. Needless to say, many people who have helped me will go unrecognised.

But others I can recognise. Thanks, therefore, go to Professor Alwyn Davies FRS at UCL Chemistry Department for originally suggesting I write a paper for the Royal Society, for materials he has supplied and for his continued interest in the Le Rossignol story. To Professor Margit Szöllözi-Janze, University of Munich, for answering my many questions regarding her biography of Haber. To the Le Rossignol family for fully engaging with this work, to Hede Hauser for her expert translations of various texts, to Mr. Christopher White, former Chairman of the Penn-Pennsylvania Trust for materials supplied, his knowledge and his enthusiasm and to Mr. T. B. Powell for many hours of discussion regarding Le Rossignol. Thanks also to Dr. Ben Marsden and the Royal Society editorial team for their expert guidance during the submission of my paper, the eventual publication of which was hugely influential in promoting an interest in the full biography.

Thanks also to Professors Bretislav Friedrich and Gerard Meijer at the Fritz-Haber-Institut der Max-Planck-Gesellschaft, Berlin for their interest in the Le Rossignol story and their invitations to speak to the Haber family and friends in Berlin on 16 May 2017, and later at the sesquicentennial of Haber's birthday at Harnack House, Berlin on 10 December 2018. And, to Professor Friedrich in particular, thank you for kindly introducing my work to Springer Publishing. I must also mention my editors (in chronological order!), Sabine Lehr and Angela Lahee, whose guidance, patience and understanding made the preparation of the text as stress free as possible.

The issue of copyright ownership regarding material published in this book has been a major concern for both Springer and myself. Consequently, strenuous efforts have been made to trace ownership wherever possible, and to attribute such material accordingly. In most cases, I have succeeded, but sometimes—where material dates back to the turn of the last century, appearing on the web as 'copyright unknown'—it has proved impossible. Such material has only been included when it has direct relevance to the story.

Finally, becoming a biographer has long been a dream for this author, but as a novice, I could hardly have picked a more difficult subject in Robert Le Rossignol! Over the 8 years or so it took to research and write his story, everyone involved played their part. I thank each one of you, whether recognised or not, for helping make a dream come true.

Cowbridge, UK   Deri Sheppard
2020

Contents

Part I

The Story of the Discovery of 'Fixation'

Part I

1

Robert Le Rossignol

Robert Le Rossignol.
Courtesy of the Le Rossignol family and the Royal Society of
Chemistry, London.

The early years, Victoria College, UCL and Ramsay.

To begin, at the beginning. It is spring ...

The opening lines from Dylan Thomas', 'Under Milk Wood', 1954.

1.1 Introduction

In Karlsruhe, southern Germany, there is a monument. A twelve-metre steel tube pointing skywards at the intersection of Engesserstraße and Fritz Haber-Weg. Rising among the buildings of the university, it commemorates the creation of the most influential technology of the 20th century, a technology to which many people on the planet today owe their existence. The monument is an old ammonia reactor belonging to the nearby BASF chemical works at Ludwigshafen and it stands in celebration of the 'fixation'

© Springer Nature Switzerland AG 2020
D. Sheppard, *Robert Le Rossignol*, Springer Biographies,
https://doi.org/10.1007/978-3-030-29714-5_1

of nitrogen.[1] 'Fixation' was conceived to nurture life. But when young, and growing innocently towards its destiny, it was 'radicalised' to support a munitions industry that has since sustained countless wars and conflicts.

The monument also stands as testament to the work of the men who discovered the technology, the 'fathers of fixation'. One of these was Fritz Jakob Haber; 'chemist, Nobel laureate, German, Jew'.[2] Even today, over three quarters of a century after his death, the reactor is sometimes found daubed in blood-red graffiti spelling the word '*Mörder*' (Murderer) as young German students recall Haber's role in the first World War.[3] Fritz Haber's legacy has polarised opinion in his own country. On the one hand 'the good German', on the other the 'father of chemical warfare', and whichever side we fall it is undeniable that his story has dominated the history of 'fixation'.

But there is another for whom this monument should equally stand. A young British 'Channel Islander'[4] whose work in partnership with Haber made a profound contribution to the discovery of 'fixation'. He was Robert Le Rossignol, and this is his story. It is the story of a kindly man with a simple uncomplicated philosophy of life. The grace of his nature was such that he rarely had a bad word to say of anyone. He was a man who walked alongside the scientific giants of his day. He knew triumph, but also deep tragedy which he suffered with a quiet dignity. And although his 'fingerprints' are to be found everywhere in the history of the discovery of the technology, history has abandoned his story, painting him in the shadows on a broad canvas of Haber's life. We cannot decant Haber from Le Rossignol, the two men were partners and friends, and what impacted Haber often shook Robert too. But if we take our time and gaze deeply enough, standing in the shadows we find another life, and another quite remarkable man. So, who was Robert Le Rossignol?

1.2   Early Years and School Days

Robert Le Rossignol was born in springtime, on 27 April 1884 at 17 David Place, St. Helier, Jersey. Robert was the third son of Augustin and Edith and the youngest of their four children (Plate 1.1).

When Robert was born his family had lived in Jersey for almost four hundred years but a common misconception[5] is that they were of Huguenot descent. 'Huguenot',[6] was the name given in 16th century catholic France to the protestant Calvinist minority which had penetrated all ranks of society,

Plate 1.1 17 David Place, St. Helier, today. Photograph courtesy of Google Earth Pro, street view, 2019

especially the nobility and the literate craftsmen. Given religious liberty by the 'Edict of Nantes', the Huguenots eventually established themselves as loyal subjects of the Crown. Later under Louis XIV, their position became insecure because they were increasingly perceived as a threat to the authority of the monarch. Gradually, their privileges were eroded and in 1685 Louis exiled all protestant pastors, at the same time forbidding the laity to leave. To the surprise of the government however, many did leave, often at great risk. Eventually, about 200,000 Huguenots escaped, settling largely in non-catholic Europe. About 50,000 came to Great Britain and the Channel Islands, with perhaps about 10,000 eventually moving on to Ireland. So, there are many inhabitants of these islands who have Huguenot blood in their veins, whether or not they still bear one of the hundreds of French names of those who took refuge here—incidentally, bringing the word 'refugee' into the English language. The link between the Huguenots and Jersey is strong. The Le Bailly's for example were Huguenots and they fled to Jersey in about 1749. Their story[7] has been handed down there, in various forms, and illustrates the dangers families faced when fleeing France.

Thomas Le Bailly, dressed as a peasant,[8] took a horse with a pannier of apples on each side, the Bible in one pannier, his baby son in the other, his wife behind him, and went through the gates of Caen to escape in a fishing boat to Jersey. As they passed through the gates a suspicious guard thrust his sword into one of the panniers, and struck the one with the Bible—not the child!

By now, as 'enemies' of France, the Huguenots were welcomed here. However, they could not escape the accusations always levelled at immigrants —that their presence threatened jobs, standards of housing, public order, morality, hygiene and that they ate strange foods! For at least half a century they remained a recognisable minority, but they made their presence felt in many professions such as banking, commerce, industry, the book trade, the arts, the army, the stage and teaching. Although many retained their Calvinist organisation and worship, by about 1760 they had ceased to stand out as 'foreign', even following the path of Anglican conformity in religion which some had taken from the very beginning.[6]

The Huguenot story 'hangs well' on the Le Rossignol family, and many Huguenots named 'Le Rossignol' may well have arrived in Jersey via the French exodus. A family of Huguenot provenance was one of French lineage —possibly nobility—of protestant heritage, of literacy and a history of hundreds of years. Much of this also fits Robert's family, but that branch of the 'Le Rossignols' to which Robert belonged had an entirely different provenance. Family members[9] who still in live in Jersey vigorously dispute the Huguenot line and point to their family tree in S. J. Le Rossignol's *Historical Notes with special reference to the Le Rossignol Family*,[10] published in 1917. The book begins with the following quotation from Falle's *History of Jersey*[11];

> In the Island there are many ancient families, not only among the Seigneurs and Gentlemen of the first rank, but even amongst those of inferior quality, several of whom can reckon a descent which in some other countries very good gentlemen would be proud of.

Indeed, Robert's family were both ancient and of 'proud descent'. Their name, 'Le Rossignol'—literally 'the nightingale'—clearly establishes their French provenance, the prefix 'Le' often being regarded as 'aristocratic' in Jersey. G. R. Balleine[12] in *Some Jersey Surnames their Origin and Meaning* classifies this name as 'descriptive', originally being applied to someone who may have been a 'sweet singer', or indeed a 'chatterer', for 'rossignolerie' in old French meant 'chattering'. According to the *Historical Notes*[10] the earliest recorded Le Rossignol of Robert's family—Guillaume or 'Guille' ('William'), appeared in Jersey around 1500, way before any Huguenot exodus and well before the first recorded influx of French Huguenots in 1548. Settling around St. Ouen in the west of the island, Guille may have been related to a 'black sheep' member of a family of minor French nobility on the borders of Brittany and Normandy (possibly 'William' Le Rossignol, who appeared about 1480), and down through the generations this branch of the Le Rossignols—just like the Huguenots—achieved a powerful presence in Jersey and indeed internationally, with family members prominent in banking, the judiciary, medicine, foreign service, business, academia and the military.

Mistaken provenance or not, what is certain is that Robert Le Rossignol was born into an established, anglicised, professional, protestant family. His father Augustin bore the given name of many generations of 'Le Rossignols'. Born in 1842,[10] he was educated firstly in Jersey, but subsequently in France at Rennes and Coutances, obtaining the French Bachelor of Science (B. és Sc.) in 1862. He then entered the London Hospital as a student of medicine becoming a Member of the Royal College of Surgeons in 1866 and gaining his Licentiate of the Royal College of Physicians in 1867. He took his M.D. Aberd at Marischal College Aberdeen in 1868, a choice which Robert later described as predicated on the fact that at the time there were just two Universities in England but *four* old Universities in Scotland.[23] He subsequently established a successful medical practise in St. Helier later the same year. Robert's mother Edith[13] (*née* Sorel, b. 1845), as befitted the wife of a professional man in Victorian times, ran the household.

Robert's birth would have brought great joy to Augustin and Edith, but for those islanders beyond the immediate Le Rossignol family, 1884 was an unremarkable year, 'enlivened' only by a visit from the UK President of the Board of Trade and a particularly severe winter's storm. The Reverend Alban E. Ragg describes aspects of Jersey at the time of Robert's birth.[14]

> The population of the island … was about 53,000, nearly 30,000 of whom resided in St. Helier. Its exports during the year were … bulls, 93; cows and heifers, 1,516; total, 1,609. Butter, 958 cwt. Fruit (raw) of all kinds, 19,613 cwt. and potatoes, 49,296 tons … 1884 was enlivened by a visit to the Island of the Right Hon. J. Chamberlain, M.P., then President of the Board of Trade, accompanied, in the Trinity House yacht Galatea, by Sir W. Vernon Harcourt, on June 3rd; The close of the year, however, proved disastrous, in that the Island was swept with one of the most terrific storms it has experienced, with resultant damage, not to houses and shipping alone, but to the coast and harbour works generally, accompanied on December 20th with the loss of the

Guernsey trader Echo off the Corbiere, when six lives were lost, and on December 22nd by a breach in the pier at Greve-de-Lecq.'

Robert's father was a pillar of Jersey's society.[10] As M.D., Augustin held a highly prestigious position and naturally participated in the intellectual and social life of the Island. For example, in 1872 he took a Commission in the Town Regiment eventually retiring with the rank of Lieut-Colonel Surgeon. He was a founder member of the Societé Jersiaise,[15] established in January 1873 for the study of Jersey's archaeology, history, language and the conservation of the environment. He was a member from 1874 to 1910, honourary Secretary from 1881 to 1896, and President in 1897. During his career he became Honourary Surgeon to the local general hospital, and to the Jersey Dispensary. He was Medical Officer of the Jersey Orphan's Home for Girls and Inspector General of the Medical Staff at the Jersey Royal Militia. He was also a founder of the Jersey Medical Society; he acted as its secretary for some years and ultimately became its president. On 21 March 1903, he was elected a 'Jurat' of the Royal Court of the island. (through French from mediaeval Latin, jurat, 'he swears,' the name given to a public official, in this case a magistrate). He was also an 'old Victorian', the name given to former pupils of Victoria College St. Helier, the island's only public school founded in 1852. In the same year as he became Jurat, he also became a member of the College committee (Plate 1.2).

Jersey at the time of Robert's birth was rural with farming and fishing the primary industries. For the ordinary population wages were low, £20 was about the annual income for the unskilled working class in relatively secure positions. Robert's provenance therefore ensured he enjoyed a privileged young life and the Le Rossignol home(s) reflected this. At the time of Robert's birth in David Place, in addition to Augustin, Edith, and siblings, Austen Clement (b. 27 October 1878), Herbert, ('Bertie'), Sorel (b. 07 November 1879) and Elsie Edith (b. 02 March 1881)[16] this busy little household included Margaret Brophy an Irish cook, Anne Billot and Anne Mollet—both quite elderly nurses—and Mary Ulrick a young housemaid. At the time David Place was a hub of Jersey commerce—many of the congregation in the local church of St. Mark's being described as 'well to do'. Today, David Place remains 'gentrified' with many medical and dental practises together with a still flourishing business community including many hotels. Even when Robert was 17 and then living at nearby Caesarea Place[17] St. Saviours Road, the household still employed servants. Rachel Jane Le Richie and Anne Louisa Way were nurse-domestic and housemaid respectively. By then however, Austen had left to study at Exeter College Oxford, 'Bertie' was employed as a

Plate 1.2 A magnificent photograph of Robert's father, Jurat Augustin Le Rossignol, circa 1903. Photograph courtesy of the Société Jersiaise.[15]

banker's Clerk at the Capital and Counties Bank in St. Helier, whilst Robert was shortly to leave for University College London.[10]

The education of all three Le Rossignol sons followed a familiar pattern viz., that of their father's at Victoria College.[18] In August 1847 the States of Jersey decided to purchase grounds from W. Le Breton, Esq. for £5,070 to erect a building capable of accommodating 300 pupils, using a financial provenance that dated back to Charles II who left land to Exeter, Pembroke, and Jesus Colleges at Oxford for the benefit of scholars from the Channel Islands. Victoria College (for boys only) opened on 29 September 1852 with the school motto *Amat Victoria Curam* (loosely, 'Victory favours those who take pains'). Augustin, then aged 10, entered as part of the first intake of 109 pupils, with the Victoria College 'index number' (VC_i) of 73—a number through which all previous pupil's provenance can still be traced.[18] Although French was the official language in Jersey, the new college was consciously patterned on the English public schools. Anglicisation of the Island was an objection raised to the building of the college, but the Bailiff of Jersey at the prize day on 31 July 1901 stated unequivocally, 'Well it has anglicised the Island, and in doing so it has done an excellent thing!'. From the beginning then, the medium of instruction here was English and this was one of the causes for the subsequent decline in the use of French, as Jersey's élite families sent their sons to the new college. The decline was compounded by the fact that the college eventually assimilated the existing grammar schools of St. Mannelier and St. Anastase which had previously provided quality education for scholars from the east and west of the island respectively.

Dr. W. G. Henderson was appointed as the College's first Principal. He was just 34 years old and he was Principal during Augustin's stay, during the course of which he laid the foundations of the future prosperity of the College. Originally, the school was divided into Classical and Commercial sides. In 1855 a grant of £50 per year was made out of crown revenues for the establishment of a (Channel Island) 'Exhibition' (scholarship) to the University of Oxford. The Natural Sciences were introduced to the school in the mid 1860s but they struggled for a place in the curriculum. A laboratory was started but it met with little support, and by 1881 Natural Sciences were briefly discontinued only to be re-introduced shortly afterwards as the school tried to move away from the classics to a wider outlook. Even so, as time went on, the teaching at the College became rather stereotyped; many of the staff were getting on in years, and the demand for a more up-to-date education became more insistent. A committee was appointed to enquire into the administration of the College and subsequently, in 1892 G. S. Farnell—educated at the City of London School and Wadham College Oxford—was appointed to take the college further. Farnell became the College's fifth Head. He was an able scholar but this ability was firmly rooted in the classics and although determined to build on previous progress, this provenance may not have benefited the development of the sciences had he progressed. His career however, was terminated by an unfortunate accident. In November 1895 the College had a half-term holiday. Farnell went out to Plemont on the ruggedly beautiful north east coast of the island to join two of his staff who had started earlier. The day was foggy, and Farnell, who was rather short-sighted, in searching for his colleagues, missed his footing on the cliffs and was found dead at their base. Subsequently, in January 1896, L. V. Lester-Garland, Fellow of St. John's College, Oxford, became Principal. Lester-Garland's stewardship of the College lasted until 1911 and it would have influenced the education of all three Le Rossignol children[19] but none more profoundly than Robert (Plates 1.3, 1.4).

Like his predecessors, Lester-Garland was strongly in favour of classics as the foundation of education, but equally he had long been pressing for the development of the natural sciences in the College. His insistence eventually led to a new block of buildings containing laboratories and classrooms which

Plate 1.3 An early engraving of Victoria College, published in the Illustrated London News, 29 September 1852

Plate 1.4 Left to right, G. S. Farnell and L. V. Lester-Garland, headmasters of Victoria College during Robert's stay.[10]

opened in 1911, but this came far too late for Robert. Even so Robert had developed an early interest in chemistry. It was the inspired appointment of William H. Pryce Jones as Science Master[20] by Lester-Garland in 1898, and encouragement by an elder brother—certainly Austen, that fostered Robert's interest.

The difficulties that the natural sciences had encountered in the college—electric light and power were not available to the laboratories until 1931—meant that the proper practice of science must have been difficult. Robert therefore would have shown considerable initiative to have flourished. However, any inadequacies of Victoria College were more than compensated by the kindness of Frederick Woodland Toms, Official Analyst to the States of Jersey, who permitted Robert to carry out exercises in his laboratory—a kindness that allowed him to acquire great skill in practical chemistry while still a schoolboy, a skill which remained with him throughout his life. Woodland Toms and Robert were acquainted through Robert's father Augustin,[23] who—like Woodland Toms—not only would have moved in the same elevated social circles of Jersey's 'great and good', but as Woodland Toms' son Humphrey was in medical practice in England, they would have shared a common experience. Understanding the work of Woodland Toms provides an insight into the influence he must have had on young Robert. What follows is largely due to Bernard Dyer's Obituary of Frederick Woodland Toms which appeared in 'The Analyst', in October 1938. Dyer was a contemporary of Woodland Toms with whom he maintained a lifelong friendship.

Woodland Toms was educated at the City of London School and rapidly developed a formidable chemistry pedigree. On leaving in 1875 he became a pupil assistant to R. V. Tuson, Professor of Chemistry at the Royal Veterinary College where he remained for two years, after which he went to the Royal College of Science for three years, studying chemistry under Professor Edward Frankland, to whom he acted as personal assistant, chiefly in matters relating to water analysis. On Frankland's recommendation he then went to Dr. (later Sir) William Perkin[21] best known for his discovery of the first aniline dye, 'Mauveine', and in whose organic research laboratory at Sudbury he served for two years. Following this he went to Guys Hospital London as senior assistant in the physiological research laboratory where he remained until 1884, when he was appointed as the first resident Official Analyst to the States of Jersey. The duties of which appointment he carried out with distinction for 47 years.

Woodland Toms was a versatile chemist exposed to the very best practises of his time. The Jersey appointment allowed him to build a laboratory that dealt with a wide variety of problems for which durable and practical solutions were required. This afforded him room for considerable innovation, for example in connection with the laws relating to the adulteration of food. The UK milk standards in force at the time—not having regard to the natural richness of the milk of Jersey cattle—were problematic and he was eventually successful in the official adoption of his own local standard which proved more effective in preventing and detecting adulteration. Food and drug analysis however, formed but one part of his work. His early experience at the Veterinary College naturally led him to take an interest in those branches of science that influenced agricultural practice. He had already contributed a series of articles to '*The Field*' on the chemistry of ensilage, at the time when its novel introduction into farm practice was attracting attention both home and abroad. Under the auspices of the Jersey Royal Agricultural Society he found ample scope for investigations relating to soils and fertilisers used in the intensive system of farming and market gardening prevailing on the island. The consumption of artificial fertilisers in Jersey was then, as it still is, very large in relation to the area of ground under cultivation. Toms found that local knowledge of the origin, composition and properties of fertilisers was often vague and their use empirical and often illogical, while the prices paid for them were sometimes disproportionate to their productive value. Toms initiated field experiments on the manuring of the staple crops of the Island, mainly potatoes and tomatoes, and the lessons derived from them were disseminated with a clarity and simplicity which led to a demand for his advice, as well as to an insistence on proper guarantees of composition by the vendors of fertilisers and on the checking of these by analysis. Later, such 'quality assurance' became legally compulsory.

His attention as States Analyst was also demanded in relation to water supplies, gas, building materials and other miscellaneous directions. As an instance of his versatility it should be mentioned that as long ago as 1878 he carried out an investigation for '*The Field*' into the comparative composition of various explosives used for 'sporting' purposes, finding that the excessive violence of certain brands of wood powder was due to a faulty proportion between two varieties of nitrocellulose (or 'gun cotton') used in its composition. Later, it was found that a mixture of gun cotton, nitro-glycerine and Vaseline ('cordite') yields a much more predictable propellant.[22] Woodland Toms was the consummate professional. He was elected a Fellow of the Chemical Society in 1875 and an Associate of the Institute of Chemistry at the time of its formation in 1878, becoming a Fellow in 1883. He was an original member of the Society of Chemical Industry, and joined the Society of Public Analysts in 1884 at the time of his Jersey appointment.

There are many 'echoes' of Frederick Woodland Toms in the career choices that Robert subsequently exercised. The fact that R. C. Chirnside, who wrote Robert's obituary[23] in 1976, makes fulsome reference to the 'kindness' of Woodland Toms bears testament to his influence. For example, whereas many pupils in Robert's position would have been content with their school chemistry laboratory, Woodland Toms permitted him the use of the Laboratory of the Official Analyst in St. Helier, an indulgence which culminated in Robert's successful analysis of a sample of cement and, we suspect, a quiet determination to become a professional chemist. We shall also see that it is quite ironic that Robert's early career was influenced by a man who not only made significant contributions to the science and application of fertilisers but also [to a very much lesser extent] to that of munitions. Practical innovation was Woodland Toms' forte and Robert's career displayed the same characteristic in abundance. Frederick Woodland Toms died in Jersey on 20 July 1938 at the age of 82.

1.3   University College London and Matriculation

Robert completed his secondary education during the 1900–1901 term but very little remains about his time at Victoria College. With regard to extra curricular activities, he seems to have been an unexceptional student. The school notes in the '*Victorian*'[24] magazine however show that he achieved a number of distinctions in the Science and Art Examination of 1900. The notes from November 1901 record that 'R. Le Rossignol obtained a First Class pass in Practical Inorganic and Theoretical Chemistry, and a Second Class pass in Model Drawing'. Robert subsequently left Victoria College for the University of London[25] where he matriculated in June 1901. The school notes for November recording that 'R. Le Rossignol has passed the London Matriculation Examination in the Second Division'. The same notes also record that Dr. Augustin Le Rossignol and his three sons presented the College with a complete copy of 'Encyclopaedia Britannica'.

'Matriculation' is an unfamiliar term these days but earlier it was commonly used to describe a student's formal entry into a University—'the act of placing a student's name upon the matricula or roll of members'. Matriculation was permitted on passing the University's matriculation examination, or on approval of a student's existing qualifications as being of an equivalent standard. Examinations are always stressful, but the London matriculation was a particularly brutal process. Just prior to the time Robert sat the examination, matriculation at the University of London had been the subject of scrutiny in the *British Medical Journal*[26] where an article raised a number of concerns regarding the age and health of the candidates, the syllabus, and the conduct of the examination.

London accepted as inevitable that candidates at examinations for University degrees were 'nervous' (or as the 'correspondent' describes them 'sicklied o'er with the pale cast of thought'), but questions were raised as to whether the matriculation process paid due care to 'prevent injury to the health of candidates.' Candidates of course were young, at the beginning of their careers, and the product of their school training. The 'evident ill-health' of a fair proportion of the candidates—'especially perhaps the girls'—suggested problems with the syllabus of the examinations of the schools which prepared the students, or with the matriculation examination itself, or both. Many candidates (again the girls) were 'anæmic, stooped markedly, or sat at their desks in a crooked attitude'. The average age of students sitting the matriculation examination was 17, the percentage 'manifestly ill' was 20–30%, and there were 27 h of written work spread over 9 subjects. It was felt that one cause of this 'evil' was that the candidates were too young—especially when compared to 'Oxbridge' entrance examinations. There were also concerns regarding the syllabus of the examination which was felt to have become 'more practical' in recent years. The compulsory subject of elementary science in particular included a 'multiplication of subjects' (mechanics, chemistry, heat, light and electricity) of low standard which encouraged 'cram'. This gave students a 'smattering of everything' which was regarded as beneath the dignity of a University. Questions in some subjects were criticised as not demanding 'thinking power' but simply memory recall. Finally, the conduct of the examination was criticised. The lack of ventilation during the summer examinations led to 'impure air' which caused some students to faint. Many were strained, suffered 'stage fright' and were siezed with diarrhœa. To alleviate this it was suggested, the simpler subjects should be examined on the first day. Although this strain applied to both sexes, the plight of the girls caused particular dismay. Girls were often markedly 'anæmic or choreic' and a school that 'urged [*such a*] girl to go up for examination had mistaken her vocation'. It was even suggested that with the help of a medical Inspector or Inspectress, the markedly neurotic might be discouraged from undertaking the strain of examinations and induced to take up a life of open air labour. Unfortunately, many of these problems are a consequence of an 'on the day' examination system and they remain with us. Brutal examination or not, R. C. Chirnside in his obituary of Le Rossignol in 1976[23] records that Robert 'always spoke affectionately' of University College. Even so, by 1900 serious concerns were emerging regarding the effect of university examinations on the health of some students, and many years later such a system was to play its part in undoubtedly one of the most tragic episodes of Robert's life.

Robert's choice to study at the University of London however, occurred more by chance than one would have first thought. Five years earlier, his brother Austen won a Victoria College scholarship to become a Channel Island 'Exhibitioner' at Exeter College Oxford where he studied Natural Sciences. However, the *British Medical Journal* of 22 August 1896 records

Plate 1.5   William Ramsay in his private laboratory at UCL. Photograph courtesy of the archives of the chemistry department of University College London, with special thanks to Prof. A. G. Davies, FRS

that Austen was successful in the Biology component of the preliminary science examination at the University of London, probably a condition laid down by Exeter prior to 'going up', as Victoria College had no track record in the sciences. Robert's father Augustin had of course studied at the London Hospital in the 1860s and Austen was shortly to become a houseman there. But according to Chirnside,[23] it was Woodland Toms' connections with a member of staff at UCL (one, 'Dunlop', who later left for Chile) that was the deciding factor. Fortuitous though Robert's choice may have been, it turned out to be inspired because London—through William Ramsay—had one of the world's foremost experimental chemists and after matriculating, Robert—by then an aspiring practical chemist himself—indeed studied chemistry under William (later *Sir* William) Ramsay at the 'Godless College'.[27] (Plate 1.5).

Ramsay was the charismatic head of Chemistry at UCL.[28] Born in Glasgow on 02 October 1852, he was a nephew of the geologist, Sir Andrew Ramsay and until 1870 he studied in his native town. But his love of travelling and languages eventually led him to Professor Rudolf Fittig's laboratory at Tübingen, Germany. While there, his thesis on orthotoluic acid and its derivatives earned him his Ph.D. and on his return to Scotland in 1872 he became assistant in chemistry at the Anderson College in Glasgow, two years later securing a similar position at the University there. In 1880 he was appointed Principal and Professor of Chemistry at University College, Bristol, and in 1887 moved on to the Chair of Inorganic Chemistry at University College London. Ramsay was an accomplished pianist, composer, and poet but above all he was an outstanding experimentalist. He was a chain smoker too, rolling his own cigarettes—machine-made ones were unworthy of an experimentalist—he fashioned his own glass cigarette holders and he kept a platinum spatula on his watch chain to examine his students' precipitates!

Ramsay's earliest works were in the field of organic chemistry. It was however in inorganic chemistry that his most celebrated discoveries were made. As early as 1885–1890 he published several notable papers on the oxides of nitrogen and followed those with the discovery of argon, helium, neon, krypton, and xenon. Yet another discovery of Ramsay's, in conjunction with Professor Frederick Soddy the importance of which it was impossible to foresee, was the detection of helium in the emanations of radium (1903). In 1904, and during Robert's stay at UCL, Ramsay was awarded the Nobel Prize in Chemistry 'for his discovery of the inert gaseous elements in air, and his determination of their place in the Periodic system'. As a result, Ramsay became a considerable celebrity in London and was cartooned both by 'Spy' for *Vanity Fair* and by Henry Tonks, Head of UCL's Slade School of Art. But in time, he had his Nobel medals melted down and the proceeds given to charity, what has been treasured at UCL during the years are therefore merely duplicates (Plate 1.6).

Prior to his move to the University of London Ramsay had left a small modern and well-equipped laboratory in Bristol where he had begun to build up a research group. Student numbers were modest, around a dozen. In London he had a larger but inferior laboratory where none of the staff were engaged in research and where student numbers approached fifty. Teaching was therefore more limited and what was possible in Bristol did not translate to London. Even so, Ramsay expected his students to know what he was doing, setting scholarship examination questions such as; 'Describe what you know of any researches in progress in this laboratory'. When in London he toured the laboratories trying to see everyone every day.

Plate 1.6 Ramsay's cartoon in 'Vanity Fair'. Photograph courtesy of the archives of the chemistry department of University College London, with special thanks to Prof. A. G. Davies, FRS

The general laboratory of the department was a long shed, behind and parallel to the north wing of the College. The western half was filled with benches called 'horse boxes' by the students whilst the eastern half, called the 'stink room', was designed for technical experiments (Plate 1.7).

It had a long slate topped table down the middle, fume cupboards along part of one side and an arrangement of flues for furnace work on the other. In earlier days, practically all the chemicals—apart from the most common ones —were made in the laboratory. Chemicals such as pure soda, potassium cyanide and methyl alcohol were all prepared by the students. Residues were preserved in jars, often unlabelled and as the department expanded, Ramsay —ever the practical man—suggested, as an exercise, that students should analyse the contents of the jars and preserve those which were useful. This was soon abandoned and a general clearance was made well before Robert joined the Department. When Robert arrived, first year ('junior') students were taught in classes in a large room next to Ramsay's private room. Conditions were primitive, for gas but no water, was laid to the benches. At first, Ramsay wanted to put Robert in the junior class but Robert

Plate 1.7 The 'horse boxes' at UCL during Robert's time. Photograph courtesy of the archives of the chemistry department of University College London, with special thanks to Prof. A. G. Davies, FRS

'protested'.[23] Having secured some initial 'seniority' because of his work under Woodland Toms, Robert was instead pleased to be allocated a place in the 'horse box' laboratory. Today, it is difficult to conceive of an undergraduate achieving any such position in a department, but the route towards Robert's degree would be unrecognisable by students nowadays. Morris W. Travers,[29] Ramsay's biographer, describes an undergraduate's lot:

> At the time, students followed no very regular course. After a certain amount of routine in qualitative and quantitative analysis and in preparations, they were set to repeat operations described in recent literature, or to deal with simple problems such as the variations in analytical processes.

In this environment, tuition was shared, and 'seniors' helped the juniors. Robert's time at Woodland Toms' laboratory gave him an advantage over other students and Ramsay recognised this. Indeed, even though by our standards still an undergraduate, by 30 March 1904 Ramsay's laboratory notebook[29] has the entry; 'Gimingham and Le Rossignol (research students) have been measuring the rate of decay of emanation from [a Thorium] filtrate …' elevating Robert and a colleague to an even higher position, a description well deserved however as the two men later published their findings in the prestigious '*Philosophical Magazine*'[30] all of which we discuss in Chap. 4.

Robert's first instruction at the laboratory was in glass-blowing, with only a foot bellows available in those days,[23] but in 1901 Ramsay put Robert to work on the 'velocity of reactions', (reaction kinetics). As a result of 'some observations' he made,[23] he won the UCL Gold Medal at the end of his first year and was moved into the research laboratory to work with F. G. Donnan who had joined Ramsay in 1898. Donnan was educated at Queen's College Belfast from 1889 to 1893 taking a BA in 1892 and his MA from the Royal University of Ireland in 1894, after which he left to work in Germany. Almost all young chemists who intended to follow a career in chemistry after graduating at a British University spent two years in Germany engaging in research and often taking a Ph.D. degree. Donnan worked there until 1897 studying under Wislicenus, van't Hoff and Ostwald, and helped to introduce the new subject of physical chemistry into Britain. On his return to the UK he joined Ramsay at UCL where he preached it zealously. Robert therefore would have been at the forefront of this new discipline although at the time there was no separate Physical Chemistry Department at UCL and even as late as 1926 the subject was still widely examined as part of the Inorganic paper.[23] (Plate 1.8).

Ramsay firstly asked Robert to tackle the reaction between hydrogen peroxide and potassium iodide,[23] moving on under Donnan to the reaction between 'potassium ferricyanide' and potassium iodide. This led to the publication of a paper, viz, G. Donnan, R. Le Rossignol, 'LXXI The velocity and mechanism of the reaction between potassium ferricyanide and potassium iodide in neutral aqueous solution'. *J. Chem. Soc., Trans.*, **83**, 703–716,

Plate 1.8   Frederick George Donnan, Robert's tutor at UCL. Photograph courtesy of the archives of the chemistry department of University College London, with special thanks to Prof. A. G. Davies, FRS

(1903). Robert also read the paper before the Chemical Society—an event of remarkable maturity for one so young—recorded in *The Victorian* of November 1903, probably because Robert's father had by then become a member of the College committee. By the time Robert graduated in 1905—with a Second-Class Honours Degree in Chemistry—he was published and had won the UCL Gold Medal, an award for practical excellence which survives to this day as the 'Ramsay Medal'.[31] (Plate 1.9).

> Records held at the Royal Society of Chemistry, Thomas Graham House, Cambridge, also shed light on Robert's performance in some of the ancillary subjects he took alongside his main diet of Chemistry. In this respect, these records show that he studied *Physics* from 1901-06, performing 'very creditably', that he studied *Mathematics* from 1903-04, performing 'satisfactorily', and that he also chose to study *Mechanical Engineering and Graphics* for two years from 1901-03, performing 'creditably'. As it turned out, Mechanical Engineering – a life-long interest of his[35] – was to be of fundamental importance to the successful solution of the 'ammonia problem' he tackled alongside Fritz Haber (Plate 1.10).

1.4   Austen Clement

1905 was therefore a memorable year for Robert but it was also memorable for another, far less welcome reason. Robert's long life was never free from personal tragedies and on 06 April 1905 the first of these occurred with the sudden death (from diphtheria) of his elder brother Austen Clement at the London Hospital. Austen was just 26 years old and had begun what should have been a glittering career in the medical profession. A popular and distinguished scholar[32] at Victoria College he left for the University of London. From here (Autumn 1896) he moved to study at Oxford where he graduated in 1889 with a second-class degree in Natural Science. He followed this with the Sutton Scholarship in 1902 and the Duckworth Nelson Certificate in 1903 becoming a Senior (science) Scholar at Exeter and taking the degree of Batchelor of Medicine and Surgery. His M.R.C.S and M.R.C.P followed and he progressed to the London Hospital where his father had also received his medical tuition. There, Austen became Pathological Assistant and House Physician in 1903, then Receiving Room Officer in 1904.[10] (Plate 1.11).

The death of a child is not a burden that any parent should be expected to bear, and Austen's untimely death dealt a severe blow to Augustin and Edith

Plate 1.9 A fragment of a photograph taken on the steps outside UCL in 1904. William Ramsay is shown seated, second from the left, in the second row alongside other academic staff. Robert appears seated on the chair at the end of the same row, confirming his 'seniority'. This is the earliest known photograph of Robert. Photograph courtesy of the archives of the chemistry department of University College London, with special thanks to Prof. A. G. Davies, FRS

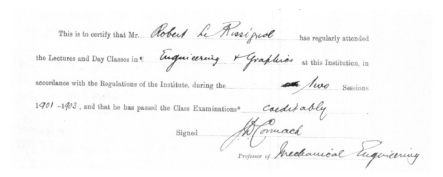

This is to certify that Mr. *Robert Le Rossignol* has regularly attended the Lectures and Day Classes in ¶ *Engineering + Graphics* at this Institution, in accordance with the Regulations of the Institute, during the *two* Sessions 1901 –1903, and that he has passed the Class Examinations* *Creditably*.

Signed *JM Cormack*

Professor of *Mechanical Engineering*

Plate 1.10 Professor Cormack's assessment of Robert's efforts in mechanical engineering. It shows that he studied the subject for two-years at UCL, even though in conversation with Chirnside in 1976,[23] he 'could not remember' if he completed the second year of the course. Photograph courtesy of Royal Society of Chemistry archives, Thomas Graham House, Cambridge, UK. With special thanks to Rob Stiles

Plate 1.11 Robert's brother Austen Clement, from 'Historical Notes … 1917'.[10]

who, along with the whole family, received it with overwhelming grief. Austen's Obituary appeared in the local Jersey 'Evening Post' and was fulsome in its praise, but a powerful testimony to his character and work appeared in the '*London Hospital Gazette*', April 1905. Describing not only the circumstances of Austen's death, but also the esteem with which the young man was held, this tribute was preceded by a brief record of Austen's life.

> … Seldom have words seemed so empty and cold as they do when we attempt to add a few lines to the above record. So terribly short was his illness that even yet we can hardly realise the sad truth. Most of our readers knew him as a house

man, and for teaching it will be hard to find an equal. In each appointment his enthusiasm prevented his instruction becoming lifeless and uninteresting, and his personal knowledge enabled him to carry out the ambition of every teacher – to add to the stores of knowledge from which he freely drew. To die at the post of duty is ever the brave man's wish and certainly in this case had its fulfilment. The disease was contracted while attending a child at the hospital at the very end of his appointments and proved only too malignant for one in his tired state. A brief 36 h of powerless watching was the lot of his friends, and then the end came. Surely, if the end must come, it brings a hero's reward and a martyr's crown. From his entry at Oxford onwards, there was one motto that dominated his whole career – work. For him it made success possible, but for others it spelt knowledge skill, confidence and recovery. His capacity for work, too, brought him the admiration of all with whom he came into contact. Always willing to help, his personality will be greatly missed from 'the London' and we must mourn the great price paid for an attempt made to save a child's life. One who knew him intimately writes as follows – 'Everyone who worked with Le Rossignol recognised that he possessed abilities of a high order. His memory was extremely good and his facts well arranged. His most conspicuous feature as a 'resident' was his power of organisation; nothing was forgotten and nothing overlooked, no matter how great the pressure. Although never hurried, his capacity for work was great. Notwithstanding the uniform success which had followed his career, his modesty was notable, and only equalled by his loyalty and amiability.' The suspense with which his illness was watched, and the gloom which overspread the Hospital when its fatal determination became known, constituted an eloquent testimony to his popularity. The sacrifice of so valuable and a promising life in an attempt to alleviate the sufferings of one of the most miserable specimens of humanity naturally raises thought as to the eternal fitness of things, but an influence and example such as Le Rossignol has left behind will live in the hearts and minds of all who knew him.

Who the *miserable specimen of humanity* was and whether s/he survived we shall never know, but as a mark of respect, the 'London' subsequently founded the *A. C. Le Rossignol Medical Scholarship* of £25 for one year, tenable at the London Hospital Medical School, in memory of Austen[33] who 'sacrificed his life to save that of a patient'. Austen's death was also marked at Victoria College. The report of the prize day in the '*Victorian*' magazine for November 1905 includes a passage from the external examiner from Exeter College Oxford, offered before the school which sent Austen to them to express the sorrow they felt at his loss. Austen was clearly a compassionate man, but the same caring thread wove itself throughout the Le Rossignol family, with Augustin and Robert both later leaving us tangible evidence of devotion to those in their care and consideration of their fellow man.

1.5 Postgraduate Studies

Robert's career so far may not have been quite as glittering as his elder brother's, but he was now a graduate—including the 'cutting edge science' of physical chemistry—and as such he would probably have been encouraged to go to Germany to study further. But instead he chose to remain at UCL for another year to sit the examination for the Associateship of the Institute of Chemistry[34] of Great Britain and Ireland—and in *organic* chemistry. The 'Institute was', he said in later years, 'the professional body, and I wanted to become a professional chemist'.[23] At this point we have to ask why he initially declined the 'conventional' route and chose this—more unusual—alternative? The answer in part probably lies in his experience with Woodland Toms. Eventually, Robert would travel to Germany, but it was clearly important to him that he would do so first and foremost as an established professional. Also, Robert never really embraced academia, preferring to remain in industry indulging his love of practical science. The professional route therefore was probably much more attractive to him and it seemed 'the right and proper thing to aim at their qualifications'.[23]

Robert passed the examination for the Associate ship in 1906 and he subsequently published two more papers with his supervisor Samuel ('Sammy') Smiles who went on to become remembered for the eponymous rearrangements of substituted aromatics. The title of these papers indicates this provenance viz., 'LXXIV.—*Aromatic sulphonium bases*', *J. Chem. Soc., Trans.*, **89**, 696–708, (1906), and 'LXX.—*The sulphination of phenolic ethers and the influence of substituents*', *J. Chem. Soc., Trans.*, **93**, 745–762, (1908). In the same year (February 1906) Robert was elected Fellow of the Chemical Society of London, an institution which he patronised throughout his life and where he maintained his contacts with both F. W. Toms and Bernard Dyer (Plate 1.12).

Robert left UCL in July 1906 and by then his credentials were noteworthy. He was a prize-winning graduate who had studied under a Nobel laureate, he was an Associate of the Institute of Chemistry and a Fellow of the Chemical Society, he was also published in physical, organic and radio-chemistry. Germany beckoned, although at the time Robert had thought of spending half a year in France beforehand.[23] Ramsay's connections with the European scientific community were extensive and he used these to place his men judiciously. Two chemists in particular stood out for him, Richard Abegg at Breslau, and Fritz Haber at Karlsruhe—both practitioners of the new physical chemistry that Robert had studied under Donnan. According to Morris

16 JAN 1906

73

CHEMICAL SOCIETY.

CERTIFICATE OF A CANDIDATE FOR ELECTION.

The attention of the candidate in whose favour this certificate is made out is specially directed to the fact that, if elected, he will be required to sign the following obligation prior to his admission into the Society:—

OBLIGATION.—I, the undersigned, do hereby engage that I will endeavour to promote the interests and welfare of the Chemical Society, that I will observe its laws, and to the utmost of my power maintain its dignity, as long as I shall continue a Fellow thereof.

Name in full   *Robert Le Rossignol*

Usual Place of Residence   *2 Caesarea Place. Jersey.*

Designation or Occupation   *Research Student.*

Qualifications.*

B S^c Hons Chem.

The velocity reaction between Potassium Ferricyanide & Potassium Iodide; in conjunction with Prof Donnan J Cs *Trans Chem Soc 1905*

Rate of decay of Thorium Emanation; in conjunction with S.T. Ginning *Phil mag 1904*

being desirous of admission into the CHEMICAL SOCIETY, *we, the undersigned, propose and recommend him as a proper person to become a* FELLOW *thereof.*

Dated this   *fifteenth.*   day of   *January*   1906

From Personal Knowledge.

*J Norman Collie
J. K. H. Inglis.
R J M Wilsmore
Samuel Smiles
Edward C. Cyril Baly*

From General Knowledge.

Proposed on   18 JAN 1906

To be Balloted for on   15 FEB 1906

Elected   15 FEB 1906   190

R. Meldola.

President.

* *Directions for filling up the Certificate are given on the other side.*   5090

◄**Plate 1.12** Robert's application for a Fellowship of the Chemical Society dated 15 January 1906. He was balloted and elected on 15 February 1906. His second publication appeared in the 'Philosophical Magazine in 1904'. His papers with Smiles were published later. Philosophical Magazine was at the time the Journal of choice for many scientific luminaries. He curiously misdates his first paper as 1902, he was to repeat such slips in later life. Courtesy of Royal Society of Chemistry archives, Thomas Graham House, Cambridge, UK. With special thanks to Rob Stiles

Travers,[29] Ramsay's own son 'Willie' was to eventually work with Haber, but even so, Ramsay wanted Robert to study under Abegg.[23] Although 'half-minded' to do so, Robert had heard that 'there were already too many Englishmen there'[23] and together with what he later described as a developing interest in 'chemical technology'[23] he wrote instead to Haber seeking a position at the Technische Hochschule—and was accepted. Now one would have thought that it was Haber's burgeoning international reputation at the time that attracted Robert, for the fine group of physical chemists he had collected at Karlsruhe produced an eclectic mix of work ranging from the theoretically fundamental to the eminently practical. But it seems there were other factors influencing his decision, factors which Robert revealed much later in conversation[35] at his retirement home in Buckinghamshire with Haber's (then) biographer, Johannes Jaenicke. Certainly, Robert wished to work in physical chemistry, but it was *not* Haber's fame that attracted him. It seems that after leaving UCL, Robert, his father Augustin and sister Elsie Edith took a holiday in Switzerland. Here he met with a friend who described Haber as 'a very nice man', and as Karlsruhe was close to the 'Black Forest' Robert found that 'a great attraction too …!' And so, it was, that he travelled to Germany to spend 'three glorious years' with Haber, simply to be able to take a walk in the Black Forest!

But Robert's contact with German science was not as one-way as this account may suggest. Although it was the 'norm' for British postgraduate students to attend German universities, German students often travelled to Britain—specially to improve their English—but equally influencing British students to practise their German. Dr. Otto Hahn had joined Ramsay's Laboratory in 1904 for precisely this reason—the laboratory already employing another very able young German chemist Otto Sakur—and these two were Robert's contemporaries at UCL. Hahn was put to work on radium but soon discovered a highly radioactive 'new metallic element' indistinguishable from Thorium he called 'radio thorium', the half life of whose 'emanation' had already been established[29] by Robert and Gimingham when working as a 'research students'. 'Radio thorium' was later found to be an isotope of thorium (^{228}Th). Hahn also worked with Sakur to show that two other new radioactive elements were in fact identical, all of which help lay the foundations of the concept of isotopes later expounded by Frederick Soddy. Whereas Hahn went on to become a giant in radio chemistry taking the Nobel prize for chemistry in 1944, Sakur left to join Haber and died in an explosion in the Kaiser Wilhelm Institute in Berlin in 1915 (Chap. 13).

1.6   And Finally … Controversy at UCL

Robert's chemistry education at UCL was of the highest standard, studying a 'blue skies' discipline under a Nobel laureate, in a laboratory internationally renowned for practical innovation and learning, alongside others, many of whom were to go on to achieve world wide scientific recognition themselves. But his education also allowed him to experience a common consequence of 'life at the top of the scientific tree' … controversy. During Robert's last years at UCL, Ramsay became involved in what were to prove the thoroughly discredited processes of extracting gold from sea water and then later the 'transmutation' of one element into another by purely chemical means. Robert Le Rossignol was not involved in either of course, but he was able to observe for the first time the significant consequences to one's reputation of getting things scientifically wrong.

Ramsay's excursions into controversy began in 1905. His high standing in scientific circles led to his unfortunate endorsement of the *Industrial and Engineering Trust Ltd.*, a corporation with a supposed secret process to extract gold from sea water. The man behind the process was H. J. Snell, a well-known stained glass and ceramic painter, playwright, songwriter, supposed 'scientist 'and inventor who had apparently made 'experiments' into methods of extracting gold from seawater. He was a larger than life character. The son of a gentleman's servant, he was married to 'the girl next door' at twenty-one, with whom he had seven children. He started the first of a string of 'creative' businesses at the age of thirty-two and also wrote several books on stained glass painting, together with many plays and books of poetry, which can be found in the British Library's rare books collection. Today it seems utterly bizarre that a figure as eminent as Ramsay could associate himself with —let alone endorse—the work of someone like Snell, but we have to understand the conditions under which scientists such as Ramsay worked in order to rationalise his behaviour. The days of government research grants had not yet arrived, and unlike the German Universities, British counterparts were not encouraged to seek and accept industrial support.[23] Ramsay had to earn such monies through 'private practice'. He could hardly have carried out his research, much less travelled as widely as he did and established friendships with a vast number of scientific men, without the money earned through his private practice. It was through this mechanism that Ramsay was able to recommend people like Richard Abegg to students such as Robert. There seems no doubt therefore that in Snell's work Ramsay had spotted 'a nice little earner'. Ramsey had apparently 'made experiments' himself and stated in a formal report that;

> there is no doubt Snell has proved that gold can profitably be obtained from sea water on a large scale, and the amount of the gold obtained is so large that whether the cost of the treatment is £1 a ton or even the outside figure of £8 a ton, which it could not exceed, it would not make very much difference.

Ramsay was retained by the Industrial and Engineering Trust to develop Snell's work. Shareholders included Lord Brassey, Lord Tweedale, the Hon. Alban Gibbs, several manufacturers and Albert Sandeman, foremost owner of the Bank of England. The syndicate had the modest capital of £3,000 in £1 shares. Such was Ramsay's standing that his endorsement even crossed the Atlantic when on 09 April 1905 the *New York Times*[36] published an article roughly calculating the amount of gold available from sea water and declaring the process as 'patented', but 'bearing a certain resemblance to the treatment adopted in the mines of the Witwatersrand'. Even some American provincial papers such as the *Minnetonka Record*[36] carried the story on 03 March 1905. The corporation bought property along the English coast to implement the gold-from-seawater process, but it quickly faded from public view and never produced any gold. Morris Travers' biography[29] of Ramsay declined to tarnish his reputation by mentioning this episode, Snell however was undeterred and in 1905 wrote that he was hoping to 'submit to the scientific world a far more wonderful discovery, the process of obtaining gold from the air that we breathe.' But the 'gold-from-seawater' concept didn't die with Ramsey and Snell, it was resurrected by Fritz Haber in the early 1920s in an attempt to alleviate Germany's wartime reparations (Chap. 13).

Ramsay's endorsement of Snell was a foolishness driven by the need to secure funding for his research, and it led to a degree of embarrassment. But his later claim that he and his University College colleagues had achieved the *transmutation* of one element into another by purely chemical means astonished the scientific world. This of course was the stuff of alchemy and those such as Rutherford, who *did* succeed in doing this by physical methods in

1919, were far from impressed. Indeed, Professor Arthur Stewart Eve in his *Life of Rutherford*[37] writes: 'Sir William Ramsay made incursions into radioactivity which were singularly unfortunate …'

Ramsay's observations on transmutation—ultimately completely unsuccessful—were firstly expressed in his laboratory note book dated 06 March 1906.[29] A series of experiments on the effects of the action of radon[38] on water and on solutions of salts of copper and other elements led him to believe that he had actually achieved the transmutation of elements of relatively high atomic weight with the formation of elements of much lower atomic weight belonging to the same Periodic Group. Typically, in a *glass* apparatus, copper (a later Group Ib element) as aqueous copper(II) sulphate or nitrate was thought to transmute to lithium (a later Group Ia element) in much the same way as radon was thought to transmute to helium in Group VIII. In coming to this conclusion Ramsay ignored the Periodic Law of the elements as it was understood at the time, and instead followed the Principle of J. W. Döbereiner, an empirical generalisation discovered in 1817 relating triads of elements with similar chemical properties to relationships between their atomic weights.

Soon after the publication of Ramsay's findings, a paper by Mme. Curie[39] appeared stating that when using a *platinum* apparatus, no lithium could be found in the aqueous solution resulting from the treatment of a copper salt with radon. Mme. Curie was so concerned that she communicated her results directly to Ramsay advising him to repeat his experiment.[29] Ramsay however remarked, 'In spite of Mme. Curie's advice, I am not going to repeat the experiment on the Cu–Li transformation. All I can say is that we succeeded in bringing about this transformation and she didn't!'.[29]

To be fair to Ramsay, the Periodic Table was far from understood. Six years were to elapse before the principle of atomic numbers by Moseley, and the existence of isotopes by Soddy brought order out of chaos and made it possible to classify the elements on a sound scientific basis. On the other hand, Ramsay had already realised that enormous energy—'such as that present in the hottest stars'—was involved in what we now recognise as nuclear transformation, even speculating that the world may have at its disposal a 'hitherto unsuspected source of energy'.[29] But such energy is simply not available in chemical reactions and so it is difficult to see how he could have believed he brought about nuclear transmutations at temperatures available to aqueous solutions. The most probable explanation of Ramsay's experiments is provided by Ramsay's biographer Morris Travers.[29] The glass apparatus used by Ramsay had a stop cock sealed to it. Tap grease was kept in an open wide-mouthed bottle standing on a bench in Ramsay's private laboratory. Ramsay smoked cigarettes constantly, knocking the ash off onto the floor, or onto the bench itself. Tobacco ash contains about 0.5% lithium. Any fine ash deposited on the surface of the tap grease would inevitably be transferred to the barrel of the tap and hence to the experiment. Cigarette ash therefore 'did for' Ramsay's theory of transmutation whilst the cigarettes themselves finally claimed him on 23 July 1916 when he died of nasal cancer at Hazlemere, High Wycombe, Buckinghamshire.

Robert first became aware of the controversy of international science through Ramsay, but controversy was to raise its head in a much more personal sense when he joined Haber. There, he was drawn into a spat between Haber and Walther Nernst. Their disagreement—a relatively trivial matter—set in motion a chain of seemingly innocuous events that changed the world forever. But at the time, who would have thought it? By all accounts Robert was a modest, uncomplicated man who 'rose without trace'. Leafing through the fragmentary public record of his early life one finds it hard—in all honesty—to find any more than just 'glimpses' of great promise. Researching this book, I found that in the chemistry department of UCL his echoes are feint … 'he lies in the dark corners of our past' one told me. Today, in the place of his birth, the Société Jersiaise,[15] Victoria College[18] and the Jersey Heritage Trust[40] are largely ignorant of this man. Little remains of his school history, there are no records of any sporting achievements in the school's main activities (at the time) of swimming or cricket. The debating society has no reference to Robert and he was not inclined to his school's military attachments. Little too remains in Ramsay's personal papers.[41]

But very few of us make an impact in our early years, most being content to be progressively 'moulded' by those around us. In this sense many people contributed to Robert's development. His father and eldest brother nurtured a caring side which was to manifest itself in later life. Pryce Jones fostered his love of chemistry, and Woodland Toms' professionalism was a profound influence on him. Exposure to Europe was provided by Ramsay, and his innovative laboratory at UCL together with Robert's life-long interest in mechanics,[35] may have led to Robert's inclination to chemical technology. Donnan too had a role to play, teaching him the new science of physical chemistry, the painstaking practise of which by Robert was later to complement Haber's driving ambition. Even so, one cannot escape the conclusion that Robert's early life was uneventful, maybe dull and destined for obscurity. Robert too recognised this, always maintaining that he owed a lot to chance in that he went to Ramsay, and then to Haber, not Abegg.[23] Indeed, without Haber, the world would probably have never heard of Robert Le Rossignol, his would have been a life as 'quiet as a domino'. So, one has to ask what was it about Fritz Haber that stimulated a young, accomplished, but otherwise unremarkable British chemist to achieve what he did, helping to make 'brot aus luft', 'bread from air' … and change the world.

Notes

1. The term 'nitrogen fixation' was introduced into technical language in the latter part of the nineteenth century. Although the world had no shortage of *elemental* nitrogen—the atmosphere contains almost 80% nitrogen in this form—nitrogen bound to, or *fixed* chemically, to other atoms in organic or inorganic compounds was far more useful. The process of achieving the chemical binding of atmospheric nitrogen was therefore termed 'fixation'. Today it is widely accepted that 'fixation' was the most important techno-logical advance of the 20th century, allowing the population of the world to expand from 1.6 billion in 1900 to today's 7 billion. The statistics are truly staggering, around 50% of the nitrogen in our bodies can be traced to factories employing the fixation process and one third of the world's popu-lation is dependent on nitrogen fixation.

2. A description of Haber by Dietrich Stoltzenberg whose family knew him well. *Fritz Haber. Chemist, Noble Laureate, German, Jew.* Chemical Heritage Press, Philadelphia, Pennsylvania, (2004). ISBN-0-941901-24-6.

3. Described by Klaus Nippert, Archivist at Karlsruhe University, in Chris Bowlby's BBC Radio 4 production, *The Chemist of Life and Death*, broadcast on 13 April 2011. Nippert goes further, describing how students often cover over Haber's name on Fritz Haber-Weg in Karlsruhe and replace it with 'Clara Immerwahr', his first wife, herself a chemist and deeply opposed to chemical warfare.

4. Strictly, the Channel Islands are not part of the United Kingdom, but dependencies under the Crown. But islanders are British nationals.

5. See for example, Fred Aftalion and Otto Theodor Benfey, A *History of the International Chemical Industry,* University of Pennsylvania Press, (1991). ISBN-10 0812282078, ISBN-13 9780812282078, p. 88.

6. *The Huguenot Society of Great Britain and Ireland* to which I am indebted for my modest understanding of the Huguenots. http://huguenotsociety.org.uk/history/.

7. A common story amongst those interested in Jersey history. Follow this typical link; http://archiver.rootsweb.ancestry.com/th/read/channel-islands/2000-03/0952382225.

8. Notice Le Bailly, 'dressed as a 'peasant', suggesting that he was really much more grand than that!

9. Communication by email with Anna Baghiani, education officer at the Société Jersiaise op. cit. (note 15) in 2009, who kindly spoke to family members regarding a query of mine.

10. S. J. Le Rossignol, *Historical Notes (Local and General) with special reference to the Le Rossignol Family (and its connections in Jersey)*, Trowbridge, (1917). The mere existence of this book indicates the importance of Robert's family in Jersey. http://catalogue.jerseyheritage.org/collection/Details/archive/110303 472/.

11. From Philip Falle's, *History of Jersey*, published about 1734.

12. G. R. Balleine, *Some Jersey Surnames: Their Origin and Meaning*. Societé Jersiaise Bulletin, (1940), http://members.societe-jersiaise.org/whitsco/balleinenam.htm. A number of the following references also have the root address 'members.societe-jersiaise.org/whitsco/' which is the home page of Mr. Tony Bellows who maintains the site for historians of Jersey and genealogists. I am grateful to Mr. Bellows for his help.

13. Edith however had a substantial provenance herself. She was connected by marriage to Mr. James Marcus Major who, as a Partner to Mr. Charles Godfray, was connected to the 'Old Bank' in Jersey, a merchant banking business run from a house in Gorey Harbour in the sailing ship days when the harbour was one of Jersey's main trading centres. Le Rossignol, op. cit. (note 10), p. 135. Today 'Old Bank house' is a hotel. http://www.oldbankhousejersey.com.

14. The reverend Ragg's account is here in more detail; Alban E. Ragg, *A Popular History of Jersey*, (1896). http://members.societe-jersiaise.org/whitsco/wragg0.htm.

15. *The Société Jersiaise*, 7 Pier Road, St. Helier, Jersey, JE2 4XW. http://societe-jersiaise.org/index.php.

16. According to '*The Historical Notes …*', Le Rossignol, op. cit. (note 10), p. 128, Austen's name, 'Clement', derived from Augustin's great aunt Mary, who married one 'P. Clement'. Apart from Robert, the children all had middle names derived from either Augustin or Edith's family. Robert, like his father, had just one given name, but he was later to introduce the 'middle name convention' to his own children.

17. Over the next twenty years the Le Rossignols are recorded at 2 Cassiana Plaisance and 2 Caesarea Place both on St. Saviour Road, the modern day A7 from St. Helier. Some say the address derives its name from the Roman Antonius, who in AD 300 called the island '*Caesarea Insula*' ('Caesar's Island'). These addresses were probably one and the same and no longer exist having been demolished and re-developed. According the Societé Jersiaise the site where they once stood is now an extension (car park) to a hotel. Those properties which remain have been largely converted into apartments, however they show that the Le Rossignol house was a fine Victorian property. Many other members of the Le Rossignol family were also living off St. Saviour Road at the time.

18. E. C. Cooper, *A Short History 1852–1928* to which I am indebted for my understanding of the early history of Victoria College. http://members.societe-jersiaise.org/whitsco/vicinfo3.htm. Victoria College, Le Mont Millais, St. Helier, Jersey. JE1 4HT. http://vcj.sch.je. The Victoria College Register Index (1852–1929) can be found at; http://members.societe-jersiaise.org/whitsco/VCIndex1.htm.

19. Austen entered in 1889, (VC$_i$ 2268) and left in 1896. 'Bertie' entered in 1890, (VC$_i$ 2295) and left in 1898, whilst Robert entered in 1894 (VC$_i$ 2470) and left in 1901—part of an intake of just 37.

20. Who, despite his fine Welsh name, was born in England(!) in 1865.

21. A popular account of Perkin's life is here; http://en.wikipedia.org/wiki/Sir_William_Henry_Perkin.

22. A. Dronsfield, 'Chemical Warfare', *Chemistry World*, Letters, **6**, 37, (February 2009).

23. Ralph C. Chirnside's sensitive Obituary of Robert appeared in; R. C. Chirnside, 'Robert Le Rossignol, 1884–1976', *Chemistry in Britain*, **13**, 269–271, (1977). Chirnside was a long-time friend and colleague of Robert from their time together at the GEC laboratories. The Obituary was based upon a tape-recorded interview he conducted with Robert on 29 March 1976, a few weeks before Robert died. The interview was not meant to form the basis of an obituary however, but rather was conducted due to a request from the Royal Institute of Chemistry—of which Chirnside was a committee member—to help celebrate its coming centenary by recording the achievements of some of its oldest and most distinguished members. For Chirnside, it was also an opportunity to 'put the record straight' regarding Robert's contribution to 'fixation'. A transcript of the interview exists which passed firstly to Robert's nephew, Clement, and subsequently—on Clement's death in November 2014—to his next of kin. The transcript remains within the family and I am grateful to them for allowing me to include its contents in this book. The transcript is some twelve pages of annotated typewritten text on A3 paper. The first three pages are Chirnside's notes and recollections which form the basis of the structure of the interview. The remaining nine pages detail the interview itself. I have used the transcript freely in this book as it contains material which did not appear in the Obituary. It is discussed it in some detail in Chap. 18.

24. '*The Victorian*', has been published annually since the foundation of the College. The magazine seeks to celebrate the achievements of the past year, and also gives an insight into the everyday life of the College. Reports include Sports—at both School and House level—Music concerts, Drama productions and much more. The Victoria College library contains copies of all 'Victorian' magazines since 1852. Contact the LRC Manager at http://vcj.sch.je.

25. The VC_i at the time showing Robert living at 7 St. John's Road, Harrow, a property that was to remain in the family certainly until the early 1950s.

26. The issue of 24 November 1900.

27. The phrase, 'That Godless Institution in Gower Street …' was coined by Thomas Arnold describing the (then) 'University of London', established in 1826 but which later became UCL. It was 'Godless' because it was completely secular, no Minister of religion was allowed to sit on the College Council.

28. Much of what follows is due to University College London from their website at http://ucl.ac.uk/ramsay-trust/life. I am grateful for their permission to reproduce it.

29. Morris W. Travers, *A Life of Sir William Ramsay K.C.B., F.R.S.*, Edward Arnold London, (1956).

30. C. T. Gimingham was a friend of Robert. The two men published this work as; C. Le Rossignol and C. T. Gimingham, 'The Rate of Decay of a Thorium Emanation'. *Phil. Magazine*. ser. 6, **8**, 43, 107–110, (1904), communicated by Prof. F. T. Trouton, FRS. Unfortunately, Robert was attributed as 'C. Le Rossignol'. The Philosophical Magazine was the Journal of choice for many notable scientists at the time. Le Rossignol and Gimingham therefore published alongside Lord Rayleigh, Ernest Rutherford, J. J. Thompson and Trouton himself.

31. Robert's success was reported in *The Victorian*, op. cit. (note 24), March 1906.

32. Reported as passing 'in the first division' from Victoria College according to the local Jersey/Guernsey newspaper, '*The Star*', on 20 July 1895.

33. England and Wales Probate Calendar for 1905 shows Austen left £164 5s. 2d., several times the average annual wage at the time.

34. Subsequently the Royal Institute of Chemistry and now the Royal Society of Chemistry. He took the exams on 3–6 July 1906, as listed in; *Proceedings of the Institute of Chemistry of Great Britain and Ireland*, **30**, 29, (1906). His address at the time was listed as 2 Caesarea Place, Jersey.

35. Jaenicke's and Robert's letters from this period and the transcript of their conversation are held at the Max Planck Gesellschaft, Archiv der MPG, Va. Abt., Rep. 0005, Fritz Haber. Haber Sammlung von Joh. Jaenicke. Nr. 253 (Jaenicke's letters) and Nr. 1496 (conversation transcript). Jaenicke's English is often clumsy and I have had to interpret it as best I can. My thanks to Berndt Hoffman at the MPG for providing the documentation by email, (Mar–Apr 2013). See also Chap. 18. Page 2 of Jaenicke's transcript records Robert declaring; '*I always was interested in mechanics …*'. The same sentiment was expressed later in Robert's 3-page hand-written account of his career, supplied to me by Prof. Alwyn Davies, FRS, from the Archives at UCL Chemistry Department. Here Robert says; '*My aim when I came to the College was to study under Prof. Ramsay and at the same time to learn as much as*

possible about engineering. I have always been, and still am, interested in that subject'. The account is anonymously annotated, 'written in 1961', and probably accompanied Le Rossignol's gift to UCL at that time. See Chap. 18.

36. The New York Times, '*GOLD FROM SEA WATER; English scientist said to have endorsed method for extracting it*', (09 April 1905). Also, the Minnetonka Record, Excelsior, Minnesota, '*Company formed in England to extract metal from water—once tried in America*', (03 March 1905).

37. Professor Arthur Stewart Eve, Macdonald Professor and Director of Physics, McGill University, Montreal. *Rutherford, being the life and letters of the Rt. Hon Lord Rutherford, OMI.* Cambridge University Press, (1939).

38. Ramsay believed that the transmutation of the elements could be achieved by utilising the 'concentrated energy' of the radium 'emanation', radon. Travers, op. cit. (note 29). The terms 'transmutation' and 'emanation' were coined by Ramsay.

39. Mme. Curie and her husband Pierre had been jointly awarded the 1903 Nobel Prize in Physics. She considered that she had taken possession of the science of radio-activity and did not take kindly to Ramsay's intervention.

40. *Jersey Heritage Trust*, Clarence Road, St. Helier, Jersey. JE2 4JY. http://jerseyheritage.org.

41. Most of the men and women who had been Ramsay's students, or had collaborated in his researches or had been members of his staff remained, at best, 'acquaintances'. Travers, op. cit. (note 29).

2

Fritz Haber and Karlsruhe

Fritz Haber, complete with Prussian 'pince-nez' spectacles. Photograph courtesy of Archive der Max-Planck-Gesellschaft, Berlin-Dahlem.

Breslau to Karlsruhe … an unexpected journey.

' … a lousy product …'.

Fritz Haber's description of his young-self in a letter to his friend Max Hamburger in 1889 (Charles[6]).

2.1 Introduction

At Karlsruhe, Haber and Le Rossignol formed one of the most important partnerships in chemical history, albeit an arrangement accommodating two quite different characters. Both men were middle-class and thoroughly privileged, but their early lives were very different. Robert was the youngest of his family. His youth may have lacked flair but it was one of stability, of quiet love, support, and gradual achievement. Determined, but 'kindly', with a

© Springer Nature Switzerland AG 2020
D. Sheppard, *Robert Le Rossignol*, Springer Biographies,
https://doi.org/10.1007/978-3-030-29714-5_2

'simple uncomplicated philosophy of life', his was a calm character, able to accommodate both triumph and disaster alike. Haber on the other hand, was the eldest child in his family and unlike Robert he suffered tragedy at the earliest possible time, his mother dying just three weeks after his difficult birth. Young Fritz was precocious, characterised by restlessness, creativity, an enthusiasm for all things new and a burning ambition to do well. But he was also sensitive to personal criticism and failure, made manifest through physical illnesses. To some, it seems the specious notion of the 'attraction of opposites' may have been at play here.

To concentrate on the differences between these two men is easy but what they had in common helped bind them. Both were from 'old families', comfortable with the values such a provenance provides. Neither showed any 'spectacular' early academic promise but both became professional physical chemists via organic chemistry, each eventually developing an interest in chemical technology. The two were also complementary characters; Fritz disliked the minutia of experimentation[1]—his personality was ill suited—but for Robert, the laboratory was 'king'. Both men also felt that they came to Karlsruhe entirely by chance and they remained friends until Haber's death in 1934. But don't be fooled, this partnership was eventually 'consummated' not by common provenance, nor the attraction of opposites … but by mutual respect, and the prospect of *money*.

Only through an understanding the life of Fritz Haber are we able to properly interpret Robert Le Rossignol's contribution to the fixation of nitrogen, but the world so nearly lost Haber's story. Because of his role in the first European war, Haber—judged by the standards of the time—was condemned as a social and scientific outcast. His last months were characterised by wanderings across Europe. With no permanent home to contain the story of his life and because of the catastrophe of the second European war, his family, friends and the documents that defined him were scattered across the continent or simply lost through the gas chambers and the destruction of the German archives. The detail of his life could therefore have easily disappeared and much of Le Rossignol's along with it. A number of friends and colleagues of Haber published short accounts his life, J. E. Coates,[1] Richard Willstätter[2] and Rudolf Stern[3] being foremost amongst these. But one of Haber's students —Johannes Jaenicke[4]—attempted a biography assembling a total of 2290 items relating to Haber's life. Jaenicke began his work in 1954 and spent the next thirty years tracking down friends and family, copying letters and documents that had survived. We therefore have to thank Jaenicke for assembling the critical mass of biographical evidence upon which our understanding of Fritz Haber is based. The interpretation of the evidence however was ultimately beyond him and by the late 1980s he delivered all his files to the archives of the Max Planck Society for the Advancement of Science in Berlin[5] who in turn made them available to researchers. These files were a 'treasure trove' for historians and they were subsequently used to build biographies of Haber and none more carefully than Vaclav Smil,[6] Daniel Charles,[6] Dietrich Stoltzenberg[6] and Margit Szöllösi-Janze of the University of Munich, whose monumental, *Fritz Haber 1868–1934: Eine Biographie*[6] remains—for this author at least—the seminal work.[7] This Chapter however adds little new of Haber, my brief story of his early years simply serves to 'contextualise' aspects of Le Rossignol's life.

2.2 Young Fritz Jakob

Fritz Jakob Haber was born in Breslau,[8] Prussia on 09 December 1868 into a large, close, Jewish family whose provenance could be traced back to the beginning of the 1800s. Fritz was the only child of first cousins Paula and Siegfried who, when they were young, had lived in the same house; a chaotic and noisy residence filled with fifteen children. Fritz's birth however proved too much for Paula who never recovered. She died just three weeks later on New Year's Eve. Twenty-seven-year-old Siegfried Haber was devastated by Paula's death, for many years later he could hardly face the world and threw himself into building up his business becoming a wealthy merchant dealing in dyes and pharmaceuticals with far-reaching business connections. It was seven years after Paula's death before Siegfried allowed himself to love again marrying nineteen-year-old Hedwig Hamburger. Three daughters followed in the next five years. The female element in Fritz's childhood was therefore provided by his step-sisters and by Hedwig who—according to Charles[6]—was a loving mother. Siegfried however found it difficult to accept the son whose birth had caused so much pain, preferring instead to dote upon his daughters.

Like Robert Le Rossignol, Fritz Haber received a largely classical education. At the St. Elizabeth Gymnasium, an elite high school closely affiliated with the largest protestant church in Breslau—half of whose pupils were Jewish[9]— Fritz grew into an energetic and talkative student, but not a spectacularly gifted one. He absorbed every thing available to an upper-middle class boy in Breslau at the time. Theatre, literature, philosophy, friendly debate and endless drinking in the city's beer cellars. As often becomes the eldest child, Fritz grew into a leader within the circle of his family and friends, at the same time becoming a great favourite due to his kindness and the pleasure he took in helping others. He soon became the dominant figure in his home and this inevitably led to confrontations with his father. The two men were radically different. Many of Siegfried's brothers and sisters were adventurous and talkative but within his family Siegfried was regarded as 'dour', not particularly social, careful with money and a pessimist. Young Fritz however was a reckless spendthrift, an eternal optimist who lost all track of time when engaged in conversation and creative in all he did. Balancing this aspect of his personality however was an emerging interest in things more structured, disciplined and academic, firstly through a knack for developing rhyming verse which he used to entertain the family at special occasions and then through an early interest in chemistry. This may have been sparked by his father who possessed some chemical expertise, but it was probably the limit of Siegfried's influence over his son. More influential was his uncle Hermann— brother of his mother Paula and cousin to his father—a liberal who ran a local newspaper to which Fritz later contributed. Uncle Hermann also provided space in his apartment for Fritz's early chemical experiments, some of them quite dangerous. By the time Fritz graduated from the Gymnasium—largely undistinguished in his written examinations except for history and mathematics but with a masterful performance in the oral aspects (Charles[6])—he decided that he had enough of Breslau and indeed of his father.

Fritz was stifled. At the time, he wrote that Breslau was an 'intellectual swamp'. 'Nothing, absolutely nothing; nothing satisfying to do, no stimulation only irritation and tedium … I'm so disgusted with my entire life here that I could burst!' (Charles[6]). Here he was idle, 'as idle as a painted ship on a painted ocean', Fritz therefore decided to go to university. To Siegfried Haber however, a university education was expensive and unfamiliar, no-one in the family had aspired to it and Jews were in short supply in the upper ranks of Germany's academic elite, (Charles[6]). Siegfried respected those modest achievements[10] that were attainable for a Jew in Breslau and he was not inclined to offer his son financial support. More acceptable to him would have been an apprenticeship that would have prepared Fritz to take over the family business. Another confrontation loomed, but uncle Hermann mediated and broke the deadlock. Siegfried and Hermann had grown up together but unlike Siegfried, Hermann was an optimist and modernistic. Things had changed remarkably for Jews in Germany but Siegfried was locked into the old ways, (Charles[6]). But through Hermann's advocacy Siegfried relented and Fritz became the first of his family to go to University. In the autumn of 1886 aged 18, he entered Berlin's Friedrich-Wilhelms-Universität (now the Humboldt University) to study chemistry and physics, drawn to both fields by the towering figures of August von Hofmann and Hermann von Helmholtz.

2.3   His University Years

Berlin at the time was dynamic, quite different to Breslau or Germany's traditional university cities. It epitomised the 1880s; it was expanding, re-developing and attracting workers and professionals from all over Europe. Berlin 'fizzed' and Fritz Haber fell in love with it. But his experience at the University fell far short of his expectations. Looking for intellectual stimulation he found only frustration and disappointment. Helmholtz was a particular culprit. His mumbled lectures failed to inspire whilst the chemistry lectures were filled with practical demonstrations of experiments that Haber found unchallenging. The repetition of the minutia of these established 'recipes' was never to attract the exuberant Fritz Haber and his first semester was a complete disappointment. Being eclectic, seeking stimulation, and 'jumping like a catfish on a pole', Haber decided to exercise the freedom of German students to move to another University to continue their studies returning to their 'alma mater' only to take their final examinations. Such a freedom was unheard of in Great Britain or the United States and it led to a maturity amongst German students that the other Western countries could only envy. Ironically, British universities have only recently embraced the concept.

Disappointed with Berlin, Fritz Haber spent the summer semester of 1887 at Robert Bunsen's Institute in Heidelberg, attracted there by Bunsen's formidable reputation and possibly by the fact that he had previously been a faculty member at Breslau University. Today Robert Bunsen is remembered by generations of chemistry students for 'his' burner,[11] when his other contributions are vastly more significant. These embraced organic chemistry, the study of arsenic compounds, gas measurement and analysis, the galvanic battery, elemental spectroscopy and geology. His study of dangerous arsenic compounds in the 1830s—for example 'cacodyl', $C_4H_{12}As_2$ (from the Greek *kakodes* meaning 'stinking') a spontaneously flammable toxic, garlic smelling compound[12]—established his name in the international scientific community. But arsenic poisoning almost killed him and nearly cost him his sight when a distillation of cacodyl cyanide with potassium acetate exploded sending a sliver of glass into his right eye. His discovery of the use of Iron(III) oxide hydrate as a precipitating agent is still the best known antidote[13] against arsenic poisoning, and it was this kind of chemical pedigree that drew Fritz Haber to the (then) 76 year old legend. Bunsen's professional and personal life centred around his laboratory and his students. He often took on introductory courses that were shunned by other colleagues. During the many hours of lectures presented each year, Bunsen emphasized experimentation and meticulous tabulated summaries, patiently introducing students to the world of analytical chemistry. Many principal players in the history of chemistry can trace their chemical roots back to Bunsen, two of the more famous were Dmitri Mendeleev and Lothar Meyer, but for young Fritz Haber the aged Bunsen was *stifling*.

Haber interpreted Bunsen's approach to teaching as emphasising proper laboratory procedure(s) over that of inspiring his students to scientific discovery. Haber of course never found any deep interest in experimentation, any benefit he accrued at the time was more in terms of a recognition of proper procedure in others rather than its practise by himself. But one also has to remember that Haber was young, unappreciative of responsibility and the role of proper procedure in establishing professional credibility. In later life, Haber's appreciation of the skills of those such as Robert Le Rossignol can certainly be traced back to Bunsen's rigour but as he embraced the world of professional chemistry, he was soon to realise that he too had to embrace rigour. Indeed, it was to become a hallmark of his work. Bunsen's distancing of matters theoretical also irritated Haber, who at Berlin—out of boredom—had already shown an interest in the philosophical issues of the day by attending the lectures of Wilhelm Dilthey. Heidelberg therefore, pleased Haber no better than Berlin and although still ambitious he began to suffer a crisis of confidence. Lacking stimulation, progress in his chosen field became difficult. At this point in his life he really had no idea what he wanted to do, describing himself to friends as 'a lousy product', uninspired by the obvious career paths of industry or academia. In contrast, at roughly the same period in his life, Robert Le Rossignol was far more assured. His early experiences with Pryce Jones, Woodland Toms, and then at UCL convinced him to pursue a career as a professional chemist, but for Fritz Haber his uncertainty began to show itself through a physical condition that would remain with him throughout his life. He complained of 'nervousness' and 'anxiety' and even at this young age he would go on long walks to relieve his symptoms.

Wracked by indecision Haber's career was about to receive another blow. In 1888, towards the end of his stay at Heidelberg, he was approaching his 20th birthday and compulsory military service—normally three years in uniform. The prospect of such a long time away from his studies daunted Haber, but for those who had attended university, a one-year option was available—as long as all his own costs could be met. The privileged Haber exercised this option and, funded by his father, he joined a field artillery regiment at his home town of Breslau. This was a convenient way to discharge his responsibilities, Haber even managing to keep up his studies by attending philosophy classes at Breslau university. But in Germany—especially in Prussia—the kudos of military rank was unsurpassed and the rigour of this year left its mark on Haber who, for the rest of his life, conducted himself with the authority that only the Prussian military could bestow. The prestige of military rank soon captured Haber and he tried to become a reserve[14] officer, but although recommended as suitable material he failed to be elected, probably because of his Jewish background—a powerful reminder of his father's maxim 'know your place'. Haber remained at the rank of a non-commissioned officer. Once more disappointed by his progress in life, in the autumn of 1890 he left Breslau for Berlin to study under Carl Liebermann at the Technische Hochschule Charlottenburg (now the Technical University Berlin). Here Haber renewed his interest in philosophy (again under Dilthey) and organic chemistry—the boom subject at the time.

Under Liebermann, Haber produced his doctoral dissertation[15] on the reactions of derivatives of piperonal,[16] 'Ueber einige Derivate des Piperonals', *Berlin G. Schade*, 1891, but he found no pride in this 'wretched' work preferring not to publish fearing that competent chemists could easily prove that 'I don't know what I'm talking about!'. At about this time however a number of German sources credit an undergraduate student named 'Fritz Haber' at Charlottenburg, Berlin with the first synthesis of 3,4-methylenedioxy—methamphetamine ('ecstasy'). He reportedly synthesised this drug in 1891 just four years after the young Romanian doctoral student Lazar Edeleano's first synthesis of 'phenisopropylamine', (amphetamine[17]). However, the synthesis does not appear anywhere in Haber's doctoral dissertation, so interesting as it may be, and bearing in mind that Edeleano also studied in Berlin, the case is hardly proven. Certainly, if the story is true and Haber had sampled his product, surely an entirely different life would have ensued!

During his final year at Berlin however, a chance event began to sow the seeds of progress in Haber's mind. He met and became friends with a fellow student named Richard Abegg, the talented son of a wealthy Berlin banker who introduced him to the new science of 'physical chemistry'. At last, here was an aspect of his chosen subject that captivated Haber. Understanding and applying the mathematical laws that governed the feasibility, spontaneity and progress of chemical reactions and which simultaneously translated physical properties in terms of atomic/molecular structure attracted him but there was still the small matter of his degree to be attended to and he returned to Friedrich-Wilhelms-Universität to sit his final examination in chemistry, physics and philosophy on 29 May 1891. Haber's performance in the examination was mixed; his philosophy was very good—he was examined by Dilthey, a teacher he respected—his chemistry was adequate but his physics was quite poor. Overall, he was an unremarkable student, but because of Dilthey's assessment he graduated *cum laude*[18] and returned to Breslau to ponder his future. At the same time, Richard Abegg also received his degree from Friedrich-Wilhelms-Universität and both he and Haber wrote to Ostwald asking to study at his laboratory. Abegg was accepted, Haber was not.

2.4   Takin' Care of Business

Fritz Haber was the only member of his family to *want* to attend University. The award of his Dr. Phil. degree both vindicated his chosen path and afforded him the distinction of being called 'Dr Haber', but it was always the intention of Siegfried that his son, in due course, should carry on the family business. Dyes and pharmaceuticals of course were chemicals, but Fritz had only rudimentary knowledge of the industrial aspects of their production. Still plagued by indecision and with no obvious career path, Siegfried used his influence to allow Fritz to gain technical experience of the chemical industry and over the next few months he spent time at three different factories; a cellulose plant in Feldmühle near Breslau, a distillery in Budapest and a modern 'soda' (sodium carbonate) factory in Kraków, Poland.[19] These placements had a number of effects on Fritz Haber. Firstly, he realised that there were large gaps in his understanding of chemical technology, secondly, he realised that the chemical industry afforded him little personal satisfaction, but thirdly at the soda plant, he saw at first hand what massive industrial investment in a crucial raw material for so many of Europe's industries could achieve, and he was impressed.

Both disillusioned and stimulated by deficiencies in his education, he decided to return to university and in 1892 he attended the Eidgenössische Technische Hochschule Zürich (the Swiss Federal Institute of Technology) to study 'cutting edge' chemical technology under a family friend, George Lunge. But he stayed for a just single semester, blaming an authoritarian teacher for his demise. By now all of Haber's dreams were crushed. Every route that he followed within his chosen subject led to mediocrity. He had been permitted to study widely but he had achieved little. His best efforts had opened few doors and just one remained, the same one as when he started. Depressingly, he had to return to Breslau to face a lifetime in his father's trade. In the spring of 1892, he began his apprenticeship.

Fritz Haber's volatile, eclectic nature was ill suited to Siegfried's sedate business. The two clashed constantly over the direction the company should take and within six months a poorly thought out business decision[20] by Fritz convinced Siegfried of the complete failure of their collaboration. Removed by his father as a 'danger to the business', Haber once more had the luxury to determine his direction in life. This freedom pointed him to academia, probably influenced by the dullness of his early experiences of industry and by the success of his friend Richard Abegg who had earlier secured a position with Wilhelm Ostwald at Leipzig, progressing from there to work with Nernst and Arrhenius. Ostwald's field of physical chemistry was by then widely regarded as the basis of chemistry and chemical technology and it was hugely engaging for Haber, but with an unremarkable degree and little to show for his studies, the time was not yet right to enter this field. Haber's experience of organic chemistry however, convinced him that synthetic chemicals would eventually replace the naturally occurring dyes that his father's business was founded upon. Possibly driven by this goal, he chose to continue his study of organic chemistry, but this time under Ludwig Knorr at Jena university, and for the next two years he became an unpaid assistant in Knorr's laboratory.

2.5   Jena, First Steps on the Academic Ladder

Even if his decision was driven by the possible discovery of synthetic dyes, his lot at Jena was to investigate the structure of 'diacetosuccinnic ester' (the methyl ester of butanedioic acid). Knorr helped Haber to write his first paper on the subject but once again, as under Bunsen, the minutia of experimentation and the meticulous application of orthodox laboratory procedure(s) left Haber unsatisfied and impatient. Jena was a decidedly third-rate German university—although it had produced one or two people of note such as Gottlob Frege the logician—but it was all Haber could aspire to at the time. If he took anything from his unhappy stay there it was that he now finally decided what he wanted to do with his life. Whatever the outcome, he would pursue a career as an academic scientist. Disillusioned with his progress in organic chemistry he summoned up the courage to apply (again), for a research assistantship with Ostwald. Despite several attempts, and a single disappointing interview, he failed to get the job—Ostwald never 'warming' to Haber as a person. Haber remained at Jena.

Once again failure visited Haber's life, but so much failure it seemed, had to have a reason and he felt that reason to be his Jewishness. Haber had suspected as much when his military ambitions were stifled, thoughts of his father's life and the modest achievements available to a Jew flooded his mind. Germany's academic structure seemed closed to non-Christians so in November 1892 at the age of 24, and maybe as an act of desperation, he stood before the baptismal font of the largest Lutheran church in Jena, recited the Apostle's creed and converted to Christianity. This aspect of Haber's life is complex and historians have tried to analyse his motive. Was it a reflection of his failure? A kick against his father? An awareness that progress in Germany was more rapid amongst those Jews who had converted, or just a cynical trade-off by Haber—his Judaism for (eventually) a Chair? The Habers were not particularly Jewish in practise (the names 'Fritz,' and 'Siegfried' for example are pure Germanic) and so Fritz may well have thought that the move to Christianity would be seamless but even after his conversion he remained as Jewish as ever, to others, and even in his own eyes.

Haber's conversion changed very little and he may well have stagnated at Jena but for the sad demise of a friend which finally gave him the opening he craved. In later life Haber was often regarded as 'Geheimrat'[21] and a raconteur par excellence. He delighted in telling the long story of how chance took him from Jena to Karlsruhe and launched him on his career but his young life was always dignified by the pleasure he took in helping others, and in Jena he had a friend whose suffering during a long and painful illness he had helped to mitigate by every means in his power. Despite his best efforts his friend eventually died. At the funeral, his friend's brother—learning of Haber's desire to leave Jena—suggested the Technische Hochschule at Karlsruhe where he said he had an influential relation who would, as a mark of gratitude, recommend him most warmly. Haber took up the suggestion but on visiting Karlsruhe he found that the relative was a porter[1] in the chemistry department! Even so, this contact was sufficient for Haber to get an interview with the Professor who—like Ostwald—received him coldly, but eventually offered him a very junior position and in the Spring of 1894, he left Jena for Karlsruhe not being entirely certain of what awaited him. Haber was always a great story teller, so how much of this is true and how much invention one will never know but on 16 December 1894, in his mid twenties and by a twist of fate, he became assistant to Professor Hans Bunte in the department of Chemical and Fuel Technology at Karlsruhe where he was to stay for the next seventeen years. With duties to perform but ample time for research, he had finally achieved what he wanted, a foot on the first rung of the academic ladder.

2.6 On Coming to Karlsruhe

Karlsruhe, capital of the state of Baden, was a prosperous city on the river Rhine in southern Germany near Stuttgart and a focus of liberalism in Germany. The university's young chemistry institute was hardly the most prestigious but it was unfettered by tradition—the kind of German tradition that would have held back a Jew's progress, (Charles[6]) It also had close relations with industry—and not just any industry, but the world renowned Badische Anilin & Soda-Fabrik (the BASF) at nearby Ludwigshafen. Haber therefore considered it doubly fortunate that chance led him to Karlsruhe. Karlsruhe was also a good place to be a Jew[22] in that only a small proportion of the Baden electorate had ever voted for the anti-Semitic parties in the elections of 1893 and 1898, (Charles[6]). In addition to its industrial links the Institute also attracted solid local government funding. Couple all of this with Haber's understanding of industry after working in Breslau, Hungary and Poland and he had for the first time a recipe for success. Here Haber found the space to set his mind free and he threw himself into his work. Over the years he amassed an amazingly eclectic collection of respected work largely in physical chemistry, the subject that had captivated him in his final year at Berlin. But what was even more amazing was that Haber claimed he was self taught never having attended a lecture on the subject (apart from his own) in his life![23] As always however, Haber was a great tease and there was more to this claim than 'met the eye'.

Having come from Jena, where he had worked in organic chemistry, to the department of Chemical and Fuel Technology at Karlsruhe, Bunte quite naturally suggested that Haber tackle a problem in organic chemistry which impinged on the fuel industry viz., the thermal decomposition ('pyrolysis') of hydrocarbons, an exercise which began Haber's life long interest in flames and the effect of heat on substances. Some thirty years earlier the eminent French chemist Marcellin Berthelot had proposed a general theory of hydrocarbon pyrolysis based on experiments with C_1 and C_2 hydrocarbons only. Now, Haber's path to Karlsruhe had been tortuous, but he had never lost ambition or confidence in his ability and when he saw Bertholet's academic jugular he ripped it out, sharply criticising the great man's work as arbitrary, confused and insufficiently supported by wider experiment—proceeding of course to prove it. The simple hydrocarbons studied by Bertholet were unrepresentative of the decomposition of larger molecules, so therefore were his inferences. Haber began *his* studies by adopting a 'higher' more typical molecule, the 'straight chain' *n*-hexane or C_6H_{14}. This choice was dictated by experimental convenience, the molecule's (then perceived) relationship to 'benzene' (C_6H_6), and its importance in oil and gas manufacture. To Haber, very much an orthodox organic chemist, the work was new yet he devised the experimental procedures involved and determined *all* of the decomposition products both qualitatively and quantitatively adapting and improving existing methods. For one so impatient with laboratory minutia this was a triumph! Haber was able to move beyond Bertholet and propose a more generalised decomposition mechanism for higher straight chain hydrocarbons introducing for the first-time concepts of chemical thermodynamics influenced no doubt by his introduction to physical chemistry by Abegg. He was able to summarise the difference between 'aromatic' and 'aliphatic' decomposition in terms of carbon-to-carbon (C–C) and carbon-to-hydrogen (C–H) bond 'strengths' (dissociation energies) through what became known as '*Haber's Rule*', concluding that the thermal stability of the C–C bond was stronger than that of the C–H bond in aromatic compounds and weaker in aliphatic compounds, and unlike Bertholet, he properly accounted for the appearance and role of 'acetylene' (ethyne) within the pyrolysis process.[1]

The pyrolysis of hydrocarbons was Haber's first independent investigation, yet it rapidly became well regarded in a little understood area. The episode also encapsulates Haber's approach to his science at the time. He was ambitious, but to avoid controversy, ambition had to be coupled with accuracy and reproducibility. He was therefore thorough, intolerant of vagueness, superficiality and theory that had no practical application, and he always tried to address the *root* of the problem. Sensitive to criticism—to the extent that it caused him physical illness—he developed an approach characterised by meticulous attention to detail thereby making it difficult for any potential critic to break him down. At last, he was making sense of Bunsen's training. His work on pyrolysis together with later studies on gas combustion was published in 1896 as a book entitled *Experimental Studies on the Decomposition and Combustion of Hydro-carbons*,[24] and in the same year it led to his appointment as *Privatdozent*[25] at the Institute at the age of just 27. At long last Haber had success, and the best kind of success too, success through his own efforts. Thrilled, he sent a copy of the dissertation to Ostwald who reviewed it positively in print but failed to respond with an invitation to his laboratory. If that was Haber's intention it was the third time Ostwald had rejected him and understandably so too. By this time physical chemistry had become an established discipline and talented scientists from all over the world were fighting for places with Ostwald, Nernst and Arrhenius. This 'establishment' regarded Haber as mediocre.

2.7 Hans Luggin and Electrochemistry

Haber determined that he was going to have to make his own way in physical chemistry and his use of bond dissociation energies was his attempt to modernise orthodox organic chemistry by placing it on a firm theoretical basis. 'Haber's Rule' had exceptions and it was only a broad generalisation, but it was one based on rigour and it permitted a degree of inference. In later years Haber was to admit that his poor grasp of the mathematical concepts limited his initial understanding and application of physical chemistry but this was about to change with the arrival at Karlsruhe in 1896 of a young able Austrian chemist named Hans Luggin. Luggin came straight from one of the most dynamic centres of European physical chemistry, the laboratory of Svante Arrhenius in Stockholm. He and Haber became firm friends and Luggin provided Haber with the stimulation of the 'shock of the new', becoming his guide to his reading in physical chemistry and casting doubt on the notion that he was *entirely* self taught. Luggin's main interests were in chemical thermodynamics and electrochemistry and he and Haber discussed the concepts involved endlessly.

Soon after Luggin's arrival, Haber began work as an electrochemist, his first major project being an examination of the electrolytic reduction of 'nitrobenzene', $(C_6H_5NO_2)$.[1] Haber's choice of nitrobenzene stemmed from his background in organic chemistry and it was natural for him to begin on familiar ground. But there was a second reason. Recent studies had shown that the reduction led to a number of products, the nature and relative proportions of which appeared to depend on a number of factors such as pH, current density, current duration and the nature of the metal used as an electrode, none of which were remotely understood. Furthermore, reduction was supposed to be effected by 'nascent'[26] hydrogen but there seemed to be great variation in the reducing power of the hydrogen leading to the different products. To the established scientific community this behaviour was distinctly odd. Haber was keen to interpret the reduction process in terms of modern theoretical knowledge and this he achieved brilliantly when he explained the appearance of the various products by introducing, and then explaining, the importance of the 'electrode potential'.

Walther Nernst had recently proposed that the potential of a gas electrode was determined by the 'concentration' of the gas in, or on the electrode, and Haber realised that this potential, varying under the conditions of the various experiments, determined the reducing power; higher potentials being equivalent to using more powerful reducing agents. By careful variation of this potential he was able to imitate the effect of using a series of chemical reducing agents of increasing activity, leading to the production of an overlapping series of primary reduction products. The work also brilliantly accounted for the apparent 'varying activity' of nascent hydrogen. Luggin too had some influence in this work, suggesting to Haber the use of a device ('Luggin's capillary') which permitted the reference electrode to sit in a separate chamber to the electrode under investigation, allowing its potential to be established precisely.[27]

Despite Haber's inexperience in physical chemistry,[28] his work on nitrobenzene was widely acclaimed and his performance at the fifth annual congress of the German Electrochemical Society in Leipzig in 1898 announced his presence on the national and international scene. Haber's military bearing saw him stride to the podium to deliver a dynamic presentation that fixed him in the minds of the audience, (Charles[6]). Some however were sceptical of this *privatdozent*, an unknown with mediocre qualifications who was prepared to take on the establishment. The following day Professor Jacques Löb from Bonn seriously criticised his work but Haber defended himself vigorously and although some felt his grasp of the subject was not as sure it should have been, his energy, quick wit and enthusiasm saw him through—earning him some influential friends on the way, (Charles[6]).

Fritz Haber soon became the 'new kid on the block' and his position was further enhanced by his new revolutionary textbook on electrochemistry; '*Grundriss der technischen Elektochemie auf theoretischer Grundlage*' (Outline of Technical Electrochemistry based on Theoretical Foundations, Munich: R. Oldenburg) published in 1898. Senior staff at Karlsruhe initially discouraged its publication realising that Haber's provenance in the subject might lead to embarrassment but the opposite occurred and its novel synthesis of the practical and the theoretical established Haber within the German chemical community earning him promotion at Karlsruhe to Professor Extraordinarius (außerordentlicher)[29] in 1898 just four years after joining Karlsruhe.

There is no doubt that Fritz Haber was deeply indebted to Hans Luggin at this time and the two could well have gone on to produce some spectacular work, but any future collaboration was cut short when on 15 December 1899 aged just 36 years, Luggin died after a brief illness. Here we see many parallels with Robert Le Rossignol who, at roughly the same age as Haber, lost his elder brother Austen Clement. Although Hans and Fritz were not related there was a tremendous respect between the two and just like Austin Clement, Luggin should have gone on to a glittering career. Such was the affinity between the two men that after Hans' death his father permitted Haber to attend to his son's scientific affairs. Fritz completed a series of experiments that Hans had started and published the results. Throughout the rest of his career Haber always recognised his professional debt to Luggin and who knows, in later years both he and Robert may well have talked about their loss.

2.8   Clara Immerwahr

The loss of Hans Luggin was both a professional and a personal blow to Haber but there was another relationship in his life that had been developing quietly over the years. Clara Immerwahr[30] had grown up along side Fritz in Breslau's close-knit Jewish community; a comfortable world held together by marriages between 'well-to-do' families with overlapping friendships.[31] Both Fritz and Clara were the children of wealthy merchants. Like Fritz, Clara was not particularly attentive to her religious provenance and by 1891 there is some evidence that their acquaintance began to develop romantically. At the time however, Clara's mother had recently died and the family had the upheaval of moving from their estate just out side Breslau to a new home on Breslau's central square. Even if it had been proposed, marriage would have been out of the question as Fritz was probably in the most uncertain years of his life with little prospect of supporting a wife. Both he and Clara therefore 'moved on' but neither really forgot each other, (Charles[6]).

Over the next decade these two young people achieved remarkable things. Fritz of course embarked on his career in chemistry which by 1901 established him as professor Extraordinarius at Karlsruhe. Clara's achievement(s) over the same period might be regarded as even more remarkable. A Gymnasium education was not available to girls in Breslau but she achieved the same level of education through private tuition and, utilising a loophole in Breslau's University admission policy towards women, gained admission to pursue a degree in chemistry—just like her father. At about the same time

Clara converted to Christianity—probably for the same reasons as Fritz Haber—and with little or no objection from her family. In 1899 Haber's good friend Richard Abegg was appointed as chemistry lecturer at Breslau and he became Clara's academic advisor. Clara's progress towards her degree was difficult in a world dominated by men. But supported by Abegg and just a few days before Christmas in 1900, Clara defended her doctoral dissertation in the *Aula Leopoldina* auditorium of Breslau university. Like Haber she was sure-footed and quick-witted in support of her research and the packed auditorium saw the dean of the faculty eventually present the first woman from a German university with her *magna cum laude*[18] doctorate, (Charles[6]).

Dr. Clara Immerwahr remained at Breslau and became Abegg's assistant, but in the spring of 1901 Fritz Haber wrote to 'two old friends and fellow chemists, Prof. Dr Abegg and Frauleine Dr Assistant', inviting the two to join him at the next meeting of the German Electrochemical Society at Freiburg just outside Karlsruhe. Soon after the invitation Fritz Haber confided in a friend that he had spent the last decade 'diligently but unsuccessfully' trying to forget about Clara. The invitation therefore had a hidden agenda and when Clara travelled to Freiburg to meet Fritz he managed—over the course of a few days—to convince her to spend the rest of her life with him and on 03 August 1901 he and Clara were married (Plate 2.1).

No doubt Fritz was overwhelmed by Clara's acceptance of his proposal but for him marriage was also just another step on the road to social and academic stability. Clara too was ambivalent. As a woman, she felt that without a marriage[32] 'a new page arising in the book of my life, and a chord in my soul, would lie prostrate and barre'. Equally, marriage meant the abandonment of her independence and her unique scientific career and it provoked a crisis of

Plate 2.1   As pretty as a picture? The young Clara Immerwahr. Photograph courtesy of Archiv der Max-Planck-Gesellschaft, Berlin-Dahlem

identity. Clara eventually had to settle into the role of a professor's wife but their relationship was constantly troubled and in 1915 it all ended tragically.

2.9 Haber, the Eclectic Electrochemist

Soon after his marriage, Haber threw himself into his work with even more vigour and his reputation grew. The next five years were the most productive of his life both in terms of the number of his publications (over fifty[24]) and the variety of topics studied. The extent to which he was becoming respected within the German scientific community can be judged from the fact that he was chosen by the *Deutsche Bunsen Gesellschaft* (the German Bunsen Society) to be their delegate to the annual meeting of the American Electrochemical Society in 1902, with a commission to report on chemical education and the electrochemical industry in the USA. On 18 August Haber undertook a fully funded sixteen-week tour. The Americans were impressed by Haber's vitality, and his subsequent report published in the *Zeitschrift für Elektochemie* in 1903 was acclaimed both in Europe and the States, but some in America regarded his trip as industrial espionage in that many things told to him in confidence were disclosed. Haber was also building on the rigour he brought to his work, developing a technique of criticism *and* suggestion that began to create for him a reputation for solving problems and getting things done. It was precisely this kind of reputation that was soon to attract the best students to Karlsruhe.

Haber's work on nitrobenzene was followed by a number of other studies viz., the electrolysis of hydrochloric acid, examination of the phenomenon of 'Zerstäubung' where large clouds of electrode metal suddenly appear at the cathode, and the proposal of a general theory of irreversible and reversible electrochemical reduction, the latter in terms of the quinone–quinol system —simultaneously developing the theory of the quinhydrone electrode later used by Einar Biilmann to determine hydrogen ion concentration, 'pH'. Like many other electrochemists of the period, Haber was also attracted to the problem of 'fuel' cells where the current-producing process was provided by the atmospheric oxidation of carbon or carbon monoxide at quite low temperatures. Fuel cells were revolutionary at the time for it was realised that almost all of the 'free energy' of oxidation (i.e., the energy of the reaction available to perform *work*) was accessible making the cells highly efficient. Haber's efforts here led not so much to the development of a practical fuel cell but rather to the novel measurement of the 'free energy of a number of important reactions, viz., the oxidation of hydrogen, carbon monoxide and carbon, not readily accessible by conventional methods.[1]

Haber then took up the question of the electrochemistry of crystalline salts, and the most important outcome of this work was the construction and understanding of new types of electrode such as the *silver|silver chloride* electrode, involving the phase boundary potential between a solid salt and its saturated solution. Further studies on boundary potentials led to the development of the 'glass' electrode and its application in determining the *p*H of aqueous solutions. This was part of a general theory of the boundary potential between an aqueous solution and a second phase permeable to water and its ions e.g., glass, or solutions of water in organic liquids like benzene. Later, in 1909, Haber wrote of the glass electrode that it was 'likely to prove of much practical value' and indeed generations of chemists have since used the glass electrode in *p*H meters to accurately measure the *p*H of aqueous solutions.[1]

Throughout his Karlsruhe period Haber was also fascinated by the chemistry and electro-chemistry of iron. Indeed, his first paper on the subject of electrochemistry (in 1897, just after Luggin's arrival) dealt with electrodeposited iron plates used for the printing of banknotes.[1] But it was iron's passivity[33] that was to capture most of his attention and his efforts here played an important part in the development of the subject. There was a practical dimension of this work too in that Haber for many years investigated the corrosion of (iron) gas and water mains due to stray currents from tramway systems which in those days were operated by direct current. Haber looked at the problem in Karlsruhe and Strassbourg. With characteristic thoroughness he isolated the relevant factors; the composition and conductivity of the earth, the direction and magnitude of small earth currents and the behaviour of iron in the earth where chemical conditions prevented protection by passivity. He developed a general theory, and corrosion was always found to occur where his theory predicted it but the practical significance of his work was largely lost with the introduction of alternating current tramway systems.[1]

2.10   The Margulies Brothers

Haber reluctantly tackles the ammonia problem

This brief résumé of Haber's early work shows that it was wide ranging and thorough. It contributed to the understanding of a number of phenomena but it was hardly fundamental. As an electrochemist however he was inevitably drawn to the study of chemical thermodynamics and in 1905 he published a truly remarkable book, *Thermodynamik Technischer Gasreactionen,* (The Thermodynamics of Technical Gas Reactions).[34] According to J. E. Coates,[1] this book holds an important place in the history of chemical thermodynamics and for three reasons viz., 'its influence on teaching, its attack on the problem [*of entropy change at absolute zero*] that Nernst solved a year later, and its timely provision of the first systematic critical survey of all the thermodynamic data necessary for the calculation of the free energy changes in the most important gas reactions'—all of which we come to shortly. But important as this book was, it was not the seminal moment of Haber's life. That occurred about one year earlier—although Haber would never have recognised it. This moment occurred somewhere around 1903/4. Haber received an unexpected request from the Margulies brothers of the Osterreichische Chemische Werk in Vienna. They had detected small but noticeable traces of ammonia in their chemical plant and they wondered if they had somehow stumbled upon a viable method of making this valuable chemical feedstock. Otto and Robert Margulies asked if there would be any chance of finding a metal which could be used for the continuous production of ammonia by passing a mixture of nitrogen and hydrogen over it, making firstly the nitride then reducing the nitride with hydrogen to form ammonia, the solid metal phase acting as catalyst.

There had already been many attempts to promote a reaction between these two gases (Chap. 3) none of which had succeeded and a widely held belief arose that it was probably impossible to achieve such an elemental synthesis. As a result, Haber was noticeably uninterested in this opportunity and indeed wrote to Wilhelm Ostwald offering him the work. Haber was aware that Ostwald had studied the problem some years earlier and thought that he had taken it as far as it was technically feasible. But puzzled by Ostwald's disinterest in the matter and persuaded by the brothers persistence[35] and their generous financial offer, he and his young student Gabriel van Oordt spent some time in the summer of 1904 investigating the process suggested by the brothers. Preliminary experiments on the reduction and regeneration of calcium and magnesium nitrides indicated little hope in the use of these metals due to the high temperatures required. Haber then decided to examine the thermal synthesis of ammonia from its elements using (unsurprisingly) iron as a catalyst. Although this was not part of the original brief, his clear understanding was that whatever the pathway to ammonia production, the yield was governed by the position of the ammonia equilibrium in what was possibly a reversible process;

$$N_{2(g)} + 3H_{2(g)} \rightleftharpoons 2NH_{3(g)}$$

In this approach he built upon the work of William Ramsay and Sidney Young who in 1884 had examined the decomposition of ammonia at atmospheric pressure and 800 °C using iron as a catalyst.[36] They found that the decomposition of ammonia was never quite complete, but when they reversed the reaction, by trying to form ammonia from nitrogen and hydrogen under the same conditions, no detectable ammonia was formed. The persistence of ammonia in the *decomposition* process clearly suggested an equilibrium between nitrogen, hydrogen and ammonia but it also meant that the equilibrium position for this (formation) reaction lay far to the left, but Haber detected a degree of uncertainty here and felt the process worthy of re-examination if only to determine once and for all the feasibility of the elemental synthesis of ammonia.

Like Ramsay, Haber was going to look at the reaction from both sides of the equation but this time sequentially in the same apparatus—a schematic and description of which can be found in 'The Thermodynamics of Technical Gas Reactions' (Fig. 2.2).[34]

He used a gaseous flow method at 1020 °C and atmospheric pressure because of the simpler apparatus required. He placed two reaction tubes in series in the same furnace each followed by an ammonia absorber. By slowly passing ammonia into the system, the *decomposition* equilibrium was established in the first reactor tube viz.,

$$2NH_{3(g)} \rightleftharpoons N_{2(g)} + 3H_{2(g)}$$

The residual ammonia was absorbed on leaving the first tube and the nitrogen and hydrogen[37] entered the second. The ammonia *formation* equilibrium was then established in the second reactor.

$$N_{2(g)} + 3H_{2(g)} \rightleftharpoons 2NH_{3(g)}$$

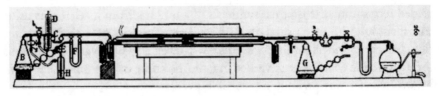

Fig. 2.2 Haber and van Oordt's apparatus from the Thermodynamics of Technical Gas Reactions[34]

On passing out of the reactor tube, the ammonia was once again absorbed and in each case the ammonia remaining/formed was determined by volumetric analysis using titration with methyl orange as indicator. Haber and van Oordt found that the amount of ammonia formed in the second reactor was virtually equal to the amount of undecomposed gas leaving the first reactor. For the first time then, the equilibrium position was confirmed and from both sides of the equation. The yield of ammonia however was predictably low varying between 0.005 and 0.0125%. Haber originally favoured the higher value as more representative but later work showed the lower value to be nearer the true equilibrium position, the higher yield eventually being traced to a special effect of the iron catalyst when fresh, viz., the presence of nitride in the iron.[38]

The results obtained by Haber and van Oordt are shown below—the values at temperatures below 1020 °C being calculated by the (integrated) van't Hoff equation (see Appendix A).

$Temp\,°C$	27	327	627	927	1020
$\%NH_3\,@\,equil.$	98.51	8.72	0.21	0.024	0.012

Overall however, the result was unequivocal and Haber's eventual conclusion was that the tiny amounts of ammonia available at equilibrium made the process impractical and therefore uneconomical. Indeed, at the time he commented;

From dull red heat upwards, no catalyst can produce more than traces of ammonia under ordinary pressure; and even at greatly increased pressure the position of the equilibrium must remain very unfavourable. To attain practical success with a catalyst at normal pressure then, the temperature must not be allowed to rise much above 300°C. ... The discovery of catalysts which would provide a rapid adjustment of the point of equilibrium in the vicinity of 300°C and at normal pressure seemed to me quite unlikely

Haber's efforts here however, were not entirely wasted. In addition to iron he had also experimented with calcium and manganese as catalysts and found that they had allowed the gases to combine at lower temperatures. Neither was he deterred—unlike many of his colleagues—by low yields, as he realised that the circulation of gases in a flow process accompanied by the removal of ammonia shifted the equilibrium and made it possible to gradually convert large volumes of gas. His results also established that the major proportion of a nitrogen/hydrogen mixture could theoretically convert to ammonia at room temperature—provided a suitable catalyst could be found—and that such a conversion would not require the generation of 'nascent' nitrogen as had previously been thought by some. He was also fully aware of the importance of high pressure,[39] and although he had the ability and technical means to proceed with experiments that optimised all these variables, he supported the widely held view at the time that the technical realisation of a gas reaction at the beginning of red-heat and under high pressure was impossible.

Haber published his results in 1905,[40] he then dismissed the whole episode and moved on expecting that to be the end of the matter. But his choice to publish at the higher[41] ammonia yield at 1020 °C was to set in motion a series of events that began his collaboration with Robert Le Rossignol, that defined his career as a physical chemist and led to the greatest technological development of the 20th century. And bye-the-way, had he published at the lower figure of 0.005%, the 'Great War' may well have been 'over by Christmas'.

2.11   Haber, Thermodynamics and the 'Heat Theorem'

In 'moving on', Haber attended to his new book which he dedicated to another chemist;

> To my dear wife Clara Haber Ph.D., in gratitude for her silent co-operation

The book was written as the result of a series of evening lectures delivered to colleagues and research students at Karlsruhe in February 1905, and it illustrates Haber's profound grasp of the theoretical aspects of chemical thermodynamics at the time.[42] Written in German, Haber's text was made available to the wider English-speaking world by Arthur B. Lamb[43] in 1907, who also incorporated the progress made in the subject in the intervening years. The book was an outstanding international success, a classic treatise of high significance. Apart from its influence on teaching, and its systematic approach to the collection of thermodynamic data, this book was revolutionary in a number of other ways. Firstly, it introduced chemical equations with fractional numbers of molecules e.g., $H_2 + \frac{1}{2}O_2 = H_2O$. Secondly, it coined the expression 'equilibrium box' for van't Hoff's device that helped explain how thermodynamics could be used to derive the 'Law of Mass

Action'.[44] But it was in its treatment of gaseous equilibria that its impact was most felt.[1]

Until the publication of Haber's book, gaseous equilibria had been discussed quantitatively in terms of the 'Law of Mass Action' and qualitatively through 'Le Châtelier's Principle',[45] but as early as 1888 Henri Le Chatelier had pointed to the importance of heat capacity in calculating equilibrium positions over a wide temperature range. Chemists generally had failed to recognise the importance of heat capacity, but writing primarily as a chemist —and referring to well known industrially important examples—Haber brought this aspect of chemical equilibria to the forefront of international teaching and research.[1]

By understanding the role of heat capacity, (see Appendix A), the equilibrium positions of gaseous reactions over a given temperature range could be determined through the integrated form of the van't Hoff equation; viz.,

$$\log_e\left(K_{p2}/K_{p1}\right) = -\Delta H/R(1/T_2 - 1/T_1)$$

where ΔH represents the average enthalpy change between the two temperatures, and K_p the corresponding (partial pressure) equilibrium constants. Now, given the arguments in Appendix A, the reader will see how it was possible to calculate the average value of ΔH over any temperature range from simple heat measurements, but given ΔH alone, the equation could only determine the *ratio* of the two equilibrium constants over that range. What the equation left unresolved was how to determine an equilibrium constant from changes in thermal data alone. This problem became the 'burning' issue of the day because thermal data was far more conveniently obtained than individual equilibrium constants. It was this fundamental problem that Haber addressed in his book.

Haber realised that the key to solving this problem lay in the free energy change (ΔG) of a chemical reaction. It was well known that ΔG was determined by the equation[46];

$$\Delta G = \Delta H - T\Delta S \qquad (2.1)$$

or under standard conditions,

$$\Delta G^\theta = \Delta H^\theta - T\Delta S^\theta$$

Once the standard free energy (ΔG^θ) was known, the equilibrium constant could be determined from the equation;

$$\Delta G^\theta = -RT \log_e K_p$$

and if the reaction took place under conditions other than standard, then the *Reaction Isotherm* applied, viz.,

$$\Delta G = \Delta G^\theta + RT \log_e K_p$$

so, it follows that if the free energy change can be measured under whatever conditions are convenient, the equilibrium constant under those conditions can also be calculated.

However, Eq. (2.1) shows us that to determine the free energy change for a reaction we need to know the enthalpy *and* entropy changes, ΔH and ΔS. Determining the enthalpy change was straightforward, it was purely a First Law problem, but there remained the problem of the entropy change. The Second Law of thermodynamics provided no guidance on the calculation of ΔS from thermal data because entropy change is defined in terms of thermodynamic reversibility and a chemical change is generally irreversible in this sense. All we can say here is that, for a change from A to B, the entropy change is related to the total heat absorbed irreversibly only by the inequality[47] below, so some other kind of third law was required to allow us to calculate the entropy change for a chemical reaction in other ways.

$$\Delta S > \sum (q/T)_{\text{irrev}}$$

By 1904 Haber and others had already suspected certain characteristic features of entropy change, in that for chemical reactions between solids at least, (e.g., in solid galvanic cells) the entropy change at 'absolute zero' must have a value of zero. In his book he developed this theme for gaseous reactions in terms of the heat capacities and entropies of the reaction components. Here too Haber suspected that as the temperature approached absolute zero the entropy change must converge to a limiting common value. He also realised that this value must be quite small—*possibly* zero—but characteristically, he could not bring himself to accept a purely speculative value entirely unsupported (as he saw it) by experiment. Even so he came tantalisingly close to identifying the behaviour that was later used as the basis of the '*Third Law*' of thermodynamics.[1]

But Haber was not alone with his thoughts because the experimental origin of the Third Law involved a number of physical chemists. By the early 1900s, van't Hoff, Nernst and the Americans Gilbert Lewis and Theodore W. Richards had also concerned themselves with the problem of changes at absolute zero. In 1902, whilst working on a different problem, Richards had collected the heats and the free energies of a number of reactions as a function of temperature. Unfortunately, Richards had a fundamental misunderstanding of thermodynamics and he was unable to 'see' what his data told him. But just down the hall at Harvard's chemistry department was Gilbert Lewis, a competent thermodynamicist who had tackled the problem of the relationship between free energy and chemical equilibrium in his doctoral thesis, which Richards had supervised but did not understand. Because he knew that the free energy change determines the point of chemical equilibrium, Lewis had proposed a general equation relating the two. Richards therefore had the data and Lewis the understanding, but the two men had fallen out, Richards accusing Lewis of 'appropriating' his data (in a separate area) and by 1904 Lewis left Harvard for the Philippines. There is no doubt that had Lewis seen Richards' data he would have spotted the 'pattern' and solved the problem. But he was now 'out of the loop' and the field was left open to van't Hoff, Haber and Nernst. van't Hoff's stab at a solution postulated two possible behaviours for the variation of the data with temperature, each missing the 'obvious' in the actual data. Haber's approach to the problem developed an over-complicated solution which, although probably quite correct, was difficult to prove.

Early in 1906 however, Nernst published his 'Heat Theorem' *based* on the pattern in the experimental data that all the others had missed. Using reliable published sources for his data, Richards had already noticed that as the temperature fell, the change in the enthalpy and the free energy for a reaction tended to converge to a some common value. At the time, this behaviour was perfectly evident from a number of perspectives. For example, Eq. (2.1) always applies, and in the limit as T tends to zero ($T \rightarrow 0$), the equation becomes,

$$\Delta G = \Delta H \text{ or } (\Delta G - \Delta H) \rightarrow 0$$

This is illustrated[48] very simply in the diagram below (Fig. 2.3); Similarly, the Gibbs Helmholtz relation[47];

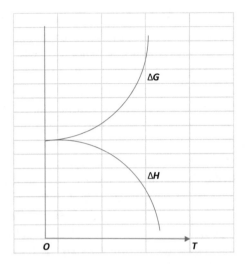

Fig. 2.3 Schematic variation of the enthalpy and the (Gibbs) free energy change with temperature

$$\Delta H = \Delta G - T(\partial G)/\partial T)_p \tag{2.2}$$

yields the same result when $T \to 0$. From Eqs. (2.1) and (2.2) any competent thermodynamicist would easily realise that the slope of the free energy curve with temperature is related to the entropy change by;

$$(\partial(\Delta G)/\partial T)_p = -\Delta S$$

whilst we have already shown that for the enthalpy curve (Appendix A);

$$(\partial(\Delta H)/\partial T)_p = \Delta C_p$$

But what Richards, van't Hoff and Haber all missed was that the slopes of the free energy and enthalpy curves always become tangential to one another as absolute zero is approached, and hence in the limit as T approaches zero so too must ΔS *and* ΔC_p. The bold step that Nernst then took was to look beyond just the published data and extend his observations to *all* chemical reactions. Nernst therefore assumed that this tangential behaviour for the variation of ΔH and ΔG with T was the 'norm' and he expressed the consequence of this in his Heat Theorem published in 1906, viz.,

$$\lim_{T \to 0} \Delta S = 0 \text{ and } \lim_{T \to 0} \Delta C_p = 0$$

implying that *all chemical reactions occurring at a temperature of absolute zero take place with no change in entropy or heat capacity*, precisely the result Haber had suspected and been trying to achieve.

To be strictly true, the heat theorem only applies to reactions involving (pure, crystalline) solids, although its principles can be applied to gas reactions too, but the hypothesis suggests that for some simple reaction A → B (say) at absolute zero[47];

$$\Delta S^{\circ} = 0 \text{ or } S_B^{\circ} - S_A^{\circ} = 0 \text{ or } S_A^{\circ} = S_B^{\circ} \qquad (2.3)$$

implying that if there is an entropy *change* at absolute zero, individual entropies S_A° and S_B° must exist. Equation (2.3) also suggests that if a substance has an entropy at absolute zero (whatever that may be) it must also have an entropy at *any* temperature $T > 0$, so that for any substance A we can also define some term S_A^T and that we further define the *absolute* entropy ($S_A{}_{abs}^T$) of a substance A at some temperature T (say) as[47];

$$S_A^T abs = S_A^T - S_A^{\circ}$$

Although the entropy change during an irreversible process cannot be determined directly from thermal measurements made during the change itself, the Heat Theorem—by identifying an absolute entropy and by defining entropy change at absolute zero as *zero*—allows the change to be calculated if the initial and final states are known and if these states can be linked by any convenient reversible path through absolute zero. For example, consider again the simple reaction[47];

$$A_{(g)} \to B_{(g)}$$

representing the conversion of one mole of A into one mole of B at some absolute temperature T. The Heat Theorem allows us to calculate ΔS for this process. Figure 2.2 shows a thermodynamically reversible path from A to B at temperature T. The 'natural' reaction is shown by path 1 to 2 (Fig. 2.4).

Path 1–3 represents one mole of A being *cooled* from T to absolute zero, path 3–4 represents the *reaction* at absolute zero and path 4–2 represents one mole of B being *warmed* from absolute zero to T. Because entropies are *state*

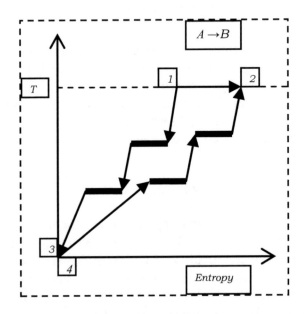

Fig. 2.4 The figure is a gross simplification. For representational purposes it has been assumed that both A and B are (ideal) gases at temperature T, so that the horizontal portions of the curves 1–3 and 4–2 represent fusions and evaporations. At absolute zero, points 3 and 4 are coincident ($S_A° = S_B°$). Because the point of coincidence makes no difference to the calculation of ΔS it can occur anywhere along the S axis not necessarily at the origin as my diagram shows. But it is counter intuitive to assume any other position. It has also been assumed that $S_B^T > S_A^T$ but there is no reason why this should always be so and the argument is the same if the opposite is the case

dependent only, the sum of the changes from 1 to 3, 3 to 4 and then 4 to 2 must be the same as the change from 1 to 2. Therefore;

$$\Delta S = ST_B - ST_A = \left(S_B^T - S_B°\right) + \left(S_B - S_A°\right) + \left(S_A° - ST_A\right)$$
$$= ST_{B\,abs} + \left(S_B° - S_A°\right) - S_{A\,abs}^T$$

Because $(S°_B - S°_A) = 0$ the equation reduces to

$$\Delta S = S_{B\,abs}^T - S_{A\,abs}^T$$

Here then lay the answer that Haber and others were seeking. We can determine the free energy change and hence the equilibrium constant of a chemical reaction by determining ΔH and then *calculating* ΔS if we can find reversible path through absolute zero to the temperature of the reaction. In the simple case of the reaction A $\rightarrow$ B, $S^T_{B\,abs} - S^T_{A\,abs}$ can be calculated if the variation of the molar heat capacities with temperature and any necessary latent heats are known—or can be estimated with confidence. The interested reader is referred to Appendix B where a calculation of the equilibrium constant for the isomeric transformation of *n*-butane into *iso*-butane from thermal data under standard conditions illustrates what Haber and the others were trying to achieve.

$$CH_3 \cdot CH_2 \cdot CH_2 \cdot CH_{3(g)} \rightarrow CH_3 \cdot CH(CH_3) \cdot CH_{3(g)}$$

Nernst's Heat Theorem ultimately led to the formulation of the '*Third Law*' of thermodynamics. The importance of thermodynamics to chemistry lies in the use of thermal data, measured over a range of temperatures, to calculate the equilibrium composition of reactions. The Heat Theorem and the subsequent '*Third Law*' provides a key to this application which could not have easily been achieved if the entropy change were not determined at absolute zero. Nernst eventually received the Nobel Prize (in 1920) but Haber's attack on this problem contained all the right ingredients. He understood the importance of entropy and of heat capacity and he suspected the answer that Nernst eventually proved. Indeed, J. E. Coates[1] observed that;

> to his eternal credit Haber was quick to acknowledge the step that Nernst had taken … in the history of this problem Haber must equally be accorded an honourable place.

Haber had been much influenced by Richards' work but Richards' misunderstanding both of mathematics and thermodynamics may have contributed to Haber's over-complicated approach to the problem. At the time however, Haber felt that his book was a great disappointment in that he had let the *Third Law* 'slip through his fingers'. Nernst too was influenced by Richards but only by his data, Nernst never really admitting that Richards' work contributed to his success. This led to bitterness between the two men and Richards never really forgave Nernst, claiming that he alone was the intellectual father of the *Third Law*. After the publication of the Heat Theorem Nernst busied himself by applying its ideas to published data and this led to another confrontation, this time with Haber over his results from the Margulies investigation. Because of this and the fact that Haber was 'beaten to the post' by Nernst, their relationship too became increasingly bitter for a while, but ironically through the Heat Theorem both men were to help each other to the ultimate triumph.

2.12 Professor Fritz Haber

By the time Robert arrived at Karlsruhe in September 1906, Haber had established a formidable international reputation for himself. Indeed, in some ways Haber *was* Karlsruhe and he was at the 'top of his game'. Through his achievements, his natural confidence and his military training he gave the impression that he was a 'great man' and a great man he eventually became when early in 1906 Max Julius Le Blanc—the incumbent professor of physical chemistry at Karlsruhe—left for a position at Leipzig. Le Blanc and Haber were not the best of friends. Haber coveted Le Blanc's position and wasted no opportunity to torment, criticise and belittle him within the Institute. When Le Blanc left, Haber called on all his friends to support his candidature for the vacant chair, and on August 10th 1906—just before Robert's arrival—Grand Duke Friedrich, Monarch of Baden, signed the official order appointing Dr. Fritz Jakob Haber to the lifelong position of (*ordentlicher*) Professor in the civil service of the German state.

Aged just thirty-seven, and taking up residence at 14 Weberstrasse[49] on the western side of the city, Haber easily slipped into his new role. The life of a full professor was what he had aspired to since leaving his parental home. Previously he had been rejected by the 'establishment' of physical chemists yet he had fought back through intelligent work to reach a respected position. Indeed, he became one of German chemistry's 'young bucks'. Now he had money to spare, money which he freely spent and often shared, and he had a scientific 'kingdom' that covered half of one floor of the yellow brick building that housed the Institute's chemistry laboratories. Haber also developed a characteristic unlike that of most professors of his time—a rapport with his students. Students galvanised Haber and he genuinely relished the role of scientific patron, shaping the professional lives of his youngsters. Nowhere was 'off limits' for him, in the streets, in the bars and in the laboratory, students were likely to encounter Haber's advice, and difficult as some of his questions could be, they loved him for his attention nevertheless.

Haber also used his home to be amongst his students. Frequent dinner parties saw him entertain his guests with an endless series of stories, jokes and rhymes composed on the spot. But home life also meant Clara, and much of Fritz's behaviour distressed her. His inability to economise both in his time and money generated constant friction between the two but Fritz also frequently failed to discharge his parental[50] responsibilities preferring instead to bury himself in his work. Clara remarked that '*if I didn't bring him to his son once in a while, he wouldn't even know he was a father!*' By this time Haber had also developed a pace of work that wore him out. 'Resting on his laurels' did not sit well with him but his approach to his work had its effect in that it aggravated a nervous condition that had been with him since his youth. It led to intestinal problems that drove him to seek the frequent advice of doctors and the seclusion of sanatoriums for rest and recuperation all of which was a strain on his marriage and on Clara. Haber's ambition inflicted a heavy price on his family.

This then was the man Robert Le Rossignol met when he arrived in 1906, and Haber could not have hoped to receive a more able research student. Through Coates' memorial lecture we see exactly why Haber came to regard Robert so highly. Yes, his formal academic credentials were impeccable, but his practical prowess—like Robert's medal from UCL—was 'gold' to Haber. Coates[1] writes;

> … [Haber's] scientific work was never lacking in thoroughness and attention to detail. It is true however that he found no deep interest in the minutiae of experimentation as such. He liked his men to be clever experimenters and greatly appreciated good work by them, but he expected them to work out the finer details.

In Robert Le Rossignol Haber certainly found a 'clever experimenter'—after all he came from Ramsay's innovative laboratory, the same laboratory that had influenced the design of Haber's ammonia experiment for the Margulies brothers. But Haber had already collected a formidable group of physical chemists at Karlsruhe who produced some outstanding work and in doing so built an excellent reputation for his Institute of Physical Chemistry and Electrochemistry. His group investigated diverse areas such as flame reactions and the process of combustion, the emission of electrons in chemical reactions and the reactions of mercury in which amalgams were formed. But there were also areas of more practical significance such as the determination of the quality of glass used in the manufacture of wine bottles. Haber also collaborated with the Zeiss works in Jena to develop a refractometer for gas analysis[51] and this in turn led to improvements in measuring instruments used in many areas of science, technology, public health and industrial safety. This device was almost certainly used for the ammonia experiments with Le Rossignol later.

Many of Haber's physical chemists too were to go on to careers of distinction. Paul Askenasy had been *Assistent* to Viktor Meyer in Heidelberg before being persuaded to join Haber, becoming Associate Professor in 1909. As early as 1907 Haber's laboratories had become known as a centre of high-pressure research, and when Friedrich Bergius conceived the idea of hydrogenating coal under high pressure it was to Karlsruhe he came, (1908)[1] working under Haber for one semester in Haber's personal laboratory. This work led to the original coal-to-oil conversion preceding the better-known Fischer-Tropsch process—which eventually saw him achieve the Nobel Prize in 1931 (jointly with Carl Bosch). Haber and Zsygmunt Klemensiewiscz were to invent the glass electrode in 1909 and J. E. Coates was to become Professor of Chemistry at University College Swansea in 1920. Paul Krassa—a distant relative of Haber's wife—was eventually appointed as Professor of Physical Chemistry and Industrial Chemistry at the University of Santiago Chile, and Hermann Staudinger a pioneer of macromolecular chemistry went on to receive the Nobel Prize in 1953. But capable as all of these men were, we also have to mention his universally admired 'servant', the technician Friedrich Kirchenbauer whom Haber had 'inherited' from Le Blanc and whom Haber described as "an unusual employee of special value". It was thanks to Kirchenbauer that many of the technical innovations of Haber's group functioned as intended. Clearly then, to make an impression within such an elite group any newcomer would have to be quite exceptional.

Robert arrived at the T. H. Karlsruhe in the autumn of 1906, living initially in a flat with Professor Bunte and his wife who—according to Robert —was 'very friendly' and helpful towards him.[52] Although matriculating as an unpaid guest instructor or 'Hospitant'—a special status reserved for those who had already graduated—no personal record for Robert remains at the university nor is it possible to determine which lectures he gave (if any at all). A search of the Generallandesarchiv[53] at Karlsruhe similarly provided no information regarding his stay suggesting, unsurprisingly, that this 'impecunious student'[54] never owned a property there. However, what does remain, is his hand written entry (student no. 89) in the Hochschule register[55] for the winter semester of 1906 showing that he eventually took up residence at "Nowakanlage 13[IV]"[56] where he apparently remained for the duration. In the same entry he declared his religion—one of few who bothered on that particular page—as 'Protestant' and his father's occupation as a Doctor in 'Medecine'. At Nowakanlage, Robert lived with a Professor Knorr. She had two daughters and someone once asked his mother … was she not afraid of her son living there? She apparently replied, '*Not at all … you haven't seen them!*'[52] (Plate 2.5).

Karlsruhe at the time of Robert's arrival was a city of about 100,000 souls built around the magnificent Karlsruhe 'Schloss', (palace) which dated back to 1715. The Schloss sat in the centre of large circular grounds—the Schlossgarten—and from it emanated 32 roads like the spokes of a wheel.

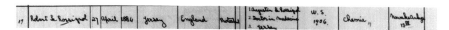

Plate 2.5 Robert's hand-written entry in the Hochschule register for 1906. Courtesy of KIT-Archives

The German nickname for Karlsruhe was therefore Fächerstadt or 'fan city'. The Schloss itself consisted of a south facing central section with the 'Schloss tower' at the centre of the 'wheel'. The wings ran out at 45° angles from the tower pointing SE and SW. The two roads running parallel with these wings formed a sector containing the old city centre and included Nowakanlage. The old city was sophisticated, refined, and offered markets, cafés, bars, museums, libraries and of course the Schlossgarten's beautiful walks. Added to this, the colleges of the Hochschule were only about 1 km away 'as the crow flies'. Karlsruhe, therefore was an ideal city in which to study, relax and of course meet new friends. Being close to the French border, 'Le Rossignol' too must have felt—culturally at least—quite at home.

When Robert first met Haber at the Hochschule, they discussed the work he had published at UCL. Robert had recently attended a short course on electrochemistry there and Haber naturally decided to ask him to repeat some work in that area.[52] Over the next few months he did 'various things'[52]—including beginning a study of the dissociation of carbon dioxide in the carbon monoxide flame, an exercise consistent with Haber's life long interest in the chemistry of flames. Working in a small laboratory alongside Krassa and Koenig,[52] Robert described these first few months as a *very happy time for everybody*,[52] and Haber as just a charming *'big boy'*[52]—but not of course in any kind of Mae West sense! Robert also came to Karlsruhe at probably the happiest time in Haber's life, J. E. Coates[1] later recalling his own time there with him;

> One remembers summer evenings round a punch bowl in the Stadtgarten, or walks through the woods to some favourite Gasthaus, when his lively spirit, his stories and charming ways were a great delight …

Robert too was smitten by Haber and they clearly got on very well together; he 'could not have been nicer', he says, he was *never* the 'Herr Professor' he was down-to-earth and he could enjoy things thoroughly, he was also extremely kind and he would find time to help everybody.[52] Robert was often invited to Haber's home to help with his German, and there he met Clara. She was 'very shy' and that gave a 'wrong impression', but even then it was clear that the couple were not suited.[52] Haber could be rude to Clara and that was uncomfortable for everyone present, but equally Haber would often bring students home at very inconvenient times for Clara and that may well have contributed to her lack of commitment to her guests. Robert also remembered that when they were first engaged, they often 'walked out' together, but after their marriage Haber 'forgot his bride', so even as early as 1906, all of this was taking its toll. Robert had also seen a photograph of Clara when she was young (possibly Plate 2.1) she was as 'pretty as a picture', but now he hardly recognised the lady he met, and of Clara's depression, Robert says he 'knew nothing'.[52]

Like Coates, Robert also recalls many walks with Haber into the Black Forest. There were often lots of people on these trips, sometimes there was 'tom-foolery'—Robert once recalling someone 'mimicking' Bunte—and there were always lots of speeches but with Haber monopolising the whole event. Robert also supports Coates' opinion of Haber's laboratory skills ... '*he was not a good experimentalist*'. Robert said, '*we used to laugh ... he could be just doing something and it would go off with a bang*!!'[52] But Haber could sometimes become angry with his students, Robert recalls a time when he asked a group about 'Ohmsche Gesetz' (Ohm's Law). '*Nobody knew what it was, I was there just as a witness*', but Haber said furiously '*I never take any people until they can answer these questions!*'[52] But such was Haber, sometimes a storm, sometimes a sunbeam.

After just a few months in the Hochschule, Robert's standing and his command of the German language[57] were well enough regarded by Haber for the 'great man' to ask him to contribute to Lamb's translation of his internationally acclaimed book on technical gas reactions,[34]—an accolade in itself. Eventually published in October 1907, his translation of the Appendix to Chap. 5 (*Some examples of reactions involving a change in the number of molecules*) was acknowledged by Lamb in the Translator's Preface, so even in this short period of time Robert had clearly made an impact. And of all the capable physical chemists at Karlsruhe it was Robert who Haber 'asked'[54] to work with him and eventually contribute to his 'Magnum Opus' viz, a successful solution to the problem of the 'fixation' of nitrogen. But why was 'fixation' *so* important to Germany, and why was Robert so well suited to help achieve it? We pursue these questions in Chaps. 3 and 4.

Notes

1. An observation taken from J. E. Coates' 'Haber Memorial Lecture'. After a brilliant career as a student at Bangor, London and Karlsruhe, followed by distinguished service during the First World War in the Royal Naval Air Service, J. E. Coates was called to the Chair of Chemistry at the newly established Swansea College of the University of Wales in 1920. At Karlsruhe, Coates was a student of Haber's and a contemporary of Le Rossignol, and on 29 April 1937 he delivered the Haber Memorial Lecture before the Chemical Society (J. E. Coates, 'The Haber Memorial Lecture', *J. Chem. Soc.*, 1642–1672, (1939)). The text is available from the on-line archive of the Royal Society of Chemistry at http://www.rsc.org. The reference to Haber's dislike of the minutia of experimentation appears on page 1671. The paper is referred to throughout this chapter and indeed throughout the whole book.

2. Richard Willstätter, *Aus meinem Leben*, Ed. A. von Stoll, Weinheim Verlag Chemie, (1958).

3. Rudolf Stern, *Fritz Haber, Personal recollections*, Leo Baeck Institute Yearbook **8**, 1, 70–102, (1969).

4. Johannes Jaenicke was a student of Haber's during the 'Gold from sea water' episode of Haber's life, (Chap. 13).

5. Max Planck Society for the Advancement of Science, Berlin. http://www.mpg.de/english/aboutTheSociety/.

6. Vaclav Smil, *Enriching the Earth*, MIT Press, (2001). http://mitpress.mit.udu. ISBN-978-0-262-693134. Here, a partial biography of Haber helps to contextualise the technology of 'fixation'. Daniel Charles, *Between Genius and Genocide*, Jonathan Cape, London, (2005). ISBN-0-224-06444-4. Dietrich Stoltzenberg, *Fritz Haber. Chemist, Noble Laureate, German, Jew.* Chemical Heritage Press, Philadelphia, Pennsylvania, (2004). ISBN-0-941901-24-6. Margit Szöllösi-Janze, *Fritz Haber 1868-1934: Eine Biographie.* C. H. Beck, Munich, (1998). ISBN-3-40643548-3.

7. I have relied heavily on the authors in notes 1 and 6 for my interpretation of Haber's early life. I hope I have provided adequate attribution. I apologise for any omission I may have made.

8. Today, Polish Wroclaw.

9. The family did not practise Jewish ritual however. According to Haber's (half) sister Else it would be a mistake to *over*-emphasise her brother's Jewishness. Christmas, for example, was more important to the family than Yom Kippur, but the tree, presents and the Christmas lunch were the focus, *not* the infant Jesus. Charles, op. cit. (note 6).

10. For example, he became an Alderman of Breslau.

11. The Bunsen burner had its genesis when Bunsen was working with Gustav Kirchhoff, a young Prussian physicist, who had the brilliant insight to use a prism to separate (spectral) light into its constituents. Thus, the fledgling science of spectroscopy was born. In order to study the spectra, however, a high temperature, non-luminous flame was necessary. The burner was quickly dubbed the 'Bunsen' burner. The concept to premix the gas and air prior to combustion in order to yield the necessary high temperature, non-luminous flame belongs to Bunsen, but credit for the actual design and manufacture of the burner goes to Peter Desaga, a technician at the University of Heidelberg.

12. Haber was to be introduced to 'cacodyl' himself in 1915, it was in part responsible for the death of Otto Sakur at the KWI. See Chap. 1 (Sakur) and Chap. 13 (cacodyl).

13. Even today, Chinese scientists have confirmed that some iron-rich porous materials can remove arsenic from drinking water in under 2 h(!). See for example the Royal Society of Chemistry's, *Chemistry World*, **8**, 3, 21, (2011).

14. Becoming a reserve officer did not mean a full-time military career, but one had to make oneself available for national duty in time of war. Whatever career one pursued the officer's uniform was a mark of social distinction in Germany and often more valuable in terms of career progress than wealth or degrees.

15. At this time Haber decided that he wanted a doctorate and not simply the diploma which would have been granted by the Hochschule. Under the rules that allowed German science students to move between universities however, his thesis had to be submitted and defended at the Friedrich-Wilhelms-Universität to achieve his doctorate.

16. 1,3-benzodioxole-5-carboxaldehyde, used in perfumery and in flavourings such as cherry and vanilla.

17. The term 'amphetamine' was not coined until 1938. The British newspaper '*The Observer*' however on 20 February 2011 claims 3–4. MDMA was first synthesised by a German chemist, Anton Kollisch of Merck.

18. *Cum laude* with praise, *magna cum lauda* with great praise, *summa cum lauda* with the highest praise. Haber's degree was therefore undistinguished.

19. Ironically, this factory was just a short distance from the site of the future Auschwitz camp where many of Haber's own family were to die in the 1939–45 world war.

20. This part of Haber's life is vague and poorly documented but according to some accounts when cholera broke out in Hamburg in 1892—eventually killing 8,600 people—Fritz impetuously bought large amounts of 'chlorate of lime' which at the time was used to treat the disease by being tipped into water supplies in order to purify them. Unfortunately (for Fritz!} the outbreak was short lived and the company was left with large amounts of a chemical that they could not 'shift'. Siegfried was *not* pleased.

21. *Geheimrat.* An 'old fashioned' German term, a title originating in feudal times describing a role as advisor to the throne—a privy councillor. In modern terms an erudite elder, one whose opinion should be respected. The actual title of 'Geheimer Regierungsrat' was eventually bestowed on Haber by the Kaiser a short time after he had received a full honorary professorship at the philosophical faculty in Berlin University in December 1911, the same university from which he had obtained his doctorate twenty years earlier. Also, Stoltzenberg, op. cit. (note 6), p. 114.

22. Conversion or not, Haber retained his sense of Jewishness. He continued to associate with Jewish friends or those of Jewish descent and non-Jews often disparaged him for this, regarding it as hypocritical and conversion merely a clever trick.

23. According to the papers Johannes Jaenicke left to the Max Planck Society for the Advancement of Science, Haber admitted this fact with glee! See, Johannes Jaenicke, '*Fritz Haber, 1868–1934*', Fridericiana, Heft 35, Universität, Karlsruhe, a reference quoted by Bretislav Friedrich of the Fritz Haber Institut der Max-Planck-Gesellschaft in *Fritz Haber, 1868–1934*. http://www.fhi-berlin.mpg.de/ ∼ brich/Friedrich_HaberArticle.pdf.

24. A complete list (in English) of Haber's publications appears in M. Goran, *The Story of Fritz Haber*. University of Oklahoma Press, Norman, Oklahoma. See also Margit Szöllössi-Janze, (note 6), where the definitive list appears in German.

25. In Germany, the title *Privatdozent* (for men, *Privatdozentin* for women), was conferred on academics who had earned their doctorate then written another thesis (or possibly a book) at a level of scholarship substantially higher than their doctorate. It was an unpaid position which allowed the carrier to lecture to students using their fees as compensation. The title was also conditional upon the success of a lecture before the relevant department or faculty of a university. The status is roughly equivalent to an associate professor and permits its holders to supervise doctoral research by students. The post is still widely held on mainland Europe but in the USA and in the UK a Ph.D. is sufficient for doctoral supervision.

26. An unusually reactive form of hydrogen produced in situ in the reaction mixture. 'Nascent' (newly born) hydrogen can reduce compounds that do not readily react with normal molecular hydrogen gas. The reactive capability of nascent hydrogen is not clear. It may be in the form of short-lived atomic hydrogen. It may also be in the form of an unstable hydride compound. It may also be possible that 'normal' hydrogen molecules are indeed formed as they are released from the reaction mixture, but in excited states that react before they relax to their ground state. Nascent hydrogen is usually produced by reaction of metals with acids, for example, zinc and sulphuric acid.

27. See, Andreas Sella, 'Luggin's capillary', *Chemistry World*, (July 2012). An eminently readable account.

28. But Haber's approach to his work was to try and master a subject 'overnight'. He would work tirelessly absorbing the experience and knowledge of others until he 'got it'. Haber had barely set foot in physical chemistry, yet he was prepared to take on the establishment and make his mark. Inexperienced or not, his work on nitrobenzene was still being referenced almost 70 years later, see for example Harwood, U.S. Patent No., 3,338,806, 1967.

29. 'Professor', but without a Chair, subordinate to the main professor who runs the department.

30. 'Immerwahr' (roughly) translates literally to *'always true'*. This may not have been the case for Haber however, see for example D. Sheppard, 'Haber's legacy', *Letters, Chemistry World*, **5**, 11, 34, (2008).

31. The same kind of marriages were also common to Jersey's elite families. S. J. Le Rossignol, *Historical Notes (Local and General) with special reference to the Le Rossignol Family (and its connections in Jersey)*, Trowbridge, (1917) shows that down the years the Le Rossignol family bonded in the same way.

32. From Stoltzenberg, op. cit. (note 6), p. 174, although see Charles, op. cit. (note 6), p. 51, who has a slightly different version.

33. *Passivity*, an element's tendency to become un-reactive often due to the formation of a tight (oxide) layer that protects it. Coates, op. cit. (note 1), p. 1649, describes Haber's explanation of iron's passivity in detail.

34. *'Thermodynamics of Technical Gas Reactions'*. By the term *'Technical'* Haber was referring the technical *realisation* of a gas reaction through chemical technology or engineering. Published in 1907, the whole book is now out of copyright but is currently available 'on-line' in English and PDF format from the internet archive at the University of California Digital Library; http://www.archive.org/details/thermodynamicsof00haberich, (also note 43).

35. Even though Haber had pointed out to the brothers the low cost of obtaining ammonia as a by-product from coking plants.

36. W. Ramsay and S. Young, Decomposition of Ammonia by Heat, Trans. *Chem. Soc.*, **45**, 88, (1884).

37. By performing the decomposition *first*, the tiny amount of ammonia left guaranteed that a virtually stoichiometric ratio of $3H_{2(g)}:1N_{2(g)}$ entered the second tube. The importance of this ratio in maximising the yield of ammonia was later established by Haber in his *'The Thermodynamics of Technical Gas Reactions'*, Haber, op. cit. (note 34).

38. See Chap. 3, Ostwald was already well aware of this problem but he did not communicate it to Haber.

39. Haber made this point in his Nobel acceptance speech in 1920 ... *"it needed only a slight modification of the pressure oven, such as that used by Walther Hempel 15 years earlier to carry out nitrogen absorption in the case of indirect ammonia synthesis under pressure of up to 66 atmospheres. But I did not think it worth the trouble ..."*. Fritz Haber, Nobel Lecture 2 June 1920; 'Fritz Haber —Nobel Lecture: The Synthesis of Ammonia from Its Elements'. *Nobelprize.*

org. Nobel Media AB 2014. Web. 9 March 2016. http://www.nobelprize. org/nobel_prizes/chemistry/laureates/1918/haber-lecture.html.

40. F. Haber, G. van Oordt, 'Über Bildung von Ammoniak aus den Elementen', (Formation of Ammonia from the Elements), *Z. Anorg. Chem.*, 44, 341–371, (1905).

41. '... *for various reasons ...*' according to Coates, op. cit. (note 1), and also Haber's opinion that, '*some discrepant values seemed to me to point to the upper limit as the probable value*', Haber, op. cit. (note 39), p. 334.

42. Haber's mastery of theory and his dislike of experimentation was to dictate his working relationship with Le Rossignol, a point we return to in later Chapters.

43. Arthur B. Lamb, at the time Assistant Professor and Director of the Havemeyer Chemical Laboratory at New York University, USA. See the link; https://www.harvardsquarelibrary.org/biographies/arthur-becket-lamb/.

44. The law of mass action is universal and applicable under any circumstance. However, for reactions that are *complete* (virtually 100% conversion from reactants to products), the law may not be very useful. We introduce the mass action law by using a general chemical equation for a *reversible* reaction in which reactants A and B react in the *forward* reaction to give products C and D, and where C and D themselves are simultaneously capable of reacting in the *reverse* reaction to reproduce A and B.

$$aA + bB \rightleftarrows cC + dD'$$

Here, *a, b, c, d* are the coefficients required for a balanced chemical equation and where the symbol $\rightleftarrows$ indicates the establishment of a chemical equilibrium, a *dynamic* state where the forward reaction ($\rightarrow$) rate equals the reverse reaction ($\leftarrow$) rate and where the composition of the mixture over time remains unchanged. In its simplest form the **mass action law** states that if the system is at equilibrium at a given temperature, then the following ratio;

$$[C]^c[D]^d / [A]^a[B]^b = K_{eq}$$

is a *constant* where [A], [B], [C] and [D] represent the *concentrations* of the reactants and products at *equilibrium* and where K_{eq} is called the *equilibrium constant* for the reaction. The *composition* of an equilibrium mixture can therefore vary depending upon the conditions under which it was achieved, but at all times K_{eq} must be maintained. For gaseous reactions, equilibrium partial pressures replace concentrations. This 'law of chemical equilibrium' can also be derived from thermodynamic methods. In 1886 J. H. van't Hoff used a hypothetical device to derive the result for the general reaction above involving ideal gases. The device comprised two large chambers, at the same temperature, in each of which the four species A, B, C and D are always in equilibrium. Each of the four walls is permeable to only one of the substances and by transferring reactants and products reversibly and isothermally between the two boxes through the walls van't Hoff was able to deduce the same equilibrium expression (see Appendix C). It was Haber who named the device the 'equilibrium box'. Needless to say, the law of mass action is a much simpler derivation but interested readers can find the detail of van't Hoff's argument in older physical chemistry text books e.g., Samuel Glasstone, *Textbook of Physical Chemistry*, 2nd Ed, Macmillan, (1966).

45. Le Chatelier's Principle, an empirical generalisation that describes what happens to the equilibrium *position* (i.e., the composition of the equilibrium mixture of a chemical reaction) when the reaction conditions are changed. The Principle was originally stated as;

Every change in one of the factors of an equilibrium occasions a rearrangement of the system in such a direction that the factor in question experiences a change in the sense opposite to the original change

A more approachable definition might be;
'when a *constraint* is applied to a chemical system at equilibrium, the system responds in such a way as to *minimise* the effect of the constraint'
Thus, in the general reaction, if *at equilibrium* we add more of A (and, or B), then the system will respond by counteracting (minimising) the effect of the increased concentration(s) by using A *and* B to form more of C and D. The equilibrium is then said to *move to* the right of the chemical equation. This behaviour follows directly from the definition of K_{eq} in that if [A] (and, or [B]) are suddenly increased the only way K_{eq} can remain constant is to use them to increase [C] and [D]. The reverse is true if we add more of C (and, or D) *moving* the equilibrium to the left.
Similarly, if we were to increase the *temperature* of a system at equilibrium, we would favour the reaction that *absorbed* heat (i.e., the *endothermic* reaction) and vice versa. The effect of *pressure* on a *gaseous* reaction too follows the Principle. If we were to *increase* the pressure then the equilibrium will react to favour the reaction (forward or reverse) that results in a *smaller* volume. In the important ammonia reaction;

$$N_{2(g)} + 3H_{2(g)} \leftrightarrows 2NH_{3(g)}$$

increasing pressure will move the equilibrium to the right resulting in more ammonia formation because the forward reaction results in *four* volumes of gas reducing to *two*. For this reaction, the decomposition of ammonia into nitrogen and hydrogen (the *reverse* reaction) is *endothermic* hence according to Le Châtelier ammonia formation is theoretically favoured by *high* pressure and *low* temperature. Simple? Not quite, reaction rates are *slower* at lower temperatures and so it would take *longer* to reach the equilibrium position. To achieve reasonable conversions to ammonia we need to use high pressures, reasonably high temperatures (with a catalyst to avoid too high temperatures) and if we *remove* the ammonia as it is formed, the reaction's drive to maintain the equilibrium constant will result in more nitrogen and hydrogen being used to replace the ammonia. For gaseous reactions where no change in volume is achieved, pressure simply affects reaction rates. High pressures mean faster reactions, low pressures the opposite.

46. The Gibbs free energy term ΔG, (after Josiah Willard Gibbs 1839–1903), reflects the energy that is available to perform *non-expansion work* on the surroundings once the demands of enthalpy (internal energy and expansion) and entropy have been attended to, thus;

$$\Delta G = \Delta H - T\Delta S$$

Any process can be regarded as *feasible* in a thermodynamic sense if it leaves the total energy of the universe unchanged, but a process is only *spontaneous* at constant pressure (i.e., occurs without work being done *upon* it) if ΔG is *negative*. The more negative this value the greater is the energy available to do work, and only reactions that 'go forward' and proceed substantially towards completion could ever achieve useful work. We therefore suspect that there must be a relationship between ΔG and K the reaction equilibrium constant.

47. M. H. Everdell, *Introduction to Chemical Thermodynamics*, English University Press Ltd., (1965).

48. In the diagram, ΔG is shown as being greater than ΔH away from absolute zero, but the reverse state of affairs is also possible. Similarly there is no intention to suggest that ΔH and ΔG approach zero as absolute zero is approached. ΔH and ΔG remain (positive or negative) *finite* quantities.

49. An address given in some of Haber's later patent applications e.g., Patent application No. 14,023 at the United Kingdom Patent Office, '*Improvements in the Manufacture of Ammonia*', dated 09 June 1910.

50. Young Hermann Haber (notice, another *Germanic* not *Jewish* name) was born to Clara and Fritz on 01 June 1902. At the time, the Habers lived on Moltkestrasse, in western Karlsruhe, where little Hermann grew up, Stoltzenberg, op. cit. (note 6), p. 176.

51. *Testing Gaseous Mixtures*, United States Patent Office, No. 830,225, Patented 04.09.1906.

52. From Jaenicke and Le Rossignol's letters and the transcript of their conversation (on 16 September 1959) held at the Max Planck Gesellschaft, Archiv der MPG, Va. Abt., Rep. 0005, Fritz Haber. Haber Sammlung von Joh. Jaenicke. Nr. 253 (Jaenicke's letters) and Nr. 1496 (conversation transcript).

53. Courtesy of Dr. Rainer Brüning, Generallandesarchiv, Karlsruhe. Personal communication by email, (O2 May 2011).

54. From Ralph C. Chirnside's sensitive Obituary of Robert; R. C. Chirnside, 'Robert Le Rossignol, 1884–1976', *Chemistry in Britain*, **13**, 269–271, (1977). Chirnside was a long-time friend and colleague of Robert from their time together at the GEC laboratories. The Obituary was based upon a tape-recorded interview he conducted with Robert on 29 March 1976.

55. Courtesy of Dr Klaus Nippert, Archivist at K.I.T. Personal communication by email, (19 April 2011).

56. Nowakanlage is still an area of Karlsruhe, but because much of the city was reduced to rubble in the second world war it has been redeveloped. This address was recorded in a number of the patents he and Haber applied for over the next few years, suggesting that it was his permanent residence in Karlsruhe.

57. UCL must have also offered Robert ample opportunity to learn the language. Many staff such as Donnan and Ramsay were fluent and the presence of German students such as Hahn and Sakur must have contributed enormously to the overall environment. Many of the important papers of the time were also in German and students were *expected* to translate them. Even the author of these lines remembers doing that in the late 1960s and early 1970s with the help of Scientific German lectures and a German-English dictionary!

3

Germany and 'Fixation'

Photo der "Preussen" unter Vollzeug beim Auslaufen aus dem New Yorker Hafen

The 'Prussian', a fully rigged clipper or wind-jammer leaving New York on 27 May 1908.[1]

Of the 'Prussian', Chile, nitrogen and Haber

'The fixation of atmospheric nitrogen is one of the great discoveries awaiting the genius of chemists!'

Sir William Crookes, part of his presidential address to the British Association for the Advancement of Science, 1898.

3.1 Iquiqui, Tocopilla and Taltal

On 31 July 1902, the German five-masted, steel clipper *Preussen*[1] (the 'Prussian'), left Geestemünde on the German north sea coast near Bremerhaven on her maiden voyage. *Preussen* remains the largest pure sailing ship ever to be built—partly on a whim of the autocratic German Kaiser

© Springer Nature Switzerland AG 2020
D. Sheppard, *Robert Le Rossignol*, Springer Biographies,
https://doi.org/10.1007/978-3-030-29714-5_3

Wilhelm II who insisted that Germany had the most prestigious vessels afloat —but principally because of Germany's need to maintain an acutely important trade route with South America. On 18 June 1899 the German Kaiser visited the F. Laeisz shipping company at Hamburg and was shown around the five-masted barque *Potosi* by the legendary Captain Hilgendorf. At the end of the tour the Kaiser turned to Carl Laeisz and reportedly asked '*Na, Laeisz, wann kommt denn nun das Fünfmastvollschiff?*'[2] Over the next three years, following the Kaiser's 'advice', the *Preussen* emerged becoming the pride of the German five-masted South American fleet. She was huge, immensely strong, incredibly well equipped, able to cover thousands of sea miles at a steady 11–13 knots and she could transport a colossal cargo of 8000 metric tonnes.

On her outward maiden voyage under the command of Captain Boye R. Petersen, *Preussen* carefully navigated her way down the English Channel and then moved westward. Off 'Start Point' in Devon, she began her journey proper heading south-west for Cape Horn on the southern-most tip of South America. Beating into the headwinds of the Cape she crossed into the Pacific then sailed north along the Chilean coast reaching the port of Iquique on 08 October just 65 days after passing Start Point. Here—using her two 'donkey' engines—she loaded 7729 tonnes of a brown mineral powder which had been poured into 62,000 sacks, and on 24 October she sailed on the return journey to Hamburg. Lizard Point (the southern-most tip of Cornwall) was passed 79 days out, and she entered Hamburg on 29 January 1903. This was the first of thirteen trips over the next ten years that was to see *Preussen* visit the northern Chilean ports of Iquiqui, Tocopilla and Taltal. Each time she loaded almost 8000 tonnes of the mineral to become the 'Queen of the seas', part of the commercial armada that plied the toughest trade route in the world—a trade that would bring a tide of the brown mineral to Europe and the United States.

The mineral was sodium nitrate ($NaNO_3$) or 'Chile saltpetre', a primary source of 'fixed' nitrogen used across the world to replenish[3] cultivated soil and to produce explosives.[4] Sodium nitrate for example could be used directly to produce blasting powder (typically 74% $NaNO_3$, 16% C and 10% S) or it could be converted to potassium nitrate, KNO_3, for use in gunpowder (75% KNO_3, 15% C and 10% S) because sodium nitrate, being deliquescent, was unsuitable. This conversion initially relied on potassium carbonate but later it was produced by 'reacting' $NaNO_3$ with 'Sylvite' (potassium chloride) large deposits of which were discovered in Germany near Stassfurt in 1857.[5] The 'double decomposition' reaction was followed by fractional crystallisation; the sodium chloride being less soluble, the potassium nitrate remaining in solution;

$$KCl_{(aq)} + NaNO_{3(aq)} = KNO_{3(aq)} + NaCl_{(aq)}$$

The quantity of nitrate *Preussen* delivered to Germany on each of her voyages was sufficient to fertilize 40,000 hectares of land or provide gunpowder for a whole German army corp.

The Chilean ports were remote but their skies were pin-pricked by a bristle of masts from sailing ships from all over the world, each drawn there by the three major deposits that lay fifty miles inland. These deposits, discovered in 1821 by Mariano Eduardo de Rivero, lay in arid plateau of the Atacama Desert a few thousand feet above sea level and each was served by its own port from Iquiqui in the north, through Tocopilla, to Taltal in the south. The crude mineral of these deposits was caliche[6] a conglomerate of insoluble material cemented by soluble salts, including nitrates, sulfates and chlorides together with borates and iodates. Miners blasted open the deposits and carried the rock on their backs to refining plants, but the process was utterly wasteful with much of the caliche remaining in place. However, the extracted mineral was then crushed and soaked in hot water to release the soluble nitrate. The solution was concentrated in tanks and the water evaporated leaving the brown nitrogen rich mineral composed of 94–96% $NaNO_3$, 1–1.5% NaCl and 1.5–2.25% water. The overall available nitrogen content was therefore about 15%, roughly thirty times as much as common 'barnyard' manures.[5]

These mineral deposits were clearly of immense economic importance and were the cause of warfare between Chile, Bolivia and Peru, a conflict ultimately resolved in favour of Chile in 1883. Their export also eclipsed an earlier source of nitrogen—guano, or sea-bird droppings—left on the islands off Peru. This material was rapidly depleted and varied in nitrogen content from 5–15%. Annual exports of the saltpetre began to rise rapidly after 1880; they surpassed 1 million tonnes (1Mt) in 1885 and were nearly 2.5Mt by 1913.[5] Concerns naturally arose regarding the long term sustainability[7] of this resource and these concerns addressed two strategic goals; a nation's ability to produce adequate harvests of food to sustain its population, and its ability to defend itself in a protracted war through a lack of munitions caused by the disruption of nitrate imports or by their outright loss.

These fears were particularly acute in Great Britain and in Germany, who together were the largest importers of fixed nitrogen. Germany's population had grown from 25 million in 1800 to 55 million in 1900 and of all the imperial European nations, she was probably the most susceptible to a naval blockade which could have easily deprived her of her most essential imports in wartime. Of all the nations capable of achieving such a blockade the British were supreme with the most powerful fleet in the world, and Anglo-German tension had risen considerably at the turn of the century because of the Boer War[8] in South Africa. For Germany therefore, investment in vessels such as the *Potosi* and *Preussen* was a national necessity and as a consequence, the beginning of the 20th century saw Germany overtake Great Britain to become the world's largest importer of Chilean nitrate. It's not surprising therefore that Germany's efforts to find new sources of fixed nitrogen were acute.[5]

3.2   Haber's Introduction to the Nitrogen Problem

The maiden voyage of *Preussen* probably 'passed Haber by'. In just two weeks time he was to leave Germany on his sixteen-week tour of the USA (Chap. 2) and he was probably far too preoccupied with his preparations to take much notice of the event. Indeed, Haber had already proved to be conspicuously relaxed on the matter of Germany's nitrogen vulnerability. In May 1899 Carl Engler, rector of Karlsruhe's university, opened a new building at the Technische Hochschule and dedicated it with a speech that reflected not only the concerns that were emerging in Germany but also the clarion call of an international giant of science, Sir William Crookes[9] who, in his presidential address to the British Association for the Advancement of Science a year earlier had exhorted science to come to the aid of the world's populations and avert a human catastrophe. Crookes did not mince his words;

> England and all civilised nations stand in deadly peril of not having enough to eat. As mouths multiply, food resources dwindle. Land is a limited quantity, and the land that will grow wheat is absolutely dependent on difficult and capricious natural phenomena… I hope to point a way out of the colossal dilemma. It is the chemist who must come to the rescue of the threatened communities. It is through the laboratory that starvation may ultimately be turned into plenty… The fixation of atmospheric nitrogen is one of the great discoveries, awaiting the genius of chemists!

Professor *Extraordinarius* Haber was probably present when Engler delivered his speech, a speech which permeated the Institute and which reflected all of Crookes fears. But Haber at least was not to be exhorted in this matter. Nitrogen was stubborn and its fixation was not a candidate for the attention of a journeyman academic. Other goals were far more attainable and much less subject to failure. Even by 1904 when he was tempted to approach the fixation problem by the Margulies brothers he did so reluctantly, and by 1905—with regard to ammonia at least—he had made his position perfectly clear.

'The peculiar inertness characteristic of nitrogen which appears so prominently …. will always place a serious handicap on the cheap technical preparation of ammonia particularly so long as the tremendous amount of combined organic nitrogen with which nature has provided us lasts. Nature brings about the transformation of elemental nitrogen into ammonia and nitric acid in her slow way with the greatest facility and on a most colossal scale. While she has accumulated nitrates at but one place on the earth's surface in sufficient quantities to mine—and its exhaustion is imminent—she has provided for us a much richer and more varied supply of combined nitrogen easily converted into ammonia'.[10]

Haber it appeared, was relaxed about the continued availability of naturally fixed nitrogen—whether organic or inorganic. At the time, the mines of Chile provided the most convenient form of nitrate and although these were considered all but 'exhausted' he maintained other sources would always be available because the enormity of the natural nitrogen cycle would guarantee a continued supply of this most critical of chemicals.[11] Maybe in this sense Haber was ahead of his time, realising that recoverable resource estimates were always far too conservative. Haber then was not going to voluntarily embark on the epic quest of finding a 'cheap technical fix' for the nitrogen problem. Nitrogen was stubborn and nature knew best how to resolve its fixation.

Even so Germany remained vulnerable, and during his 1902 tour of 'the States' Haber was drawn to the problem for the first time when he visited the 'Atmospheric Products' factory at Niagara Falls which employed one of the most modern and technologically advanced approaches to fixation, viz., the oxidation of atmospheric nitrogen in electric arc furnaces according to;

$$N_2(g) + O_2(g) \rightleftharpoons 2NO(g); \Delta H^\theta = +90.4 \, kJ \, mole^{-1}$$

Haber of course duly reported progress in this area on his return to Germany and despite his public scepticism and his reluctance to get involved, the problem—in one form or another—occupied him for the next few years.[11]

3.3   Early Commercial Progress in Nitrogen Fixation

The oxidation of atmospheric nitrogen was probably the most technically advanced commercial attempt to bind elemental nitrogen. The process had a long history. It was well known to occur during thunderstorms and in the laboratory by 'sparking' mixtures of nitrogen and oxygen. But by 1890 researchers realised that using an electric arc discharge concentrated huge amounts of energy in a small space resulting in much higher yields. However, even at 2000 °C very little conversion occurred, and temperatures above 3000 °C were necessary to produce commercially useful amounts of the oxide (but still only 1.5–2% by volume). The highly endothermic nature of the formation required large amounts of energy and the electric arc process could only be commercially viable in countries where sustained supplies of inexpensive electricity were readily available. The arrangement of the arc(s) too was important and several novel solutions arose to maximise conversion. A further technical difficulty concerned the conversion of the oxide to useable product. Nitric oxide was stable at high temperatures but cooling favoured the reverse reaction causing the equilibrium to shift to the left, decomposing the oxide back into nitrogen and oxygen. To avoid decomposition of the expensively obtained oxide, rapid cooling, or 'quenching' was necessary to bring the it to a manageable temperature (around 800–1000 °C) without losing too much yield. Subsequent condensation of the gases produced nitric acid which was reacted with 'lime' to produce calcium nitrate $Ca(NO_3)_2$, a compound containing 17% nitrogen used directly as a fertiliser.

$$2HNO_3 + CaO = Ca(NO_3)_2 + H_2O$$

Two serious commercial attempts were made to 'fix' atmospheric nitrogen in this fashion. The first was by Bradley and Lovejoy who founded the 'Atmospheric Products Company' launched in 1902, inspected by Haber during his visit to the 'States'. They relied on the power generated by the newly built Niagara Falls hydrostation but the process proved to be uneconomic and the Company folded in 1904. The second attempt by Birkeland and Eyde opened as a prototype in Norway in 1903. Norway was blessed with naturally high heads of water which removed the need for the construction of expensive dams making hydroelectricity particularly attractive. Plants with increased capacity were built there in 1905 and again in 1911 but even when these achieved a maximum of 70,000t[5] calcium nitrate per annum (during 1913), it was equivalent to only about 3% of the Chilean fixed nitrogen exports and less than 5% of what was available by other means. Low yields coupled with an energy input of around $180-270GJt^{-1}$ of fixed nitrogen made this method unsustainable in the long term[5]—especially for Germany.

Other commercial processes that returned 'fixed' nitrogen were based on ammonia by-product recovery obtained from coking plants, and the 'cyanamide' process. 'Coke' (impure carbon) was needed for 'pig-iron' smelting and for the generation of 'coal gas' used in gas lighting. Coke in turn was produced from coal by heating it in the absence of air. Now, coal contains around 1–1.5% nitrogen, arising from the decomposition of proteins present in the original biomass. When the coal is heated in the absence of air some of the nitrogen is released as ammonia. For decades the ammonia was an unwanted by-product and released to the atmosphere. With a growing understanding of the importance of ammonia, coking ovens were modified to achieve by-product recovery, which for ammonia commonly meant its conversion into ammonium sulphate $(NH_4)_2SO_4$. Coke by-product recovery was largely confined to Western Europe and the United States and by 1900 the total annual global output was around 0.5 Mt ammonium sulphate. By 1913,[12] German by-product recovery output alone accounted for almost two thirds of the global output of fixed nitrogen by this means, but Chilean saltpetre exports were still larger accounting for almost twice that amount. By 1910, the German output of ammonium sulphate had exceeded the British total but this was still only about half of what she needed. During the first world war the demand rose sharply. The extract below taken from the *Manuals of Chemical Technology—III,*[12] (p. 45), describes an advanced by-product recovery process in operation widely in Germany by 1915 and illustrates the investment made here (Fig. 3.1).

An earlier means of fixing atmospheric nitrogen was discovered in 1860. The original process employed barium carbonate at 1200 °C according to;

$$BaCO_3 + N_2 + 4C = Ba(CN)_2 + 3CO$$

the cyanide subsequently being converted to ammonia and barium hydroxide in the presence of water vapour. Later versions used much cheaper calcium carbide which produced 'cyanamide' rather than the cyanide. The carbide was produced by the fusion of lime and carbon ('coke') in an electric furnace;

In the **Otto-Hilgenstock Ammonia Recovery Process** (Fig. 15) the old condensing plant is entirely dispensed with, the tar being removed from the entering gases by a tar spray at A at a temperature above the dew-point of the liquors.

After depositing the tar in B the gases pass directly through an exhauster E into the saturator F, where the whole of the ammonia is caught by the sulphuric acid. The gases coming from the saturator are hot, and contain all their moisture in the form of steam; the gases are, therefore, passed forward to the oven flues, and the troublesome and offensive waste liquors are thus got rid of. C is the tar-spray pump, D the tar-spray feed pipe, G is the acid-spray catch box, H the mother liquor return pipe, J the tar store, K the tar-spray overflow pipe, L the condensing tank, M the pump delivering tar to railway trucks, N the pump delivering condensers to the saturator F.

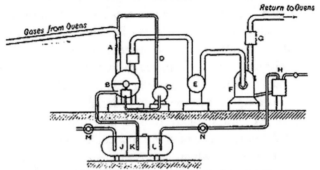

Fig. 15.—The Otto-Hilgenstock Ammonia Recovery Plant.

This process, by abolishing condensing plant, liquor tanks, ammonia stills, lime mixers, pumps, etc., effects a great saving, since less floor space is required, and nearly the whole of the steam required to distil the ammoniacal liquor made by the condensing process is abolished. Also no ammonia is lost, as often arises in a distilling plant. The ammonium sulphate produced contains 25-25.5 per cent. N and contains less than 0.1 per cent. of tar.

Fig. 3.1 *An advanced German ammonia by-product recovery system, from Manuals of Chemical Technology[12] 1915*

$$CaO + 3C = CaC_2 + CO$$

The molten carbide was removed, cooled, ground and reacted with pure nitrogen[13] in iron retorts at about 1000 °C;

$$CaC_2 + N_2 = CaCN_2 + C$$

Hydrolysis followed producing ammonia.

$$CaCN_2 + 3H_2O = CaCO_3 + 2NH_3$$

the whole being referred to as the 'cyanamide' or 'Frank-Caro' process, discovered by Fritz Rothe in 1898 and adapted for industrial use by Adolf Frank and Nikodem Caro. The process became the first commercial method of

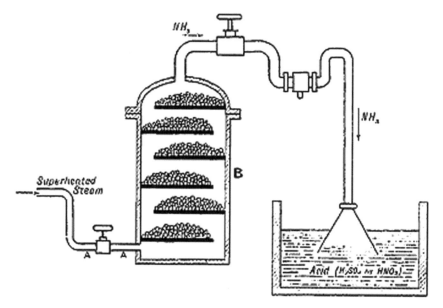

Fig. 3.2 *A schematic Illustrating the cyanamide process, from Manuals of Chemical Technology*[12] *1915*

fixing atmospheric nitrogen but it's energy costs again were high—at best 130 GJ t[−15]—due to the electricity requirements and the production of the coke. By 1915, the cost[12] of producing ammonium sulphate from cyanamide in the UK was just £4. 13 s ton[−1]. Even so, Germany became a major player and by 1918 thirty-five cyanamide plants produced 325,000t fixed nitrogen per annum[5] dominating the world's production (Fig. 3.2).

3.4 Early Academic Progress Fixation by Synthesis: The Combination of Elemental Nitrogen and Hydrogen

Both the electric arc and the cyanamide processes provided fixed nitrogen, but both were expensive. The simplest chemical route to nitrogen fixation was long known to be through the synthesis of ammonia according to;

$$N_{2(g)} + 3H_{2(g)} \leftrightarrows 2NH_{3(g)}$$

but academic progress to harness the apparent simplicity of this reaction was laboured, and even in the earliest history[14] of this problem it became apparent that to tease these two elements to combine, temperature, pressure, and catalysis were all important.

Nitrogen was known to combine with hydrogen to form ammonia as early as 1784—Claude-Louis Berthollet even determined the approximate ratio of the two constituents in the compound. By 1788 high temperature synthesis (by passing nitrogen and hydrogen—or water vapour—over red-hot iron or glowing coals) had been attempted and then repeated by more than a dozen chemists during the nineteenth century. The effect of pressure was firstly examined in 1811 by George Friedrich Hildebrand who, after a failed attempt to combine the elements at atmospheric pressure, submerged a sealed flask containing a mixture of the gases in the sea to a depth that achieved a pressure of 3.1 MPa (about 30 atmospheres)—with little effect. The first recorded use of a catalyst dates from 1823 when Johann Wolfgang Döbereiner experimented with platinum, and by the 1870s many other different metals and compounds had been tested and indeed patented. Henri Saint-Claire Deville apparently also managed some form of fixation in 1865.

The first 'modern' examination of the reaction was undertaken by William Ramsay and Sidney Young in Bristol in 1884 who studied the decomposition of ammonia at atmospheric pressure in the presence of an iron catalyst at 800 °C, the template later used by Haber and van Oordt. By the term 'modern' we mean one which understood that a chemical equilibrium was probably involved and which benefited from the supreme practical expertise of the workers. The clear conclusion from their experiment however was that the equilibrium position for the synthesis of ammonia lay far to the left. Numerous attempts at ammonia synthesis followed during the 1880s and 1890s but all were fruitless and then, at the turn of the century, the problem was effortlessly solved by a giant of German physical chemistry, Wilhelm Ostwald.

Ostwald was not only a perceptive scientist but also a perceptive statesman in that he long realised that the naval supremacy of the British would deny Germany fixed nitrogen for her fertilisers and munitions if the antagonism of the Boer War were to escalate into a full-scale conflict—fears which were later realised in 1914 but for different reasons. So, in 1900 he announced a solution to this national threat when he notified Fabwerke Hoechst and BASF that he had succeeded in synthesising ammonia from its elements using an iron catalyst at atmospheric pressure. The low cost of this process both in terms of materials and energy would clearly revolutionise the supply of Germany's fixed nitrogen making her completely free of the saltpetre trade—

even though her supply of hydrogen was not entirely certain. Such was Ostwald's standing—especially as an expert in catalysis—that German industry took notice and BASF immediately made an agreement with him to investigate his process. This they did through a young chemist named Carl Bosch. Bosch replicated Ostwald's machine at BASF and charged it with a fresh sample of his catalyst. At first ammonia was generated but it soon stopped. Bosch was puzzled and so he repeated the experiment several times always using fresh samples of the catalyst. Each time the result was the same. Bosch had only joined BASF in 1899 but his report of April 1900 demolished Ostwald's infant, tracing the small mounts of ammonia formed to the hydrogenation of iron nitride present in all commercial iron catalysts at the time. Even at a pressure of 0.5 MPa (~ 5 atmospheres) catalysis failed to produce ammonia. But Ostwald's reaction to Bosch's report was arrogant and dismissive, …

> when you entrust the task to a newly hired, inexperienced chemist who knows nothing, then naturally nothing will come of it.

But he was soon forced to accept Bosch's opinion and the patent application he had submitted was withdrawn. Even so this application contained all the ingredients for the successful industrial production of ammonia viz., elevated temperatures, an appreciation of the role of high pressure, a metal catalyst and circulation of the gases. Consequently, in his autobiography[15] he salvaged some pride when he maintained that his efforts had at least established him as the 'intellectual father' of the ammonia industry, although ….

> I have certainly not become its real father, for all the difficult and varied work needed to create a technically and economically viable industry from the right ideas was carried out by those who took on the abandoned infant.

Over the next few years Ostwald withdrew from the ammonia problem. When Haber approached him in 1904 with regard to the request of the Margulies brothers, his lack of interest in the problem puzzled him but Ostwald's humiliation at the hands of the young Bosch explains why. Even so, Ostwald did not inform Haber of the effect of the hydrogenation of the nitride present in the iron catalyst, had he done so we may well not have had a story to tell. 1901 saw a further attack on the ammonia problem by another giant of the chemistry world viz., Henri Louis Le Chatelier. By 1900 the physical chemistry of the ammonia synthesis was sufficiently well established for Le Chatelier to calculate the necessary temperature, pressure and quantity of iron catalyst needed. In 1901 he built a small high-pressure apparatus along the necessary principles but it blew up and he abandoned his research. Le Chatelier was an expert metallurgist, well practised in high temperature studies and the behaviour of gaseous mixtures and therefore ideally placed to tackle the ammonia problem, but his findings were recorded in an obscure French patent taken out under a foreign name and lost to the mainstream thrust of research. Indeed, Haber was later to acknowledge that he only became aware of Le Chatelier work after he had successfully, and independently, completed his own experiments.[16]

This then was the' state of play' by 1905. Fixed nitrogen was available, but expensively and in limited amounts. The ammonia synthesis was well understood from a thermodynamic perspective but arguably less so from a catalytic one. Many giants of chemistry had tackled the problem and failed, giving rise to a prejudice that deterred further investigation. But as Ostwald observed, all the ingredients for a solution were there and it became inevitable that at some time a physical chemist would provide 'very special equipment'[16] that would blend these ingredients and tease nitrogen and hydrogen to combine in commercially viable amounts. But no-one it seemed, was about to volunteer and take up this poisoned chalice.

However, the impasse was broken when Walther Nernst, along with co-workers K. Jellinek and F. Jost, used his new 'Heat Theorem' to re-examine the published results for various gaseous equilibria. They noticed that only in the single case of Haber's figures for the Margulies ammonia investigation, were published experimental results seriously at odds with his predictions. In particular, Haber's choice for the percentage of ammonia present in the equilibrium mixture at atmospheric pressure and 1020 °C (i.e., 0.012%) was far too high for Nernst who subsequently wrote to Haber in the Autumn of 1906 informing him of his concerns. Nernst was Germany's pre-eminent physical chemist and Haber—a freshly appointed professor—was disturbed by his observations. The letter was a private communication—although Nernst had already voiced some concerns about Haber's figures in his Silliman[17] Memorial Lecture at Yale[18]—but Haber realised that if the issue were to go unchallenged then it had immense potential to damage his hard-earned reputation. Consequently, he was forced back to address the ammonia problem.

In his Nobel lecture on 02 June1 920,[16] Haber recalls that;

> ... in 1906... a new determination of the ammonia equilibrium became necessary. In the course of his investigations... Nernst succeeded in finding an approximate formula which permitted a prediction of the [position of

chemical] equilibria. In the case of ammonia this gave a deviation from the values obtained at my first measurements… This deviation led to fresh determinations of the equilibrium which Nernst carried out at his Institute in a pressure oven [designed] by him while I, in collaboration with Robert Le Rossignol, repeated the determinations at normal pressure with greater care than before.

Because of the difficulty of the ammonia problem it was unlikely that Haber's results would ever have been seriously challenged by further experimentation. Indeed, if anyone other than Nernst had criticised Haber's work, he may well have dismissed them, but Nernst was universally respected and his concerns had to be addressed. Nernst too, had the advantage of his 'Heat Theorem' which he adapted to apply to gaseous systems *via* his 'approximation formula'[19] which meant that estimates of equilibrium compositions could be made without the bother of actual experimentation. Haber needed the best to help him tackle Nernst. He chose Robert. But what talent(s) did this young award-winning chemist bring to Haber that made him so suited to help solve the ammonia problem?

3.5   But Finally… What of Preussen?

The quality of this vessel—designed specifically for the nitrate trade and the rigours of the Cape Horn passage—allowed her to complete thirteen uneventful voyages to the north west coast of Chile. On each occasion she delivered German exports and brought back a treasure trove of fixed nitrogen. On 31 October 1910 she left Hamburg on her fourteenth voyage, headed for Valparaiso loaded with a general cargo that included sugar, cement, bricks, railway materials and one hundred pianos. There was no reason for Captain J. Heinrich H. Nissen to expect his transit along the English Channel to be anything other than normal, but on the evening of November 6th there was some concern regarding the autumn weather which was particularly dark and foggy. Logging a sedate 4 knots, *Preussen* erred on the side of caution and paid due respect to the conditions. However, the same could not be said for the Captain of the Newhaven-Dieppe cross-Channel steamer *SS Brighton* which was making seventeen knots under the same conditions. Just before midnight, the *Brighton,* entirely mis-judging the speed of the big sailing ship, turned to cross in front of her but struck her full abreast between the foremast and the bowsprit. *Preussen* sheered off one of the steamer's two funnels and ripped a hole in her hull, but with her own bow stove in *Preussen* had to be taken in tow by the steam tug Alert. Eighteen miles from Dover, Nissen tried to anchor in

Plate 3.3 *Preussen, broken and aground in Crab Bay November 1910.*[1]

the lee of Dungeness, but the ship's anchor chains parted in a squall and he was forced to run for Dover. Standing into Dover escorted by three tugs Alert, Albatross and John Bull, *Preussen's* top hamper created so much windage that the tow lines parted and she began to drift. Setting sail in an effort to back out of the shallows, *Preussen's* bow snagged on a reef in Crab Bay. The East Cliff coastguards managed to get a line across the ship by rocket but the crew (and two passengers) refused to leave. For the next two days the crew worked tirelessly to save their magnificent ship but all attempts to free the huge five-master failed. Crew and passengers were finally taken ashore. Later that day her back broke and she ended her days where she lay (Plate 3.3).

The sad sight of this magnificent machine, dismasted and broken, probably symbolises the beginning of the end for the German nitrate trade. An end caused not by this event in particular, but by a much more modest machine perfected by Robert Le Rossignol in Karlsruhe a year earlier. To this day, and at the lowest of spring tides, the remains of *Preussen* can still be seen in Crab Bay.[20]

Notes

1. Information regarding the Preussen is widely available on the internet, but I am particularly indebted to the web sites http://www.wrecksite.eu and http://www.taubmansonline.com/AERPREUSSEN.htm, together with reports in the Dover Express, (East Kent News) from 1910. Copyrights are unknown.
2. Now Laeisz, when does the five-masted [*i.e. clipper*] ship arrive?
3. Patrick Coffey, *Cathedrals of Science.* Oxford University Press, (2008). ISBN 978-0-19-532134-0. Populations of the major European countries had almost tripled in the 19th century, but the amount of arable land available had increased very little. To feed the increased populations intensive farming was

required to increase the yield per acre of cereals such as wheat, maize, oats and rye. But these in turn deplete the soil of its essential nutrients, especially nitrogen. Replacement of the nitrogen was traditionally achieved by crop rotations and by spreading barnyard manures. Crop rotation involved planting legumes (such as beans or peas) which were able to utilise nitrogen from the air and convert it to a mineral form readily assimilated by plants. This helped replace the nitrogen, but reduced the production of the more profitable cereal crops. Spreading manure only replaced a small potion of the nitrogen removed from the soil. As cities grew across Europe, food was exported to them from the farms and nitrogen was lost in sewers and privies and not returned to the soil, Coffey, op. cit. (note 3), p. 83.

4. The ready availability of the saltpetre sources was a major factor in the invention and the commercialisation of a number of modern explosives. In 1900, the USA used almost 50% of its imports of $NaNO_3$ to make explosives. Indeed, but for the discovery and exploitation of the Chilean deposits, the explosives industry as we know it today would have been impossible.

5. Vaclav Smil, *Enriching the Earth,* MIT Press, (2001). http://mitpress.mit.udu. ISBN-978-0-262-693134.

6. Sodium nitrate—like all nitrates—is highly soluble and therefore can only accumulate in arid areas on the planet. The Atacama Desert had not seen any rain of significance for centuries and these peculiar circumstances allowed nitrates and the other minerals to accumulate in huge quantities as caliche.

7. As is often the case with recoverable resource estimates, these fears were far too conservative. An estimate of the country's resources around 1908 put the nitrate reserves at 920 Mt, and at an annual rate of 2.7 Mt—the usage per annum just before WW1—the reserves would have lasted for 340 years. But 920 Mt of $NaNO_3$ contains less nitrogen than is now fixed in just two years by the synthesis of ammonia, Smil, op. cit. (note 5), Chap. 3. Even by 1920, '200 Years Supply in Chile' was reported in *The Times* in an article entitled 'Nitrates for War and Peace', (Tuesday 31 August 1920).

8. The Boer of course were of Dutch-German extraction.

9. (Sir) William Crookes was a man whose opinion had to be respected. He was born in London on the 17 June 1832. His scientific career began when, at the age of fifteen, he entered the Royal College of Chemistry in Hanover Square, London, studying under August Wilhelm von Hofmann. From 1850 to 1854 he filled the position of assistant in the college, and soon embarked upon original work, on certain new compounds of the element selenium, the selenocyanides. Leaving the Royal College, in 1854 he became superintendent of the meteorological department at the Radcliffe Observatory in Oxford, and in 1855 was appointed lecturer in chemistry at the Chester training college. From this time his life was passed in London, and devoted mainly to independent work—journalistic, consulting, and academic. In 1859 he founded the *Chemical News*, which he edited for many years and conducted on much

less formal lines than is usual with journals of scientific societies. Crookes' life was one of unbroken scientific activity. He was never one of those who gain influence by popular exposition; neither was he esoteric. The breadth of his interests, ranging over pure and applied science, economic and practical problems, and psychical research, made him a well-known personality, and he received many public and academic honours. He was knighted in 1897 and he died in London 04 April 1919. I am grateful to Oxford University's Chemistry Department earlier web page: http://www.chem.ox.ac.uk/icl/heyes/LanthAct/Biogs/Crookes.html.

10. From, '*Thermodynamics of Technical Gas Reactions*', Haber, Lecture Five. http://www.archive.org/details/thermodynamicsof00haberich. Later, Robert could not have been more aware of the nitrogen problem and Haber's attitude to it, after all he was the translator of the Appendix to the chapter in which these words appeared.

11. Indeed Haber once calculated that the binding of nitrogen to oxygen during thunderstorms in the upper atmosphere deposited several tenths of a gram of fixed nitrogen per km^2 to the earth's surface. Haber had also studied the oxidation of nitrogen in an electric arc as early as 1907 in collaboration with BASF and Adolf Koenig and had proposed a mechanism for the reaction and a process which retarded the decomposition of the oxide—see Chap. 7 However, by the time the ammonia synthesis had been perfected he had abandoned this route.

12. Geoffrey Martin and William Barbour, *Manuals of Chemical Technology III, Industrial Nitrogen Compounds and Explosives*. Crosby Lockwood and Son, 7 Stationers' Hall Court, Ludgate Hill, E. C., and 5 Broadway, Westminster S. W., LONDON, (1915).

13. Obtained by the 'Linde' process, *i.e.*, the liquefaction of air.

14. See, Smil, op. cit. (note 5), Chap. 4, for a 'potted history' of all the pre-1900 attempts at the synthesis of ammonia, or his reference, *Gmelins Handbuch der anorganischen Chemie, Stickstoff*. Berlin: Verlag Chemie, pp. 327–328, (1936).

15. Wilhelm Ostwald, *Lebenslinien: Eine Selbstbiographie*. (Lifelines: An autobiography). Volume 2, Chap. 12, Berlin: Klasing, (1926).

16. Fritz Haber, Nobel Lecture 2 June 1920; 'Fritz Haber—Nobel Lecture: The Synthesis of Ammonia from Its Elements'. *Nobelprize.org*. Nobel Media AB 2014. Web. 9 March 2016. http://www.nobelprize.org/nobel_prizes/chemistry/laureates/1918/haber-lecture.html.

17. The Silliman Memorial lecture series was established in 1888 in memory of the mother of Benjamin Silliman, one of the first professors of science at Yale University. The University began publishing the lectures of distinguished science guest speakers in 1901.

18. Diana Kormos Barkan, *Walther Nernst and the Transition to Modern Physical Science.* Cambridge University Press, 1999. ISBN- 0 521 44456 X.
19. We introduce the approximation formula in Chap. 5.
20. Accounts differ. Some say Crab Bay, others Fan Bay, some say the collision occurred on 05 November others 06 November and some even maintain that the captain of the Brighton was not really to blame. Even so, he was convicted and lost his 'ticket'.

4

The Golden Chemist

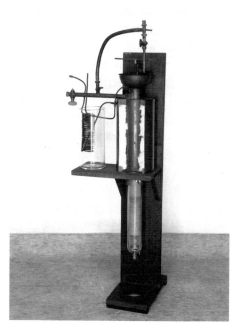

*William Travers' original hydrogen liquefier in continual use at UCL
from 1899–1903. Courtesy of the Science and Society Picture Library,
National Science Museum, London.*

Robert's 'ammonia apprenticeship' at UCL

' I was always interested in mechanics ...'

Robert Le Rossignol in conversation with Johannes Jaenicke in 1959.[1]

© Springer Nature Switzerland AG 2020
D. Sheppard, *Robert Le Rossignol*, Springer Biographies,
https://doi.org/10.1007/978-3-030-29714-5_4

4.1   Introduction

Throughout his professional life and beyond, Robert Le Rossignol maintained an active interest in the development of the ammonia synthesis. He derived great satisfaction from the contribution that the Haber-Le Rossignol[2] Process had made to the feeding problems of the world—particularly in the developing countries. But as a young man he was only drawn to the ammonia problem through his contact with Haber. Dietrich Stoltzenberg[3] suggests that Robert had worked on the ammonia problem with Ramsay before joining Haber, but Ramsay's only investigation in this area was published in 1884— the year of Robert's birth. No doubt the problem was discussed at UCL during Robert's time there but it was hardly Ramsay's mainstream concern. Equally however, it would be naïve to presume that Robert travelled to Haber without any kind of provenance in matters of technical engineering. Indeed, his early interest in 'chemical technology' was one reason for him choosing Haber over Abegg. Until his death in 1976 Robert was always grateful for his time at UCL, but as he left us only a modest account of this period,[4] one can only speculate that it was the knowledge and skills he gained there that laid the foundations for his later success at Karlsruhe. Although we can call on little biographical evidence to support this conjecture, Morris Travers in '*A Life of Sir William Ramsay*'[5] provides ample evidence to suggest that—even before Robert arrived there—engineering knowledge of the type he eventually applied at Karlsruhe already existed at UCL.

4.2   Travers' Liquefiers

In the Kensington Science Museum's Large Object Store at Wroughton, Swindon[6]—a second world war maintenance airfield in Wiltshire—lies a largely forgotten machine with coils, a glass jar and two vacuum vessels. This machine, the first hydrogen liquefaction apparatus, may well have helped to change the world. From 1898 to 1903 Ramsay was concerned with the isolation of neon, krypton and xenon. His laboratory was a foundry of experimental innovation. The isolation and physical determination of the inert gases required bespoke equipment much of which is described in Travers' biography, but 'cutting edge' technology too penetrated the laboratory. By 1899 Ramsay needed to make liquid air in order to accumulate krypton and xenon-rich residues which were later separated by fractional distillation. The college had a Hampson air liquefier—minus its compressor and motor—so it could only be operated by compressed air or oxygen, very expensive and inefficient. The only hope of progress in this matter lay in acquiring an air compressor and a motor to drive it. Money became available[7] and a small compressor of the type made by the Whitehead[8] Torpedo Company at Fiume[9] on the Adriatic for charging torpedoes—together with an electric motor—were ordered and delivered by Easter 1899.

Like Haber, Ramsay was in no way 'mechanically minded' but having received a letter from Ostwald who had installed similar apparatus, he passed instructions to Travers to supervise the installation. Consequently, the compressor and motor were bolted to a couple of wooden planks in the 'compressor room'—a disused lavatory with the usual apparatus removed.[5] Ostwald however did not tell Ramsay what had happened at *his* laboratory when his machine started up … it chased the operators across the floor and only stopped when the electrical leads to it broke away! Travers was much more fortunate, and after a nerve racking start the 'lab boys' got quite used to running it at 180 atmospheres. On occasions, when the 'boys' were more concerned with reading their 'penny dreadful', the pressure often rose way above the danger limit but no accidents ever occurred.[5]

Soon after the commission of the air liquefier further problems arose regarding the separation of the inert gases by fractional distillation. Air contained helium, and liquefaction of some of the gaseous mixtures could not be achieved at liquid air temperatures. Liquid hydrogen was therefore required as a cooling agent, and for financial reasons—even though colleagues believed hydrogen could not be compressed without a danger of the operator being blown through the skylight of the compressor room—the apparatus had to be built within the department. The laboratory mechanic Mr. Holding had little time to spare for this adventure—other than modifications to the compressor—so Travers had to design and build the liquefier himself.[10] The progress and manner of this development is interesting, and although it took place just before Robert joined UCL, it illustrates the kinds of engineering skills available within the department, skills which Robert, in time, surely would have become exposed to—and wanted to master.

The first step was to purchase a steel gasometer to hold about 100 ft^3 of hydrogen. This was set up in an area outside the laboratory and a pipe from the gasometer led to the compressor. A (wooden) beer barrel was bought for the generation of hydrogen from zinc and sulphuric acid, and after several coats of paint it became sufficiently gas-tight. The design of the apparatus was fairly straightforward. Hydrogen from the gasholder was compressed at (up to) 180 atmospheres and cooled by passage through a coil immersed in an alcohol/solid carbon dioxide bath. The cooled pressurised gas then passed through another series of coils surrounded by a jacket of liquid air. An exhaust pump was used to reduce the pressure in the jacket and allow the air to boil thereby further cooling the hydrogen in the coils *via* the Joule-Thompson effect. Finally, the cold hydrogen was passed through an expansion valve which caused further cooling by the same effect[11] to produce liquid hydrogen collected in a vacuum flask. Remaining gaseous hydrogen was returned to the compressor to be re-circulated.

Apart from the vacuum flasks[12] and the expansion valve which was built by Brin's Oxygen Works Horseferry Rd., Westminster, (later the British Oxygen Company, BOC) the materials and skills were provided locally or 'in house'. The apparatus itself was made from brass tubing and blanks easily obtained from metal dealers in Clerkenwell, London. Making the apparatus required little more than some skill in soldering and the only accessories required were the exhaust pump and a motor to drive it. Fortunately, a student had a blowing pump in his private workshop, and with its valves reversed, it served as the exhaust pump. The motor was borrowed from the Engineering department. The apparatus was literally made from 'odds and ends', it was built in just six weeks and cost about £50. The only design difficulty faced was the problem of connecting the receiving glass vacuum vessel with the metal body of the liquefier. This was overcome by a compressed rubber ring whilst the only operational difficulty was caused by the blockage of the expansion valve and its seizure at such low temperatures. By soldering a cross bar to the head of the valve, enough 'purchase' could be achieved to force the valve to open, and blockages were cleared by allowing the pressure to rise.

The apparatus was used for the first time on Saturday 30 June 1899. Mr. Holding and a team of willing senior students prepared the hydrogen and the liquid air, and after initial cooling with the solid carbon dioxide and the liquid air, the compressor and exhaust pump were started. When the pressure rose to 180 atmospheres the expansion valve was opened and liquid hydrogen collected in the receiving vessel. However, this was short lived as the valve both seized and blocked. The next day, having soldered the cross bar to the head of the valve, the experiment was repeated and when the valve blocked again Travers allowed the pressure to rise, much to Ramsay's distress who cried '*It'll burst Travers … it'll burst!*' The valve however cleared and the apparatus remained intact for the duration of the experiment and indeed for the next four years being used continually by Travers and his colleagues until he left the department in 1903.[5]

By the time Robert arrived in the chemistry department at UCL the liquefier had been mounted in a prominent position on the wall outside the compressor room. For a young man 'interested in mechanics' and developing an inclination towards 'chemical technology', this machine would surely have been an immense attraction and it possessed so many aspects of the engineering Robert was later to employ at Karlsruhe. It had a compressor that could achieve a working pressure of almost 200 atmospheres, it circulated gas back to the compressor—the same gas that Robert was to later encounter with Haber. It also incorporated cooling and separation systems and customised valves to control gas flow. It was operated too by 'senior students'[5] and we know that Robert was certainly one of these because of his time with Woodland-Toms (Chap. 1). A detailed schematic of the apparatus had also been published[10] in the prestigious *Philosophical Magazine* for the 'mechanically minded' to pore over, and Travers' liquefier was of course in constant use during Robert's time at UCL (Fig. 4.1).

4.3   Beyond Engineering …

But in addition to these engineering aspects Robert had also acquired a thorough experimental and theoretical background in reaction kinetics and chemical equilibria culminating in the paper he produced with Donnan in 1903 *viz.*, G. Donnan, R. Le Rossignol, 'LXXI The velocity and mechanism of the reaction between potassium ferricyanide and potassium iodide in neutral aqueous solution'. *J. Chem. Soc., Trans.*, **83**, 703–716, (1903), referred to in Chap. 1. Here, Donnan and Robert observed that when potassium 'ferricyanide' and potassium iodide react in neutral aqueous solution, potassium ferrocyanide and free iodine are gradually produced. Conversely, a solution of iodine in potassium iodide oxidises ferrocyanide to ferricyanide. In both cases they observed that a definite state of equilibrium was attained expressed by the equation;

$$2KI + 2K_3Fe(CN)_6 \leftrightharpoons 2K_4Fe(CN)_6 + I_2$$

and a rough preliminary experiment, in which the amount of free iodine was estimated by rapid titration of the cooled and diluted ('quenched') reaction mixture, showed that the equilibrium could be approached from either side— a technique employed by Ramsay in the ammonia study of 1884 and subsequently used by Haber.

This was a really complex chemical equilibrium, far more so than the ammonia problem Robert was to tackle later in Karlsruhe. Their study produced a substantial paper—fourteen pages in length—packed with reaction kinetic theory and the kind of experimental detail that was to later

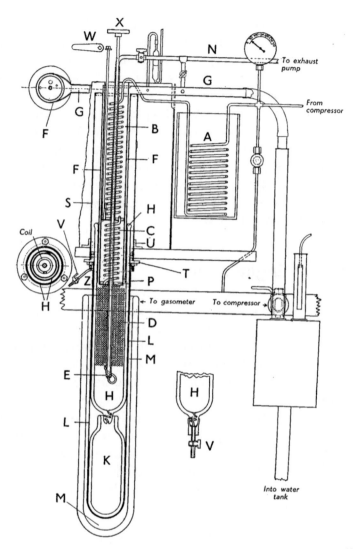

Fig. 4.1 A schematic of Travers' liquefier from the philosophical magazine.[10] Courtesy of Taylor and Francis Ltd., www.tandfonline.com

characterise Robert's professional life. Here, Donnan and Robert proposed[13] 'a quinquemolecular' reaction, inasmuch as the reaction velocity is well represented by the equation;

$$-dc_1/dt = kc_1^2 c_2^3$$

where c_1 represents the concentration of ferricyanide and c_2 that of the iodide. The experimental results however were best explained by assuming the reaction to proceed as;

$$2Fe^{+++} + 3I^- = 2Fe^{++} + I_{3^-}$$

the ferric ions resulting from the instability of the ferricyanide complex. So, just two years after entering UCL, through this detailed experimental investigation—soon to be read before the Chemical Society—Robert demonstrated a firm grasp of experimental design and observation, of reaction kinetics and mechanisms, and of dissociation equilibria—the latter through the stability constants of complex ions. And although all the experimentation was performed by Robert, the mathematics was due to Donnan[4]—a harbinger of Robert's later role with Haber. This paper was rapidly followed by a second, co-authored with his friend C. T. Gimingham, this time in another—quite different—'cutting edge' area *viz.*, 'The Rate of decay of Thorium emanation', the first page of which is shown in Plate (4.2).

At the time of this investigation, Ramsay had become interested in Uranium and Thorium, and he had obtained ores from the Thomas Tyrer Company of Stratford, East London. Robert fused the ores with sodium bisulphate (flux) before carrying out the investigation, releasing around 1000 litres of helium which Ramsay hoped to sell for a large sum. Published in the much-admired *Philosophical Magazine* and endorsed by no less than F. T. Trouton[14] himself, this paper ventured into the exciting world of radioactivity where of course Ramsay was soon to make a 'singularly unfortunate incursion' (Chap. 1). Here, charged with the confidence of youth, these two young men 'took on' Rutherford and declared that after a 'series of experiments' the 'true value [*of the half life of the Thorium emanation is*] 51 s,[15] and not 60 as found by him', proceeding of course to 'prove' it. (Chap.1) The paper also demonstrates Robert's inclination towards things mechanical, in that the electrometer was modified to produce a virtually 'dead beat' instrument, i.e., one which indicated a steady true reading without excessive oscillation of the needle. It was for work such as this that Robert had become the 'golden boy' of chemistry at UCL.[16]

Unknowingly then, Robert's time at UCL may well have prepared him to work on the ammonia problem. He would have known about the hydrogen liquefier and its engineering, indeed he probably passed it most days outside the compressor room. All the skills needed to fabricate such machines were available within the department and his work on reaction kinetics with Donnan would have formed the basis of his understanding of chemical

this conclusion still remains undisturbed, even though the
shape be not one of revolution.

Whether the conditions of the limit can be sufficiently
attained in experiment is a question upon which I am not
prepared to express a decided opinion. From the logarithmic
character of the infinity upon which the argument is founded,
one would suppose that there might be practical difficulty in
reducing the section sufficiently. Even if an adequate reduction were possible mechanically, the conductivity of
actual materials might fail. We must remember that in the
theory the conductivity is supposed to be *perfect*.

Terling Place, Witham,
    June 12, 1904.

XI. *The Rate of Decay of Thorium Emanation.*
By C. LE ROSSIGNOL *and* C. T. GIMINGHAM[*].

DURING the investigation of a substance, giving an
emanation which was suspected to be that of thorium,
we had occasion to determine the time taken for known
thorium emanation to decay to half-value, and finding that
the value obtained was rather less than Prof. Rutherford's
(Phil. Mag. 1900, xlix. p. 1), we carried out a series of
experiments, which go to prove that the true value is 51
seconds, and not 60 as found by him.

Before going on to describe our method of measurement,
it might be advantageous to give a short account of the
apparatus employed (fig. 1).

The electrometer was of the Thomson-White pattern, with
the bifilar suspension removed and replaced by a quartz fibre;
the electrodes were insulated by means of ebonite coated

Fig. 1.

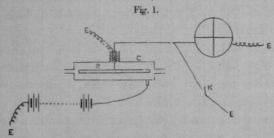

with shellac, and the instrument was practically dead-beat.
A transparent celluloid scale was used, divided in millimetres;

* Communicated by Prof. F. T. Trouton, F.R.S.

◄ **Plate 4.2** The first page of Robert and Gimingham's paper showing their modification to the standard Thomson-White quadrant electrometer. From the Philosophical Magazine. C. Le Rossignol and C.T. Gimingham, 'The Rate of Decay of a Thorium Emanation'. Phil. Magazine. ser. 6, **8**, 43, 107–110, (1904), communicated by Prof. F.T. Trouton, FRS. Unfortunately, Robert was attributed as 'C. Le Rossignol'. Page courtesy of Taylor and Francis Ltd., www.tandfonline.com

equilibria. By winning the gold medal he had clearly demonstrated outstanding practical ability and all that was needed was the good luck to be introduced to a problem that needed all this expertise. By choosing to go to Haber he unwittingly achieved this. It is arguable whether any of the talented people Haber had assembled at Karlsruhe would have been capable of bringing the rich mix of theoretical understanding and practical expertise that Robert did, and maybe this explains why Haber chose Robert to help him. A much less generous view would assume that no-one at Karlsruhe wanted to grasp the 'ammonia nettle' and as a new recruit, Robert 'drew the short straw'. Haber however, was a good judge of his students and he would never have tolerated mediocrity for this task. The letter from Nernst was potentially damaging to Haber and he would have demanded the best available collaborator to confirm the credibility of his work for the Margulies brothers. The best it seems was Robert, and both he and Haber began their re-investigation of the ammonia equilibrium around February 1907.

Many years later, R. C. Chirnside[4] observed;

> [Le Rossignol] was essentially a chemist, with strong engineering skills and understanding. He had an amazing grasp of the inter-relation of fundamental chemical, physical and electrical concepts and could work out on paper complex problems from a few basic relationships. Throughout his life he maintained his practical mechanical aptitude.

By Robert choosing Haber and then Haber choosing Robert, serendipity was to provide a solution to the ammonia problem. But firstly, Nernst's criticism had to be addressed and it was the painstaking, professional, practical work of Robert Le Rossignol that Haber 'clothed himself in' to meet the challenge that came at Hamburg in May, 1907.

Notes

1. Jaenicke's and Robert's letters from this period and the transcript of their conversation are held at the Max Planck Gesellschaft, Archiv der MPG, Va. Abt., Rep. 0005, Fritz Haber. Haber Sammlung von Joh. Jaenicke. Nr. 253 (Jaenicke's letters) and Nr. 1496 (conversation transcript).

2. One of the few references to the '*Haber–Le Rossignol*' Process that ever appeared in the literature, this one from Chirnside's sensitive Obituary of Robert. R. C. Chirnside, 'Robert Le Rossignol, 1884–1976', *Chemistry in Britain*, **13**, 269–271, (1977).

3. Dietrich Stoltzenberg, *Fritz Haber. Chemist, Noble Laureate, German, Jew.* Chemical Heritage Press, Philadelphia, Pennsylvania, (2004). ISBN-0-941901-24-6.

4. Robert's Obituary, Chirnside, op. cit. (note 2), was based upon a tape-recorded interview Chirnside conducted with Robert on 29 March 1976, a few weeks before Robert died. A transcript of the interview remains with the Le Rossignol family. Robert's modest account of his time at UCL is contained here.

5. Morris W. Travers, *A Life of Sir William Ramsay K.C.B., F.R.S.*, Edward Arnold London, (1956).

6. Science Museum at Wroughton, Swindon. UK. SN4 9LT. Inventory number 1931–1158.

7. £100 from a private benefactor according to Travers, op. cit. (note 5).

8. Robert Whitehead was a brilliant Victorian engineer who, whilst honoured by many countries world-wide, received little or no recognition from the land of his birth. He was the world leader in torpedo development and production up to the First World War.

9. At the time Fiume was part of Italy but now it belongs to Croatia.

10. Incidentally, the first one to be described in the scientific literature for liquefying hydrogen; Morris W. Travers, XXXVII 'The liquefaction of hydrogen'. *Philosophical Magazine*, Ser 6, **1**, 4, pp. 411–423, (1901). Liquefaction of hydrogen had also been achieved by Sir James Dewar, however Dewar and Ramsay were not on friendly terms at the time and there was no chance of UCL 'borrowing' Dewar's equipment. Travers' apparatus was later donated to the Science Museum, South Kensington, LONDON.

11. Gases such as nitrogen—the major constituent of air—cool upon expansion at room temperature (a $+ve$ Joule-Thompson effect) but hydrogen warms up (a $-ve$ Joule-Thompson effect). For hydrogen gas to cool upon expansion its temperature must be below its pressure dependent inversion temperature T_{J-T} when internal interactions allow the gas to do work on expansion. This requires pre-cooling of the gas to 78 K (about -195 °C) which the vapourisation of the liquid air achieves.

12. According to Travers, op. cit. (note 5), these flasks were obtained from Berlin.

13. There is a distinction to be made between reaction '*order*' and reaction '*molecularity*' which may not have been developed at the time? A '*quinque-molecular*' reaction tends to suggest reaction *order*, i.e., the number of atoms or molecules whose concentrations determine the kinetics of the process. *Molecularity* refers to the number of atoms or molecules taking part in each act leading to chemical reaction—and especially the Rate Determining Step. Today, 'five' would be regarded as 'suspicious'.

14. F. T. Trouton F.R.S. was Quain Professor of Physics at UCL during Robert's time there. He is best remembered for 'Trouton's Rule', a relationship between boiling points (K) and the molar 'latent heats' of vapourisation (at atmospheric pressure).

15. Today, the value is accepted as 54 s, closer indeed than Rutherford's value.

16. Robert's success on achieving the departmental gold medal (Chap. 1) was reported in *The Victorian*, the annual publication of Robert's school, Victoria College, in March 1906.

5

Hamburg, 12 May 1907

Herr Geheimrat Regierungsrat, Prof. Dr. Hermann Walther Nernst.

The Bunsen Society Meeting[1]

I don't like to say too much. Haber missed the equilibrium by just a little bit.
Nernst was not a pleasant gentleman. [In a letter] he told Haber all his work
was wrong. And Haber was afraid to publish it …

*Robert Le Rossignol in conversation with Johannes Jaenicke in 1959 regarding
Haber's confrontation with Nernst at Hamburg.*[2]

5.1   Approaching the Ammonia Equilibrium

After writing to Haber with his concerns, Nernst and his assistant Fritz Jost
spent the next six months examining the ammonia equilibrium themselves
using a small pressurised electric oven specially designed by him and capable
of being heated to 2000 °C, the entire unit being immersed in a large water
bath. They did so from the perspective of the decomposition of two moles of
ammonia, according to;

© Springer Nature Switzerland AG 2020
D. Sheppard, *Robert Le Rossignol*, Springer Biographies,
https://doi.org/10.1007/978-3-030-29714-5_5

$$2NH_{3(g)} \leftrightarrows N_{2(g)} + 3H_{2(g)}$$

the partial pressure equilibrium constant (atm^2) being expressed as;

$$K_p^{\text{Nernst}} = \left(p_{H_2}^3 \times p_{N_2}\right)/p_{NH_3}^2$$

Conversely, Haber's re-investigation of his work for the Margulies brothers was expressed in terms of the 'development' or synthesis of one mole of ammonia according to,

$$\frac{1}{2}N_{2(g)} + \frac{3}{2}H_{2(g)} \leftrightarrows NH_{3(g)}$$

the partial pressure equilibrium constant (atm^{-1}) now being expressed as;

$$K_p^{\text{Haber}} = p_{NH_3}/\left(p_{H_2}^{\frac{3}{2}} \times p_{N_2}^{\frac{1}{2}}\right)$$

The two approaches of course are complementary so that;

$$K_p^{\text{Haber}} = \left(K_p^{\text{Nernst}}\right)^{-\frac{1}{2}}$$

or

$$\log_{10} K_p^{\text{Haber}} = -\frac{1}{2}\log_{10} K_p^{\text{Nernst}}$$

However, the ammonia equilibrium is the ammonia equilibrium, whichever way it is reached.

5.2 Nernst's Presentation

Nernst's presentation[3] was clinical and mathematically elegant—his subsequent publication contained only two pages—but it was far from comprehensive in that he reported just seven results. Nernst probably felt that this was all that was necessary to discredit Haber's work and support his own 'Heat Theorem'. Nernst made a critical design decision early in the investigation viz, to study the equilibrium at 'high' pressure to maximise the percentage ammonia in the equilibrium mixture and thereby minimise any error in the quantitative determination(s). To this end, experiments were performed at 50–70 atmospheres by moving a mixture of hydrogen and nitrogen into the oven using the 'streaming method'[4] where the gas mixture was funnelled through a long tube. The gases were held in the tube and the temperature kept constant until equilibrium was reached. On release of the mixture, the temperature was rapidly reduced so that the reaction was immediately halted ('quenched') at the equilibrium position. Quite rightly then, Nernst has to be recognised as the first to synthesise ammonia at 'high' pressure[5] and to determine an equilibrium under these conditions (Fig. 5.1).

These experiments used just one catalyst, (viz., platinum in the form of a foil, 'Als Katalysator wurde Platinfolie benutz') and Nernst devised a novel volumetric determination of the ammonia present in the equilibrium mix. When the final mixture emerged from the oven, it was passed through 5–10 cm^3 (depending on the expected %ammonia) of 0.01N[7] sulfuric acid containing methyl orange as indicator. Using a measuring glass, the volume of gas required to produce an 'end point' was observed, so that the %ammonia in the mixture at any equilibrium temperature and total pressure could be calculated leading to the evaluation of K_p^{Nernst} at T. Nernst's early results however fluctuated wildly, the effect eventually being traced to residual oxygen which generated water, but the problem he believed was soon resolved. Using the experimentally determined K_p^{Nernst} values at various temperatures, Nernst then adjusted the equilibrium %ammonia in terms of a stoichiometric ratio of $3H_{2(g)}:1N_{2(g)}$ at one atmosphere total pressure[8] to provide a direct comparison with Haber and van Oordt. This led to the simple expression;

$$K_p^{Nernst} = \frac{(0.75)^3 \times (0.25)}{x^2};\qquad\qquad(5.1)$$

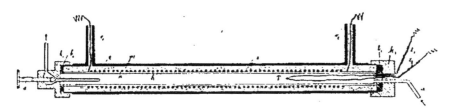

Fig. 5.1 The rather crude diagram presented by Nernst and Jost in their paper[3] illustrating their 'specially designed' oven, with an iron reaction tube encased in burnt magnesia containing a porcelain tube wound with a heating coil.[6] Gas entered at the right and exited at the left through the fine capillary tube connected to what appears to be a 'conical' valve

with the assumption that the partial pressure of the remaining ammonia (x) at equilibrium was so small that the sum of the partial pressures of the product gases constituted the total pressure. The observed (beobachtung or 'beob.') %ammonia at equilibrium was then simply $100x$.

Having determined the %ammonia experimentally at various temperatures, all that remained was to provide a theoretical estimate based on the 'Heat Theorem'. Nernst used his 'approximation' formula for gases to achieve this viz.,

$$\log_{10} K_p = -Q_{p(T)}/4.571T + \sum v\, 1.75 \log_{10} T + \sum vI$$

and interested readers are referred to Appendix C for a derivation of this expression. Here, $Q_{p(T)}$ represents the 'heat of reaction' for the decomposition of ammonia at pressure P and temperature T, i.e., the enthalpy change ΔH, but the symbol Q_p was in use at the time. Nernst however simplified matters further by deciding that the heat of reaction at 'ordinary' temperatures would suffice here because Q_p varied little with temperature. $\sum v$ represented the change in the number of moles during the reaction and $\sum v\, I$ the change in the 'conventional chemical constants' for the reaction (Appendix C again)— essentially the difference in various integration constants generated during the development of the expression. Nernst provided a value of 1.3 for this term. Given that the accepted 'ordinary' heat of reaction (Q_p) at the time was $+12,000$ cal mole^{-1} over a wide temperature range and that the change in the number of moles of course was $\sum v = +2$, Nernst's equation[9] for the decomposition of two moles of ammonia became;

$$\log_{10} K_p = -\frac{24,000}{4.571T} + 3.5 \log_{10} T + 2.6 \qquad (5.2)$$

Now this expression—already an approximation and with fragile values for the 'conventional chemical constants'[10]—was simplified further. Over the temperature range Nernst employed (958–1313 K), the term $3.5\log_{10} T + 2.6$ varied very little, so he decided to replace it with a single average value, and although this value was about 13.31, he found that 12.86 (97% of the 'true' value) proved 'more suitable in practise'. Nernst also queried the value $+12,000$ cal mole^{-1} found by Thomsen and Bertholet which, he decided, was 'too low' since their method of estimation provided 'unsteady values' leading to an appreciable error. Nernst subsequently favoured $+14,010$ cal mole^{-1} (an increase of almost 17%) and all this led to the expression;

$$\log_{10} K_p = -\frac{28{,}020}{4.571T} + 12.86 = -\frac{6130}{T} + 12.86$$

Now any physical chemist reading these lines today would regard Nernst's modification of these values as, frankly, 'over-accommodating'—a point we return to later—but substituting for K_p from Eq. 5.1, Nernst finally arrived at an equation of imperious mathematical austerity;

$$\log_{10} x = 3065/T - 6.918 \qquad (5.3)$$

which gave the partial pressure of ammonia (x) at equilibrium and one atmosphere total pressure at any temperature T. The calculated (berechnet or 'ber.') %ammonia was again simply $100x$. Nernst and Jost presented their results in the following table, with $\sqrt{K_p^{\text{Nernst}}}$ being included to facilitate an easy comparison with Haber and van Oordt (Table 5.1).

Nernst also noted[11] that for $x = 0.00012$ (i.e., 0.012% NH_3), Eq. 5.3 gave $T = 1023$ K; in 'far better agreement' with the value 893 K theoretically found via Eq. 5.2 than Haber's published figure at 1293 K (see Chap. 2). In this respect Nernst made pointed reference to page 187 of Haber's book[12] where his work with van Oordt appeared. He could of course have referred to the original paper[13] so his choice seems rather spiteful and calculated to discredit a book which had previously been received with some acclaim. Nernst also noted that if the enthalpy term in Eq. 5.2 was a 'little larger', the agreement between theory and observation would be 'practically perfect'. Finally, Nernst defended his figures (in a retrospective note—'Nachträglicher Zusatz') against deviation from ideality at such 'high' pressures. He repeated his observations at 12–15 atmospheres and found experimentally that at $T = 1188$ K, $100x = 0.0046$, whilst the equations delivered the value

Table 5.1 Nernst and Jost's results adjusted to one atmosphere

t (°C)	T (K)	$\sqrt{K_p^{\text{Nemst}}}$	$100x$ (beob.)	$100x$ (ber.)
685	958	1830	0.0178	0.01960
809	1082	3783	0.0087	0.00820
836	1109	4460	0.0072	0.00702
876	1049	5900	0.0055	0.00561
920	1193	7560	0.0043	0.00480
1000	1273	10,200	0.0032	0.00308
1040	1313	12,170	0.0026	0.00261

0.0048, suggesting that no significant deviation occurred. With this 'rider' he concluded his presentation.

Nernst was only four years older than Haber but he was firmly established as an eminent German physical chemist, indeed he had achieved everything that Haber could ever have dreamed of aspiring to, there was even talk of the Nobel Prize. Nernst's short presentation therefore 'showcased' his 'Heat Theorem', and he was intent on using the Bunsen Society meeting to defend its integrity. To Nernst, Haber was just another 'young buck' from a minor provincial German university, Haber's work it seemed had to be discredited. Now the professional humiliation of others to confirm or promote one's own position is commonplace within the academic community, and Haber could have anticipated it from the day he received Nernst's letter. In Hamburg therefore, he must have known that he was about to receive some of the 'medicine' he had handed out to Max Julius Le Blanc at Karlsruhe, and to make matters worse Le Blanc had arrived from Leipzig to chair the discussion. Academic vultures it seemed were gathering, so the tension in the Karlsruhe group[14] that day must have been palpable. Haber, a freshly appointed professor, had a hard earned reputation that could yet be easily broken; Robert, his young co-worker, was responsible for the experimental integrity of their presentation. Although biographical history has concentrated on Haber's position in this meeting, the pressure on Robert too was considerable for it was his efforts that had to return the evidence that prevented Haber going 'naked into the debating chamber'. But sound as Robert's efforts were … Jaenicke later recorded (1959, Chap. 18) that Haber was afraid.[2]

5.3   So How Should Haber Respond?

Two approaches were available to Haber that day, one was to attack. For example, the reproducibility of Nernst's results could be questioned; there were very few of them, and his approach to the theoretical predictions made heavy use of approximations. These ranged from the conventional chemical constants to values that 'proved more useful in practice' such as the modified heat of reaction and the average value for the constant in his equation. But Nernst's approach seemed to work well for many other reactions. This, coupled with his reputation and the near universal acceptance of his 'Heat Theorem', made this a difficult route to follow.

Haber therefore would have preferred a second route, one which he had been developing ever since his first independent investigation of the pyrolysis of hydrocarbons at Karlsruhe. It was characterised by meticulous attention to detail making it difficult for any opponent to break him down, and in Robert Le Rossignol he had the ideal experimental partner.[15] Indeed, the stamp of Robert Le Rossignol, 'professional chemist', should have been writ large on the work presented by Haber that day because the two men had laboured for months on the revised investigation and had submitted (29 April 1907) a substantial paper[16] to the Bunsen Society shortly before the meeting began. But the transcript of Haber's subsequent 'presentation'[17] shows that it was curiously defensive and not at all representative of the careful work done by the two men. So we firstly 'reprise' this paper because what Haber was able to convey in response to Nernst on that day—and subsequently reported in several biographies[18]—did both himself and Le Rossignol an historic justice.

5.3.1   On the Ammonia Equilibrium

Their paper began by describing 'Die älteren Versuche', the 'older experiments' conducted with van Oordt, and they immediately withdrew the (frankly *quite* indefensible[19]) value of 0.012% equilibrium proportion of ammonia at 1293 K. This value, they felt, may have arisen either because of a special effect of the iron catalyst when fresh (i.e., the presence of nitride) or because during the decomposition of the ammonia, insufficient time had been allowed to permit the equilibrium position to be reached, resulting in the presence of a higher percentage of residual ammonia than should otherwise have been the case.[20] Overall however, Haber and Robert declared themselves 'content' with the basic design of the van Oordt experiments (Chap. 2). Their new investigation therefore proceeded along the same lines as before but with refinements that established the reproducibility of their results which in turn, they maintained, showed that they must represent the true equilibrium position.

   The design of the Haber-Le Rossignol experiment was more robust than Nernst's, examining as it did the equilibrium from the decomposition and the 'development' (or synthesis) sides and the whole process involved a considerable range of practical skills practised entirely by Robert. Measured volumes of dry ammonia at atmospheric pressure were passed over different catalysts at temperatures between 1000 and 700 °C. Unchanged ammonia was 'washed out', the product gases[21] dried then passed over a second share of the same catalysts. The newly formed ammonia was again 'washed out' and both washings determined volumetrically. In their paper Haber and Robert presented a table containing the results of 49 separate attempts at six different temperatures. Their description of the experimental setup and preliminary work showed just how exhaustive their approach had been. They firstly varied the source of ammonia. Experiments 1–36 used commercially available solutions of ammonia together with ammonium nitrate, whist for 37–49, gaseous ammonia was taken from a cylinder. Different sources of ammonia generated different flow rates but this was regulated by overcoming the resistance of a manometer of controllable height.

The decomposition and development reaction tubes were always side by side in the oven. These tubes, (60 cm in length) were therefore always at the same constant temperature (within 5° at 1000 °C) and it was here that the catalyst resided, so that the gases never passed over catalyst in cooler parts of the oven. The gases were also allowed to remain in these tubes for hours to ensure that equilibrium was reached. Various materials were also used for these tubes such as quartz and china, the general idea here being to make the tube less permeable to the diffusion of gases that they felt was a fault[22] with the Nernst set-up. Temperature measurement was via a thermocouple and galvanometer, one end of the thermocouple lay in the oven—pushed into both tubes—the other lay in ice. This device was calibrated using the gold melting point at 1062 °C giving an accurate galvanometric determination of oven temperature.

Four different catalysts were used viz., iron, nickel, chromium and manganese and these were prepared in the laboratory to control their purity.[23] All the catalysts, except nickel, were 'raised' on iron free asbestos. Prior to use, the asbestos was, for 'day after day', boiled with concentrated hydrochloric acid and heated in a hydrogen current for at least 10 h, followed by renewed boiling with the acid. The asbestos was impregnated with the catalyst, and the amount used varied depending on the experiment. In the case of iron, between 10 mg and 1 g were examined. Between experiments, the apparatus was cleansed of any undecomposed ammonia by flushing it overnight through auxiliary washing flasks containing 20% sulphuric acid with glass wool filter plugs to prevent any acid vapour remaining in the apparatus afterwards. The remaining part of their description of the experimental setup concerned quantitative aspects of the ammonia determination, and this aspect of the paper was critical because Nernst was unhappy with 'Haber's' earlier experimental accuracy when determining the low values of ammonia present at 1 atmosphere equilibrium pressure.

Immediately after the decomposition reaction tube were *two* sequentially placed Volhard flasks modified so that they required only 15 cc of titration acid to fill them. In this respect the flasks were originally charged with 10 cc of 0.02N HCl to which 5 cc of distilled water[24] were added. Two such flasks were also placed just after the development tube. In each case, the second flask was to act as a control, absorbing any traces of ammonia not 'held back' by the first. Robert noted that no ammonia was ever absorbed in the second flask. The ammonia determination could therefore proceed with confidence on the first. The ammonia 'held back' in the absorption bottles was determined by titration. The bottles were washed out with distilled water to provide a final volume of 40 cc. 5 cc of ethers containing 2 mg per litre of 'Jedeosin' (Erythrosin B) were then added as indicator. This solution was titrated with 0.02N aqueous ammonia (according to the established method of Mylius and Foerster). The titration therefore required 3.4 mg of ammonia for complete neutralisation of the *original* acid and Haber and Robert declared that the end point for the residual acid was sharp enough to recognise 0.01 mg ammonia (formed or undecomposed), 'very clearly'.

The equilibrium constants and %ammonia at equilibrium found experimentally by Robert are shown in Table 5.2.

Robert also confirmed by analysis, that once the decomposition equilibrium had been reached and the residual ammonia washed out, the 'remainder' gas always contained 25% nitrogen[25] before passing into the development tube, so that stoichiometric amounts of the gases were present during the formation of ammonia. Having determined the equilibrium constants experimentally with meticulous accuracy, all that remained was to provide theoretical estimate(s). Once again in this respect, the Haber-Le Rossignol paper was robust in that two methods were employed. In the first, the equilibrium constants were calculated using what could be described as the 'conventional' approach. In the second they used Nernst's novel 'approximation' formula for gases. The 'conventional' approach had already been developed for the ammonia reaction in Haber's book, (see pp. 203–204 of Lamb's translation[12]) and although this book had been a critical success, Haber also referenced Bodländer's[26] respected work here, probably to add

Table 5.2 The experimental equilibrium constants and % ammonia at equilibrium and atmospheric pressure for the Haber-Le Rossignol investigation.[16] Subsequently, this was the only table Haber offered the audience in his Hamburg presentation

t (°C)	T (K)	$K_p^{\text{Haber}} \times 10^4$	$\%NH_3$ @ equil.
700	973	6.80	0.0221
750	1023	4.68	0.0152
800	1073	3.34	0.0109
850	1123	2.79	0.0091
930	1203	2.00	0.0065
1000	1273	1.48	0.0048

credibility. Employing the 'approximation formula' as a double check may have been an appeasement to Nernst, after all who can resist the self satisfaction when one's ideas are used to confirm another's work?

At the time, the 'conventional' approach was developed in terms of the 'reaction energy' A, the decrease in A representing the maximum reversible work (of all kinds) available in any isothermal process. The general form of the equation[27] used by Haber was;

$$A = \left(Q_0 - \sigma_p T \log_e T - \sigma' T^2\right) - RT \log_e K_p + \text{const.}\, T$$

and for interested readers, Appendix D contains a more modern but equivalent derivation of this equation—and indeed all the equations used by Haber from 1907 to 1914.

At equilibrium, $A = 0$ because no useful work can be done by the system. Appendix D also shows that the term in parenthesis is simply the 'heat of reaction' at absolute zero (i.e., the enthalpy change), adjusted to the temperature of the experiment by including the variation in the heat capacity between reactant(s) and product(s). Haber conventionally assigned the symbol $Q_{p(T)}$ to this term resulting in the expression;

$$0 = Q_{p(T)} - RT \log_e K_p + \text{const.}\, T$$

Converting to common logarithms, substituting for $Q_{p(T)}$ (using the accepted 12,000 cal *evolved* mole^{-1} of ammonia formed—but see Appendix D again), rearranging in terms of $\log_{10} K_p$ then simplifying gave;

$$\log_{10} K_p = \frac{12,000}{4.57T} + \frac{\text{const}}{4.57}$$

With the constant determined experimentally at—26.93 by Robert, Haber presented Table 5.3 for the 'conventional' calculated ($K_p^{\text{Conv eqn}}$) equilibrium constants at the various temperatures, to which we have added the corresponding %ammonia present at equilibrium.

Haber's theoretical foundations—being more conventional than Nernst's —were therefore difficult to criticise, but he proceeded to show that by using Nernst's own 'approximation formula', the *same* results could be obtained. Given Nernst's expression;

$$\log_{10} K_p = -Q_{p(T)}/4.571T + \sum v\, 1.75 \log_{10} T + \sum vI$$

and now with $\sum v = -1$, $I = 1.3$ and $Q_{p(T)} = -12{,}000$ cal mole^{-1}, he arrived at;

$$\log_{10} K_p = 12{,}000/4.571\, T - 1.75 \log_{10} T - 1.3$$

However, Haber decided to take into account the average specific heats of the ammonia and 'permanent' gases from absolute zero to the temperatures of the experiments, a point which Nernst had failed to address.[10] By adding the term $0.000651\,T$ to the right hand side, the average specific heat of ammonia at constant pressure became $9.5 + 0.004157\,T$ whilst at the same time the increase in the average specific heat of the 'permanent gases' became $0.0006\,T$. With the Nernst approximation now in the form;

$$\log_{10} K_p = 12{,}000/4.571T - 1.75 \log_{10} T + 0.000651\, T - 1.3$$

Haber presented the 'calculated' equilibrium constants $\left(K_p^{\mathrm{Nernst\ eqn}}\right)$ to which we have again added the corresponding %ammonia at equilibrium, Table 5.4.

Tables 5.2, 5.3 and 5.4 provide comprehensive agreement between the experimental and calculated results with no need for the 'adjustments' made by Nernst. Indeed, Robert's experimental results were in far better agreement with Nernst's predictions than his own figures, which because of his approximations tended to favour a lower equilibrium %ammonia. The consistency and sheer attention to detail of the Haber-Le Rossignol investigation should have convinced any audience, but could Haber convey enough of it to convince Nernst?

Table 5.3 The calculated equilibrium constants and % ammonia at equilibrium and atmospheric pressure using the conventional equations

t (°C)	T (K)	$K_p^{\mathrm{Conv\ eqn}} \times 10^4$	$\% NH_3$ @ equil.
700	973	6.30	0.0205
750	1023	4.72	0.0153
800	1073	3.53	0.0115
850	1123	2.79	0.0091
930	1203	1.94	0.0063
1000	1273	1.48	0.0048

Table 5.4 The calculated equilibrium constants and % ammonia at equilibrium and atmospheric pressure using Nernst's 'approximation' formula

t (°C)	T (K)	$K_p^{\text{Nernst eqn}} \times 10^4$	%NH_3 @ equil.
700	973	6.34	0.0206
750	1023	4.62	0.0136
800	1073	3.48	0.0113
850	1123	2.69	0.0087
930	1203	1.88	0.0061
1000	1273	1.44	0.0047

5.4   The Discussion

In all honesty, the detail of their investigation did not come out in the meeting,[28] and one has to accept that the simplicity of Nernst's approach was probably more convincing. Nernst too was a rather abrasive character. Having already ruffled Theodore Richard's feathers with regard to the Third Law he was not on particularly good terms with Svante Arrhenius the Swedish physical chemist, and he was about to become bad friends with Haber because Robert's results of course, were consistently higher than his own. J. E. Coates[29] adjusted Nernst's figures to Haber's experimental temperatures to illustrate this for us (Table 5.5).

After Haber's 'presentation' the discussion began with Nernst being characteristically blunt.

'If one operates with yields of fractions of milligrams [*of ammonia*] gained over long periods, then very little can be learned. I would recommend that you centuple the quantity of ammonia. Indeed, with our simple arrangement, we get so much[30] in 5 min that we can titrate exactly. Nevertheless, the present difference in our figures is very small in comparison with earlier numbers, but I would like to propose that Professor Haber now uses - instead of his earlier method that gave *such* unreliable values - a method at higher pressure that returns really precise values'.

'But', replied Haber, 'using your apparatus the equilibrium was approached from only one side, I approached it from both sides'.

Table 5.5 Experimentally determined %ammonia at equilibrium and atmospheric pressure for both sets of workers

Temperature (°C)	700	750	800	850	930	1000
Nernst-Jost	0.0174	0.0119	0.0087	0.0065	0.0043	0.0032
Haber-Le Ross.	0.0221	0.0152	0.0108	0.0091	0.0065	0.0048

'That's not a problem!' declared Nernst. Belittling Haber, he announced to the audience, 'You can do that in *my* apparatus, as much as you like!'.

Haber was taken aback by this jibe enquiring only, 'Do your seals[31] permit that?' to which Nernst replied with confidence 'Yes … why not?' Haber however, had realised that at the heart of Nernst's attack was the question of the accuracy of the ammonia determination at such a low pressure. With total confidence in Robert's ability he leapt to his defence.

'Then I want to say the following. Being able to determine down to one-hundredth of a milligram of ammonia through our titration method is simply not a problem. The earlier difficulties emerged because such precise measurements were not available to us.'

To Nernst this still wasn't good enough and he now felt he had the upper hand, so he summed up with a final cruel attack.

'Then perhaps may I just note one fact that is of general technical interest. It is regrettable that the equilibrium tends even more to the side of much reduced formation [*of ammonia*] than has been assumed up to now because of Haber's highly erroneous[32] [*previously published*] data, for one could really have thought of producing ammonia synthetically from hydrogen and nitrogen. But in fact the conditions are even less favourable. The yields are about one third of what was previously expected.'

Haber remained silent for the remainder of the session and the audience was stunned at his public humiliation. In a charitable act, one which Haber's previous behaviour towards Le Blanc hardly deserved, Chairman Prof. Dr. Le-Blanc moved to deflect the course of the discussion away from Haber towards a more general appraisal of the ammonia synthesis. He enquired;

'I have also seen, with some regret, that the gains become unfavourable at high temperatures, for my laboratory is also busied with attempts to develop ammonia, - but in different ways. Recently we worked with dark electric discharges at high temperatures and the question now naturally arises whether at these temperatures the discharge itself may have some influence. Is anything well known about that? Nernst however was not to be drawn to an area where he had no experience simply declaring, 'We did not work with electric discharges', and by failing to stimulate discussion in another direction, Le Blanc probably felt that Haber had been humiliated enough. He drew the session to a close noting that in view of the highly unfavourable equilibrium ratios discussed, the fixation of nitrogen via this route was 'doubtful' and the nitrogen-oxygen route may yet prove more successful.

Now Nernst's outburst was really quite puzzling from a number of perspectives. Firstly, Haber had publicly withdrawn his old figures, and Robert's new figures were in better agreement with Nernst's approximation formula than Jost's, which relied on considerable modification of the constants to achieve an acceptable 'fit'. Indeed without the adjustments, the basic formula would have returned much higher figures for the percentage of ammonia. Nernst also intimated that Haber had concluded that the ammonia synthesis was feasible, but that was not the case. Haber was already on record (Chap. 2) saying that '… *even at greatly increased pressure the position of the equilibrium must remain very unfavourable.*' One therefore tends to hold a very uncharitable view of Nernst's motive(s) that day, in spite of some authors[33] opinion that Nernst was 'a fair and honest competitor', able to 'reach an agreement' with Haber and Le Rossignol. Well maybe, but any 'agreement' appears to have been entirely on Nernst's terms. After the meeting Svante Arrhenius unsurprisingly endorsed Haber's findings, but the damage was done, and not only in Hamburg but at an international level as well, for the full text of the discussion was soon published in the journal *Elektrochemische Zeitschrift*.[3] As a result, Nernst's summary was generally accepted. He regarded the position of the ammonia equilibrium as substantially settled in his favour. Haber's figures were wrong while his own were in sufficiently good agreement with his 'Heat Theorem', and this remained the position over the next year or so. Although Nernst had caused considerable damage to Haber's reputation, his conclusions were not without significance to himself. Between 1901 and 1907 the theoretical foundations of the ammonia synthesis were laid by three giants of physical chemistry, Le Châtelier, Ostwald and Nernst. But as Smil[34] says;

> anyone of these chemists could have been the actual inventor of a technically viable ammonia synthesis but for unique circumstances – the explosion in Le Châtelier's laboratory, Ostwald's erroneous interpretation of experimental results and Nernst's opinion that reaction yields were far too low … and these led all three scientists to abandon their ammonia related work.

In a bizarre twist of fate therefore, Nernst's vitriolic outburst at Hamburg had the effect of removing[35] himself and alienating many other competitors from the ammonia problem leaving the road open[36] to Haber and Robert, who, driven by the damage to their reputations, took up the challenge again. Another road however, often leads to another rut, and even without the competition the path to commercial ammonia synthesis was to prove a hard road indeed.

Notes

1. There is confusion in the literature here. Some authors such as Kormos Barkan, *Walther Nernst and the Transition to Modern Physical Science*, Cambridge University Press, (1999), p. 130, and Thomas Hager, *The Alchemy of Air*, Three River Press, New York, (2008), p. 65, refer to a meeting at 'Hannover'. Margit Szöllösi-Janze, *Fritz Haber 1868–1934: Eine Biographie.* C. H. Beck, Munich, (1998), Vaclav Smil, *Enriching the Earth*, MIT Press, (2001) and Timothy Lenoir, *Instituting Science. The Cultural Production of Scientific Disciplines*, Stanford University Press, (1997), however, identify the meeting at Hamburg on 11–12 May, 1907. This author agrees with the latter authors.

2. Jaenicke's and Le Rossignol's letters and the transcript of their conversation on 16 September 1959 are held at the Max Planck Gesellschaft, Archiv der MPG, Va. Abt., Rep. 0005, Fritz Haber. Haber Sammlung von Joh. Jaenicke. Nr. 253 (Jaenicke's letters) and Nr. 1496 (conversation transcript).

3. W. Nernst and F. Jost, 'Über das Ammoniakgleichgewicht', (On the Ammonia Equilibrium), *Elektrochemische Zeitschrift*, **13**, 521–524, (1907). Jost was a research student at the time, his dissertation being '*Über die Verwendung Eines Elektrischen Druckofens bei Behandlung Chemischer Gleichgewicht*', (On the use of an electric pressure furnace when treating chemical equilibrium), dissertation, Berlin, 1908.

4. W. Nernst, 'Experimental and Theoretical Applications of Thermodynamics to Chemistry', *Silliman Memorial Lectures*, University of Yale, p. 31, (1907). The 'streaming' method was first applied by Langmuir for 'very accurate' determinations of *dissociation* equilibria for water and carbon dioxide, Kormos Barkan, *op. cit.* (note 1), p. 129. It is curious that for his Hamburg presentation, Nernst employed a technique used for the study of dissociation equilibria, proceeded to discuss the equilibrium in terms of ammonia dissociation, then appears to have used a mixture of nitrogen and hydrogen to establish the equilibrium via the ammonia *synthesis*(?). It would have been simpler to dissociate ammonia and avoid the need for a stoichiometric mix of the two gases. The paper covering the presentation contains no detail regarding the mixture composition nor its provenance (although see Chap. 6) and simply refers to the electric pressure oven 'in which a mixture of nitrogen and hydrogen is under suitable pressure'. But it is not entirely clear to this author if this mixture was introduced initially, or formed by the decomposition of ammonia. However, forming ammonia at high pressure led to much greater yields and therefore made the quantitative analyses much easier.

5. In Coates' Haber Memorial Lecture, delivered before the Chemical Society in London on 29 April 29 1937, he says, "Nernst was thus the first to synthesise ammonia *under pressure* … ". Coates, *op. cit.* (note 29), p. 1651. But on this basis, Dronsfield and Morris, 'Who really discovered the Haber process?'

RSC, Education in Chemistry, (May 2007) fabricated a mischievous and mildly sensational argument to credit Nernst with the discovery of what was to become known as the 'Haber Process'. Predicated on being 'the first to synthesise ammonia at high pressure' in his electric pressure oven, these authors' claims can be dismissed. Nernst's oven existed simply to support his 'Heat Theorem', but the 'Haber Process' is universally accepted as forming the basis of the *industrial* synthesis of ammonia. Later chapters will show that 'Haber's' 'process' was a complex affair at the cutting edge of the understanding of technical gas reactions. Nernst's apparatus made no pretensions whatever to represent a 'process', nor even an application, a point which he acknowledged in the discussion at Hamburg and which he later repeated before the Patent courts in 1912 when he declared that his interest had been of a 'purely theoretical nature'. (See Chaps. 6, 12 and Dietrich Stoltzenberg, *Fritz Haber. Chemist, Noble Laureate, German, Jew.* Chemical Heritage Press, Philadelphia, Pennsylvania, (2004), p. 176.) Indeed, Chap. 6 shows that he did not even achieve a correct equilibrium. If one wishes to meddle with history, one should get one's facts right. Authors would be better advised if they referred to the 'Haber Process' for what it really was, the 'Haber – *Le Rossignol*' process.

6. Kormos Barkan, *Walther Nernst and the Transition to Modern Physical Science*, Cambridge University Press, (1999).
7. 'Normal' (*N*) solutions were in use at the time and even through to the 1960s. The concept of 'normality' rendered the calculations of volumetric analysis very simple. Normality was centred around the idea of 'equivalent weight'. For example, a 'normal' (1 *N*) solution of sodium hydroxide contained the molecular weight of NaOH in grams (40.0 g) per litre, whilst a 'normal' (1 *N*) solution of hydrochloric acid also contained the molecular weight of HCl in grams (36.5 g) per litre. One litre of 1 *N* NaOH therefore reacted completely with one litre of 1 *N* HCl so that 40 g of NaOH was *equivalent* to 36.5 g HCl. However, a 1 *N* solution of an acid such as sulphuric, contained just 49 g per litre, (*half* the molecular weight in grams) since a 1 *N* solution of sulphuric acid would then react completely with a 1 *N* solution of say NaOH or NH_4OH or $NH_3(aq)$. The 'equivalent weight' of sulphuric acid w.r.t these titrations was therefore just 49 g.
8. For an 'ideal' system, K_p is independent of pressure.
9. Incidentally, even after all these years this equation has often been mis-quoted in the literature. Poor type-setting in the original publication led to the minus sign being *incorporated* into the quotient line, the first term appearing as $24,000/4.571T$ i.e., +ve not −ve. As an example, see; p. 9, J. R. Jennings, *Catalytic Ammonia Synthesis: Fundamentals and Practice*, Springer, (1991).

10. The 'conventional chemical constants' were calculated using an assumed value for the specific heat of ammonia which had never been measured accurately at high temperatures, Jennings, *op. cit.* (note 9), p. 10.

11. Even after his 'adjustments', he maintained that his predictions were still in better agreement with Eq. 5.2—the 'benchmark'—than Haber's.

12. F. Haber, '*Thermodynamics of Technical Gas Reactions*', pp. 204–206, University of California Digital Library, translated by Arthur B. Lamb.

13. F. Haber, G. van Oordt, 'Über Bildung von Ammoniak aus den Elementen', (Formation of Ammonia from the Elements), *Z. Anorg. Chem.*, **44**, 341–371, (1905).

14. History has not recorded whether Robert was present at the meeting, but most likely he was. Even so, the division of work between the two men was becoming established; Robert the practician, Haber the theoretician.

15. Haber made clear reference to his working 'partnership' with Robert ('*in Gemeinschaft mit Herrn Le Rossignol* …) in the subsequent presentation, emphasising the equal role Robert played in designing their experiments and applying the 'most possible care' to their investigation.

16. The Haber-Le Rossignol investigation is contained in the paper; F. Haber and R. Le Rossignol, 'Über das Ammoniak-Gleichgewicht', (On the Ammonia Equilibrium), *Ber. Bunsenges., Phys Chem.*, **40**, 2144–2154, (1907), received on 29 April 29 1907, and published shortly after the meeting.

17. The transcript of Haber's 'presentation', appears as part of the discussion which followed immediately after Nernst's paper, Nernst, *op. cit.* (note 3). However, submitting his paper barely two weeks before the meeting meant that it could hardly have been subjected to detailed scrutiny before Haber got to his feet. Being 'afraid', maybe that was the intention, but equally it failed to convey the enormity of the work the two men had undertaken.

18. Vaclav Smil, *Enriching the Earth*, MIT Press, (2001), Daniel Charles, *Between Genius and Genocide*, Jonathan Cape, London, (2005), Dietrich Stoltzenberg, *Fritz Haber. Chemist, Noble Laureate, German, Jew*. Chemical Heritage Press, Philadelphia, Pennsylvania, (2004) and Margit Szöllösi-Janze, *Fritz Haber 1868–1934: Eine Biographie*. C. H. Beck, Munich, (1998).

19. Robert's *new* values confirmed this figure as indefensible. One would have expected Nernst to have been pleased by this retraction and let the matter rest. But on that day, he was the *Über* pedant.

20. Haber's 'presentation' in Hamburg also began with the same retraction and it seems to have taken up almost half of his allotted time.

21. As with the van Oordt experiment, performing the decomposition *first*, guaranteed that a virtually stoichiometric ratio of $3H_{2(g)}:1N_{2(g)}$ entered the second tube. The proof of the optimal mix of nitrogen and hydrogen for maximum ammonia yield (a stoichiometric ratio of 1:3) had already been

established by Haber in, '*Thermodynamics of Technical Gas Reactions*', Haber, *op. cit.* (note 12), pp. 204–206.

22. The diffusion of the small hydrogen molecule into iron was later to become of paramount importance in the history of the synthesis of ammonia (Chap. 11). At the time neither Haber nor Le Rossignol realised this.

23. To generate the iron catalyst, the purest commercial iron oxide was obtained and reduced in a stream of hydrogen at 1000 °C. There was, therefore, no possible formation of nitride. Occasionally, ferrous oxalate was used instead. For the nickel catalyst, the purest commercial nickel nitrate was treated with silicic acid. The solid was dried and the metal produced by heating to 350–400 °C in a stream of hydrogen, Haber noting that at this temperature 'a catalytically very effective product is received'. Chromium was prepared electrolytically from a solution of the sulphate. It was easily removed from the platinum electrode then ground down in a mortar. Manganese was obtained electrolytically as a mercury amalgam. The mercury was removed in a vacuum and the manganese ground as before. Little did the workers know that the eventual successful industrial catalyst was much more of a chemical 'mongrel' (Chaps. 10 and 11).

24. The distilled water used to 'top up' and 'wash out' was weakly alkaline against the indicator 'Jedeosin' (Erythrosin B), but by the addition of 1.4 cc 0.02 N HCl to a litre of the water it was made neutral to this indicator.

25. Between 25.1 and 25.2% nitrogen was usually confirmed—probably by using Haber's patented refractometer, (Chap. 2). *Testing Gaseous Mixtures*, United States Patent Office, No. 830,225, Patented 04 September 1906.

26. Bodländer, *Zeitschrift für Elektrochemie.* **8**, 833, (1902).

27. See R. Luther, *Zeitschrift für Elektrochemie*, 12, 97, (1906), for Haber's thermodynamic terminology, also quoted in Louis De Vries, *German-English Scientific Dictionary*, 3rd Ed., McGraw-Hill. One might also refer to Lamb's translation of Haber's book, Haber, *op. cit.* (note 12).

28. Haber only referred to the Nernst 'approximation formula' and presented just one table, (viz, Table 5.2). The whole report of Haber's 'presentation', Nernst, *op. cit.* (note 3), was barely a page long—and much of that was taken up with the retraction of his previous result at 1020 °C in 1905, Haber, *op. cit.* (note 13).

29. J.E. Coates, 'The Haber Memorial Lecture', *J. Chem. Soc.*, 1642–1672, (1939).

30. Nernst was beguiled by the large yield of ammonia his experiments provided; he didn't question the possibility that his equilibrium could be poor—a fact established later by Le Rossignol. Although he criticised Haber, *his* results were at fault. Le Rossignol dealt with tiny yields, but the work was fastidious, accurate and reproducible.

31. Nernst was operating at 50–70 atm. Examining the equilibrium from both sides consecutively would have meant a much longer period in the oven, hence Haber's concern regarding the seals at such 'high' pressure.

32. Nernst used the term '*stark unrichtigen Zahlen*', which translates to '*strongly inaccurate*' data or even '*severely inaccurate*' data. This was 'hard talk' indeed.

33. Kormos Barkan, *op. cit.* (note 6), pp. 130–131.

34. Smil, *op. cit.* (note 1).

35. As consultant to the Griesheim Chemical Works, he had considered ammonia synthesis feasible. But some time after Hamburg they lost confidence and he walked away from the problem, Stoltzenberg, *op. cit.* (note 5), p. 87.

36. But they probably didn't know it at the time.

6

Bestimmung Des Ammoniakgleichgewichtes Unter Druck

Determination of the ammonia equilibrium under pressure.[1]

We found out that Nernst's equilibrium [for Hamburg] was very bad ....

Robert Le Rossignol in conversation with Johannes Jaenicke in 1959.[2]

6.1   The Legacy of Hamburg

After the horror of Hamburg, it was clear that Haber was deeply offended by Nernst's opinion of his work and mindful of its effect on his perception by others. Quite naturally, he took it as a personal slight and an injury to his reputation causing him to fall into sombre mood. As usual, this began to affect his health and a few weeks later Clara wrote to their friend Richard Abegg expressing her concerns[3]

> .... Fritz is again suffering from his old sickness - stomach, digestion, nerves, skin problems – all of which weaken him ....

Quite naturally too, biographical accounts of the period have concentrated on Haber, and there is no doubt that both on the day, and biographically, Haber 'took the hit'. But Robert too must have been hurt by Nernst's criticism because it was his experimental practise that Nernst had publicly denounced.

© Springer Nature Switzerland AG 2020
D. Sheppard, *Robert Le Rossignol*, Springer Biographies,
https://doi.org/10.1007/978-3-030-29714-5_6

Like Haber, Robert had a reputation to protect, that of 'professional chemist', and although yet modest, it was a reputation he had long aspired to, one which he had proudly brought to Karlsruhe and one which he was minded to preserve. The Hamburg paper was Robert's first real professional test and he could hardly have been thrown into a more difficult arena. On the one hand he faced a persistently obstinate chemical problem and on the other a brilliant opponent of world renown—and a curmudgeon. History has shown however, that this young man's work was equal to the challenge, indeed later, he even regarded himself lucky to be presented with the challenge.[4]

Prior to Hamburg, Robert could not have been more aware of Haber's attitude to the ammonia synthesis. In Lecture Five[5] of the 'Thermodynamics of Technical Gas Reactions'[6] Haber had announced ....

> The peculiar inertness characteristic of nitrogen which appears so prominently .... will always place a serious handicap on the cheap technical preparation of ammonia ...

However, Haber's scepticism did not deter Robert from providing a comprehensive and professional experimental treatment of the problem, a treatment that both men felt must settle the matter. But after their public humiliation in Hamburg they were forced to return to it. Convinced of the efficacy of their procedures and once more in partnership, they began another detailed investigation but this time—constrained by Nernst's criticism—at a modest elevated pressure of 30 atmospheres. This pressure was achieved by utilising the 'high pressure cylinders' used for the commercial supply of nitrogen, hydrogen and ammonia gases. The result of this investigation was the paper 'Bestimmung des Ammoniakgleichgewichtes unter Druck' published in April 1908, and here—without the respect shown towards Nernst in Hamburg—they provided closure *and* retribution.

If ever one aspired to become a professional chemist at the time, one needed to look no further than '*Bestimmung* ...' for the standard required. In surely one of the most exquisitely[7] crafted pieces of early 20th century chemical practise and theory, Haber and Robert comprehensively responded to Nernst's criticisms of their Hamburg presentation. Their paper stood as an open letter to the international chemical community proclaiming simply 'we told you so ... here's the proof .. ' But exquisite or not, the paper didn't improve the prospects for the industrial synthesis of ammonia one jot—at least from the perspective of the equilibrium. But then again, that was not its purpose. Initially this paper should be seen as an act of pure scientific self indulgence by the two men, an indulgence aimed squarely at restoring their reputations and not at producing an application. And it did so by undisputedly establishing the most accurate profile of the ammonia equilibrium to date, a profile which remains essentially the standard today. But even so, what could that 'bring to the table'? After all, the Hamburg meeting had already given the whole scientific community all they needed to reasonably anticipate the (ideal) behaviour of the equilibrium at various temperatures and pressures. Experimentally determined equilibrium constants[8] were available from 700 to 1000 °C and using the integrated form of the van't Hoff equation, these could be extrapolated with confidence down to temperatures of 200 °C, or even lower (as Haber and van Oordt had already done). Here, with ammonia now forming a much larger percentage of the equilibrium mixture, there had to be an adjustment in the expression for the partial pressure equilibrium constant. But Haber, in his book 'The Thermodynamics of Technical Gas Reactions' had already described K_p under these circumstances, so that figures of the type shown in note 9 would have been easily constructed by the two men—and indeed by anyone else who was interested enough to bother.[9]

And such was the fearsome reputation of the ammonia problem, that Haber and Robert by restoring their reputations had only really confirmed what everyone generally suspected ever since Nernst's outburst .. the ammonia synthesis was stubborn; at temperatures amenable to industrial production catalysis was impossible, at higher temperatures yields were vanishingly small, so what if anything did this 'exquisite' paper really contribute to the ammonia problem?

6.2 Well Firstly …

With complete confidence in the procedures for their Hamburg presentation, Haber and Robert must have been puzzled by the fact that their figures were so still much higher than Nernst's—usually about 33% higher, (Table 6.5 later). The two men therefore decided to reproduce Nernst's work, and in doing so they showed that the reason for this was that Nernst's equilibrium was quite simply, *wrong*.[10] Those choosing to work on the ammonia synthesis would naturally want to start from a conventionally accepted position, and in this respect Haber and Robert claimed '*Bestimmung* …' as 'de rigueur'. The paper therefore, did not simplify the problem, but it certainly set the standard for subsequent work. Haber and Robert achieved this by showing that the results they presented at 1 atm were entirely consistent with their new figures at 30 atm. Continuing on from Hamburg, Robert's experimental accuracy and attention to detail was stunning; every avenue was explored and any potential criticism was anticipated and addressed. The two men were now 'playing hard-ball' and in a substantial (16 page) paper, Nernst was questioned on almost every page.

They began by announcing that in view of their new results, '*Herr Nernst was not entitled to his previous objection*'. Nernst's values for the equilibrium, they claimed, lay *so* far to the nitrogen:hydrogen side that they could not be justified. Much of this they felt could be attributed to Nernst's experimental arrangement which he had already admitted was 'simple'.[11] Their arrangement however, was 'substantially' different. Haber and Robert continued by criticising Nernst's use of an internal heating element ('Innenheitzung') in his oven. This, they argued, produced a non-uniform temperature profile; in their experiments the pressurised 'ammonia bomb' was uniformly heated,[12] externally, by immersion in an oil bath, and even with determinations being made from both sides of the equilibrium at 30 atm, they obtained the same results as at 1 atm. The presence of impurities too, they felt, dramatically influenced the results. Nernst had revealed that early inconsistencies were observed in his experiments because of the presence of small amounts of oxygen in the ammonia which formed water. Haber and Robert found that with an iron catalyst, any oxygen was removed as the oxide. The catalytic effect however was not impaired[13] and so for iron this impurity was not a problem. For other metals however oxygen had to be removed. But Robert also found that in the commercial gases used by both himself and Nernst, residual carbon dioxide[14] was equally problematic because it was an unexpected partial pressure and unaccounted for in he calculations. Finally, Robert simulated Nernst's experiments by using platinum as catalyst, but when this metal was finely distributed on an asbestos bed he found that 'platinsilicid' formed on it's surface. As a result the catalytic effect was then 'very slight'. None of this had been addressed by Nernst, so his figures were always going to lie on the side of reduced yields of ammonia at equilibrium.

Once the shortcomings of Nernst's experimental design had been addressed the two men had to establish the integrity of their own investigation. The apparatus[15] and quantitative determinations they described were essentially the same as that used in their Hamburg paper with of course the addition of the pressurised reaction chamber, a clear quartz tube about 60 cm long with walls 3 mm thick, which unlike Nernst's metal tube, was less permeable to diffusion. Gases from commercial high pressure cylinders[16] were brought to 30 atm by means of a combined pressure reducing valve/mercury manometer[17] from the firm J. Amsler-Laff. The Mechanical Institute of the University was well experienced in the use of this technology and they also provided a Schaffer-Budenberg manometer by means of which the experimental pressure in the reaction chamber was measured. The apparatus was firstly exhaustively tested for leaks. The Schaffer-manometer was capable of measuring 0–60 kg cm^{-2} excess pressure (i.e., up to $\sim$60 atm). Its scale was divided on a circle of 24 cm diameter where each 'stroke' accurately measured pressure differences as small as 0.05 kg cm^{-2} ($\sim$0.05 atm). Even with such accuracy, no detectable loss was observed over prolonged periods at 30 atm. To maintain precision, the manometer was re-calibrated between every experimental determination.

The quantitative determinations so heavily criticised by Nernst in Hamburg were confidently retained by the two men, ammonia now being absorbed by 0.05 N hydrochloric acid and residual acid being titrated with 0.0523 N potassium hydroxide using methyl orange as indicator. Predictably, as reflected by the higher concentrations of acid and alkali used, typical ammonia yields at the higher pressure were now of the order of milligrams—depending on the temperature and volume of gas—rather than the fractions of a milligram obtained at atmospheric pressure. With yields now in the region of 10–100 times larger, Nernst's major criticism was finally laid to rest.

6.3   Then Secondly …

'*Bestimmung* …' probably introduced a prototype of what was to eventually become an essential component of high pressure chemistry viz., a special 'conical valve'—designed entirely by Robert—and with which he could achieve precise control of gas flows through the apparatus.[18] By means of this device, together with a 'gas clock' Robert was able to pass accurately measured volumes of the pressurised gases through the reaction chamber for 30–300 min depending on the volumes used and the temperatures of the experiments. Once equilibrium was achieved the valve allowed the gas to be moved effortlessly on through the absorption bottles. For the later form of this invention he was to obtain a lower form of German patent or 'Gebrauchsmuster' and eventually an award equivalent to £500—'a lot of money at that time for an impecunious student!' he was later to comment.[4] Indeed, it amounted to about ten years salary at the time.

Having established the integrity of their apparatus and now with precise control of the gas flow, the two men conducted 56 pressurised experiments reporting a series of 27 of these in temperature groups of '700 °C', '800 °C', '900 °C', and '974 °C'. Here they examined the formation of ammonia using stoichiometric mixtures of nitrogen and hydrogen obtained from the decomposition of commercial ammonia, and mixtures produced after obtaining the two gases from commercial sources. They also reported the decomposition of commercial ammonia itself. The first group of (8) results was reported at '900 °C' with 10 mg iron ('eisen') as catalyst. The second group of (6) results was reported at '800 °C' with 1 g manganese as catalyst and the third group of (6) results at '700 °C' again 1 g with manganese as catalyst.

'Tabelle 1', (Table 6.1) below—taken from 'Bestimmung ...', shows Robert's published results at '900 °C'. Those presented at '700 °C' and '800 °C' employed a similar format.

Here the columns correspond to the EMF (voltage) across the thermo-couple, the corresponding calibrated temperature (°C), the flow time in minutes, the gas volume (in litres) used, that volume corrected to 0 °C and 760 mm ('standard' 'STP' conditions at the time), the volume (cm^3) 0.0523 N KOH solution used to neutralise the excess acid, the yield (in mg) of ammonia at equilibrium and the % ammonia in the equilibrium mix. The first two groups of rows correspond to the 'development' or formation of ammonia using gases obtained from the decomposition of ammonia, and from a mixture of (atmospheric) nitrogen and hydrogen respectively. The third group row shows results from the decomposition of ammonia itself.

Now, the partial pressure p_x of any component x in a gaseous mixture can be expressed in terms of that component's % volume in the mixture (x) as;

$$0.01(x) \cdot P = p_x$$

where P is the total pressure. So, using the final column in these tables (% ammonia by volume), the equilibrium constants at the various temperatures were easily calculated by the two men. Assuming once again;

$$\frac{1}{2} N_{2(g)} + \frac{3}{2} H_{2(g)} \leftrightarrows NH_{3(g)} \quad \text{and}$$

$$K_p = p_{NH3} / \left(p_{H2}^{\frac{3}{2}} \times p_{N2}^{\frac{1}{2}} \right) \quad \text{atm}^{-1}$$

by substituting into K_p for each individual partial pressure, Haber showed that;

$$K_p \cdot P = 100 \cdot (NH_3)/(N_2)^{\frac{1}{2}} \times (H_2)^{\frac{3}{2}}$$

Given that the overall pressure in their experiments was always over-whelmingly due to hydrogen and nitrogen in the ratio 75 to 25%, the

Table 6.1 'Tabelle 1' taken from 'Bestimmung ....'. Experimental results recorded by Robert for the '900 °C' group of experiments

Tabelle 1. Versuche bei 900 °C mit 10 mg Eisen

	EMF in Millivolt.	Temperatur in Grad C.	Zeit in Min.	Gasvolumen in Liter	Gasvolumen In red. Litern.	0.0523 N KOH in Kubic zentimeter	NH_3 in Milligramm	NH_3 in Vol-Proz.	
1	8,34	897	75	2,16	1,97	3,62	3,22	0,215	aus zersetztem
2	8,39	902	71	2,15	1,97	3,45	3,07	0,205	Ammoniak
3	8,38	901	140	2,21	2,02	3,55	3,16	0,206	aus
4	8,38	901	62	2,72	2,46	4,37	3,89	0,208	Luft-Stickstoff
5	8,38	901	–	2,25	2,03	3,54	3,15	0,205	und Wasserstoff
6	8,38	901	60	2,00	1,82	3,30	2,94	0,213	aus Ammoniak
7	8,38	901	78	2,22	1,99	3,43	3,05	0,202	
8	8,38	901	80	2,31	2,07	3,62	3,22	0,205	

denominator in the expression could be regarded as a 'constant' equal to $(25)^{\frac{1}{2}} \times (75)^{\frac{3}{2}}$ or 3247, so at 30 atm K_p was expressed[19] as;

$$K_p \cdot 30 = 100 \cdot (NH_3)/3247 \tag{1}$$

Therefore, given the average % ammonia at equilibrium from '*Tabelle* 1' as 0.2071% for example, K_p was easily calculated as 2.13×10^{-4} atm^{-1}. This calculation was repeated for each temperature group. However, in the last group of (7) results at '974 °C'—the most contentious temperature given the dispute in Hamburg—and with 20 mg iron as catalyst, Robert varied the composition of the gas mixtures around the optimal 75% hydrogen to 25% nitrogen to determine the maximum and minimum values to be expected for K_p. For the first four experiments the ratio was 25.9% nitrogen to 74.1% hydrogen (const. *eqn 1* = 3246.1) and in the remaining three experiments the ratio was 75.2% hydrogen to 24.8% nitrogen (const. *eqn 1* = 3247.7). Therefore, the average K_p for the first group could confidently be expected to be a maximum value whilst the average K_p for the latter was regarded as minimum. Using the same calculation as before, the range for K_p at 974 °C was determined as 1.48×10^{-4} to 1.56×10^{-4} atm^{-1} and this was expressed by Haber and Robert in their final table for the equilibrium constants at 30 atm viz, where the middle column provided the % ammonia at equilibrium (Table 6.2). With precise control over the flow of the gases and meticulous attention to detail in their experiments, the two men claimed that these figures were the most accurate ever recorded and a true representation of the equilibrium at 'high' pressure.[20] But because of Nernst, there remained the question of the need to establish demonstrable concordance of practise with theory.

Table 6.2 Experimentally determined equilibrium constants from 'Bestimmung …'.[1] The notation > separates minimum and maximum values at 974 °C

Grad C.	Prozent NH_3 bei 30 Atm. ($N_2 : H_2 = 1:3$)	$K_p \cdot 10^4$
700	0,654	6,80
801	0,344	3,56
901	0,207	2,13
974	>0,144	>1,48
	(0,152)	(1,56)

6.4   So Thirdly

'*Bestimmung* …' wedded Robert's precise experimental practice to contemporary theoretical prediction and Haber and Robert extended the arguments from their Hamburg paper to establish that their calculations applied with a high degree of accuracy across the pressure and temperature ranges employed. In this respect, and even though the ammonia synthesis had a fearsome reputation, they showed that there was nothing unusual about the reaction and that the equilibrium could be described perfectly well by the standard thermodynamic equations of the day—without recourse to 'titivation'. Just as Nernst had 'showcased' his 'Heat Theorem' at Hamburg, Haber now confidently referred to his book[21] throughout this part of '*Bestimmung* …' and in doing so – just as in Hamburg - Haber demonstrated once again that the kind of modifications made by Nernst were unnecessary to explain the results.[22] The implication of course was again obvious.

Haber and Robert firstly summarised their results at 30 atm and 1 atm comparing them to the expected K_p values using the same equation they used in Hamburg viz,

$$\log_{10} K_p = \frac{12,000}{4.57T} - 5.8927$$

and although not every temperature corresponded to measurements made at both 30 atm and 1 atm, the consistency of the figures and their agreement with conventional theory was compelling. Table 6.3 below, (taken from '*Bestimmung* ….') shows the calculated ('berechnet' or ber.) and found ('gefunden' or gef.) equilibrium constants reported by the two men along with the % difference between their observed values and theory (the final column).

Table 6.3 A summary of the equilibrium constants presented by Haber and Robert[1]

Grad C.	$10^4 K_p$ ber.	$10^4 K_p$ gef. (30 Atm.)	$10^4 K_p$ gef. (1 Atm.)	Prozent Abweichung des gef. von ber. Werte
700	6,4	6,8	ca. 6,8	+6,2
750	4,72	–	4,68	−0,9
801	3,57	3,56	3,33	−0,3 and −6,7
850	2,79	–	2,79	0,0
901	2,21	2,13	–	−3,6
930	1,95	–	2,00	+2,6
974	1,63	>1,48	–	<−9
		(1,56)		−(4,3)
1000	1,48	–	1,48	0,0

Table 6.4 Results from the 'best fit' equation provided by Haber and Robert[1]

Grad C.	$10^4 K_p$ ber.	Grad C.	$10^4 K_p$ ber.
700	6,58	901	2,11
750	4,76	930	1,85
801	3,53	974	1,53
850	2,71	1000	1,38

Ignoring the −9% value at 974 °C—which was really quite speculative—the experimental values all fell within about ± 6% of the theoretical prediction, and with very minor modification[23] to their formula, viz,

$$\log_{10} K_p = \frac{12,800}{4.57T} - 6.06$$

the agreement, they declared, became 'somewhat better'; (Table 6.4).

In stark contrast however, using Nernst's equation with his 'over accommodating values' and with the equation now written to describe the formation of one mole of ammonia viz.,

$$\log_{10} K_p = \frac{14,010}{4.571T} - 6.43$$

the difference between his calculated values for K_p and Haber and Robert's experimentally determined values was disturbing. Table 6.5, again taken from 'Bestimmung ...', shows Nernst's values as consistently lower, resulting in a significant difference in the %ammonia at equilibrium. Nernst's prediction being lower than Haber's by about 30% to almost 50%. The ammonia synthesis was therefore yielding substantially higher product than Nernst was willing to concede in Hamburg.

Table 6.5 A comparison between Nernst's predictions and Haber and Robert's experimentally determined values[1] i.e., 'our values' (unser wert). The final column gives the % difference

Grad C.	$K_p .10^4$ (Nernst)	$K_p .10^4$ (Unser Wert)	Unser Wert liegt über dem Nernstschen um Prozente Desselben.
700	5,26	6,80	29
801	2,66	3,56	34
901	1,52	2,13	40
974	1,06	>1,48	>40
		(1,56)	(47)

Table 6.6 Comparison between Haber and Robert's experimental results[1] and those predicted by his equation from Chap. 5 of his book[6]

Grad C.	$10^4 \, K_p$ ber.	$10^4 \, K_p$ gef.
700	6,84	6,80
750	4,89	4,68
801	3,60	3,56
850	2,77	2,79
901	2,17	2,13
930	1,90	2,00
974	1,60	>1,48
		(1,56)
1000	1,45	1,48

Finally, using the data and equations from Chap. 5 of his book, Haber presented the following expression for K_p which took into account the change in the heat capacities of ammonia and the 'permanent gases' (nitrogen and hydrogen), which they had by now measured themselves;

$$\log_{10} K_p = 2215/T - 3.626 \log_{10} T + 3.07 \times 10^{-4}T + 2.9 \times 10^{-7}T^2 + 4.82$$

and interested readers are referred to Appendix D for its derivation. Using this final equation, Haber and Robert compared the calculated equilibrium constants to those found experimentally and expressed their results in Table 6.6 below—again taken from '*Bestimmung …*'.

Once more the agreement between theory and practise was compelling. But Haber and Robert had already shown in Hamburg that by both a 'conventional' route and by making entirely appropriate modifications to Nernst's novel 'approximation' formula their experimental results were consistent with theory, and '*Bestimmung…*' simply served to enforce this view. Judging by their annoyance at Nernst's criticism of their earlier work the two men probably already knew this so, what did they get out of '*Bestimmung …* ' that they didn't already suspect?

6.5 Well Finally

Haber and Robert saw for themselves the effect of pressure on the equilibrium and they were entirely at ease with the additional technical complexity this presented. Haber's original investigation for the Margulies brothers was conducted at normal ('usual') pressure to simplify experimental procedure, but in summing up '*Bestimmung* ... ' the two men now declared that they had;

> ... devised an apparatus which permits pressurised experiments at 30 atm and almost 1000 °C by external heating and with streaming [flowing] gases, with the same comfort with which they are feasible at usual pressure ... .

and after the horror of Hamburg they even permitted themselves a little indulgence;

> Determination of the ammonia ... [at high pressure] ... by means of this arrangement at 700 °C, 801 °C, 901 °C and ... 974 °C ... confirmed our values obtained at usual pressure ... making Herr Nernst's statements regarding the situation of the equilibrium ... irrelevant' [!].

an observation which Nernst never publicly acknowledged until some years later. Indeed, at the time, Nernst's public utterances on the matter continually polemicised against Haber's figures.[24] To make matters even worse between the two sets of workers, Haber had earlier received a preliminary copy of a paper by Jost regarding the ammonia equilibrium entitled 'Die Lage des Ammoniakgleichgewichtes'[25] in which Jost tried to salvage something from the sorry mess of his experiments by claiming his 'numbers' were in line with those of Haber and Le Rossignol, and although Jost's paper was eventually published in the same edition[26] of *Zeitschrift* as '*Bestimmung* ...', Haber and Robert would have none of it. *Zeitschrift* therefore included an additional paper by the two men—also entitled 'Die Lage des Ammoniakgleichgewichtes'[27]—which followed on from their main paper; this put Jost firmly in his place. With the 'bone of contention' between he two sets of workers being the efficacy of external versus internal heating and the attendant accuracy of temperature measurement, Haber and Robert's repost began quite politely.

> The note with the same title which Herr Jost kindly let us read in manuscript form makes us add a few words to the report which we gave previously about the topic. Herr Jost regards it as unacceptable that we didn't install the thermocouple in free gas, but for reasons of convenience, between the pipe of the experiment and the pipe of the stove ..

but after a vigorous and incontestable exposition of the evidence, the two men announced that;

> We cannot find that Herr Jost is right when he thinks that our numbers correspond on the whole with his …

With both Nernst and Jost comprehensively chastised, the position of the ammonia equilibrium at pressure was now unequivocally decided in favour of Haber and Le Rossignol and these workers had produced more synthetic ammonia than anyone else before—despite Nernst's continued rantings.

6.6 The Legacy of 'Bestimmung …'

Now revenge is sweet, but '*Bestimmung* ..' provided something else really quite surprising—for Haber at least. It wasn't the fact that he was right about the position of the equilibrium, nor was it the ease with which he achieved agreement between theory and practise, but rather it was the 'comfort' with which pressurised gases could be controlled and made to move through an apparatus with an innovative but otherwise entirely conventional level of technical engineering. This engineering design was provided of course by Robert, and Haber was overwhelmed by his ability to plan and build apparatus calling him 'a true engineer!'[28] Even Kirchenbauer, the department's most accomplished mechanic well versed in the construction of an eclectic range of apparatus, stood in awe of Robert's skill. But to put things into context, Robert must have already known that the transport of gas(es) at high pressure was fairly easily obtainable in engineering terms. After all, the hydrogen liquefier at UCL was just a 'Heath Robinson' device constructed from beer barrels, a steel tank, brass tubing, a student's pump, and a compressor used for charging torpedo tubes. Yet this concoction easily circulated hydrogen gas at 180 atm for many years. To Robert then, 30 atm must have seemed an entirely achievable pressure and thanks to his time at UCL he achieved it entirely.

By 1908 the ingredients for something spectacular were coming together at Karlsruhe. The experimental and theoretical profiles of the ammonia equilibrium had been established beyond doubt and with Robert's skills the seeds of a viable technical solution had been sown. But who or what was to provide the spark that was to ignite these workers to provide the most important technological development of the 20th century? Well, surprise, surprise … it was Nernst.

Notes

1. F. Haber and R. Le Rossignol, 'Bestimmung des Ammoniakgleichgewichtes unter Druck', (determination of the ammonia equilibrium under pressure), *Zeitschrift für Elektrochemie und Angewandte Physikalische Chemie*, **14**, 15, 181–196, (1908).

2. Jaenicke's and Le Rossignol's letters and the transcript of their conversation on 16 September 1959 are held at the Max Planck Gesellschaft, Archiv der MPG, Va. Abt., Rep. 0005, Fritz Haber. Haber Sammlung von Joh. Jaenicke. Nr. 253 (Jaenicke's letters) and Nr. 1496 (conversation transcript).

3. Dietrich Stoltzenberg, *Fritz Haber. Chemist, Noble Laureate, German, Jew.* Chemical Heritage Press, Philadelphia, Pennsylvania, (2004), p 85. Source; Clara Haber to Richard Abegg, 23 July 1907, MPG, Dept. Va., Rep. 5, 921–926.

4. R. C. Chirnside, 'Robert Le Rossignol, 1884–1976', *Chemistry in Britain*, **13**, 269–271, (1977).

5. He was 'aware' of course, because he had translated the appendix to that chapter.

6. '*Thermodynamics of Technical Gas Reactions*'. available 'on-line' in English and PDF format from the internet archive at the University of California Digital Library; http://www.archive.org/details/thermodynamicsof00haberich.

7. Coates, described the paper as 'a beautiful piece of work …'. J.E. Coates, 'The Haber Memorial Lecture', *J. Chem. Soc.*, 1642–1672, (1939).

8. Even though there was clearly some disagreement over the K_p values it was not enough to spoil the overall picture.

9. In 1928, in his paper 'Zur Geschichte der Herstellung des synthetischen Ammoniaks', (the history of the manufacture of synthetic ammonia), Robert published the table of % ammonia at equilibrium that he and Haber expected from their results at various pressures and temperatures (1–200 atm and 200–1000 °C). He said, '*With these results one was now able to see what sort of amount one could obtain under different pressures and temperatures …*'. Figure 6.1 is based on such a table. At the time, anyone could construct such figures if so inclined. We refer to this paper throughout Chap. 8, R. Le Rossignol, 'Zur Geschichte der Herstellung des synthetischen Ammoniaks', *Naturwissenschaften*, **16**, 50, 1070–1071, (December 1928).

10. '*Oh what a tangled web we weave when first we practice to deceive …*'. (From, *Marmion, A tale of Flodden Field*, Sir Walter Scott). Indeed, it appears that Nernst's paper too had been 'exquisitely crafted', crafted to accommodate Jost's poor experimental figures.

11. Nernst freely admitted this in the discussion at Hamburg. See W. Nernst and F. Jost, 'Über das Ammoniakgleichgewicht', (On the Ammonia Equilibrium), *Elektrochemische Zeitschrift*, **13**, 521–524, (1907).

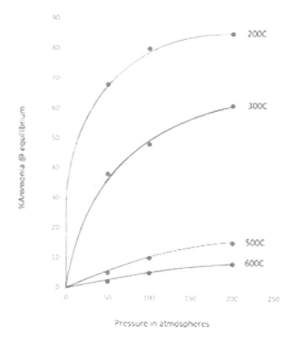

Fig. 6.1 The % ammonia at equilibrium at various pressures and temperatures

12. Temperature measurements were made as in Hamburg with an accurately calibrated thermocouple.
13. Later, during the industrial scaling of the process, the BASF, after trying thousands of catalysts, found iron oxides acceptable. (See Chaps. 10 and 11).
14. The carbon dioxide was therefore removed by 'kalilauge' ([caustic] potash solution) vessels attached to the high-pressure cylinders.
15. In '*Bestimmung* …' Haber and Robert referred to labels on a diagram ('*figurentafel*') but no diagram appeared in the paper nor indeed that issue of *Zeitschrift*. Even with the help of Wiley Interscience we have not been able to trace it. Even so, Coates, op. cit. (note 7), described the apparatus as 'simple and admirably adapted to its purpose'.
16. At the time pressures up to about 100 atm. were available from commercial cylinders.
17. Robert later observed that gas cylinders at the time were supplied with simple shut off valves that "*in no way allows an easy regulation of the out-flowing gas stream. For this purpose, complicated or reducing valves are necessary*" This observation led eventually to his special conical valve (see Chap. 8).
18. '*Bestimmung* …' was published in April 1908, the valve patent was applied for earlier in Germany on 16 March 1908, (see Chap. 8). Because the diagram is missing, we cannot be exactly certain about the use of the valve,

but the word 'ventil' (valve) crops up a few times in the paper such as *Stahlventil*—steel valve, and *ventilgewinde*—valve thread. The paper also mentions '*Weichblei*', a soft lead seal, a feature of Le Rossignol valves.

19. Incidentally confirming Le Chatelier's Principle, that the %ammonia at equilibrium was directly proportional to the total pressure at T.

20. Nernst had shown that no appreciable deviation occurred at 50–70 atm. so the two men confidently assumed 'ideal' behaviour at 30 atm.

21. Haber, op. cit. (note 6). He even referred to its title in English, i.e., 'The Thermodynamics of Technical Gas Reactions'.

22. Indeed, years later in a bizarre turnaround of events, Haber's calculations were used as evidence to support Nernst's Heat Theorem (see Chap. 15).

23. The increases in the constants here lie between 3 and 6% only.

24. Even by 1909 when Haber and Le Rossignol had completed their work on the laboratory scale synthesis of ammonia, Haber wrote to Paul Krassa saying that; '*My old friend Nernst is falling back into the old tune .... and all hell will break loose when he finds out about the synthesis! ..*'.

25. The situation (position) of the ammonia equilibrium'.

26. F. Jost, 'Die Lage Des Ammoniakgleichgewichtes', *Zeitschrift f. Elektrochemie*, **14**, 373, (1908).

27. F. Haber and R. Le Rossignol, 'Die Lage Des Ammoniakgleichgewichtes', *Zeitschrift f. Elektrochemie*, **14**, 513–514, (1908).

28. Stoltzenberg, op. cit. (note 3), p.63.

7

Taking Care of Business

Haber's contracts with the BASF and Le Rossignol

To every man is given the key to the gates of Heaven.
The same key opens the gates of Hell.
An old Buddhist saying.

7.1 Money

Scientific progress occurs in a variety of ways, but there is no doubt that throughout history, progress has always favoured the 'prepared mind'. Sometimes, progress involves a romantic 'eureka' moment when a single—often unexpected—event allows the mind to instantly 'crystallise' its understanding, and although such romance did not contribute to the success eventually achieved by Haber and Le Rossignol, a 'prepared mind' certainly did, 'prepared' by two years of painstaking laboratory work and the barbs of professional criticism *in extremis*. But the men's work did not just prepare their minds, it led to a burgeoning of their relationship. Haber of course was eclectic and theoretically astute, Robert was his '*true engineer*'[1] and crucially by now the two men shared a common language,[2] allowing the efficient communication of ideas and strategies. They complemented one another perfectly, and just as Abegg had sown the seeds of a career in physical chemistry in Haber's mind, so too did Robert's work convince him that a solution to the ammonia problem might just be achievable. But even as the thought crossed his mind, what was Haber to do about it? There were obvious barriers to nitrogen fixation via ammonia; the need for seemingly unachievable levels of pressure and the discovery of a catalyst that could work at temperatures low enough to produce industrially appealing amounts of ammonia (around 10%). Maybe Haber was not too discouraged by pressure,

© Springer Nature Switzerland AG 2020
D. Sheppard, *Robert Le Rossignol*, Springer Biographies,
https://doi.org/10.1007/978-3-030-29714-5_7

after all he and Robert must have discussed the simplicity of Travers' liquefier. But high pressures require a compressor and that costs money so there was the question of funding, and whereas a compressor may represent a relatively modest outlay, to pay someone appropriately for the time, effort and skill required to develop an application was quite a different matter. To be able to move further in this direction with his engineer, Haber would always need money and of course a reason to do so.

But Haber was not without funding. On appointment to his Chair at Karlsruhe on 01 October 1906, he received a personal salary of 5,200 Marks together with an additional 4,000 Marks for new apparatus and instrumentation. By 1907 this additional payment had risen to 4,200 Marks for assistance plus 1200 Marks for equipment and new work rooms.[3] The additional money was of course meant to serve the whole of Haber's department but he was becoming judicious and able to direct some of this funding towards those whom he felt might best progress his aims. Robert had arrived in Karlsruhe in the autumn of 1906 and matriculated as a 'guest instructor',[4] unpaid except we suspect for the 'bank of mum and dad'. His professional credentials, his efforts in adding to Lamb's translation of Haber's book and his progression of van Oordt's work all impressed Haber, and over the next two semesters (January 1907 through to January 1908) their relationship was strengthened as he became Personal Assistant[5] to Haber at an annual salary of 1,000 Marks, about £50.[6] There is no doubt that this represented a considerable outlay for Haber but to put it in perspective, Robert was only receiving what his father Augustin paid his long term senior[7] household staff in Jersey. During this tenure Robert distinguished himself, developing the conical valve and providing the experimental results for both the Hamburg presentation and 'Bestimmung …'. As assistant to Haber he had established the position of the ammonia equilibrium beyond doubt and if further progress was to be made then Haber realised that additional funding was required to keep this young man 'onboard'. But at the beginning of 1908, funds were scarce and if monies were not found Robert would once again have resorted to 'guest instructor',[8] surely a poor return for one so gifted. Advancements in science however are never made without some luck and so it was in Haber's case when an entirely different aspect of his interest nitrogen fixation came to Robert's rescue.

7.2   The Contracts with BASF

Haber was an electrochemist, and although in 1899 he had shown little enthusiasm when Engler had exhorted his staff at the Hochschule to tackle the problem of nitrogen fixation, the developing importance of the production of nitric oxide (NO) in a high-voltage electric arc had led him as a matter of professional integrity to address the problem. In 1907 he and his student Adolf Koenig published their first paper on the topic.[9] Conventional wisdom at the time (Chap. 3) demanded that it was preferable to proceed at the highest possible temperatures to obtain the highest equilibrium %NO, but Haber and Koenig argued that lower temperatures would minimise the decomposition of the oxide and thereby possibly increase the overall yield. Haber conducted experiments at different pressures, he cooled the arc from the wall as well as the anode and determined the relationship between energy consumption and frequency up to about 50 kHz. His best results gave yields of around 10% NO in air at decreased pressure with conversion efficiencies 10–15% higher than conventional methods and this was a significant conclusion.[10]

The publication of this work however, soon attracted the attention of BASF at Ludwigshafen. They had an interest in almost every kind of chemical innovation and they had already set up a research program into nitrogen fixation. During 1903 for example the young Carl Bosch had been investigating the possibility of barium cyanide synthesis from barytes, coke and nitrogen gas, but it lasted barely a year. Prior to that in 1899, BASF had established their own research program into fixation via the electric arc, and as cheap electricity was an essential component in the success of this technology, the early 1900s saw BASF begin preparations for the construction of a pilot plant at the Alz hydrostation in Bavaria. BASF were also acutely aware of developments in the electric arc process in Norway—a country blessed with high hydro-plateaux—and made an agreement with Norsk-Hydro, who owned the franchise for the Birkland-Eyde design in December 1906. As a result of this cooperation BASF set up an experimental plant at Kristianssand in southern Norway where various arc furnaces were tested. Haber's publication therefore drew immediate recognition from BASF.

Ever since Crooke's speech (Chap. 3), Karl Engler, Professor and Rector of Karlsruhe University, was aware of the importance of nitrogen fixation, he was also aware of the potential of his new young professor. But crucially, Engler sat on the Board of Directors[11] at BASF and he was also aware of the company's strategic aims in this area *and* their interest in Haber's work. Early in 1908 therefore, Engler asked Haber if he would like to enter into a binding contractual agreement with BASF to investigate nitrogen fixation further and soon after, BASF director August Bernthsen visited Haber in Karlsruhe. Now, when working for his father Siegfried, Haber's business decisions were lamentable but something must have persisted 'in the genes', and given the confidence that came from being 'courted' by BASF, he tried to drive a hard bargain with Berthnsen. Haber's vigorous and varied demands ruffled a few of Berthnsen's feathers and Engler had to subsequently write to the Board in his support, recommending Haber as;

> … a very energetic man, from whose talent and enterprising attitude I expect important results and whose solutions of disputed matters in the field of electrochemistry show that he is as much a fundamentally sound expert as he is a sharp and clever dialectician. As he is not unaware of his worth, and, like Ostwald and his pupils, he would like to earn something, he does not come naturally quite cheap ….[12]

As a result of Engler's mentoring, two agreements were concluded with BASF on 06 March 1908. These essentially marked the serious progression of Haber from civil servant to private entrepreneur,[13] a progression easily understood given Haber's earlier experience of the chemical industry, (Chap. 2) and it began a complex web of financial dealings over the years with his co-workers and BASF which have never been fully resolved. Haber subsequently became funded by BASF, doubling or tripling his already generous salary which he used to hire assistants and/or pay for equipment. Indeed, BASF became Haber's sponsor and his laboratory received more money than the whole Hochschule itself. For an academic however, the sponsorship came at a high price. The contracts placed severe restrictions on Haber's ability to disseminate his work. He was not allowed to publish without the express approval of BASF, he was not allowed to divulge technical detail, and he was not allowed to work with other firms without BASF's authorisation.[14] In short Haber had to accept a 'gagging order' and its secrecy has left a heavy legacy for scientific history.

The first contract between Haber and BASF concerned the synthesis of nitrogenous gases from nitrogen and oxygen. In this contract, BASF offered Haber a 10% share in the profits of any patentable development during the duration of the patent (15 years at the time) but Haber asked for an additional 10,000 Marks, and when eventually in 1909 BASF got their patent, Haber got his money. Haber also insisted on Koenig being employed by BASF but being allowed to continue working for him. The second contract was drawn up at Haber's insistence. He demanded that the company also support his efforts to synthesise ammonia—seen both as a vindication of his confidence in its viability and in Robert's work. By this time there were abundantly clear indicators of Haber's appreciation of Robert Le Rossignol, viz, choosing him over other talented members of his department at Karlsruhe to tackle the ammonia problem, entrusting him with the translation of part of his book and the experimentation upon which his reputation depended, recognition of all this through his appointment as personal assistant, and then negotiation of a separate contract with BASF which clearly depended upon Robert's participation. None of this carried much weight with BASF however, but they reluctantly agreed the contract, Haber noting later that it was;

> ... only out of consideration for my wishes and not out of confidence in the matter itself ....[15]

These contracts showed how eager BASF were to support arc research and how slow its leadership was to support ammonia synthesis. Haber referred to this in his Nobel acceptance speech,[16] in that the company;

> thought so highly of my efforts to obtain improved efficiency from electrical energy in the combining of nitrogen and oxygen, as to get in touch with me in 1908 and – by providing the resources – to facilitate my work on the subject: whereas they agreed with every caution to the proposal to back me in the high pressure synthesis of ammonia as well, approving it only with hesitation ....

Through this second contract, Haber secured a larger salary for Robert[17] to continue working for him, describing him to BASF as 'indispensable'.[17] He also sought assurance that BASF would look favourably on him and help him in his career.[17] Restrictions were again placed on publication and cooperation, and Haber received 6000–8000 Marks per year until February 1911, (later re-negotiated) together with a share of the earnings ('royalties') from any *patentable* application over the patent's lifetime. During this time, Robert was essentially paid by the BASF but only via Haber, and at no time did the company ever take direct responsibility for the young man.

Two other aspects of this second contract are also interesting. Haber felt guilty that previous partners who had turned his mind towards the ammonia problem, and who had paid him handsomely, were not now involved. In fact, Haber dropped any remaining commitment to the Margulies brothers in favour of the BASF—a point the brothers pursued later. He therefore wrote to the Board of Directors of BASF asking whether they would share any future discoveries or possibly license any future technology to the Margulies brothers. The idea was summarily dismissed by BASF—Engler reminding Haber that German inventions were *not* shared with 'foreign' companies. Haber never raised the point again—although the brothers did when a few years later they tried to sue him for breach of contract. Secondly, the fact that Haber insisted on the additional contract with BASF was proof enough of his regard for Robert's work, but on 01 May 1908, Haber made a remarkable private contract with Le Rossignol regarding royalties—an arrangement which Chirnside[6] described later as 'generous' and which finally 'consummated' the men's professional relationship. Only when Margit Szöllözi-Janze published her magnificent biography of Haber[3] were we made aware of this contract, which promised Le Rossignol a 40% share of Haber's BASF royalties[18] from patentable, technically useful inventions if he helped him achieve a technically feasible ammonia synthesis.[19] Other indicators aside, there could be no greater recognition of Robert Le Rossignol's importance to Haber, for the potential profit from the fixation of nitrogen in Germany and probably world-wide by license was of 'stellar' proportions. The two men also agreed to share a proportion of their royalties with the Technical Hochschule to provide a foundation endowment for the benefit of the progression of physical chemistry at the Institute.[20]

7.3   Robert's Contract. The Corn Fields or the Killing Fields?

As a result of his negotiations, Haber set up two bank accounts[21] to deal with monies received into, and payments to be made from, the two main contracts. The first of these dealt with the arc process and the second with ammonia. It was transfers from this second fund that payments to Robert Le Rossignol were eventually made some years later. Robert's engagement with Haber's private contract established him not just as an employee but as a full partner in any subsequent invention, a move easily rationalised from the perspective of an 'impecunious'[6] young student. It was a contract entered into in good faith and a just recognition of his contribution over the previous two years. Robert probably realised that a solution to the difficult problem of nitrogen fixation was within his grasp and if patented its development by BASF would eventually provide him with significant reward. The consequences of providing BASF with a possible solution to Germany's nitrogen problem should also have crossed his mind. Indeed, two years later at a meeting of the

Scientific Union in Karlsruhe,[22] Haber's previous reticence regarding fixation had disappeared and he now pronounced that in Germany there was;

> … an extraordinary need for bound nitrogen, mainly for agricultural purposes and to a much smaller extent for the explosives industry and the chemical industry ….

As a partner to Haber, Robert from the outset must have been aware of the link between his work and the '*explosives industry*' because Haber was not expounding anything new here. As early as the summer of 1903 Ostwald had already written to the newspaper *Schwäbische Merkur*[23] describing why he had become involved with the problem of ammonia synthesis in the first place and why he and his son-in-law Ernst Brauer[24] had developed a process that oxidised ammonia to form nitric acid—an essential ingredient in the production of the major explosives and propellants of the day viz, TNT (2, 4, 6 tri-nitro toluene), picric acid (2, 4, 6 tri-nitro phenol), and 'cordite'—a mixture of 'gun cotton' (nitrocellulose), nitro-glycerine and 'Vaseline'. Ostwald declared that;

> The significance of bound nitrogen … is especially high for both war and peace.

In the article he described its agricultural significance but went on to develop its importance in military terms;

> There is another agency with a vital interest, the administration of the armed forces. Without saltpetre the best military is almost helpless … Were a war to break out today between two great powers, one of which was able to prevent the export of saltpetre from Chile's few harbours, that ability alone would allow it to render its opponent almost incapable of fighting.

Ostwald would not have to look far to find 'two great powers' at the time. Europe was awash with them. Great Britain, Germany, France, Russia, Austria-Hungary, Italy, and the Ottoman Empire. The smaller but none-the-less troublesome states of Romania, Bulgaria and Serbia had caused tensions in Europe for generations and throughout the 19th century the major players had tried to maintain a balance of power through a complex but unstable network of political and military alliances. Robert would have been politically naïve not to have been aware of these tensions and the possibility of war, but he was just a minor scientist, and with Europe at peace the commercial aims of Germany's industrial giant BASF were in terms of the production of ammonium sulphate as *fertilizer*—by far the most common use of ammonia in Germany at the time. Robert's contract with Haber was also simply in terms of the production of ammonia, so one could reasonably argue therefore that even with a link to the 'explosives industry', to Robert, the whole contractual process would naturally have seemed just an opportunity for personal, financial and professional progression and nothing more sinister. After all, why should this young man's ambition be restrained by some hypothetical future, having to choose between the corn fields and the killing fields? Decades later another Robert—Robert Oppenheimer—summed up the scientist's dilemma[25];

> A scientist cannot hold back progress because of fears of what the world will do with his discoveries.

But there was surely one other reason for Robert's engagement with this contract, and that was the exciting opportunity it presented to parallel the work of Woodland Toms, whose contribution to the science of agricultural fertilisers had led to legally binding assurances of quality in the United Kingdom, and whose 'kindness'[6] had meant so much to him. Oppenheimer was fully aware of where his work was leading and surely so too was Robert Le Rossignol. But soon, by '*calling down nitrogen from the heavens to make plant growth possible* …', Robert and Haber were to be given 'the key to the gates of Heaven', but would civilization … '*prove itself worthy of the gift?* For of course … '*it is one that society may also use to its undoing* …'.[26]

7.4   Haber's Dilemma

With Koenig and Le Rossignol now 'on board' Haber was in an enviable position. He had secured the financial backing of an industrial giant, his co-workers were of the highest calibre, the problems they addressed were of international importance and whichever route were to prove the most effective, professional kudos and huge financial gains were to be made. But therein lay Haber's dilemma, where to place his best efforts? Biographical evidence suggests Haber's heart lay in the ammonia synthesis, but BASF were entirely sceptical of this route and Haber would have required some significant evidence to convince the company otherwise. In May 1908, during a trip to Berlin, the evidence dropped into his lap. Here, he ran into Walther Nernst and learned of Nernst's connection with the Griesheim Chemical Works—

major producers of compressed gases. Now, whether Nernst had read 'Bestimmung ...' by then remains unclear but Haber knew that Jost was still active regarding ammonia because of their recent exchange in 'Zeitschrift' (Chap. 6). Haber described Nernst's behaviour at the time as 'euphoric'.[27] Nernst now believed that the synthesis of ammonia from its elements was a distinct possibility and Haber was deeply disturbed for he regarded the ammonia problem as *his* territory. Subsequent to this meeting of course,[28] the Griesheim Works privately advised Nernst of their lack of confidence in this approach, and without industrial support—but unbeknown to Haber— Nernst soon walked away from the problem. But Berlin lit Haber's 'blue touch paper', and he retreated to Karlsruhe convinced that he now had to contend with determined competition from an old adversary. Here then was the reason he needed to move forward with his engineer. The thought of Nernst, like some kind of chemical 'cowboy',[29] riding into 'Habertown' on the back of a shabby equilibrium held together with theoretical sticking plasters, *then* 'robbing his intellectual bank', was too much for Haber to countenance and with funding from the BASF in place, from this point onwards, all his efforts regarding the problem of nitrogen fixation focussed on ammonia and its elemental synthesis.

Because of his work with Le Rossignol, the principal technical barriers to overcome were perfectly clear; carry out the process within apparatus capable of operating at least 100 atmospheres—but preferably double that—and take advantage of this by lowering the temperature, simultaneously finding a catalyst that could promote the reaction at around 700 °C or even lower. Of these two technical objectives, high pressure must have seemed more easily achievable. Biographers Szöllössi-Janze[3] and Stoltzenberg[30] differ in describing how Haber progressed from here, but both agree that he had by this time 'become aware'[31] of industrially important processes that proceeded at very high pressures indeed, viz, the Linde process for the liquefaction of air at ~ 200 atmospheres, and secondly the '*Formiat herstellung*' a pressurised process for making sodium formate from caustic soda and producer gas. Stoltzenberg describes Haber receiving support from BASF to purchase a compressor but provides no detail. Szöllössi-Janze on the other hand is more precise, suggesting that funds were made available to Haber as a consequence of his role as an expert witness in a dispute between the firms Elektrocemischen Werken Bitterfeld and Koepp and Co. Whatever the origin of the funds, a compressor was eventually purchased—although we know nothing of its provenance[31]—but Szöllössi-Janze insists it came with;

an additional device to achieve the liquefaction of both air and hydrogen.[32]

and in this respect, we can safely infer that it must have been capable of pressures approaching 200 atm as described by Szöllössi-Janze.[32] Even if Le Rossignol had no influence on this purchase—an opinion with which common sense and circumstance would not concur—his time at UCL must have made him entirely comfortable with its outcome. So began the race for synthetic ammonia and the march of nitrogen and hydrogen into everyone's lives. Within a year Haber and Le Rossignol had crossed the finishing line and solved a problem that had bedevilled chemical science for well over 100 years and the next chapter explains exactly how they did it.

Notes

1. Dietrich Stoltzenberg, *Fritz Haber. Chemist, Noble Laureate, German, Jew.* Chemical Heritage Press, Philadelphia, Pennsylvania, (2004), p. 63.
2. Haber's English was never that great. Even by 1933 he confided in a letter to Sir William Pope whilst at Cambridge, '*Mein englisch ist nein gut genug ...*', (My English is not good enough ...), (Chap. 16). The two men probably conversed in German with the occasional English 'thrown in'.
3. These figures are taken from Margit Szöllösi-Janze, *Fritz Haber 1868-1934: Eine Biographie.* C. H. Beck, Munich, (1998), p. 152 and p. 173.
4. See Chap. 2, supported by Szöllössi-Janze, op. cit. (note 3), p.176.
5. Replacing Gerhardt Just, who was Haber's first private assistant.
6. R. C. Chirnside, 'Robert Le Rossignol, 1884–1976', *Chemistry in Britain*, **13**, 269–271, (1977).
7. See Chap. 10 for some examples of the salaries paid by Augustin and Edith to their staff.
8. Szöllössi-Janze, op. cit. (note 3), p. 176.
9. F. Haber and A. Koenig, 'Über die Stickoxydbildung im Hochspannungsbogen', (the formation of nitric oxide in the high-voltage arc), *Z. Elektrochem,* **13**, 725–743, (1907).
10. Haber may not have been entirely convinced of the efficacy of the arc approach however. In his Nobel acceptance speech on 02 June 1920, he described an alternative method that he had considered, using a much more accessible form of energy viz, heat. The utilisation of the heat of gas flames in air seemed to Haber to be compatible with the formation of nitric oxides as occurred in the explosive mixtures of motors. Carbon monoxide, hydrogen and acetylene were contenders to provide the heat, however he failed to pursue the matter at the time as yields were again only a few percent. Fritz

Haber—Nobel Lecture: The Synthesis of Ammonia from Its Elements'. *Nobelprize.org.* Nobel Media AB 2014. Web. 9 March 2016. http://www. nobelprize.org/nobel_prizes/chemistry/laureates/1918/haber-lecture.html, p. 333.

11. As a member of the 'Surveyance Council' of the Badische factory. Szöllössi-Janze, op. cit. (note 3).

12. Vaclav Smil, *Enriching the Earth,* MIT Press, (2001) p. 76, and Daniel Charles, *Between Genius and Genocide,* Jonathan Cape, London, (2005), p. 90.

13. This progression really had its genesis earlier, on 04 September 1906, when he patented the gas refractometer with the Zeiss company, (*Testing Gaseous Mixtures*), United States Patent Office, No. 830,225, Patented 04.09.1906.

14. Szöllössi-Janze, op. cit. (note 3), p. 173.

15. Daniel Charles, op. cit. (note 12), p. 89.

16. Fritz Haber, Nobel Lecture 2 June 1920; 'Fritz Haber—Nobel Lecture: The Synthesis of Ammonia from Its Elements'. *Nobelprize.org.* Nobel Media AB 2014. Web. 9 March 2016. <http://www.nobelprize.org/nobel_ prizes/chemistry/laureates/1918/haber-lecture.html>.

17. Szöllössi-Janze, op. cit. (note 3), p. 173.

18. Later, their private contract was re-negotiated for the year 1918. See Chap. 15.

19. Szöllössi-Janze, op. cit. (note 3), p. 1740. Her reference; Vgl. HS 2067 (BASF, W I Haber, Verträge ab 1920) Aktennotiz Holdermann betr. Zahlung der BASF an Haber, 25 February1953. The existence of this contract is so important to the story of Robert Le Rossignol that I contacted Prof. Szöllössi-Janze by email to confirm my translation of her book, which she kindly did by email, 01 August 2014.

20. Szöllössi-Janze, op. cit. (note 3), p. 225. Once again, my understanding of this arrangement was confirmed by Prof. Szöllössi-Janze by email, 01 August 2014. During the war in September 1916 the amount donated by the men had reached 50,000 marks at which point the 'Haber Foundation'—as it was eventually called—was realised. Haber thought of leaving the money to the chemistry department allowing Engler and Bunte to work out the details of its expenditure. At the beginning of January 1917 Haber gave them the money, the largest part of it in 5% 'Fünfter Deutscher Kriegsanleihe' (5th issue German war bonds) but stipulating that only the *interest* should be used in furthering the objectives of the Institute. The Haber Foundation survived the time of German hyper-inflation, put itself back into good shape in 1926 and by 1936 had acquired a wealth of 26,000 Reichesmark. It existed during and after the second world war and only in 1985 when it was merged with other financially weak foundations to create the 'Vereinigten Studienstiftung der Universität Karlsruhe' (the United Study Foundation of the University of Karlsruhe) did its name finally disappear.

21. Szöllössi-Janze, op. cit. (note 3), p. 174 and p. 481. These accounts were set up at the Dresdener Depositenkasse and the Württembergischen Vereinsbank.

22. A lecture entitled 'Making Nitrogen Useable', at the Verbandlungen des Naturwissenschaftlichen Vereins in Karlsruhe, (the natural science association in Karlsruhe), 18 March 1910, later published in *Zeitschrift für Elektrochemie*, **23**, 20–23, (1909/1910).

23. Wilhelm Ostwald, *Lebenslinien: Eine Selbstbiographie*. (Lifelines: An autobiography). Volume 2, Chap. 12, Berlin: Klasing, (1926).

24. Wilhelm Ostwald, *Abhandlungen und Vorträge, 1903* (Essays and Lectures, 1903), Leipzig: Voith, p. 326, (1904).

25. Peter Goodchild, *J. Robert Oppenheimer, Shatterer of Worlds*, BBC Publications, (1980). ISBN 0 563 17781 0.

26. These lines were written by Professor H. E. Armstrong, FRS, in *The Times* on, Tuesday 06 February 1934, on the occasion of Haber's death. See Chap. 16.

27. 'Euphoric', Szöllössi-Janze, op. cit. (note 3), p. 175.

28. Stoltzenberg, op. cit. (note 1), p. 85.

29. Nernst's work for Hamburg really was a 'shoddy' job.

30. Dietrich Stoltzenberg, *Fritz Haber. Chemist, Noble Laureate, German, Jew*. Chemical Heritage Press, Philadelphia, Pennsylvania, (2004).

31. The term 'became aware' is somewhat confusing as there was ample time for scientists such as Haber to have become aware of this technology. For example, the Linde company began to make air liquefiers from about 1895. By 1897 the company had sold 14 laboratory liquefiers. In 1900 the Linde air liquefier was shown at the world exhibition in Paris and won the Grand Prix, and by 1901 the Linde workshop in Lothstrasse, Munich—not too far from Karlsruhe—had built a further 60 of these machines for laboratory use. Haber's machine was probably sourced from here. From; Ebbe Almqvist, *History of Industrial Gases*, Springer, (2003). ISBN-10: 0387978917, ISBN-13: 978-0306472770.

32. Szöllössi-Janze, op. cit. (note 3), p. 172.

8

The Big Fix …

Synthetic ammonia at Karlsruhe.

Come down … there's ammonia!
… You have to see how the liquid
ammonia is pouring out!

Fritz Haber, running through the laboratories in the T. H. Karlsruhe, March 1909.[1]

8.1 The Secret Years

Vaclav Smil in his book '*Enriching the Earth*' maintains;

> The synthesis of ammonia belongs to that special group of discoveries – including Edison's lightbulb or the Wright brothers' flight - for which we can pinpoint the date of the decisive breakthrough …

The BASF archives contain a 'down-to-earth' four-page report sent by Prof. Dr. Fritz Haber to the company's directors on 03 July 1909 describing the events of the previous day when Alwin Mittasch and Julius Krantz from BASF witnessed the continuous production of synthetic ammonia in Haber's Karlsruhe laboratory. We can therefore certainly include 02 July 1909 in Smil's 'special group'. But pinpoint this event as we might, it is a poor substitute for the personal and scientific history that was lost because of the conditions imposed upon Haber by the BASF. At the time, the development of the Haber-Le Rossignol process was probably the most demanding technical project ever undertaken in an academic setting,[2] but the principal players involved in the drama have long since left the stage and the whole

© Springer Nature Switzerland AG 2020
D. Sheppard, *Robert Le Rossignol*, Springer Biographies,
https://doi.org/10.1007/978-3-030-29714-5_8

episode has become something of a mystery, for these were years of commercial secrecy.

A few papers spread over almost 20 years, and four crucial industrial patents—some submitted jointly by both Haber and Le Rossignol—have allowed us to form an understanding of the technology they developed during the period from April 1908 to July 1909. But biographers have uncovered little regarding the 'inspiration and the perspiration' involved in the discovery of high-pressure fixation e.g., the division of work, the individual contributions and the provenance of the technical innovations they employed. According to Haber's biographer Margit Szöllözi-Janze, Haber never tired of praising the contributions of Le Rossignol and indeed Kirchenbauer, but in this regard she says; 'there is hardly any ... precise biographical information available ...'.[3] Szöllözi-Janze also maintains that these two men were largely responsible for the machine Mittasch and Krantz saw that day, but she is unable to describe in detail the nature of any laboratory work they did together. However, she maintains that Le Rossignol's contribution was 'decisive'.[3]

Even so, Le Rossignol has sometimes been treated as a 'Johnny-come-lately' in the story of ammonia synthesis, 'magically' appearing on the scene with his gift 'for solving problems of practical engineering',[4] constructing the high-pressure apparatus and then 'disappearing'. But previous chapters have shown that this is simply not true. For three years, Le Rossignol was at the heart of the ammonia synthesis in both an experimental and an engineering sense. And even though we have little understanding of their relationship, he was a much more important figure than Kirchenbauer, a conjecture we confirm later in Chap. 18. Dietrich Stoltzenberg—himself a chemist—is another biographer who has accepted that it was Le Rossignol alone who brought the apparatus to the point where the BASF finally adopted it,[5] and of course what we do know, is that unlike Le Rossignol, Haber never entered into any kind of private financial contract with Kirchenbauer, neither did he submit any patents nor publish any papers with him. Indeed, communication between the two must have been difficult, because Haber could not understand Kirchenbauer's Pfälzer dialect[6] and these facts alone surely stand as testament regarding the relative importance of these two men. Amazingly too, ever since Jaenicke deposited his 'treasure trove' at the MPG Archives,[6] evidence has existed that confirms this understanding, but it has lain dormant, and now with the appearance of Chirnside's transcript of his conversation with Robert in 1976,[7] we are able to add a little more detail. But even with this additional understanding, where does it leave us? Thomas Hager in '*The Alchemy of Air* …'[8] provides an entertaining account of what might have happened in the Karlsruhe laboratory during this time, and even though his view lacks any kind of verifiable support, one suspects that it is probably not too far from the truth. In this respect then, a sprinkling of facts coupled with conjecture based on what one could reasonably expect to have happened has to suffice, and the sad truth is that we probably know more about what happened between March and July 1909 than we will ever know about the whole of the preceding year. Neither Haber nor Le Rossignol left us a detailed, 'blow by blow', personal account of this period, and Robert's three years at Karlsruhe never led to any kind of academic dissertation which may have provided some additional insight.[9] Nevertheless, we have fragmentary glimpses into the events and relationships developed during this period. For example, in 1928[10] Robert published the only account of his recollections as part of an international appreciation of Haber's 60th birthday, and although it was lacking in detail—even misdating[11] the meeting with BASF to *June* 1908—what emerges is his immense regard for Haber at the time. Here Robert is self-deprecating, far from identifying himself as a principal player in the evolution of the ammonia synthesis, he says;

> As early as 1904 Haber had worked on the ammonia synthesis. This work was taken up by van Oordt and published in 1905. That really was the basis for the production of ammonia even if in view of the knowledge at the time, an application seemed impossible. Nevertheless, Haber devoted himself to it and in 1906 he decided to do new research work on it. This task was taken up together with the author of these lines at first from a purely scientific point of view … .

After outlining his recollections of that time, he is reverential, totally underplaying his role in the whole affair;

> At the end of these lines I should like to take the opportunity to express my deep gratitude for the friendship granted to me by Haber. It is rare for a young person such as I was at the time to be given he luck to come into contact with such an excellent teacher. A teacher whose own enthusiasm and devotion to his work translated to his students and in whom they always found a true friend and advisor. No wonder that even today, [1928] many of his pupils think of him with admiration, enthusiasm and gratitude, remembering when they were allowed to share time with him …

Haber too held Robert in high regard, for later he was fulsome in his praise, such as at the Hurter Memorial Lecture given before Robert's own countrymen at Liverpool University on Wednesday 26 November 1913 where he described events leading to the synthesis of ammonia, and then again in his Nobel acceptance speech on 02 June 1920.[12] Both of these are expanded in later chapters.[13] We know too that Robert's enthusiasm for his task translated to 'running the extra mile'. Many years later in 1966, Paul Krassa, Haber's student between 1906 and 1909 and a friend of Robert, describes[14] how he worked in the private workshop in Karlsruhe during the holidays, manufacturing the 'special valves' which could control the flow of gases and tolerate the high pressures. One other aspect also emerges regarding the activity in the Karlsruhe laboratory at the time, in that at least one of the catalysts (uranium), was prepared for use by breaking up the commercial metal 'with a hammer![15] Today of course, 'Health and Safety' would have stern words regarding the practice, but it seemed to have had no ill effects, Robert at least living to the ripe old age of 92, but these anecdotes seem to be all we have at a day-to-day level. From a technical point of view however we know rather more and so the story of synthetic ammonia at Karlsruhe begins here, but our understanding of Kirchenbauer's contribution is limited to his apparent construction of a 'special joint' which was used 'wherever possible in the apparatus ...'[15] and which was later successfully patented in his name.[16]

8.2   The Technical Realisation of Synthetic Ammonia

At the time, ammonia in the form of 25% commercial ammonium sulphate possessed a value of 89 Pf. kg^{-1} (9.5 cents lb^{-1}), whilst the combined nitrogen and hydrogen from which it was composed was valued at 20 Pf. kg^{-1} (2.14 cents lb^{-1} of ammonia).[15] In British terms at the time, this amounted to the fact that the cost of manufacturing ammonium sulphate from synthetic ammonia came to just £2. 6s. 6d per ton,[17] and of course the plant could be built anywhere, not just where cheap electricity was available. By comparison, ammonium sulfate from the cyanamide process cost twice as much per ton.[17] The economics of ammonia production were therefore compelling, but the cost of the raw material was far from negligible. Nitrogen of course was available in 'inexhaustible' quantities from the liquefaction of air (the 'Linde process'), but also from the action of air and 'producer gas' on heated copper, and as a by-product in the manufacture of 'formic' (methanoic) acid from producer-gas and caustic soda. Hydrogen was available; from the decomposition of distillation gases, by the alternate action of steam and reducing gases on iron, by the electrolysis of water, by the action of water-gas on calcium hydroxide, from the electrolytic manufacture of alkali, and the manufacture of oxalates from formats.[18] In their experiments at Karlsruhe however, Haber and Robert used nitrogen from the Linde process—which was relatively free from Argon—together with 'electrolytic' hydrogen. The approach developed by the men employed two machines, viz., an apparatus for comparing the efficiency of various catalysts, and a separate small scale continuous circulation[19] apparatus which employed the most successful catalysts from the first machine, but repeatedly returned unreacted residual gas (refreshed with stoichiometric nitrogen and hydrogen) to the 'contact substance' thereby avoiding a situation where ammonia was removed but the valuable residue allowed to go to waste.

But long before Haber and Le Rossignol had devised a practical machine to synthesise ammonia from its elements based around these ideas, the patents people at BASF demanded that the notion of a catalytic circulation process be protected. As early as 13 October 1908 therefore, Haber, together with Le Rossignol on behalf of the BASF, submitted his first patent concerning synthetic ammonia viz., Deutsches Reichspatent D.R.P. 235421, issued on 08 June 1911.[20] This 'base' patent—supplementary patents followed—subsequently became known as the 'circulation patent' and its principle is still used in every ammonia plant today. The invention that Haber and Robert were yet to realise was summed up in the patent claim;

Process for synthetic production of ammonia from its elements, whereby an appropriate mixture of nitrogen and hydrogen is continuously subjected to both the production of ammonia under the influence of heated catalysts, and continuous removal of the resulting ammonia, characterised by constant pressure and the transfer of process heat from the ammonia containing reaction gases to the ammonia-free gas mixture.

Many of the ideas behind this claim stemmed from the work presented in Hamburg and in 'Bestimmung …' but the practical realisation of the 'transfer of process heat…' and the 'continuous removal of the resulting ammonia …', however, were yet to be addressed.

8.2.1 Evaluating Candidate Catalysts

Haber's results with van Oordt, indicated that to work at atmospheric pressure, a temperature of 300 °C could not be exceeded, but at the time no catalyst was known that operated efficiently at such low temperatures. The most effective catalyst they discovered was manganese, and in practise temperatures between 500 and 700 °C had to be employed. Above 700 °C the advantage of the increased rate of reaction was more than counterbalanced by the low yield of ammonia. Below 500 °C, a much larger volume of 'contact substance and space' was required to maintain a reasonable reaction rate, and since a high-pressure apparatus is effected more easily at low volumes, temperatures below 500 °C could not be entertained.

The original apparatus for comparing the efficiency of various catalysts consisted, in its simplest form, of a strong steel cylinder with a bore of about 6–7 mm heated externally in a bath of fused saltpetre.[15] Little other technical detail is available, but this approach was soon abandoned in favour of a series of more sophisticated apparatus which employed a convenient *internal* electrical heating mechanism. Now this might seem a little odd bearing in mind Haber's criticism of Nernst, but here of course the objective was to produce a template for an industrial process and not simply the accurate measurement of an equilibrium position. Electrical heating therefore had to be the more practical option, and not only for the evaluation of the catalyst but also for the continuous circulation apparatus. Haber and Robert's original diagram[15] (Fig. 8.1) shows the apparatus they finally developed to compare the efficacy of candidate catalysts.

Here, 'Fig. 1' in the diagram shows two views of the 'furnace' as it was then called, where a strong steel outer tube enclosed a thin iron tube (9–13 mm diameter), wrapped in asbestos paper and wound with a nickel wire so that the whole could be heated electrically, the workers maintaining that 'with correct adjustment of the heating coil, a field of constant maximum temperature of 4–5 cm length can be obtained …'.[15] The catalyst, or 'contact substance', was contained between asbestos plugs in an inner glass or quartz tube, the same tube containing the wires for the thermocouple. The pressurised gases from the compressor entered the apparatus at the top and passed down through the thin iron tube, over the catalyst and then out of the apparatus at the top via a capillary tube emerging from the bottom of the inner tube and running up through the annulus between the inner tube and

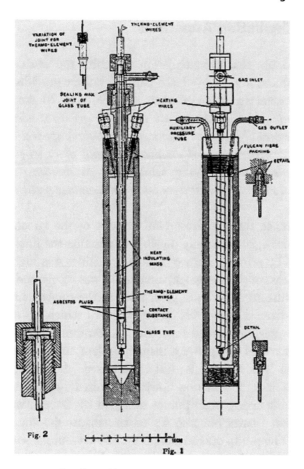

Fig. 8.1 The apparatus developed by Le Rossignol to evaluate candidate catalysts[15]

the strong steel outer tube. Because the thin tube was likely to deform at the high pressure, an auxiliary pressure tube introduced gas to the annulus thereby equalising the pressure inside and outside the thin tube. The seals in this apparatus were made gas tight using 'Vulcan fibre',[21] and 'Fig. 2' in Fig. 8.1 shows how this was achieved whilst simultaneously providing insulation for the heating wire(s). Analysis of the gas was via a Raleigh interferometer supplied by Zeiss and for which Haber had gained a patent earlier in 1906[22]—the gases being previously passed through sulphuric acid to remove the ammonia.

8.2.2 The Circulation Apparatus

There is little doubt that the circulation apparatus too was developed in an incremental fashion, and in his 1928 paper[10] Robert provided some indication of how progress was made between May 1908 and March 1909. He tells us that *'great difficulties ... arose again and again from the scientific and technical side ... [eventually] a small technical [circulation] apparatus was built but experiments made with it proved disappointing and that's why more strenuous efforts were made to find a better catalyst. ... At the same time when the experiments for a suitable catalyst were taking place a new smaller apparatus was built ...'*.

Like the furnace then, at least two versions of the circulation machine evolved, but once again it is only possible to describe the final machine with any accuracy. The primary feature of these machines was high pressure, but because of the cost of the feedstock, circulation was recognised as an essential component, returning 'unbound' gases back to the catalyst and replenishing the used gases from a high-pressure reservoir, of which no technical detail remains. Their previous work—especially 'Bestimmung ...'—had suggested that circulation would not prove a major problem and this design not only economised on raw materials but minimised the work required to re-compress the gases. Circulation however required a circulation pump, and so a small 'double acting' steel pump, designed by Robert, was built which not only circulated gases but also served to exhaust the apparatus prior to pressurisation. The pump operated at about 300 rev. min^{-1} with a maximum capacity 72 m^3 h^{-1} measured at 'NTP', and although in practice it was not quite as efficient as Robert would have liked due to the leather packing of the piston proving 'not quite satisfactory',[15] it operated well at pressures up 200 atm. 'Troublesome' as Haber later described the whole task,[23] at no time during this period did the workers report any particular difficulty in achieving and maintaining pressures between 100 and 200 atm whilst freely moving gas through the apparatus. This was due to 'Kirchenbauer's'[16] 'special joints', the use of 'Vulcan fibre' gaskets which made other joints gas-tight, and of course Robert's special 'conical' valves.

'Conical valves' however were hardly a new invention; indeed, they had been around for almost 100 years and were widely used in steam engines. Their use in a 'pneumatic' apparatus was described as long ago as 1820[24] when The New Monthly Magazine carried an article by Sig. Crivelli, Professor of Natural Philosophy in Milan who described such a valve, declaring that ' ... *the emission of the gas may in this way be regulated with great*

nicety and its retention, if required, secured in a very perfect manner …'.
A simple conical valve permits relatively fine control over gas flow by lowering
(or raising) a conical 'male' plug into (or from) a conical 'female' chamber
whose volume therefore increases or decreases only gradually because of the
taper. The plug is connected to a threaded shaft and control is via a knob or
cross bar on top of the shaft. Robert's valve however had a new and innovative
aspect to qualify it for a *Gebrauchsmuster* (Chap. 6) and in this respect it was
described contemporaneously as a 'needle' valve. The 'needle' valve uses a
conically tapered pin—the needle—inserted into the valve inlet to gradually
open a space for even finer control of flow. Robert's 'needle' was relatively
long with an extremely acute angle of inclination between 85° and 90°, and
being attached to the plug, it too was raised or lowered via the threaded shaft.
The result of using this form of valve was that by turning the hand screw,
only an extremely small and comparatively long annular space was opened
through which the gas may escape from the valve inlet to its outlet. Robert's
original diagram from his patent[25] is shown below (Fig. 8.2).

At the time, there was no simple valve which allowed gas to be withdrawn
from a high-pressure cylinder 'bubble by bubble' or in a rapid stream.

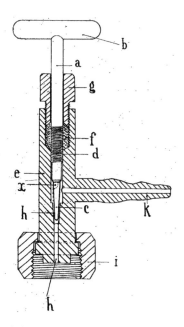

Fig. 8.2   Robert's drawing of the conical valve from his patent application of the 16th
March 1908.[25] 'c' shows the conical pin or needle, the gas input and output are via 'h'
and 'k'. 'x' shows the inclination angle of the cone taper. The photograph in Fig. 8.4
later, clearly shows six of these valves in use in the 'circulation apparatus'

Robert's valves added much finer control to gas flow and became a standard component in high pressure chemistry[26]—at the time being widely known throughout Germany as simply 'Le Rossignol' valves. Even as late as 1963 for example, a British textbook of preparative inorganic chemistry confidently declared that 'the usual Le Rossignol' valves can be used for fine control …'.[27] Today needle valves are ubiquitous, and many different designs are available tolerating pressures of many thousands of atmospheres. As for Kirchenbauer, 'his'[15] joints addressed the very serious engineering problem of working continuously with gas at high pressure without leakage. This was achieved by carefully turned screw joints with a male conical portion of 16° angle screwing into a female portion of 20° angle, and together with a conical copper gasket, perfectly gas tight connections were made.[28] 'Kirchenbauer's' joints were widely used and frequently adapted, but subsequent literature often referred to them as 'Haber-Le Rossignol' joints.

In addition to the problems of pressure and circulation, two other technical aspects of the circulation apparatus exercised the ingenuity of the men, viz., the efficient utilisation of the exothermic nature of the reaction, and the removal of ammonia. The (standard heat of) formation of ammonia from its elements releases $46.2 \ kJ \ mol^{-1}$, a 'gift' which simply had to be adopted into the process to minimise the thermal work done. Some sort of heat-exchanger —later called the 'heat regenerator'—had to be devised by means of which fresh incoming gases were pre-heated by the exhaust gases before being raised to operational temperature by the internal heating element. Cooling of the exhaust in this way also displaced the equilibrium in favour of ammonia and so with an efficient heat exchanger a 'win win' situation arose. This meant that in the circulation machine, the 'furnace' holding the 'contact substance' had to be modified to accommodate the heat exchanger; a bundle of 127 steel capillaries wound with iron wire and supported at either end by perforated hexagonal plates, the capillaries passing through the perforations. Such a device provided a huge surface area, the incoming gases passing around the tubes carrying the exhaust gas and through the interstices achieving remarkable heat transfer. No photographs of this original device exist but precisely the same Le Rossignol model was used by the BASF for the industrialisation of the synthesis, and the photograph in Fig. 8.3 shows one such heat exchanger—damaged by 'hydrogen attack'[29]—but displaying the essential features of the design.

The final development in the evolution of the circulation apparatus was stimulated by the discovery of a 'spectacularly' effective catalyst—probably uranium—dispersed on an asbestos medium, (Sect. 8.5) capable of increasing yields by as much as a factor of 20.[7] Prior to this, at low levels of conversion,

Fig. 8.3 Carl Bosch's adaptation of Le Rossignol's 'heat regenerator' after suffering 'hydrogen attack'. Photograph courtesy of the Nobel Foundation, Carl Bosch Nobel Lecture, 1931[30]

Haber suggested that removal of the ammonia could be achieved by absorption (conventional in industry), followed by recovery through a reduction in pressure.[7] But when yields rose to around 6%, according to Robert[10] Haber *'had the idea of* [continuously] *producing* [then removing] *liquid ammonia in very small quantities ... of about 1 cm³'*—an approach facilitated by the fact that the compressor they purchased came with a liquefaction capability (Chap. 7). However, in conversation with Chirnside in 1976, Robert says that he discounted Haber's absorption idea in favour of *his own* suggestion of liquefying the ammonia by cooling. By crediting Haber with the idea in his 1928 paper,[10] Robert probably avoided 'upstaging' the old man. Whichever way, removing the ammonia displaced the formation equilibrium to the right, but doing so by cooling in a circulation apparatus, returned[31] very cold residual gas to the furnace. Another heat exchanger was therefore indicated—later called the 'cold regenerator'—which used the relatively warm gas leaving the furnace and entering the liquefier, to heat up the very cold gas leaving the liquefier on its way back to the furnace. The final form of the Haber-Le Rossignol apparatus therefore consisted of a 'furnace' containing the 'contact substance', a 'heat regenerator', a liquefier and a 'cold regenerator', a circulation pump, valves to control the flow of gas, soldered copper tubing, and 'Kirchenbauer's' joints to connect all the components, the whole being supplied by a high pressure gas reservoir maintained by the compressor.

8.2.3 The Final Form of the Circulation Apparatus

The original Haber-Le Rossignol apparatus is shown in the photograph, Plate 8.4. Figure 8.5 shows the Haber-Le Rossignol schematic[15] of the modified furnace and the circulation apparatus—now incorporating a drier,[32] a pressure gauge and a liquid ammonia 'water' gauge glass. Most of the

Plate 8.4 The original Haber-Le Rossignol machine. Photograph courtesy of the Archive der Max-Planck-Gesellschaft, Berlin-Dahlem

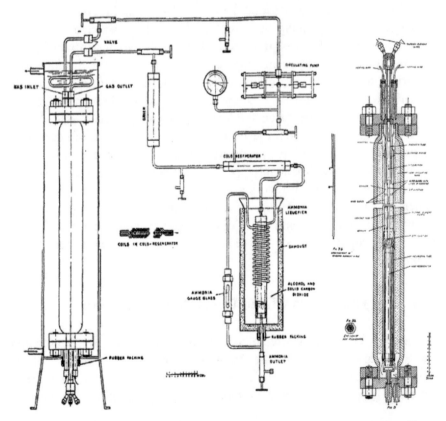

Fig. 8.5 The original schematics describing the machines built by Le Rossignol[15]

components in the schematic can be picked out in the photograph, but four of Robert's valves are laid horizontally on the bottom LHS and are not quite so obvious.

The furnace on the right-hand side of Fig. 8.5 shows the heating and thermocouple wires at the top, and the gas inlet and outlet at the bottom. The bundle of steel capillaries of the heat exchanger is shown inside the furnace in the bottom half with the 'contact substance' above that—both being part of the inner tube. Just to the left of the furnace is a plan view of the hexagonal form of the heat exchanger. In the schematic of the circulation apparatus however, the furnace is inverted with the thermocouple and heating wires at the bottom and the gas inlet/outlet at the top. Here, the two components marked '*valve*' are not 'Le Rossignol valves' but 'check valves', (*Rückschlagventil*) used to prevent gases flowing in a direction opposite to that intended. Within the overall schematic for the apparatus is a diagram showing the additional heat exchanger, essentially a steel tube containing a linear 'plait' of three copper tubes interwoven in a spiral and united at the inlet and outlet. 'Warm' gas from the furnace passed through this plait into the liquefier, whilst cold gas from the liquefier passed into the tube containing the plait and back to the furnace, the whole facilitating an efficient heat exchange.

8.2.4 Using These Machines

The beauty of Robert's design and engineering can be appreciated by understanding how these machines were used. Once a possible 'contact substance' had been identified using the simple furnace, it was transferred to the circulation apparatus. The substance was placed in the modified furnace, the whole re-assembled, and the bolts tightened. Using the circulation pump the apparatus was evacuated. The mixture of nitrogen and hydrogen from the high-pressure reservoir (gasometer) was admitted through the Le Rossignol valve shown in the schematic just above the pressure gauge. With other Le Rossignol valves opened, the whole apparatus could be brought to the desired pressure and once this was achieved the valves were closed. The liquefier was charged with a mixture of alcohol and solid carbon dioxide kept at -25 to -39 °C, but mostly around -30 °C. The circulation pump was set working and a current sent through the heating wires. From now on, circulation was controlled using Robert's valves.

Once the apparatus became fully operational the mode of action was as follows. The gas mixture was sent into the furnace and over the outer surfaces of the capillary tubes conveying the hot gas away from the 'contact substance'.

This pre-heated the incoming gas to 400–500 °C and in doing so abstracted almost all the heat from the exhaust gas. This was why the term 'heat regenerator' was coined. Passing over the electrical heating element, the incoming gas was economically raised to the desired temperature by the time it entered the 'contact chamber'. Fixation then occurred. The mixture streamed out of the furnace through the fine capillary tubes of the regenerator, simultaneously pre-heating more incoming gas, and by cooling, moving its own equilibrium position towards the formation of ammonia. By the time it reached the second heat exchanger it was practically at room temperature. Indeed, the two men reported[15] that the end of the tube nearest the furnace could be held quite 'comfortably in the hand'. The gas then passed through the copper 'plait' of the second exchanger into the liquefier where the ammonia was removed—the gauge glass indicating the amount developed. Passing out of the liquefier, the cooled 'remainder' gas entered the exchanger for a second time but now through the tube *surrounding* the 'plait'. Here, the relatively warm gas passing through the plait heated it up before the pump sent it back to the furnace. On the side closest to the liquefier, the second exchanger eventually developed a thick coating of ice whilst on the side nearest the furnace it was at room temperature, hence the term 'cold regenerator' was coined to describe it. In continuous operation, ammonia could be removed as a liquid or 'blown off' in gaseous form whilst replacement gas mixture was simultaneously added to the system via the valve connected to the high-pressure reservoir.

The 'furnace' and circulation machines were used to examine a wide variety of potential 'contact substances'. At the time little was known about catalytic activity, Robert recalling that '*we knew this was a way to help reactions to go in the right direction but the choice was often a matter of trial and error …*'.[7] The two men had already examined chromium, manganese, iron, nickel and platinum in their previous two papers. All these materials were meticulously prepared in belief that impurities would seriously impair the efficacy of the catalysts but none of these 'cut the mustard', i.e. generating reasonable reaction rates and yields at 'low' enough temperatures. Others such as palladium, iridium and ruthenium had no catalytic effect at all. But then, and not for the first time in the history of the ammonia synthesis, chance intervened.

8.3   At Last .... the Breakthrough

At this point in his career, Haber was recognised as a diligent and innovative practitioner of physical science and this led to many opportunities for consultancy. One such position was with the Auergesellschaft Company of Berlin, investigating new and novel materials as potential filaments for electric light bulbs—a highly competitive research area at the time. The company was founded by the fabulously wealthy Jewish banker and entrepreneur (*Geheimrat*) Leopold Koppel, together with the Austrian chemist Karl Auer von Welsbach. The company's research interests were in the areas of gas mantles, luminescence, incandescence, the rare earth elements and radioactivity—including uranium and thorium. In 1908 the Auer offered Haber a position as scientific/technical consultant through which—purely by chance—he had access to 'exotic' materials that he could select as potential catalysts. One such material was osmium, another was uranium. The two elements however differed markedly in provenance. Osmium was rare, just a hundred kilograms existed in the whole world, spent or broken bulbs being bought up by the company to recover the metal. The Auergesellschaft however had much more uranium, essentially a 'waste' product left over from their process for extracting radium from uranite.

Haber's desperate search for an effective catalyst had eliminated many of the obvious candidates and led him to consider the unconventional. Therefore, early in 1909 he and Robert decided to examine finely ground osmium[33] in their 'furnace'. The result was encouraging. Even better than that, osmium appeared to be far and away the best 'contact substance' they had yet encountered, converting around 8% of the nitrogen/hydrogen mixture into ammonia at about 175 atm—and at temperatures as low as 500–600 °C, about double what they have previously achieved.[7] It was at this point, and with such 'high' yields, that Robert suggested separating the newly formed ammonia by liquefaction, and he modified the circulation machine to accommodate his innovation—developing the 'cold regenerator' at the same time and probably incorporating the cooling devices supplied with the compressor.

According to Robert's account,[10] the modified circulation machine was far from its final form when in the third week of March 1909 the two men decided to move on anyway, placing finely divided osmium in the contact chamber of its 'furnace'. The 'furnace' was then reassembled and the bolts tightened. The compressor brought the gas from the gasometer to a pressure of ∼ 200 atm. The apparatus was evacuated, the pressurised gas admitted and the liquefier cooled. Finally, the heating element and the circulation pump were turned on and the men stood back, their work largely done. Time passed slowly as the machine went about its business of balancing heat and pressure. All the seals held, and nervous glances must have been exchanged as the valves were occasionally manipulated to move the gases on through the machine. Then ... ice began to form on the cold regenerator, and slowly at first, tiny beads of liquid ammonia were seen in the collecting vessel. The men stared transfixed and after a while it became clear that their machine was producing about 1 cm^3 of synthesised liquid ammonia per minute which they easily drew off in a continuous cycle. Haber instantly realised what they had achieved; the frustration of his early life, the 'Third Law' slipping through his fingers, and his humiliation by Nernst flooded over him as the ammonia appeared, and he could hardly restrain himself. Like a child opening a longed-for present on Christmas day, he had to share his joy, running wildly through the laboratories he shouted 'Come down ... there's ammonia!... You have to see how the liquid ammonia is pouring out!' Colleagues gathered around. Max Mayer, a close friend of Haber recalled this moment vividly[34] Hermann Staudinger too recounted ... 'Suddenly Haber came to me and shouted ... Come down there's ammonia! He had a ... long capillary and in it was about 1 cm^3 of liquid ammonia. I can see it still. Then Engler joined us. It was fantastic!'[35]

8.4   Brunck, Bernthsen and Bosch

At this point, the circulation apparatus was still fairly crude, and other potential catalysts had yet to be examined, but such was Haber's enthusiasm that on the 23 March 1909—and as directed by his contract with BASF—he wrote to Ludwigshafen reporting his success, suggesting at the same time that they buy up the world's supply of osmium from the Auergesellschaft and anywhere else it could be found. Haber's letter landed on the desk of August Bernthsen, the man in charge of BASF's research. Bernthsen received it with undiluted scepticism and he communicated his initial feelings to Haber, Robert later recalling,[10] that 'in a personal conversation with Haber [Bernthsen] told him that [BASF] wasn't interested in the experiments ...'. Haber was utterly dismayed and he asked Carl Engler, who sat on the board of BASF, to act as mediator. Engler by-passed Bernthsen and contacted Heinrich von Brunck – chairman of the board since 1907, and on 26 March 1909,[36] according to Robert,[10] Brunck 'travelled to Karlsruhe to find out for himself about the production of ammonia.'

Brunck, Bernthsen and Carl Bosch—the latter, *Herod* to Ostwald's infant —all suddenly appeared in Haber's laboratory. The three men examined Haber's machine and there were many technical questions. One in particular Robert recalled much later in life[6] as Bosch enquired '*What about the totraum?*'—'totraum' being the dead-space behind the circulation pump piston. Robert was surprised by this question, and he had to explain to Bosch that the dead-space played no role in the ammonia 'trap' because liquid ammonia behaved like water, thinking to himself at the same time … 'You don't know that much, great Bosch!'. But once the questions were over, Bernthsen expounded his concerns. How could BASF consider adopting a process that occurred at (at least) 100 atm when they recently had an auto-clave explode at just seven? When Haber then suggested that 200 atm—or more—would be even better, Bernthsen was horrified, but he continued. As far as the BASF was concerned, what Haber was suggesting was a process that occurred on a continuous and not a batch basis, a process that demanded high temperatures, extraordinary pressures, that to become profitable could not be allowed to stop or even fail! Nothing like it existed in the known world of technical gas reactions and coupled with that it depended on a catalyst that was rare, therefore expensive, and hardly understood. Haber was crestfallen, but then Brunck turned to the man who had been in charge of the company's nitrogen fixation research since 1903.

Bosch had been quietly assessing what had been put before him throughout the day. He had listened attentively as Haber and Robert answered questions about the process. Periodically he recalled his own efforts into fixation. He knew how important the nitrogen project was for BASF, he was their 'golden boy' but he had been working on fixation for almost seven years without success. The arc process was predicated on cheap electricity deals and was location dependent, other fixation processes were competing and he needed a breakthrough to consolidate his position within the company. After he had debunked Ostwald's experiments, he had even promised his future wife Else [*Schilbach*] that *he* would solve the nitrogen problem,[37] but confronted with the Haber-Le Rossignol machine it was obvious that he had lost the race to discover the technology. But neither could *they* solve the nitrogen problem alone. That would require an industrial process, and immediately a new direction in life opened before him.

Haber and Le Rossignol had the right mix for an industrial synthesis viz, temperature, pressure and catalysis. But osmium? Surely an equally effective catalyst could be found? Bernthsen's concerns about the high pressure too were serious but Bosch was a metallurgist, he understood metals and what they could withstand. The Haber-Le Rossignol process was a 'long shot' but it was feasible, easily understood, location independent and even though it would demand the development of a whole new technology, to Bosch it seemed the 'best show in town'. He had lost the race to discover the technology, but the opportunity to develop it commercially was there. With great courage, Carl Bosch faced Brunck and replied without hesitation, 'I think it can work. I know exactly what the steel industry can do. I think we should risk it',[38] but this would have been simply 'posturing' if Bosch had not proceeded to take on the problem personally, which of course he subsequently did.

The matter was settled and according to Robert,[10] ' … from that day on, the directors looked at the experiments with more goodwill, but still sceptical interest.' Brunck boosted the BASF's support for Haber, but only conditionally. No-one from the BASF had yet seen this miracle machine in action, and even if it proved promising it would not only have to work in Karlsruhe but in Ludwigshafen's workshops as well. There could be no crooked chemistry at Karlsruhe. Osmium too was a stumbling block. BASF realised that even if all the osmium in the world were captured, there simply wasn't enough to form the basis of a viable industrial process. And so all those present at Karlsruhe on 26 March 1909 agreed that if progress was to be made there had to be a successful demonstration of the apparatus and a credible movement towards a more sustainable catalyst. But just to be on the safe side, BASF's patent people re-appeared to protect this latest development and Haber's second patent—the 'osmium catalyst' patent—was filed on 31 March 1909. By 13 August, the 'circulation' and 'osmium' patents had been submitted in the United States[39] as well—naming Haber *and* Le Rossignol as Assignors to the BASF.

8.5 The Final Push

Over the next few months there was a spring in everyone's step and work progressed rapidly; to perfect the amendments to the circulation machine, to reduce its size, to find a replacement for osmium and to improve the cooling. In this last respect, the collecting vessel of the liquefier was adapted to include a 'water glass' gauge which provided a clear exposition of the amount of liquid ammonia being formed. Shortly after, the two men examined uranium as catalyst and this proved to be just as spectacular as osmium and although expensive, much more readily available. Experiments with the uranium catalyst indicated yields of around 10% at about 100 atm and 500 °C, undoubtedly a commercially appealing prospect. At the time the two men were enamoured by the versatility of these new elements. Osmium for example, could be used as a finely divided metal or in the form of a compound which upon being used becomes converted to metallic osmium, the men suggesting osmium oxide. Similarly, uranium could be used as the metal, as an alloy, as the nitride or in a mixture with other compounds. However, as the oxide its effectiveness was reduced so the two men suggested that water —'or bodies which give rise to water'—be excluded. Their enthusiasm however was short lived. Later, osmium and uranium were thoroughly discredited (Chap. 10) by the BASF as catalysts. But such difficulties lay in the future and as far as Haber and Robert were concerned progress towards a more sustainable catalyst had been made. By June 1909 they also felt they were in a position to address the second of the BASFs demands, a demonstration of the feasibility of the high-pressure synthesis of ammonia from its elements, with gas recirculation.

The demonstration was set for 01/02 July 1909 at Haber's laboratory in Karlsruhe, to where Carl Bosch, Alwin Mittasch (BASF's catalyst expert) and Julius Kranz (BASF's chief mechanic) were all cordially invited. The two days set aside for the demonstrations involved a number of experiments, but Haber's guests could only make the trip to Karlsruhe on the second day, by which time the machine had probably been assembled and disassembled several times over. Nevertheless, Robert recalled[10] the excitement in the laboratory as they waited for the guests to arrive. But ever the practical man and intimately bound to his machine, he was also tense and sceptical, charged with the fear of *Tücke des Objekts* (the 'spite of things'). Friday 02 July 1909 was the day that the final form of the laboratory scale technical apparatus was properly demonstrated for the first time, and although confident in his engineering and the performance of individual parts, he knew that so many things could still go wrong.

The morning of the demonstration began well enough. Preliminary checks on the gas supplies, the compressor, the cooling agents, the valves and the joints all passed without incident. The machine was disassembled and prepared by charging the furnace with almost 100 grams of osmium powder.[38] Assembling the machine was quite straightforward, after all this had been done many times before but, on this occasion, when someone tightened the bolts on the furnace, Robert's worst fears were realised and he recalls that 'Ein Bolzen des Hochdruckapparates sprang beim Festschrauben …'[10] as a thread on one of the bolts stripped. Now this set-back may have been caused by cross-threading or over-tightening the nut, or simply by wear and tear from the frequent disassembly and re-assembly. Whichever, it threw to laboratory into a panic, and just as the guests arrived. '*Tücke des Objekts*' indeed—but now with the added bitterness of perfect timing.

With a bolt stripped the seals/gaskets wouldn't hold and the apparatus could not be pressurised. There was little choice but to fix the problem, but that morning the laboratory was psychologically 'tooled up' for demonstrating and not for engineering. For Bosch in particular, time passed slowly as the repairs dragged on and he soon left for another appointment, failing to return.[40] Haber may well have seen his dreams going out of the door along with Bosch but several hours later the fix was made and he and Robert were 'back in business'. From now on things would go smoothly. Everyone knew instinctively what to do next and Mittasch and Kranz were about to witness a flawless performance by the machine as the 'furnace' was reassembled and the new bolts tightened. Once again, the compressor pressurised the gas from the gasometer. The apparatus was evacuated, the gas admitted and the liquefier cooled. Finally, the heating element and the circulation pump were turned on and Robert stood by his valves.

Not for the first time that day, time passed slowly for Mittasch and Kranz as the machine began to practise its art. At first, they saw little happening as Robert skilfully moved his gasses on. But then, at about nine o'clock at night[6] … ice began to form on the cold regenerator, pearls of liquid ammonia gradually appeared in the water glass only to disappear like tears in the rain as liquid formed. Mittasch stared enthralled as the liquid began to rise and Haber periodically drew off samples to convince him that it was indeed ammonia. Over the next few hours the machine performed faultlessly as the men chatted, asked questions, collected the ammonia periodically and calculated the yield. After a while Mittasch had seen enough and he shook Haber's hand enthusiastically, utterly convinced by the process. Whatever was to happen next in the development of synthetic ammonia would be out of Robert's hands but he realised that he was part of an historic occasion and he made sure he obtained a memento of the day. He kept it for the next 43 years, but after five hours he turned the machine off. That night they made 500–600 g of liquid ammonia.[6] The next day Haber wrote a letter[41] to the directors of BASF describing the events of 02 July 1909;

> Yesterday we began operating the large ammonia apparatus with gas circulation in the presence of Dr Mittasch and were able to keep its uninterrupted production for about five hours. During this whole time, it had functioned correctly and it produced liquid ammonia continuously. Because of the lateness of the hour, and as we were all tired, we had stopped the production because nothing new could be learned from continuing the experiment. All parts of the apparatus were tight and functioned well, so it was easy to conclude that the experiment could be repeated … The steady yield was 2 cm^3/min and it was

possible to raise it to 2.5 cm^3. This yield remains considerably below the capacity for which the apparatus has been constructed because we have used the catalyst space very inefficiently.

When Brunck read the letter and discussed the matter with Mittasch and Bosch, he became convinced that the BASF should commit its full resources to the commercialisation of 'Haber's' process. After all, the cost of producing fixed nitrogen in this way would be just a fraction of that of the arc. The patent department at the BASF subsequently made clear to all involved the need to absolutely protect every advance in this field and they asked Haber to send the documentation necessary for the appropriate patent applications. By the 21 July 1909[42] Haber had provided the BASF with the outline of a third key patent, a 'supplementary 'to the original 'circulation' patent called the 'high pressure' patent, Deutsches Reichspatent D.R.P. 238450, filed on 14 September 1909. Two further proposals were submitted later, the 'diffusion' patent (D.R.P. H48062 IV/12 g) and the 'external heat' patent (D.R. P. H47701 IV/2 k). The first of these was shortly to prove particularly apposite following BASF's discovery of 'hydrogen attack' (Chap. 10). It claimed that if a metal body—heated externally—was to be used as the reaction chamber, then it could be protected from chemical change by a protective coating. The second proposal patented the idea of the 'cold regenerator' to ensure efficient heat exchange, and both these proposals were submitted in Robert's name as well as Haber's both as Assignors to the BASF.

Other patent applications were to follow, but Haber and Robert appeared to have solved a problem—at least in principle—that had bedevilled chemists for well over a century. The BASF now had the task of scaling up the modest 75 cm high prototype and building a whole new technology and industry around it, and rarely if ever had a process been brought in an academic laboratory to such an advanced state of technical development before being handed on to industry.[43] If successful, the two men easily saw the consequences of all this. It would answer Crooke's challenge, solve the world's nitrate problem, end the dangerous Chilean trade and beat Nernst and Ostwald to chemistry's 'holy grail'. They would share in the profits too, becoming rich and famous! Haber was beside himself with the prospect and he reverted to his basic instincts, throwing a celebratory party a few days later at a local hotel, J. E. Coates recalling[43] that Haber declared 'invite whomever you like!' Everyone from the laboratory was there, together with University colleagues, friends and anyone that anybody wanted to bring along. The wine flowed freely that night and when it was all over Coates recalled that 'We could only walk in a straight line by following the streetcar tracks!' (Fig. 8.6)

Fig. 8.6 The partygoers! Haber's staff in the summer of 1909 outside the TH Karlsruhe. Robert is seated second from the left and looking particularly pleased with himself. Courtesy of the Archive der Max-Planck-Gesellschaft, Berlin-Dahlem

8.6 The End Game at Karlsruhe

This was an exciting time for Robert to be with Haber, but along with their solution of the ammonia problem came some 'loose ends' to tie up. Being scientists, their instincts were to communicate their discovery through the literature and so they quickly wrote up their account ready for publication. However, because of the commercial secrecy imposed by BASF, this was not to happen for some time yet. For Robert there was an additional task, for alongside his work on the ammonia synthesis he had still been working on the original problem that Haber set him when he came to Karlsruhe viz., the study of the dissociation of carbon dioxide in the carbon monoxide flame, and the two men wrote, then quickly published, their paper on the subject.[44] As a final 'tidying up' exercise, Robert published a single page paper[45] entitled, '*A simple precision regulating valve*' in which he described his new conical high-pressure valve to the German scientific community. In time, it was to make his name famous across Germany.

With his affairs at Karlsruhe now in order Robert would have been available to involve himself in the technical development of the ammonia synthesis, surely a 'dream come true'. But in what role? As far as Haber was concerned the development was now BASF's problem, and little—if any—of it would be done at Karlsruhe. Robert's relationship with BASF was only transitive, i.e., *through* Haber. The company itself bore no obligation towards the young man insisting that he was entirely Haber's responsibility. Robert too, of course, was a 'foreigner' and as Engler had once reminded Haber regarding the Margulies brothers of Vienna, German inventions were *not* shared with foreigners. There was no prospect of employment with the BASF even though Haber had asked them to help him in his career, and Haber's report to BASF on the 03 July 1909 had already indicated that his ammonia account was 4400 Marks in the red.[46] Robert's position and salary were therefore in doubt, with the very real prospect of having to revert to 'Hospitant'. At this time in his life, and after almost three years of hard work, such insecurity was simply unacceptable. Unsurprisingly then, in August 1909 Robert left Haber and Karlsruhe. After a short 'sojourn' in Jersey where he shared—what turned out to be—precious time with his family, he travelled on to the Auergesellschaft in Friedrichshain, Berlin where Max Meyer, a great friend and former colleague at Karlsruhe,[47] had already secured a lucrative position for him as an advisor to the 'Auer' to work on the manufacture of 'glow lamps'—a position also endorsed by Haber's recommendation.[7] Such was their friendship that for some months Meyer and Robert shared a property together in Grünau[6] in the borough of Köpenick, a South East suburb of Berlin. Now Robert's friendship with Mayer was to prove particularly convenient. Mayer was from a Jewish family in Ulm and after studying at Heidelberg he came to Karlsruhe in 1903 as Bunte's personal assistant. He received his habilitation there in 1906 eventually becoming one of Haber's closest friends, advising him on questions of patent law and professional issues. Later, as aspects of Haber's life became ever more complex, Haber granted him power of attorney to represent him in all legal affairs —including the administration of the funds from the two accounts he set up to deal with the arc and the ammonia synthesis. In this capacity he also became the authorised representative ('plenipotentiary') in matters concerning Robert Le Rossignol in Germany.[48] Eventually, he became a director of the Auer. For Robert, the move to Berlin may well have signalled the end of his involvement with the ammonia synthesis, however along with Haber, he was to remain bound to it in an advisory sense for the next few months, financially for years to come, and spiritually for the rest of his life. But for now, a new career beckoned at the 'Auer' in what was at the time another 'cutting edge' technology—electric lighting.

Notes

1. Recollections of Max Meyer and Hermann Staudinger at the time, recorded by Johannes Jaenicke in the late 1950s, (notes 34 and 35).
2. Patrick Coffey, *Cathedrals of Science*. Oxford University Press, (2008). p. 89.
3. Margit Szöllösi-Janze, *Fritz Haber 1868–1934: Eine Biographie*. C. H. Beck, Munich, (1998), p. 176.
4. See for example, Daniel Charles, *Between Genius and Genocide*, Jonathan Cape, London, (2005), p. 87.
5. Dietrich Stoltzenberg, *Fritz Haber. Chemist, Noble Laureate, German, Jew.* Chemical Heritage Press, Philadelphia, Pennsylvania, (2004), p. 63.
6. Contained in Joh. Jaenicke's and Le Rossignol's letters and the transcript of their conversation on 16 September 1959 held at the Max Planck Gesellschaft, Archiv der MPG, Va. Abt., Rep. 0005, Fritz Haber. Haber Sammlung von Joh. Jaenicke. Nr. 253 (Jaenicke's letters) and Nr. 1496 (conversation transcript).
7. Ralph C. Chirnside's Obituary of Robert appeared in; R. C. Chirnside, 'Robert Le Rossignol, 1884–1976', *Chemistry in Britain*, **13**, 269–271, (1977). The Obituary was based upon a tape-recorded interview he conducted with Robert on 29 March 1976. A transcript of the conversation remains with the family and was made available to this author.
8. Thomas Hager, *The Alchemy of Air*, Three River Press, New York, (2008). ISBN 978-0-307-35179-1.
9. At the time, the T. H. Karlsruhe was only allowed to award a doctorate for engineering, but the BASF would never have permitted such a public examination of Le Rossignol's work anyway. In later years, Haber's refinement of his theoretical work—published in 'Seven Communications' (Chap. 12)—*was* used by Haber's co-authors in *their* dissertations. See Stoltzenberg, *op. cit.* (note 5), p. 93, and his reference no. 40. See also Szöllössi-Janze, *op. cit.* (note 3), p. 866 and her references for publications between 1914 and 1915.
10. R. Le Rossignol, 'Zur Geschichte der Herstellung des synthetischen Ammoniaks', (The history of the manufacture of synthetic ammonia), *Naturwissenschaften*, **16**, 50, 1070–1071, (December 1928).
11. This was probably just a 'typo'. Even in his sixties Robert was perfectly capable of recalling important dates. For example, see his visit to the 'Ramsay Exhibition' in 1952, (Chap. 18).
12. Fritz Haber, Nobel Lecture 2 June 1920; 'Fritz Haber - Nobel Lecture: The Synthesis of Ammonia from Its Elements'. *Nobelprize.org*. Nobel Media AB 2014. Web. 9 March 2016. http://www.nobelprize.org/nobel_prizes/chemistry/laureates/1918/haber-lecture.html.
13. See Chaps. 12 and 15.

14. Vaclav Smil, *Enriching the Earth*, MIT Press, (2001), p. 78, his reference; 'Zur Geschichte der Ammoniaksynthese', (The History of the Ammonia Synthesis), *Chemiker-Zeitung*, **90**, 105, (1966).

15. F. Haber and R. Le Rossignol, 'Über die technische Darstellung von Ammoniak aus den Elementen', (The technical production of Ammonia from the elements), *Z. Elektrochem.* **19**, 53–72, (1913). There was also a more concise English version abstracted from the above viz*., The Production of Synthetic Ammonia*, Journal of Industrial and Engineering Chemistry, **5**, 328–321, (1913).

16. The 'patent' number was D.R.G.M 376.829, Haber, *op. cit.* (note 15). The German Patent Office currently records two patents by Kirchenbauer, in 1914 and 1922, but D.R.G.M. 376.829 was not strictly a patent, rather it was a 'registered design', a Deutsches Reichsgebrauchmuster, a form of basic copyright protection which lasted initially for three years, extendable to six, the same level of protection afforded to Robert's valve design. However, Chirnside's transcript contains the dialogue; [RC] *'I had always understood that your contribution to the final success of the work included the fact that you had designed and made the high-pressure joints* etc., *with your own hands?' …* [RLeR] *'That is right, I designed and made this pump …'*, which in turn *suggests* that Robert certainly had some input to what was finally credited to Kirchenbauer. The 'transcript' referred to here is of a conversation Chirnside had with Robert on 29 March 1976. The transcript remains with the family. See also Chap. 18.

17. Geoffrey Martin and William Barbour, *Manuals of Chemical Technology III, Industrial Nitrogen Compounds and Explosives*. Crosby Lockwood and Son, London, (1915), p. 56.

18. Eventually, none of these sources were to prove suitable for the industrialisation of the process.

19. A closed circulation process inevitably meant that the level of residual argon would rise. Feed-stocks with minimal %argon were therefore preferred.

20. BASF 1908. Patentschrift Nr 235421; 'Verfahren zur synthetischen Darstellung von Ammoniak aus den Elementen', (Process for the Synthesis of Ammonia from the Elements), Kaiserliches Patentamt, Berlin, (June 8, 1911), Smil, *op. cit.* (note 14), p. 79 and p. 278. According to Robert however, Haber did all the patent work. Robert simply signed the documents before the various national consuls and patent lawyers. This comment is contained in the transcript of Le Rossignol's conversation with Joh. Jaenicke on 16 September 1959 held at the Max Planck Gesellschaft, Archiv der MPG, Va. Abt., Rep. 0005, Fritz Haber. Haber Sammlung von Joh. Jaenicke. Nr. 1496.

21. Vulcan fibre is still in wide use today as an electrical insulator and in making seals. Invented in 1859 it was made from cotton and/or cellulose fibre soaked in zinc salts. It is a ductile, tough, non splintering material.

22. *Testing Gaseous Mixtures*, United States Patent Office, No. 830,225, Patented 04.09.1906.

23. This is a reference from Haber's *Hurter Memorial Lecture* in 1913. *Hurter Memorial Lecture*, University of Liverpool, Wednesday 26 November 1913. Published in *J. of the Society. of Chemical Industry*, **2**, XXXIII, (31 January 1914).

24. *The New Monthly Magazine*, **13**, (1820).

25. '*Conical screw down valve with an angle of inclination between 85° and 90°*. Submitted in Germany on 16 March 1908. Application submitted in the UK on 16 July 1908, accepted 01 April 1909, UK Patent No. 15,065. The German Patent Office records also provide details of this later application.

26. Largely because of Robert's work, from 1907 onward Karlsruhe began to develop a reputation as the centre for German high-pressure chemistry, eventually leading to high pressure syntheses of methanol and benzine from coal (Chap. 16), this in turn attracted capable physical chemists to Haber's department. When Bergius conceived the idea of hydrogenating coal under high pressure, it was to Karlsruhe he came in 1908. J. E. Coates, 'The Haber Memorial Lecture', *J. Chem. Soc.*, 1642–1672, (1939), p. 1653.

27. For example, G. Brauer and P. W. Schenk, *Handbook of Preparative Inorganic Chemistry*, **1**, Academic Press, (1963). Indeed, a personal communication with Dr. Graham Holmwood of Bayer (by email) extends the life of the 'Le Rossignol valve' to even later. When Dr. Holmwood began his career with Bayer AG Pharmaceuticals Division in Wuppertal, Germany in 1974, he discovered the name 'Le Rossignol' in the company's internal equipment catalogue. This catalogue was subsequently replaced by an online version in which his trace was lost, but in the last paper version (1995) he was still in there, with a soft lead (Weichblei) seal for a 'Rossignol valve'. Dr. Holmwood is sure that the 'Rossignol valve' was still used in the 70s and 80s for ammonia and (probably) chlorine 'bombs'. Even in 2011 some could still be found in cupboards at the Bayer Labs, a remarkable longevity.

28. Described as such for example in 1915, in *Manuals of Chemical Technology-III*, Martin, *op. cit.* (note 17), p. 55.

29. 'Hydrogen attack' was quickly discovered by Bosch. At such high pressures, decarburisation of the steel occurred leading to a brittle iron hydride, a point we discuss in Chap. 11.

30. Carl Bosch Nobel speech; '*The Development of the Chemical High Pressure Method During the Establishment of the New Ammonia Industry*', available at; https://www.nobelprize.org/nobel_prizes/chemistry/laureates/1931/bosch-lecture.html and also from, *Nobel Lectures, Chemistry* 1922–1941, Elsevier Publishing Company, Amsterdam, (1966).

31. According to the 1913 abstracted paper, Haber, *op. cit.* (note 15), the vapour pressure of the liquid ammonia was slightly higher than the tables indicated, and so returning the 'remainder gas' to the contact substance meant that the presence of residual ammonia affected the equilibrium yield slightly.

32. Removal of any water was important. If it entered the cooling apparatus it would form ice which blocked the valves. It also affected some catalysts and the equilibrium.

33. Today we know that osmium is in the same 'transition metal group' as iron. But Haber and Le Rossignol were not experts in catalysis and this confounded their progress. Their understanding—exemplified by their papers—was that *pure* substances made the best catalysts. In the event, mixtures were found to be more effective.

34. Stoltzenberg, *op. cit.* (note 5), p. 87, Johannes Jaenicke in conversation with Max Meyer, 09 November 1958. MPG, Dept. Va, Rep. 5.

35. Stoltzenberg, *op. cit.* (note 5), p. 87, Johannes Jaenicke in conversation with Hermann Staudinger. MPG, Dept. Va, Rep. 5.

36. Smil, *op. cit.* (note 14), p. 80.

37. Smil, *op. cit.* (note 14), p. 86, his reference no. 11.

38. Stoltzenberg, *op. cit.* (note 5), p. 88.

39. United States Patent Office. Serial numbers 512,679 and 512,680.

40. Timothy Lenoir, *Instituting Science. The Cultural Production of Scientific Disciplines.* Stanford University Press, (1997). ISBN 0-8047-2042-6. One anecdote from Lenoir, (p. 230, his ref. 49) is that Bosch apparently '*told Haber and Le Rossignol that he didn't need to see the conclusion of the demonstration; he knew it would work ...*' Although this is not corroborated by Le Rossignol's 1928 paper, Le Rossignol, *op. cit.* (note 10), it may have been an extension of his feelings from his earlier visit to Haber's laboratory on 26 March 1909. (Chap. 8).

41. Smil, *op. cit.* (note 14), p. 81, his reference, no, 72; Fritz Haber, (1909). A letter to the directors of the BASF, 03 July 1909. 'BASF Unternehmensarchiv, Fritz Haber, Allgemeine Correspondenz II', (BASF Corporate Archive, Fritz Haber, General Correspondence II), no 92.

42. Stoltzenberg, *op. cit.* (note 5), p. 89.

43. An observation taken from J. E. Coates' 'Haber Memorial Lecture'. Coates was a student of Haber and a contemporary of Le Rossignol, and on 29 April 1937 he delivered the Haber Memorial Lecture before the Chemical Society. (J. E. Coates, 'The Haber Memorial Lecture', *J. Chem. Soc.*, 1642–1672, (1939)).

44. F. Haber and R Le Rossignol, 'Über der Dissociation der Kohlensäure in der Kohlenoxydknallgasflamme', (The dissociation of carbon dioxide in the carbon monoxide flame), *Z. Phys. Chem.*, **66**, 181–196, (1909). See also Haber's Hurter Memorial Lecture, *op. cit.* (note 23), p. 5.

45. R. Le Rossignol, 'Ein einfaches Feinregulierventil', (A Simple Precision Regulation Valve), *Z. für Analytische Chemie*, **48**, 9, 568, (1909).

46. Szöllössi-Janze, *op. cit.* (note 3), p. 180.

47. Szöllössi-Janze, *op. cit.* (note 3), pp. 185–6.

48. Szöllössi-Janze, *op. cit.* (note 3), p. 45.

Part II

The Industrialisation of 'Fixation' and the Great War

9

The Auergesellschaft, Berlin

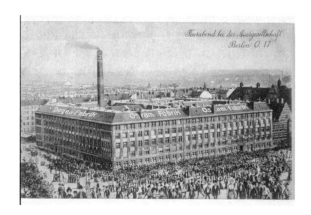

The factory buildings of the Degea-Fabrik (Deutschen Gasglühlichtgesellschaft AG/Auergesellschaft) and the Osram Factory at Rotherstrasse, Friedrichshain, Berlin. Source, historical postcard reproduced form Presglass-Korrespondenz.

Tungsten filaments and a pawn in a power game.

Genius is one percent inspiration, and ninety nine percent perspiration ....

Thomas Alvar Edison, 1847–1931.

9.1 Introduction

July 1909 was an exciting month at the physical chemistry department in Karlsruhe, but the tension, uncertainty and calamity associated with the demonstration for the BASF exhausted Haber and he began to suffer a recurrence of his old ailment—stomach cramps and pains. At the end of the month he took a holiday at the health resort of Pontresina in the Engadine district of Switzerland. Here, sweet-smelling pine and larch forests surrounded the elongated spa town perched upon a south-west facing terrace in

© Springer Nature Switzerland AG 2020
D. Sheppard, *Robert Le Rossignol*, Springer Biographies,
https://doi.org/10.1007/978-3-030-29714-5_9

the high sided valley of the Upper Engadine, just seven km from St Moritz. The first hotels were built here early in the 19th century—often in the 'Belle Époque' style—and adding to the charm of this small resort was the architectural blend of fine old Engadine houses with their 'sgraffito' (decorated walls) dating from the 17th and 18th centuries. Pontresina sat below the towering Bernina pass—the route towards Italy—and its magisterial scenery offered hundreds of kilometers of walks such as to the wild and romantic Val Roseg or the Muottas Muragl—a trip to the top of either rewarding the walker with a magnificent view over the lakes of the Upper Engadine. Here the air was fresh and cool, and Karlsruhe should have been a world away but Haber found it difficult to release himself—even for a short time—from his concerns for the ammonia synthesis.[1] From his hotel room he wrote to Bosch to check on progress, which at the time centred on the suitability of osmium and uranium as catalysts, and the discovery of a cheaper but equally effective alternative. In terms of the latter, Bosch reported that nothing outstanding had yet showed up but that some concern was developing regarding both the suitability of osmium, and the durability of uranium.

No doubt all of this worried Haber and it may well have impinged on his enjoyment of the break, but for Robert there was a more immediate concern. With Haber luxuriating in Pontresina, Robert travelled to Berlin in late August 1909. Max Meyer's offer of a position at the 'Auer' company allowed him to move to the heart of a vibrant capital city, one which had recently become the fifth in Europe to boast an underground railway (in 1902) and probably familiar to him when accompanying Haber on his search for the perfect catalyst. It's possible to rationalise Robert's move to Berlin in terms of the insecurity of his position at Karlsruhe after the BASF monopolised the industrialisation of the Haber-Le Rossignol process. But one also has to recognise the exciting new challenge that the new 'Auer' factory (opened in 1908) offered him, and the strategic aims of its controlling owner, Leopold Koppel. For these were the 'electric years', the years of Tesla, Westinghouse and Edison, and Koppel was hatching big plans for his fellow Jew Haber, plans which Mayer and Le Rossignol's presence in Berlin undoubtedly facilitated.

9.2 Leopold Koppel

Koppel, a reclusive banker and entrepreneur, was born in Dresden on 20 October 1843. In 1890 he established his private banking house, Koppel and Co., the source of much of his wealth, and in 1892 along with the Austrian chemist (Baron) Carl Auer von Welsbach he created the Deutsche Gaslühlichtgesellschaft-Aktiengesellschaft, 'DGA' or the 'German Gaslight Company', forerunner of the Auergesellschaft, (the 'Auer'). Koppel was not a scientist, but he recognised a good business opportunity when he saw it. In 1885 von Welsbach had received a patent for the 'gas mantle', a significant technological development at the time. Using a mixture of the oxides of magnesium, lanthanum and yttrium which he called 'Actinophor', he impregnated a network of guncotton which was then heated. As the cotton burnt away it left a fragile residue which glowed brightly with a green tinted light. These mantles were not commercially successful and the company he formed to sell them failed in 1889, but in 1890 a new improved mantle using thorium and cerium oxides proved far more robust and its use spread rapidly throughout Europe. von Welsbach's success attracted Koppel and he took a controlling interest in the business they subsequently founded together (Plate 9.1).

With Koppel's financial support, von Welsbach became more adventurous. He began work on more robust metal filament mantles, first with platinum and then with osmium. Both these metals were scarce and expensive but such was the importance of the technology of gas lighting at the time that he developed a method that created a paste of osmium oxide powder, rubber and sugar which when squeezed through a nozzle and 'fired', left a fine wire of osmium. This wire was originally meant to form the basis of a new gas mantle, but across Europe and the western world electrical power was being introduced and this technology competed with gas to provide lighting for domestic, commercial and public use. The history of the development of electrical power is fascinating. It parallels that of many other areas of technological discovery in the 19th century. Eccentric and rich inventors, spectacular accidents, battles over patents, the rapid decline and growth of companies, many now household names. By 1900 over 25 million electric 'light bulbs' had been sold by the Edison Electric Light Co. in the USA, and by 1907 the Brooklyn Edison Company showed off the all-electric-house of the future, with clocks, sewing machines, phonographs, fans, electric heating rings, domestic heating, burglar alarms, and doorbells. Penetrating many aspects of life as it did, von Welsbach therefore turned his attention to ways of using the osmium wire in electric lighting.

Plate 9.1 A very rare photograph of the banker and entrepreneur, Leopold Koppel, 1910. Courtesy of Archiv der Max-Planck-Gesellschaft, Berlin-Dahlem. Later illustrated by David Vandermeulen, from 'One hundred years of the Fritz Haber Institute'.[2] Courtesy of Fritz-Haber-Institut, der Max-Planck-Gesellschaft, Berlin-Dahlem

9.3   Electric Lighting

Today the electric lamp is ubiquitous, mundane, part of everyday life and noticed only when it fails, but in the early part of the last century the technology was 'white hot' attracting the best brains and intense research effort. The subject was so 'hot' it was 'cool'. By the time Robert joined the 'Auer', electric light already had a history of 100 years, first being seen publicly in 1808 at a demonstration of the 'arc lamp' by Sir Humphrey Davy. The arc lamp operated by positioning two-pointed carbon rods opposite one another, each connected to a different pole of the power supply. The tips were then brought into contact so that the electricity flowed through them causing the tips to heat up slightly. They were then pulled apart very slowly and a spark, or arc, of electricity leapt the gap, causing the temperature of the tips to rise further. When the tips were hot enough, vaporising carbon became incandescent giving an intense blue-white light. It was not a cheap method of lighting however and the intense brightness and UV radiation meant it was not a practical proposition for domestic use. The term 'arc lamp' remains in use today for bright lights often used to illuminate work spaces, but their technology is entirely different. But by 1890 there were over 130,000 arc lamps in use nightly for lighting public spaces such as 'Broadway', NY, in the United States for example, each producing around 700 'candlepower' (cp).[3] For domestic and commercial lighting, the 'incandescent' lamp was favoured, and like the arc lamp it too had a history of many years. The 'light bulb'—an evacuated glass bulb containing an electrically heated thread-like 'filament'—was made a practical reality by Joseph Wilson Swan and Thomas Alvar Edison. Almost universally in these early attempts, carbon produced by a variety of mechanisms, was the preferred choice for the 'filament' because it had the highest melting/sublimation point of any element—the key to a practical cost effective incandescent light bulb being the longevity of the filament. The breakthrough for Edison and Swan came in 1879 when they independently developed an incandescent bulb that lasted a practical length of time—at best about 13 h! But the contributions of many others were also important and two of these already had a profound influence on Robert's life.

Sir William Crookes—who in 1898 had exhorted chemists to rid the world of its dependency on Chile saltpetre—had earlier developed a carbon filament from animal and vegetable fibres, 'parchmentised' in 'cuprammonic' chloride. His bulbs were tall and narrow with straight sides and his filament was 'M' shaped. His main contribution to the light bulb however, was in the vacuum pumps that he helped develop with Hermann Sprengel[4] for removing air from the tubes. Without these pumps, the practical light bulb we recognize today could not have been developed. But for Robert, there was a much more familiar player in the form of Walther Nernst, whose engagement with the technology of electric lighting almost by itself defines its importance. Nernst's lifelong interest of course was in the interaction of heat and matter and he found various minerals that became incandescent when heated. These materials were first used for gas lamps but he quickly saw the possibility of heating them by electricity. Unlike carbon filaments, the materials he used were not subject to oxidation at the temperature of incandescence and so there was no need to enclose them in an evacuated bulb—with obvious engineering and manufacturing advantages. Any enclosing bulb was used simply to protect and isolate the hot element (the 'glower') from its environment. The glower itself was made from oxides of the rare earth elements. Nernst lamps gave a higher luminosity at a lower energy consumption than competitor technologies but at ordinary temperatures the glower materials were non-conductors. Lamp engineering was therefore complicated by the need for a 'pre-heater' which had to be switched off once the operating temperature had been reached and which also introduced a delay before the lamp became luminous. Excessive surges in current also had to be avoided as these lowered the resistance of the glower and so the whole was quite a sophisticated piece of 'micro engineering'.

In 1897 Nernst formed the 'Nernst Electric Light Company' and produced a number of variations of his basic lamp most using a ceramic of zirconium and yttrium oxides as the 'glower'. The production models were given alphabetic labels, the 'A' type had a vertical filament and operated between 100 and 250 V. The 'B' type was smaller and fitted into an ordinary lamp holder, it had a horizontal filament and gave 25–32 cp at 230 V. The 'C' type, called the 'Luna', had a horizontal filament and finally the 'D' type, which was an improved and enlarged 'B' type, gave 75cp at 110 W. In tests, the average life was between 1,000 and 1,500 h per lamp depending on the voltage, although practical use suggested much less (Plate 9.2).

By 1903 the Nernst lamp had become quite important, which in turn made Nernst wealthy.[5] Even though the Nernst lamp never became the dominant force in lighting, it was used widely across Europe—including the UK. It was used in Buckingham Palace, in the clock faces of 'Big Ben', as well as for street lighting and domestic use. It gave a fifty percent saving in current, and light equivalent to eight (16 cp) carbon lamps. The bulb was expensive but its main disadvantage was that it took about half a minute to light up as it had to get hot first. It also suffered from poor manufacturing standards which resulted in a fairly inconsistent life span. In Maidstone Kent, the bulbs were used for street lighting where the Authority recorded some lasting for two or three thousand hours but others failing after only one hundred hours, overall, they found an average life of almost seven hundred hours. Another problem

EVZ-066 Nernstlampe

Plate 9.2 A model B 'Nernst lamp'. From https://commons.wikimedia.org/

was that the light output fell off considerably during the lamp's life-time and it also varied in damp weather.

9.4   The Tungsten Filament Bulb

With the emergence of these new technologies the carbon filament's days were numbered, but they remained in use for some time because they were cheap. For twenty years however, manufacturers had been looking for more durable materials for the filaments. von Welsbach's osmium wire was an obvious contender and in 1898 he re-directed his research efforts producing an efficient osmium filament bulb. It was very fragile however and had to be burnt base up to prevent the hot filament from sagging and breaking. But he patented it, set up the 'Auer' Company to manufacture the bulb in 1899, and marketed it in 1902. In a further improvement in 1903, Austrians Franz Hanaman and Alexander Just made a tungsten lamp, tungsten being the best refractory metal. This was made by depositing tungsten onto a carbon filament. The process was not suited to mass production however, but Hans Kuzel of the 'Auer' was able to adapt the osmium squirting process to tungsten, and in 1906 the DGA introduced and marketed a tungsten filament bulb under the trademark 'OSRAM'—from osmium and the German word for tungsten, wolfram—Koppel owning the mark. When Robert arrived at the 'Auer' in 1909, things had moved on and there was now an obvious need for a more simplistic production process involving fine *drawn* tungsten filament wires, and it was on this problem that Robert began work there in 1909, later moving on to develop methods of bulb exhaustion. Robert's initial work on the tungsten ductility problem had some success using a method that heat treated the brittle filament material, but Osram were not enthusiastic, probably because tungsten—along with osmium—were not the only contenders in the filament market. In 1903 Dr. Werner Bolton and Otto Freurlin, head of Siemens & Halske's incandescent lamp works, developed a tantalum filament after Werner had found a method of drawing it out into fine wire. In 1905 the tantalum bulb was launched world-wide, but by 1907 they had developed an even better alloy of tungsten (wolfram) and tantalum, which was marketed as the 'WOTAN' bulb. It was quite successful, in use up to 1914, and the name is still used today. Tungsten filament 'OSRAM' lamps were marketed in the UK by the British General Electric Company (GEC) in 1906. In 1907 they also marketed Siemen's 'WOTAN' lamp, but the technological advance of ductile tungsten, eventually made *it* the best option. In 1910, the American William Coolidge beat Robert and the 'Auer' to the discovery of the technology when he developed a process for making tungsten that could be drawn out into long fine wires (Chap. 15). These filaments were in use world-wide by 1911, and they remain the standard filament today. From now on, technological advances were to move away from the filament to problems of economic large-scale production, and the OSRAM bulb shown in Plate 9.3 illustrates where one of the improvements was made.

Notice the 'pip' seal on the crown, the point at which evacuation of the bulb occurred during manufacture. Most electric bulbs at the time had a 'pip' (or 'tip') seal. Some examples of decorative bulbs appear to have had the pip flattened and polished smooth—but at an additional cost. In 1919 an easier method of making bulbs without 'pips' was introduced. Air was extracted from the base, and the bulb was sealed there. Over the next few years all the largest manufacturers changed to this method and it was here that Robert was to eventually make another important technical contribution (Chap. 14). By 1909, after the sale of four million Nernst lamps, the AEG discontinued its production providing only spare parts, and like all other manufacturers they began intensive research into rare earth filament techniques. However, it is widely accepted that the appearance of the Nernst lamp drove research into the filament bulb, but had the lamp's problems been successfully addressed, we might well have waited much longer for a filament successor.

Plate 9.3 An OSRAM lamp from 1910. From, https://fr.wikipedia.org/wiki/Fichier: Osram_lamp,_1910,_high_candle_power_type_(Forty_Years_of_Electrical_Progress).jpg

9.5   The Economics of Electric Lighting— Koppel's Strategic Aims

The evolution of the light bulb was not only driven by the available technology but by its economics. From 1906, power stations throughout Europe saw a steady decrease in demand as more and more people switched to efficient metal filaments. In comparison with carbon, metal filament bulbs were expensive, but in their favour the life span averaged about one thousand hours—with a maximum of six thousand being recorded—running costs were comparable with carbon, and the light was brighter. In the UK, some power companies tried to prevent customers moving to the new technology, but this was quickly deemed illegal. Eventually, the companies realised that falling demand was just the 'lull before the storm', because electricity now held a price advantage over 'dirty' gas, which would lead to an increase in the number of installations—and therefore in profit. Improvement for the consumer of course meant permanent financial losses to the power companies but they responded by reducing the price of electricity, and third parties developed more and more electrical 'gadgets' which not only made the 'electric-house' a reality, but commonplace. The electric-house soon became something to aspire to, but these were dangerous times for those using electricity. Electric cabling was often not insulated. When it was, (oiled) paper or dangerous asbestos were used and house fires and deaths by electrocution were commonplace. There were no 'trip' switches or fuses, and there was no legal enforcement requiring these. Crazy devices flooded the markets like a 'tablecloth' that had un-insulated electric wires woven into it where diners could push the prongs of a bulb into it to provide table lamps. The fact that water or salt would be used by the diners did not seem to be a consideration! There were also women's curling tongs that burned the hair and left many women bald! Refrigerators of the day used refrigerants such as di-ethyl ether, di-chloro methane and ammonia, the escape of any of these vapours being highly dangerous.

   Throughout Europe however, hundreds of power companies arose to satisfy people's aspiration, each with its own area of distribution, supply voltage and tariffs and it was the economics of electric lighting that clearly illuminated Koppel's engagement with the technology in Germany. The 'Auer' company too offered a natural progression for Robert, moving as he did from one cutting-edge technology to another. The creation of efficient and affordable incandescent lighting was a multidisciplinary activity ideally suited to Robert's engineering skills, but it was also a chemical problem[6] as well as a physical one, and today Robert might be best described as the forerunner of a new breed, the 'materials scientist'. The issue of the filament may well have been settled in favour of tungsten by 1910–11 but many other problems remained viz, of the formation of the bulbs and the composition of the glasses involved, of pumps and methods of evacuation or the use of inert gas, the problems of metal-glass seals, the balance of electrical input and light output, the engineering of mass production, and on a wider scale the wiring of homes, the generation and distribution of electrical power (AC or DC?), of meters and regulators to keep supply constant and (eventually) the safe design of all the new domestic 'gadgets' that would not shock or kill. At the time, all of this had huge financial potential and that attracted Koppel, particularly in terms of the patents and licenses involved. Koppel's 'capture' of Le Rossignol and Mayer—two of Haber's closest colleagues at Karlsruhe—would certainly have deepened the pool of experience at the 'Auer' but Koppel was after a 'bigger fish', and these two may have been just 'sprats to catch his mackerel'. The relationship between Haber and Koppel had been growing since 1908 because of Haber's role as consultant to the 'Auer'. Almost as soon as Koppel had secured Robert, he made Haber an offer of a directorship within the company (September 1909) believing that the presence of Le Rossignol (and Mayer) in Berlin would have been attractive to him. Koppel's plan was to establish a separate foundation for nitrogen research headed by Haber viz, the *Gesellschaft für Verwertung Chemischer Produkte* (GVP) (Institute for the Utilization of Chemical Products) and Koppel offered Haber quite excellent conditions. He would have his own laboratory, be able to choose his own chemists, his areas of research and decide which innovations were to be scaled up to industrial levels. His salary would be 50,000 marks, he was to share in any profit to the extent of 2.5% and he could retain his position and remuneration as scientific/technical advisor in electric lighting.[7] Stunningly attractive as this offer was, Haber was bound by certain contractual arrangements at Karlsruhe and he was still enamoured by the freedom of academic life so he declined it, but he realised that Koppel had placed him in a strong position to renegotiate his earlier contracts with the BASF and the relatively modest sums he felt they contained. Using Koppel's offer as a lever, Haber opened new negotiations with the BASF and in October 1909 presented five essential points that had to be met if he were to remain 'on-board'[7];

1. An increase in salary from 6,000 to 20,000 Marks and annual costs of 3,000 Marks.
2. Limitation of the work to the ammonia synthesis exclusively.

3. Extension of the contract to 10 years.
4. Improvements to the process introduced by BASF were not to affect his royalties[8]—which were increased—if the scientific basis was originally introduced by 'himself'. And finally,
5. To have some influence on the future direction of the technology.

These improved conditions also favoured Robert because of his private contract with Haber, but the BASF were not best pleased by Haber's new demands. Reluctantly, they agreed to all of them but noted their corporate displeasure in the new contract, 'resenting' Haber's 'suspicions' about the original royalty arrangement and concluding that 'in future we would want to set certain limits to Herr Haber's somewhat unrestrained ideas'.[9] To Haber, the 'industrialisation' of the Haber-Le Rossignol process was the BASF's problem. To Robert, the contractual dealings with the BASF were Haber's problem and so here Robert was passive. Whatever improvement(s) Haber was able to extract from the company would accrue to him through contractual 'osmosis', so he was free to concentrate on a new chapter in his life in Berlin safe in the knowledge that if the monies *did* come 'rolling in', he could claim his due entitlement. From 1909 onward, Robert pursued a largely uneventful professional career at the 'Auer'. However, he retained his contacts with many Karlsruhe colleagues and with Haber in particular who, 'back at the ranch' in Karlsruhe, was now dealing with a whole range of concerns. Matters of patents, of technical clarification(s) for Bosch and Mittasch and the irritation of industrial secrecy preventing him from publishing his work and achieving the academic kudos he felt should accrue. The next few years were tumultuous for all concerned. The BASF was to transform the basic Haber-Le Rossignol process into an industrial giant, Robert's personal life was to run the full gamut from deep sadness to unbounded joy and Koppel was to finally 'capture' Haber, bringing him to Berlin where he and Robert were re-united.

Notes

1. According to Coates, p. 1671, 'Beauty in nature made no deep impression on him'. J.E. Coates, 'The Haber Memorial Lecture', *J. Chem. Soc.*, 1642–1672, (1939).
2. *One Hundred Years of the Fritz Haber Institute*. B. Friedrich, D. Hoffmann and J. James. This paper can be found at; http://www.fhi-berlin.mpg.de/mp/friedrich/PDFs/ACh-100-Article.pdf.
3. Candle power (cp) is an obsolete unit which measured luminous intensity. Today we think of a bulb's light intensity in terms of Watts, but there is no

real connection between the two, because a bulb's wattage is a measure of the *input* energy and there is no guarantee that all the electrical input energy is converted to light. However, one cp equals 0.981 cd, which measures luminous intensity in a given direction of a source that emits monochromatic light of 1/683 W per steradian—a sphere measuring 4π (12.56637) steradians. So, one cp at 0.981 cd, equals 0.981/683 per steradian or about 0.2 W over the whole sphere. 700 cp was therefore about 140 W output. Pretty poor really.

4. Hermann Sprengel was a German born chemist who developed the pump whilst working in London in 1865. Sprengel pumps were not only used by Crookes, but by Ramsay, Swan and Edison as well. The pump was a critical component for incandescent light technology removing enough air to allow the filament to glow for a commercially useful life time. See for example: http://en.wikipedia.org/wiki/Sprengel_pump.

5. Selling his patent to the AEG (*Allgemeine Elektrizitätsgesellschaft*) in 1898 reputedly made him one million Reich Marks. Diana Kormos Barkan, *Walther Nernst and the Transition to Modern Physical Science.* Cambridge University Press, 1999, p. 100.

6. Kormos Barkan, op. cit. (note 5), p. 94.

7. Margit Szöllösi-Janze, *Fritz Haber 1868–1934: Eine Biographie.* C. H. Beck, Munich, (1998), pp. 185–186.

8. Haber's royalties were established at about one Pfennig per kg ammonia with the help of Engler in 1913. There was some discussion of a fixed payment of 7.5% profits which Haber tried to re-instate post-war to protect his income from erosion by inflation. From Szöllössi-Janze, op. cit. (note 7), pp. 482–483.

9. Thomas Hager, *The Alchemy of Air*, Three River Press, New York, (2008), p. 103, and his references.

10

Of Koppel and the Kaiser

The Kaiser Wilhelm Institutes.

One never notices what has been done;
One can only see what remains to be done ....

Marie Curie (1867–1934), from a letter to her brother, 18 March 1894.

10.1 Progress at the BASF

With Robert and Meyer Mayer ensconced at the 'Auer', Haber returned to Karlsruhe from Pontresina to be briefed by Bosch. It soon became clear that the BASF already had serious concerns regarding the suitability of both uranium and osmium as 'contact substances' and in the next chapter we discusses the difficulties involved. Consequently, Bosch and Haber visited one another regularly and reviewed different strategies regarding the pre-treatment of uranium (in particular) to achieve a longer catalytic life, but it seemed obvious to Bosch that neither uranium nor osmium would be able to serve as catalytic agents. Even so, by January 1910 after an intense effort (Chap. 11), the search for an inexpensive, effective, durable catalyst was over and the BASF wrote to Haber with their findings;

> Now that our experiments in the area of producing ammonia from the elements, which you started, have reached a partial conclusion, we take this opportunity of informing you of the results ... We have managed to find that iron works as a catalyst; with respect to ease of use and cheapness, it cannot be equalled by any other, and using it produces yields as great as those with uranium and osmium.[1]

© Springer Nature Switzerland AG 2020
D. Sheppard, *Robert Le Rossignol*, Springer Biographies,
https://doi.org/10.1007/978-3-030-29714-5_10

Iron of course was an old friend of Haber's, never-the-less he was utterly surprised by the BASF's conclusions which seemed to run counter to all that he had expected. He wrote back immediately congratulating everyone involved and added with amazement …

> Here iron, with which Ostwald first worked and which we then tested hundreds of times in its pure state, now is found to function when impure! It strikes me again how one should follow every track to its end.[2]

With this critically important development following on so closely after the successful demonstration in the previous summer, Haber began to get impatient with the BASF regarding the publication of his work. The BASF of course wanted as little as possible to reach the public regarding progress and Haber felt frustrated. The suffocating secrecy even extended to Haber's laboratory at Karlsruhe which had to remain barred to unauthorised entry and this led to tension within the Institute. January and February therefore saw much to-ing and fro-ing between Karlsruhe and Ludwigshafen as Haber repeatedly put his case for some form of public recognition of his work. The BASF eventually agreed, and a meeting was arranged for 18 March 1910 at the Scientific Union in Karlsruhe. Haber's talk was to be titled, 'Making Nitrogen Usable', but the BASF of course insisted that the detail of just how this was achieved remain secret.

10.2   18 February 1910, St. Helier, Jersey

With Haber 'taking care of business' at Karlsruhe, Robert was able to concentrate on 'getting his feet under the desk' at the 'Auer' in Berlin, but he had hardly done so before the second tragedy of his young life struck, with the sudden death of his father, Augustin, in St. Helier. In less than five years Edith had lost her eldest son and her husband, both taken from her by brief, fatal and unexpected illness. Jurat Le Rossignol had been in attendance at the State's Royal Court as recently as the Tuesday before his death, and though it was apparent that he was not in his usual health no-one dreamt for a moment that his end was so near. By the afternoon of 18 February, he had passed away quietly in his sleep at his home in Caesarea Place - family at his side. His Obituary appeared in the local *Jersey Evening Post*, *The Daily Telegraph*, the *Jersey Times* and the *British Medical Journal* for March 1910 (Plate 10.1).

𝕺bituary

AUGUSTIN A. Le ROSSIGNOL, M.D.Aberd.,
M.R.C.S.Eng., L.R.C.P.LOND.,
HONORARY CONSULTING SURGEON, JERSEY GENERAL
DISPENSARY.

By the death of Dr. Augustin Le Rossignol, which occurred on February 18th at St. Helier, Jersey, as the result of acute broncho-pneumonia, the profession in Jersey has lost one of its most senior and respected members.

From the very onset of his brief, fatal illness, Dr. Le Rossignol appeared to realise that he would not recover, but he met his end with perfect calmness and entire trust. The interment took place at St Saviour's cemetery and was numerously attended, among others, by the Lieutenant-Governor of Jersey, the Chief Magistrate, and several members of the medical profession.

Dr. Le Rossignol married Miss Edith Sorel, by whom he had a family of three sons and one daughter. The eldest son, Austin, a young medical man of marked ability and much promise, died three years since, of diphtheria, when a resident officer at the London Hospital, the others, together with Mrs. Le Rossignol, survive to mourn the loss od a most attached and devoted husband and father.

Plate 10.1 An extract from Augustin's Obituary in the British Medical Journal unfortunately mis-spelling and mis-dating Austen's name and death. Article reproduced by the author

Augustin was held in the highest regard by the Jersey Legislative Assembly. On his appointment as Jurat in March 1903 much was expected of him, and their anticipation was fully realised. Recognising him as a man of exceptional ability, confrères nominated him to many of the standing committees where, through regular attendance and enthusiasm, he became an invaluable member. His independence of character won him admiration and whenever he thought it expedient to address the House he did so with a thorough grasp of his subject - always having the ear of members. In his judicial capacity he showed integrity, impartiality, a sound judgement and an obvious care for those brought before him.[3]

The sudden death of Augustin in his 68th year stunned the family, but as a medical man well used to diagnosis, he understood his fate and with the little time left to him he put his house in order. Augustin's Will[4] was dated 27

August 1909 and in it he bequeathed—according to the laws of the island—one third of his residual estate to his wife Edith, with the remainder to his surviving children Herbert, Elsie and Robert in equal shares. However, on 17 February 1910, the day before his death, two signed codicils were added which were in keeping with his humanity. In the first of these he gave the sum of one hundred pounds to one, 'Bessie Stephenson'. The same amount was left to the Jersey Medical Society, four fifths of the annual interest from which was to assist any medical practitioner in the island who, 'as member of the society and who has practised for at least ten years, may be 'in need'. The remaining one fifth interest was added to the capital. To his coachman 'Bartlett', he bequeathed a years wages of £57 4s. 0d., to his cook and housemaid each, a further sum reflecting their annual wage, and to Rachel Jane Richie, 'who has been many years in the family', the sum of £27 6s 0d. The second codicil ensured that after the death of any one of his children, and *after* the death of their mother, the income from his children's share of his estate would pass to that child's husband or wife if they were without issue. This ensured that his bequeath would not only help his children but their partners equally. Augustin Le Rossignol left a personal estate in the United Kingdom valued at £11,872 11s.[3]—including shares such as in the Great Western Railway[5]—all of which of course passed to Edith and the children. With the average annual (mainland UK) wage at the time around £50–70, this figure in today's money translates to a purchasing power equivalent to many millions of pounds. Probate of the Will, along with the two codicils was granted to Edith, and in just a few years Robert had moved from being an 'impecunious student'[6] to being a rather wealthy young man.

But in the year his father died, much happier times were to come when a few months later he married the youngest daughter of Carl Walter of Karlsruhe, one Frauleine Agnes Emily Hedwig Walter (b. 24.1.1885). Robert and Emily were married on Tuesday 11 October in Karlsruhe, and *The Times* of London on Friday 14 October 1910 carried the announcement of their marriage which was to last for the next sixty five years. Following their marriage, Robert and Emily left for their honeymoon—something of a 'curate's egg'—for Robert was sent to the 'States by the 'Auer' to meet and stay with Coolidge in Schenectady with whom he became good friends, and this business trip doubled up as their honeymoon. Undoubtedly, these two men must have discussed the problem of tungsten ductility at length, ... but Robert later remarked of the business trip and honeymoon, 'this is not a good mixture!'[6]

By the time of Augustin's death, Robert had left Grünau and was living at 8 Marburger Strasse in the Charlottenburg district of Berlin,[7] a little to the west of Friedrichshain, and this probably became the first marital home for the couple. Today, Marburger Strasse is part of 'old Berlin' and during Robert's time there it accommodated many apartments typical of the city from 1880 to 1910. There was 'old school' elegance here. The apartments often had two entrances, grand imposing frontages for the more well off and a plain back entrance for the less well-to-do. The main thoroughfare of Charlottenburg was the elegant Kurfürstendamm, modelled to some degree by Bismarck on the grand boulevards of Haussmann's Paris. Not long after the street's urbanization in 1886 it spawned residential districts, and in the early 20th century during Robert and Emily's time it became the lively centre of Berlin's social and cultural life. By 1911 however, the couple could be found at the impressive Prinzregentenstrasse 108, Wilmersdorf, Berlin, which lay just off the Prager Platz. Wilmersdorf has always been an affluent middle-to-upper class, inner-city, villa and apartment area with beautiful 'Art Déco' quarters. The move to Prinzregentenstrasse probably reflected the couple's improved financial status … but there may just well have been another reason too.

10.3   The Scientific Union Meeting, 18 March 1910

We know that Robert and Haber were still in personal contact with one another in Berlin at least until February 1913 (Chap. 12). Even as late as March 1915 Haber remained cognisant of Robert's position in Germany (Chap. 14). Haber therefore must have been aware of Augustin's death, and condolences were probably offered to Robert at Berlin, although we can call on no biographical evidence to support this conjecture. Neither do we know if Robert was at the Scientific Union meeting to hear Haber release his frustration and announce to an unsuspecting world that the problem of synthetic ammonia synthesis had been solved. Haber's lecture, later published as a shortened version in *Zeitschrift für Elektrochemi*,[8] made clear the extraordinary need in Germany of bound nitrogen for agriculture, together with the explosives and chemical industries. Of course this represented an 'about turn' from the opinion expounded in his book (Chap. 2) but having solved this difficult problem, he was now in an entirely different place. In his talk he addressed the source(s) of bound nitrogen and included all 'the usual suspects' viz, the 'fragile' Chile saltpetre imports, the role of fixation by plants, the oxidation of nitrogen together with its formidable energy barrier, along with such as the Frank-Caro process which he regarded as a 'multistage and therefore comparatively complex process that has not yet proved its economic feasibility'. With the expert audience in despair as to what could be done regarding Germany's position, he suddenly lobbed a stick of 'technological dynamite' amongst them[9];

> In contrast, there is the preparation of ammonia through direct combination of the elements nitrogen and hydrogen ...,

pausing briefly to suggest that this was thought impossible, he continued,

> ... work carried out in the Physical Chemistry Institute of the 'Fridericiana' in Karlsruhe has disproved this assumption.

Haber proceeded to exhibit, and then describe, the high pressure circulation process developed by himself and Le Rossignol, but mischievously—although quite in keeping with the BASF's policy of secrecy, or if not, confusion - even championing osmium and uranium as 'transfer agents'! Finally he concluded;

> These experimental results appear to ensure the basis for a synthetic ammonia industry ... The Badische Anilin und Sodafabrik in Ludwigshafen has successfully continued further from this process, which I have described here, so that the high-pressure synthesis of ammonia from its elements can now be included among the processes on which agriculturalists can pin their hopes when they ... search for new sources of the most important substance they require.[9]

The audience was stunned, and over the next few days Haber was overwhelmed with offers of collaboration. The BASF too were taken aback by the sensation that Haber's lecture caused in professional circles, and when Haber asked how he should distribute preprints of his talk, they made it clear to him that they had no intention of allowing any such thing, the BASF later writing to Haber;

> The less that the process ... is talked about in the next years, the less attention that is paid by those interested as to whether the technical realisation is going to succeed, [and] the more likely it is that we can win an advantage over the competition in the technical use of the process, this will also be to your advantage.[10]

Never-the-less, Haber's lecture was reported widely including in the UK when *The Times* carried a report of the talk entitled 'The Synthetical Production of Ammonia', in their Engineering Supplement on 18 May 1910. Much of the technical side of the process was reported including the temperatures, pressures, catalysts, cooling system, circulation process and yields—which reached commercially appealing levels 'upwards of 8%'. The article also reported—probably for the first time—that Haber had conducted 'a series of experiments with the assistance of Mr R. Rossignol', however there was no mention of Robert's nationality, the inference being that this was probably a Franco-German collaboration. However, Robert was now 'tagged' to Haber and 'fixation' at a 'popular' as well as a scientific level. Because of the continued secrecy and tensions at the Hochschule, Haber began to re-direct his efforts towards other research areas, and over the months he gradually became less and less attached to his chair at Karlsruhe. Haber however had become a respected member of the Institute. His relationships with Bunte and Engler were familiar and strong. Baden's capital city too was quiet and cosy, and one less 'driven' might well have accepted the crown of 'Saviour' and spent their remaining career there. But his ambition, along with matters brewing in Berlin, were to decide his future for him. Haber was about to enter a period of his life that was to define him to his own generation and of many more to come - even down to those of today.

10.4   Of Koppel and the Kaiser

Unbeknown to Haber, the prospect of Berlin appeared just two months after his talk at the Scientific Union, and its somewhat unexpected architect was Adolf Von Harnack, a theologian and prominent church historian. Deeply religious, but highly critical of conservative theology, von Harnack was equally appreciative of the role of science in German life—even being elected as a Member of the Academy of Sciences in 1890. It was as a distinguished 'organiser of science' that von Harnack was to unwittingly, but profoundly, affect Haber's life.

Von Harnack had been a close observer of the expansion of Imperial Germany at the time, and had become increasingly concerned by its fragility. Expansion he maintained, had to be driven by economic power, which in turn depended on efficient and innovative industrial strength. As Germany acquired political influence in Europe, so too her acquisition of overseas territories made her an increasingly important player on the world stage, but this in turn demanded a corresponding expansion in the German navy, one to rival the 'big guns' of the day, such as Great Britain. In all of these concerns von Harnack realised that Germany's prestige, power and survival was umbilically connected to its science, and here he detected fatal flaws—especially so, as the growing tensions within Europe made many nations look to their war machines. Soon, their distant rumble began to be heard across the continent.

In its drive toward expansion, the German state was already overcommitted with few spare resources to dedicate to its scientists and technologists on a whim that a particular avenue might lead to a commercially viable development. It was not possible for expansion to be driven by the 'Haber model', viz., a lone professor, the 'gentleman scientist', whose considerations of 'reward' were measured in terms of honour, service, patriotism, respect and fame, but struggling to obtain funding for an important project in an institution inadequately equipped for their needs. Harnack therefore looked to America and dreamed of a similar systemic approach to science in Germany with 'think tanks' populated by the best brains in the country, with state of the art equipment, effective channels of communication and strong links with industry and commerce. Early in 1910, von Harnack had a private audience with the Kaiser and delivered a speech much along these lines, in turn pointing to the fabulously wealthy private American research institutions established by Andrew Carnegie and John D. Rockefeller. These institutions were able to tackle profound problems that lay beyond the grasp of individual university departments, and Germany had nothing to rival them. Harnack told the Emperor that the nation rose and fell along with its science, and that he too should establish a series of elite private research establishments and—appealing to the Kaiser's vanity—call them 'Kaiser Wilhelm Institutes'. Wilhelm was smitten and adopted the idea completely—a position easily rationalised by the fact that his institutes would be funded largely by the altruism and patriotism of good Germans. But first find your 'good Germans', and to this end a meeting was arranged in Berlin by the leaders of Germany's dominant State, Prussia, at Wilhelmstrasse 63, on 14 May 1910.[11]

The Moghuls summoned there that day had their fortunes built on a variety of industries such as steel, electrical equipment, chemicals and banking. Amongst the gathered was the reclusive Leopold Koppel. Koppel listened attentively to the presentation, one which asked these men to donate a proportion of their fortunes for the good of the German nation. Koppel soon saw an opportunity to curry favour with the establishment—always close to the heart of the German Jew—whilst at the same time establishing an entity that, through patents and licenses, would enhance his already powerful business empire. Koppel's financial interests lay in the establishment of an

Institute of Physical and Electrochemistry, but who would he chose to direct it? Koppel knew just the man, and soon after the presentation he asked Fritz Haber to travel to meet him in Berlin. Haber of course knew Koppel well, and the two men discussed the Emperor's plans for German science when they met at the offices of Koppel and Company on Unter den Linden, a boulevard that ran from the Brandenburg Gate to the Emperor's palace. Haber in turn had already seen for himself the powerful basis of American science on his sixteen week tour of the 'States' in 1902, so the Emperor's plans struck an immediate chord with him. But soon, Koppel moved on to his personal agenda; he was willing to fund the construction of an entire Kaiser Wilhelm Institute for physical chemistry and electrochemistry, he would largely cover its year-on-year operating costs—although its director would be employed by the Prussian State—but he would do so with just one proviso, that Haber became the Institute's founding director.

Now in September 1909, Koppel had already tempted Haber with a lucrative offer of a directorship in his company which Haber had declined, (Chap. 9) but this new offer was different. Here was the call not of just of commerce, but of *Empire*, a call which every German Jew longed for, viz., to be brought 'into the fold', and to become accepted as 'establishment'. Koppel's too had always seen much of himself in Haber, both were Jews, both had converted to Christianity, both rose from humble beginnings and reached the top of their professions by hard work. Both too had a taste for the expensive, although unlike Haber, Koppel disliked publicity and unlike the already wealthy and established Walther Nernst, Haber was still 'hungry' and genuinely appreciative of success. But Koppel's offer was not simply patronising, he realised that with such close links to Haber he would benefit from 'special access' to his professor, an access which in turn would allow Koppel and Co. to benefit, through patents and licenses, from appropriate scientific work conducted at his Institute.

This was an offer that Haber could not refuse—after all, that would confound both the Emperor and Haber's patriotism - and so Koppel moved to place his plan for an Institute of Physical Chemistry and Electrochemistry before the Prussian State machine. There was initially some dissent—the eminent Walther Nernst was preferred as director for example—but by September 1910 Haber received an invitation to meet with officials in Berlin. Koppel's gift to the nation along with his chosen director were approved. It took some time for Haber to release himself from the bonds that tied him to Karlsruhe but by July 1911, Haber, Clara and their son Hermann left for a new life in Berlin, a life that for Haber was to last until 1933.

The move to Berlin also 're-united' Haber with Robert who was still at the 'Auer', and who in turn welcomed a new life shortly after the Habers' arrival, as Emily gave birth to their first son, John Augustin, at 108 Prinzregentenstrasse on 25 September 1911, the notice of his birth appearing in *The Times* on Friday 29 September 1911. Robert probably saw John Augustin's birth as rationalising the deep sadness of his father's sudden death the previous year, and it is of course quite probable that Robert and Emily shared their joy with the Habers at this time, but no biographical record remains to support this conjecture. Meanwhile at Ludwigshafen, another of Robert's progeny was receiving some serious attention at a highly accomplished 'finishing school' but, as is often the case with a 'boarder', little of what was to finally graduate was attributable to its parents. The next Chapter in our story therefore deals with the industrialisation of the Haber-Le Rossignol process, but contrary to the popular name *'Haber-Bosch'* process, neither Haber nor Le Rossignol had much input here.

Notes

1. This work was due to Alwin Mittasch. His catalysts have hardly been improved upon even today. From, Dietrich Stoltzenberg, *Fritz Haber. Chemist, Noble Laureate, German, Jew.* Chemical Heritage Press, Philadelphia, Pennsylvania, (2004). p. 90. Source, BASF to Haber, 13 January 1910, MPG., Dept., Va., Rep 5.

2. Stoltzenberg, op. cit. (note 1), p. 90. Source, Haber to BASF, 14 January 1910, MPG., Dept., Va., Rep 5.

3. This description has been adapted from, S. J. Le Rossignol, *Historical Notes (Local and General) with special reference to the Le Rossignol Family (and its connections in Jersey)*, Trowbridge, (1917), p. 181.

4. A copy of Augustin's Will was obtained from the Jersey Archive, Clarence Road, St. Helier, Jersey. JE2 4JY.

5. From the Great Western Shareholders Register 1835–1932. Augustin invested in 1889, and Robert kept his father's shares at least until 1925.

6. A quote from Ralph Chirnside's sensitive Obituary of Robert; R. C. Chirnside, 'Robert Le Rossignol, 1884–1976', *Chemistry in Britain*, **13**, 269–271, (1977).

7. Le Rossignol, op. cit. (note 3), p. 132.

8. Stoltzenberg, op. cit. (note 1), p. 90. 'Über die Nutzbarmachung des Stickstoffs' (Making nitrogen useable), *Verbandlungen des Naturwissenschaftlichen Vereins in Karlsruhe*, **23**, (1909/1910), and F. Haber, 'Über die Darstellung des Ammoniaks aus Stickstoff und Wasserstoff', ('The synthesis of ammonia from nitrogen and hydrogen), *Z. Elektrochemie* **16**, 244, (1910).

9. Stoltzenberg, op. cit. (note 1), p. 91, taken from note 8, 'Making nitrogen useable'.

10. Stoltzenberg, op. cit. (note 1), p. 91. Source, BASF to Haber, 19 October 1910, MPG., Dept., Va., Rep 5, 2085.

11. Daniel Charles, *Between Genius and Genocide*, Jonathan Cape, London, (2005), p. 118.

11

From Karlsruhe to Oppau

An ammonia converter dating from about 1915. Photograph courtesy of the corporate archives, BASF.

Industrialisation of the Haber-Le Rossignol Process

... I should like to stress how important completely smooth running is for the synthesis of ammonia. Every stoppage at a single point affects the whole plant, and after a stoppage it takes hours until everything is back to normal. It is no exaggeration for me to say that, technically speaking, it depends entirely on smooth continuous operation whether the process is economical...

Carl Bosch, Nobel acceptance speech, 21 May 1932.[1]

© Springer Nature Switzerland AG 2020
D. Sheppard, *Robert Le Rossignol*, Springer Biographies,
https://doi.org/10.1007/978-3-030-29714-5_11

11.1 Early Progress at the BASF

With Haber and Robert progressing their careers in Berlin, Carl Bosch and his team were moving ahead with the industrialisation of the ammonia synthesis at the BASF. Very early on, Bosch had realised that there were three key problems to be addressed viz, the discovery of an inexpensive, effective and durable catalyst, an economic supply of the feedstock gases of hydrogen and nitrogen, and the construction of the high-pressure converters. These three problems were tackled in parallel but they were to prove of varying complexity, the design of the converters being by far the most difficult. The BASF had early success with regard to their first key problem. Their letter to Haber in January 1910 (Chap. 10) had reported 'a partial conclusion' to the ammonia synthesis, in that their search for an effective, inexpensive catalyst was already over. Even by September 1909 the BASF had made significant progress in the matter. They had tooled up a workshop fully adapted to make the 'pilot' apparatus suitable for the prolonged laboratory experimentation necessary for the discovery of a durable catalyst. Their starting point had been the original Haber – Le Rossignol 'furnace', but they soon found it far too fragile for an industrial environment. As a result, it was completely redesigned, retaining its essence but implemented in a much more robust fashion, and incorporating a removable 'cartridge' capable of containing about 2 g of material which allowed different catalytic candidates to be examined on a 'production line' basis. This task was appointed to Alwin Mittasch, Hans Wolf and George Stern. Twenty-four of these laboratory converters were made and they were in continuous service 'around the clock' for years, eventually (by 1922) examining over 20,000 'runs' of more than 4000 separate catalytic candidates (Figs. 11.1, 11.2).[1]

During September 1909 this investigation was in its infancy, but osmium and uranium were already causing the BASF some concern. Osmium it seemed, was easily converted to the volatile tetroxide by the presence of air—almost impossible to exclude in a high-pressure apparatus on an industrial scale—whilst uranium's sensitivity to both air and water eventually meant that there was no form in which it could be used to lay the foundations of an industrial process.[1]

Every pure metal with a known catalytic property at the time was subsequently examined by the BASF, but compounds and mixtures of compounds—especially those involving iron—were to reveal themselves as unexpectedly suitable candidates. On 06 November 1909 for example, one of Wolf's samples of impure magnetite (Fe_3O_4) from the Gällivare mines[2] in northern Sweden

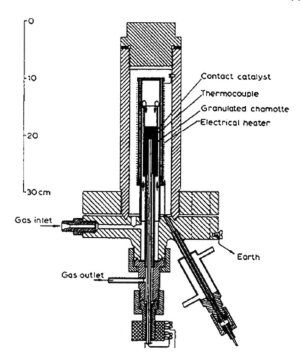

Fig. 11.1   The new robust laboratory converter designed by George Stern and used by Mittasch for the examination of candidate catalysts. The auxiliary pressure tube is now shown diagonally. Figure courtesy of the Nobel Foundation, Carl Bosch Nobel Lecture, 1931[1]

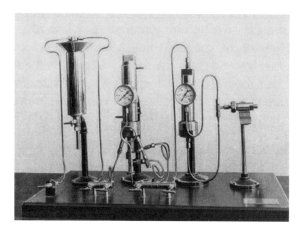

Plate 11.2   The BASF version of the Haber Le Rossignol apparatus showing the new converter in the centre (with a cooling jacket of iced-water), a drier to the right and the cold regenerator to the left. Notice however the 'Le Rossignol' valves along the bottom and how much more robust this apparatus now appears. Photograph courtesy of the corporate archives, BASF

provided an unusually high yield—a yield unmatched by an exhaustive study of other magnetites. Using his array of newly designed Haber-Le Rossignol reactors, Mittasch began a systematic study to determine which additions to the basic iron catalyst were most effective. Sodium and potassium hydroxides were the earliest tested additions, and this led to the first BASF patent regarding 'mixed catalysts' (DRP 249 447), filed on 09 January 1910 by Bosch, Mittasch, Stern and Wolf.[2] Eventually, Mittasch concluded that the presence of other elements acted as 'promoters' of the ammonia reaction and he soon developed a mixed catalyst of magnetite and bauxite, together with oxides of potassium, calcium and magnesium which according to Bosch was 'relatively fast acting, technically perfect, easily manipulable, stable and inexpensive.'[1] At the time, these catalysts were of a completely new type but even today the composition of the mixed catalysts in an ammonia plant remains virtually the same as Mittasch and his co-workers originally discovered. By January 1910 the search for a catalyst was effectively over and the BASF wrote to Haber with their findings (Chap. 10).

11.2   Nitrogen and Hydrogen

The supply of raw materials however proved more difficult. By 1909, nitrogen of sufficient purity to prevent catalytic 'poisoning' was available from the Linde process, but the procurement of hydrogen of an equivalent purity represented a prime cost, especially so since the efficiency of the high-pressure conversion of the gas mixture into ammonia was such that *its* cost was a minor factor. None of the known hydrogen producing processes at the time (Chap. 8) were suitable, and given the BASF's access to vast quantities of coal, Bosch decided[1] that the source of hydrogen had to be 'water gas', a mixture of hydrogen and carbon monoxide obtained by passing steam over red hot coke according to;

$$C(s) + H_2O(g) \leftrightharpoons CO(g) + H_2(g) + 118.7 \text{ kJ mole}^{-1}$$

This gas mixture contained about 50% hydrogen, 40% carbon monoxide, and 5% each of nitrogen and carbon dioxide. A cryogenic process was initially used to remove the carbon monoxide at −200 °C and ∼25 atm, but as the yields from the ammonia synthesis increased Bosch looked for a better solution. By summer 1911, Wilhelm Wild, one of Nernst's pupils working for the BASF, had perfected a cheaper method of hydrogen production from 'water' gas, using a catalytic[3] process that reacted the carbon monoxide with the steam to 'shift' the reaction towards the production of more hydrogen;

$$H_2O(g) + CO(g) \leftrightarrows CO_2(g) + H_2(g) - 41.2 \text{ kJ mole}^{-1}$$

In the early ammonia plants this reaction was combined with another to provide the simultaneous production of feedstock nitrogen, thereby avoiding an undue dependence on the Linde process. Here, both 'water' gas and 'coke-oven' gas (typically 62% N_2, 32% CO, 4% H_2 and 2% CO_2) were produced by separate processes, the latter from the reaction of hot coke (> 1000 °C) with air. The two gases were then washed with water, mixed, and passed over activated charcoal to remove the catalytic 'poison', hydrogen sulphide ($H_2S_{(g)}$)—always present as an inevitable consequence of the use of coal. The 'clean' gases were then passed over the catalyst and the 'shift' reaction occurred leading to the formation of more hydrogen. Any remaining CO_2 and CO was removed (respectively) by 'scrubbing' at ~25 atm followed by absorption in cuprous ammonium formate, but moisture was allowed to remain. Sometimes, this purified 'synthesis gas' contained too much hydrogen and this was corrected by the addition of nitrogen procured via the Linde process to achieve an exact $H_2:N_2$ stoichiometric ratio of 3:1. By 1912, Bosch and Wild had patented their version of the 'shift' reaction as part of the BASF's *Wasserstoffkontaktverfahren* process for the mass production of hydrogen.[4]

11.3   The Converters

The design of the industrial high-pressure converters began in late 1909 and was based on the Haber-Le Rossignol model. The circulation pump and ammonia separator were additional concerns of course, but these were initially thought to be relatively 'conventional' problems. As far as the converters were concerned, there were no examples in industry that could be followed. The Linde process operated at ~200 atm but the apparatus was made from soft soldered copper utterly unsuitable for the synthesis of ammonia at 600 °C and higher. Bosch began by using two strong steel 'contact' (reactor) tubes about 2.5 m long and 30 mm thick drawn by the Mannesmann Company. Following 'Bestimmung...' these tubes were heated externally (by gas), and mindful of the problem of fires and flarebacks which can occur when hydrogen escapes at high pressure and temperature, the tubes were housed in reinforced concrete sleeves far away from the busy centres of activity in the plant. Using these tubes at the experimental site (designated Lu398) in the main Ludwigshafen compound, the BASF first began to synthesis liquid ammonia as early as 18 May 1910. By 19 July, enough ammonia had been produced to fill a 5 kg container,[5] but there were clear problems developing with the reactor tubes. These tubes performed well within a circulation system for about 80 h, but then they failed catastrophically and burst, Bosch later observing that, 'Had these tubes been charged with osmium, the world's supply of this rare metal would have disappeared!'[1]

Examination of the burst tubes showed that they were swollen. Some change in the material of the inner wall had resulted in a loss of elasticity and this change had progressed outward until the undamaged part was so thin that it yielded to the internal pressure. A chemical attack was suspected and nitrogen was at first thought to be the culprit because the literature showed that iron nitride, a silvery brittle compound, was apt to form when hot iron and ammonia come into contact. Chemical analysis of the changed material however showed no sign of nitrogen. Now Bosch's provenance was in metallurgy, and he applied a technique little known to chemical engineering at the time viz, metallurgical 'etching'. Consequently, the following picture slowly emerged.

Because of its high mechanical strength, carbon steel was the obvious candidate to construct the reactors. This steel has a structure in which carbonaceous perlite is dispersed in a matrix of pure iron, but in the changed parts of the reactor tubes, there was no sign of the perlite, and the resulting structure had been destroyed by cracking. Bosch realised that the perlite had disappeared because of the 'de-carburisation' of the steel, but this in turn should have resulted in 'soft iron' which should not have cracked. On the contrary, the changed material was hard and brittle, rather like 'cast' iron. Desperate laboratory investigations soon revealed the true problem. Diffusion of hydrogen[6] into the steel caused de-carburisation via the formation of methane gas leaving a brittle alloy of iron hydride. The gas, trapped under extreme high pressure within an embrittled metal, added to the mechanical stress on the material, loosening the structure and contributing to a complete wall failure. Bosch concluded that all carbon steels would suffer this 'hydrogen attack' and fail in a matter of hours or days. He also concluded that even in the absence of 'hydrogen attack', external heating of the steel jackets of these converters put further stress on their walls because the temperature and pressure gradients involved led to buckling and cracking, causing heavy explosions and costly repairs. The original Haber-Le Rossignol experiments of course, had only been run for a few hours and so these problems had never arisen—although to be fair, Haber and Robert had also converted to internal heating. (Chap. 8) Overall however, it was remarkable just how little guid-ance their experiments provided regarding the uplifting of the technology to an industrial scale. Even so their design, based on a continuous circulation process was sound, and remains essentially that which is still used today.

With the problem identified but no clear solution yet in mind, the BASF moved their entire research effort to a new site at Ludwigshafen, designated Lu35. The gas preparation and cleansing system machinery, the high-pressure reactors, circulation pumps, cryogenic separators and scrubbers.[7] were all moved to build a new prototype plant which became operational on 10 August1910—but still using the suspect reactor tube design. The first three months of operation saw only 100 kg of ammonia produced, but by December 1910 a daily rate of almost 10 kg had been achieved. This rose to 18 kg day^{-1} in the first two weeks of 1911.[8] In the meantime, between July 1910 and March 1911, Bosch and his team tried a number of different solutions to obviate the problem of the burst tubes, a problem they realised jeopardised the whole project. Initially, everything they tried was either unsuccessful or unpromising and their conclusion was quite simple;

We assumed…. that the diffusion of hydrogen into the iron, the de-carburisation of the perlite, and the formation of brittle iron hydride are just unavoidable. They had to be rendered harmless by a modification in design…[1]

Bosch focussed on the role of the reactor walls. These he concluded served two purposes viz, to act as a gas tight seal, and to mechanically withstand the extreme pressure. He ingeniously decided to separate these two roles by designing a reactor made of two tubes, an outer pressure bearing steel jacket and a quite thin inner 'sacrificial' lining of soft steel in such a way that;

As the hydrogen… diffuses [and] *passes through the thin lining it is able to escape without building up pressure before it can attack the outer steel jacket at the high temperature. This is readily achieved by grooves produced on the outside when the* [inner] *tube is being turned and by a large number of holes drilled in the steel jacket through which the hydrogen is free to emerge… the thin lining tube fits tightly against the jacket under the high pressure and… when it has become brittle there is no way of expanding it… so that no cracks occur.*[1]

The early recognition of 'hydrogen attack' of course had already led to the submission of the supplementary 'diffusion' patent, (Chap. 8) whilst the first reactor based on the twin tube proposal—which provided the 'protection' mentioned in the patent application—became operational from March 1911.

Here losses due to diffusion were minimal, however this reactor still retained external heating.

Having found a solution to 'hydrogen attack', Bosch now had to address the difficulties of the buckling and cracking arising from the external heating. Internal heating with electricity was immediately discounted because of the unavoidable losses to prevent overheating of the jacket. Gas heating provided a more controllable solution[9] whereby hydrogen was burned with admitted air within the reactor to raise it to operational temperature, the water formed being seemingly harmless to the catalyst.[10] By the end of 1911, internal heating had been added to the twin tube reactors which by now had reached a length of 4 m and a diameter of 15 cm. Shortly afterwards a further innovation was added which reduced the 'hydrogen attack' on the convertor's jacket. This was achieved by continually flushing the space between the lining and the jacket with nitrogen—harmless to the metal. The nitrogen also reduced the pressure difference between the inner and outer parts of the lining tube alleviating the tendency to buckle.

The influence of the improved technology on the stability of the industrial pilot process was reflected in the steadily rising ammonia output during 1911. The first two weeks of 1911 saw production achieve around 18 kg day^{-1}. By July 1911 a daily rate of 100 kg ammonia was surpassed. By the end of 1911 —and now with twin tube reactors and internal heating—the yearly output was around 11 tonnes, with a daily average of 30 kg. Even by April 1911, the BASF realised that progress was sufficiently advanced to begin planning for a full-scale plant operating at 10 tonnes of ammonia day^{-1}, although at the time, the pilot plant was yielding less than 0.1 tonne. By the time a decision to go ahead with a new plant was eventually made in November 1911, its anticipated daily production had increased to 30 tonnes fixed nitrogen day^{-1}, with 70% of that destined for the production of ammonium sulfate fertiliser.[11] The speed of development after the original Haber - Le Rossignol demonstration in July 1909 was therefore truly spectacular.[12]

11.4   Of Compressors and Valves

During this time there were also more peripheral, but no less taxing problems to be addressed, such as the design of the circulation/compression pumps, and the valves. Initially the high-pressure compressors were a great source of worry. Contemporary systems were designed to compress or circulate air, for such as locomotives in mines and the Linde process. Little attention had been paid to the 'stuffing boxes',[13] and leaks could be tolerated. But in high pressure systems involving hydrogen, losses meant money, the risk of explosion, and stoppages. The form of the ammonia synthesis Bosch was developing was particularly dependent on continuous smooth running. Any stoppage would affect the whole plant and take hours to get back to normal, which in turn determined whether the process is economical. After years of trials Bosch eventually developed 3000 horse power compression units capable of functioning reliably without stopping for six months, after which they received their regular periodic clean. The same effort saw the development of large piston and rotary gas circulation pumps that operated reliably at 200 atm.[1]

The valves too presented a significant technical problem. In the Haber-Le Rossignol apparatus, Robert's valves had only one purpose, viz, to finely control the passage of high-pressure gas through the apparatus, standard 'check valves' being used to ensure the directional consistency of the gas. The industrialisation of the process however led to a large system of functional units interconnected by thousands of metres of flanged tubing with valves, that in addition to controlling gas flow, also had to permit the isolation and rapid de-pressurisation of parts of the plant in the event of rupture or stoppage. At least three types of valve were designed[1]; the 'quick acting' valve, which rapidly energised in the event of a tube breakage, the 'self closing' valve, automatically using the gas flow to seal off one side or another of the tube when the gas flow becomes too high due to tube rupture, and the 'slide valve', which was essentially a 'needle valve' that rapidly closed and permitted a release of pressure. Needle valves of course were modelled on Robert's original patent of 1907.[14]

11.5   The First Ammonia Plant at Oppau

With the decision to build the new plant made in November 1911, the actual construction began on 07 May 1912. The main BASF site at Ludwigshafen had insufficient space for what—because of the huge volumes of gas involved —was anticipated to be a somewhat sprawling development. A new 500,000 square metre site was chosen just north of Ludwigshafen between the village of Oppau and the river Rhine. The plant was to be completely integrated, receiving deliveries of coal and coke, generating 'water' and 'coke-oven' gases, purifying, enriching, and pressurising the synthesis gas, generating the ammonia, capturing it using the water scrubbers,[15] forming the sulfate by reaction with sulfuric acid, storing it, and accommodating the rail network for supply and distribution. Because most of the technical problems had already been solved, work proceeded smoothly and the plant produced its first synthetic ammonia on 09 September 1913, after less than fifteen months of construction,[16] a stunning achievement of contemporary engineering.

The throughput of the plant of course depended on the capacity of the converters. During and after construction their capacity was increasingly expanded. By the end of 1911 the pilot plant converters were 4 m long with a diameter of 15 cm, this soon increased to 4 m long and 23 cm diameter and then to 8 m long with a diameter of 28.5 cm, the latter each weighing 3.5 tonnes. These were the reactors initially installed in Oppau. By 1915 the converters were 12 m long with a diameter of 108 cm and weighed 75 tonnes.[1] During the first full year of Oppau's production in 1914, the plant 'fixed' 20 tonnes of nitrogen daily which in turn translated to 8800 tonnes of ammonia or 26,280 tonnes of ammonium sulfate per year. Flushed with success and with an insatiable market for their product, the BASF made plans to expand production to 40 tonnes of fixed nitrogen daily, equivalent to 150,000 tonnes of ammonium sulfate per year.[17]

At the same time Bosch, who had now been umbilically connected to the nitrogen problem for eleven years, realised another long-cherished dream viz, to open an agricultural research station to study the influence of nitrogenous fertiliser on plant physiology. This was eventually established at Limburgerhof near Ludwigshafen,[17] and one feels that as far as Robert was concerned, there could hardly have been a more fitting recognition of the technology he helped to create—one which Woodland Toms too must have entirely approved.

Notes

1. Carl Bosch Nobel speech; '*The Development of the Chemical High Pressure Method During the Establishment of the New Ammonia Industry*', available at; https://www.nobelprize.org/nobel_prizes/chemistry/laureates/1931/bosch-lecture.html and also from, *Nobel Lectures, Chemistry* 1922–1941, Elsevier Publishing Company, Amsterdam, (1966).
2. Vaclav Smil, *Enriching the Earth*, MIT Press, (2001), p. 95.
3. A catalyst of iron and chromium oxides at 250–450 °C.
4. See Smil, *op. cit.* (note 2), p. 97, his reference no. 47.
5. See Smil, *op. cit.* (note 2), p. 99, his reference no. 56.
6. Hydrogen is a vey small molecule and diffuses easily through many substances. Early airships for example deflated easily unless the buoyancy bag was coated with a substance that hindered diffusion—often aluminium coatings. The easy diffusion of the hydrogen and the 'high' temperature and pressure caused the de-carburisation.
7. In early plants, the ammonia was removed by absorption in water 'scrubbers'.

8. See Smil, *op. cit.* (note 2), p. 99.

9. The situation resembled an electric cooker hob. Turning the hob off still leaves it hot for some time. Turning off a gas hob however has an immediate effect.

10. Recall that moisture was not removed from the water-gas/coke-oven gas mix.

11. Ammonium sulfate was chosen simply because of the availability of sulphuric acid produced from sulfur dioxide obtained by 'roasting' pyrites (FeS) in air.

12. See Smil, *op. cit.* (note 2), p. 100, his reference no. 59, for the details concerning the Oppau plant production tonnages.

13. A *stuffing box* is an assembly which is used to house a 'gland seal'. It is used to prevent leakage of fluids such as water, or gases such as steam, between sliding or turning parts. It was the 'stuffing box' of Robert's circulation pump, that caused him some concern. See Chap. 8.

14. '*Conical screw down valve with an angle of inclination between 85° and 90°*'. Submitted in Germany on 16 March 1908. Application submitted in the UK on 16 July 1908, accepted 01 April 1909, UK Patent No. 15,065.

15. Later, to economise on transport costs, the ammonia was separated as a liquid by intense cooling as in the original Haber-Le Rossignol process.

16. See Smil, *op. cit.* (note 2), pp. 100–102 and his reference no. 62.

17. See Smil, *op. cit.* (note 2), p. 102.

12

Berlin 1911–1914

Of patents, publications, and the prelude to war.

You can fool all of the people some of the time. You can fool some of the
people all of the time. But you can't fool all of the people all of the time ....

Abraham Lincoln, US President, 1809–1865.

12.1 The Patent War 1911–12

As director of the Kaiser Wilhelm Institute for Physical and Electrochemistry
at Dahlem Berlin, Haber was to move from the laboratory into the 'office'.
Leaving the laboratory was never a challenge for Haber for he rarely showed
any enthusiasm or aptitude for what to him was the tedium of experimental
practise. Theoretical physical chemistry too was changing. A new picture of
the atom was emerging through quantum mechanics, and its application to
chemistry led to a new generation of mathematical tools which had largely left
Haber behind.[1] Instead, the 'great man' began to transform himself into a
'great German'. Like Nernst he soon became *Geheimrat* or privy councillor
(Chap. 2). He accepted a place in the councils of government, advising
Germany's elite in matters of science and industry. His rather pompous style
—first exhibited in Leipzig (Chap. 2)—had by now been embellished with
trademark cigar and fashionable clothes. Profiles, policies, politics, budgets
and administration were to become the order of the day. There was much to

© Springer Nature Switzerland AG 2020
D. Sheppard, *Robert Le Rossignol*, Springer Biographies,
https://doi.org/10.1007/978-3-030-29714-5_12

administer too, and not only those matters relating to the new Institute, but also the 'left overs' from Karlsruhe—especially the various BASF patent applications. Lincoln's perceptive observation may have had its provenance in politics but it had equal legitimacy in patent law and through 1911–12 the Haber-Le Rossignol applications, as 'assignors' to the BASF, had begun to generate significant challenges from 'some of the people' regarding their originality and novelty. These of course were the patent lawyers of competitor companies and they were not easily fooled. Haber's talk at the Scientific Union in 1910 had generated immense interest in the ammonia synthesis but even so, the BASF had been remarkably successful in keeping the extent of their subsequent progress hidden. However, the challenges were now 'raining in' and the BASF faced a dilemma viz, how to establish their claims without having to reveal the progress they had made.

In order that a patent be granted to protect the exploitation of some invention it was (and still is) necessary to show that the claim defining the invention must be new, not obvious, useful, and/or industrially applicable. Under such circumstances exclusive rights are granted by a national government to an inventor for a limited period of time in exchange for the public disclosure of the invention. Certainly, the companies raising the objections were no fools. Kunheim, Griesheim-Elektron, Hoechst all contested the Haber-Le Rossignol early patent applications, their claims being based largely on Nernst's 1907 paper (Chap. 5) with Fritz Jost which was quite rightly the first to examine the ammonia equilibrium under pressure. Indeed, even Jost raised an objection but he soon withdrew. Haber therefore had to expend much time and effort in addressing their concerns as his correspondence with the BASF and the German Patent Office shows. The Hoechst Dyeworks company led the charge, eventually submitting a nullity suit—a formal claim of invalidity—to the Patent Office on the grounds that the catalytic synthesis of ammonia under pressure had already been discovered. This objection primarily concerned the third key patent, no. 238450 (the 'high-pressure patent') but supplementary patents were also contested on the basis of the presumed 'invalidity' of 238450.

Haber received the Hoechst nullity claim on 09 September 1911 and he immediately contacted the BASF. They suggested a transfer of the patent business to themselves, to which Haber had little choice but to agree. In the meantime, the companies began to organise themselves for the coming court cases and expert witnesses were engaged; Ostwald for Hoechst and his former pupil, Walther Nernst, for Griesheim. Ostwald's opinion was that the process described in the patent was simply an extrapolation from low pressures to higher ones and that its result could be entirely expected. It was therefore not 'new' but scientifically 'obvious'. As far as industrial applicability was concerned, there was no hint of a discussion on how the 'eventual difficulties' i.e., the expected engineering problems, would be overcome and so one could not possibly deem it as applicable in that sense. Now Ostwald's foray into the ammonia arena in 1904 had unfortunate consequences, as the BASF's Carl Bosch could well testify (Chap. 3). Because of this weakness he was not seen as the principal opponent despite his eminence, although challenging his notion of applicability would undoubtedly lead to the BASF divulging the progress they had made and the engineering solutions employed, and so his opinion still needed to be discredited. Nernst however was a different proposition and it was the BASF's cunning manipulation of *his* opinion that led to a resolution in their favour. Although Nernst had allied himself with Griesheim, the BASF suddenly remembered that he had earlier also helped them [with their objections], so they approached him again with two questions to address[2] viz, are the declarations in patent 238450 correct, and could these have been predicted on the basis of publications known to you before the establishment of the patent?

Their approach however was 'sugar coated', as the BASF appealed to his vanity offering him an annual 'honorarium' of 10,000 marks for five years.[3] The BASF put various apparatus at Nernst's disposal in order to 'help him formulate his opinion'. After going through the motions of conducting a few experiments, Nernst suddenly pledged it a 'national duty' for him to defend Haber's patent and his observation was now that;

> I cannot but express my active conviction that patent document 238450 deals with results of a completely new type, and the declarations therein form a solid experimental foundation for an extremely important new technical process … it seems to me without a doubt the … patent under discussion is … worthy of the protection given by the granting of a patent.[4]

The opinions of the 'old master' Ostwald, and his 'apprentice' Nernst, were therefore at odds and what was surprising to contemporary observers was the subsequent astonishing union of former academic 'adversaries' such as Nernst and Haber.[5] But what a difference an honorarium makes, and how adroitly the BASF played their hand. In February 1912 the BASF informed Haber of Nernst's 'miraculous' support for his process and a date was set for hearing the counter claims to the patent(s) at the national court in Leipzig on 04 March 1912. A few days before, the BASF team of Haber, Nernst, Bernthsen and Bosch[6] met to discuss tactics, but their unanimous opinion was that their position was weak. During the morning of the hearing, claims and counter claims were presented by the patent attorneys. Late in the morning Richard Weidlich, the Hoechst attorney, presented a thorough justification of their suit to which Bernthsen had managed only a modest, unconvincing reply. Then suddenly, his principal witness Nernst, whose fundamental contribution to the ammonia synthesis he had just praised, marched into the courtroom 'arm in arm' with Haber in a tangible gesture of confidence and unity. Nernst proceeded to deliver a passionate defence of Haber in which he described his own work as of 'scientific interest' only and of 'no technical relevance'. Nernst explained how only Haber–Le Rossignol had created the 'prerequisites for a technical success by investigating new pressure ranges' and anyone who could not see that must be technically 'blind'. Nernst's evidence was devastating. So much so that soon after, Weidlich turned to his assistant and told him 'we can go home'.[7] At 3.33 p.m. on 04 March 1912, Bernthsen telegraphed the BASF management, '*Hoechst's nullity suit against our ammonia pressure patent rejected, and they are to pay costs*.'[8] The patent was 'in the bag' and little of what the BASF had achieved up until then had been divulged. It was the perfect outcome and not just for the BASF but for Haber and Robert as well, for their royalty payments from the BASF depended upon 'the profits from any *patentable* development'. Realising the advanced state of the technical progress of the ammonia synthesis, these two men could now look forward to substantial earnings over the fifteen years of a patent protection at the time.

We know too that, sometime after the patent challenges were rejected—the date is vague—Robert met Nernst at the 'Auer'. Nernst told him that even *he* could not have done the ammonia work better, to which Robert replied, '*Herr Geheimrat, es war auch mit Ihre Hilfe …*' (it was also with your help) and Nernst, he said, '*swallowed it like butter*',[9] a story Robert also re-told to Chirnside in 1976.[10] Later, when Nernst was in a 'good mood', he invited Robert to meet Einstein at a dinner given by Nernst at the Automobile Club in Berlin. '*As a young man I was glad to meet Einstein*',[9] he said. With success at the courts, dining at the high table of the European scientific elite, life was good for Robert and it may not be entirely coincidental that few months later —December 1912—Emily fell pregnant again.

12.2   Let Me Publish Be Damned!

With 'matters patent' now resolved, with Lu35 producing 11 tonnes of ammonia annually, and construction of the Oppau plant imminent, Haber felt free to turn to issues of publication. During the time he had left at

Karlsruhe, and then again on arrival in Berlin, he and his assistants[11] had completed the work on ammonia, part of which was the subject of their dissertations and which incidentally fully confirmed the accuracy of Robert Le Rossignol's experimental figures.[12] This material, together with the account he and Le Rossignol wrote in the summer of 1909 (Chap. 8), was sent to Ludwigshafen with the urgent request for permission to publish as soon as possible. So much time had passed and Haber became desperate to publicly establish his provenance in this matter. Very little of what the BASF had achieved to date had escaped into the public domain; it was still very much a secret process and the patent department at the BASF were determined to keep it so, repeatedly objecting to various passages in these papers and demanding changes. Eventually they relented and permitted aspects of his research to become public. What eventually followed was seven papers published consecutively with his assistants from 1914 through to 1915 in *Zeitschrift für Elektrochemie* under the title 'Untersuchungen Über Ammoniak: Sieben Mitteilungen', (Investigations on Ammonia: Seven Communications).[13] These publications were subsequently referred to in the Nobel Award Ceremony Speech on 01 June 1920 when Haber received the Nobel Prize. However, from the BASF's point of view these papers were largely benign; simply academic studies examining for example uranium as a catalyst or more accurate measurements for heat capacity. But the seminal work on the subject was published by Haber jointly with Robert Le Rossignol a little earlier in 1913, in *Zeitschrift für Elektrochemie* (Chap. 8), and for the first time the detail of their work in Karlsruhe was also made available to the English-speaking scientific world through the *Journal of Industrial and Engineering Chemistry* (Chap. 8), in April 1913. Even in London, *The Times* Engineering Supplement on Wednesday 28 May 1913 ran a detailed article describing this work, clearly identifying 'the Englishman', Le Rossignol as Haber's partner. However, some months earlier in September 1912, at the Eighth International Congress of Applied Chemistry in New York City, Bernthsen had already described aspects of '*his*' process for synthetic ammonia. His paper was reported in detail in the *New York Times* on 12 September 1912 under the heading;

**GIVES OUT SECRETS
OF MAKING AMMONIA**

Noted German Chemist
Ex-pounds before Congress
His Synthetic Process.

At the previous conference in London in 1909, Bernthsen had said that a solution to the synthetic ammonia problem was as far away as ever, but now he announced, 'problem solved', and he gave out the 'secrets' of making ammonia, but these were largely confined to temperatures and pressure, the candidate catalysts—osmium and uranium, thoroughly misleading—and issues of the purity of reactants, with no mention at all of the difficult aspects of technical engineering that had been overcome. He also announced that the process was patented everywhere, and that the Oppau works was under construction. In a prophetic statement he also pointed out that the *peaceful development of the various new industries for the combination of the nitrogen of the air ... is to be expected ...*. *This* then was the first real public announcement of the existence of the new technology, which he maintained had been *purposely delayed for the benefit of his brother chemists who are now assembled in this city*. From 1912 onwards then, the world should have been aware that Germany was becoming 'nitrogen self-sufficient' and the papers published from 1913 to 1915 simply confirmed this view.

12.3   The Secrets of Synthetic Ammonia Really Revealed?

The 1913 publications laid bare how the breakthrough made by the two men in Karlsruhe was achieved. Technical and engineering detail was disclosed, including the design and construction of the 'furnace', the need for a circulation apparatus, the pump and compressor, the hot and cold 'regenerators', the most successful 'contact' substances and the sources of nitrogen and hydrogen. Theoretical equilibrium yields of ammonia at various temperatures and pressures were discussed as were the details of the *practical* yields found at various flow rates through the apparatus. The % ammonia at ~ 600 °C increasing from about 3 to 5% as the pressure increases and the flow rate falls.

The papers therefore described what was ostensibly a viable process and with enough detail to allow a reconstruction by anyone interested in this most important of industrial reactions. But its whole content was of course technically obsolete. With patents already granted, and progress on the industrialisation so far advanced, the BASF's part in permitting publication can only be seen as mischievous, designed to pacify Haber but to send any new workers 'off on the wrong path'. For those who wished to follow the path however, the suggested sources of hydrogen (Chap. 8) would soon prove uneconomic and often too 'dirty', the 'contact' substances described would be unusable on an industrial scale, and the importance of the valves, the fragility of the apparatus and the inevitability of 'hydrogen attack' would all have to be discovered anew—and if they did so they would find the problems already discovered and solutions patented.

Of course, at the time of *their* experiments none of these problems had emerged, so for Haber and Robert the papers were still a true reflection of the work they had conducted—although both knew that what they presented to the world in 1913 was hardly 'de rigueur', but rather more 'passé'. Even so, just as at the Scientific Union meeting in 1910, interest was international and immense. By the end of the year, Haber was to travel to Liverpool to address the British Society of Chemical Industry and to provide the prestigious Hurter Memorial Lecture[14] which he entitled '*Modern Chemical Industry*'. For Robert, the 1913 papers together with Haber's frequent, sincere and public acknowledgement of his help, was to make his involvement in the discovery of the technology internationally known. We know that Robert kept reprints the paper (certainly the German language version) well into later life as a testament of his work (Chap. 18), and on the rare occasions that he mentioned this part of his life he did so with reference to this paper. Eventually however, his expertise in the production of ammonia in Germany was also to make him known to the Department of Munitions in the UK.

12.4 Life in Berlin, 1913

Unlike Sir William Ramsay, Haber often kept in touch with former colleagues and students, and as both he and Robert were working in Berlin, their relationship was particularly close. Robert of course had been personal assistant to Haber in Karlsruhe for some years and naturally remnants of this relationship still existed, the two men often being in contact by telephone and 'face to face'. The manifestation of this relationship can be seen in a letter Robert wrote to an old Karlsruhe colleague[15] on behalf of Haber, to whom

Koppel of course had 'special access', and relied upon to populate the Auergesellschaft with suitably talented employees. This letter also provides a rare insight into Robert's opinion of his life in Berlin at the time.

Prinzregenten Strasse,
Berlin, Wilmersdorf
2/2/1913

My dear Hodsman,

Very many thanks for your letter and my sincerest congratulations on getting Bone's Job. I suppose you are now a Herr Professor. It must be jolly nice for you to be in your native town.

I am writing to you with a special object, namely to ask you if you will come and take a job in Berlin with the Auergesellschaft. Haber telephoned me today and asked me if I thought you would be willing to take such a job, I told him that you had now left the cyanide firm and that you were in Leeds in Bones' place and so did not think you would, still he is not satisfied and wants to know if you are willing to come and what your terms would be in that case. So you had better write to the Herr Geheimerat, Faraday Weg 4, Dahlem, Lichtenfelde bei Berlin.

I don't know anything as to what they want you to do so cannot give you the slightest aid, I was with Haber this afternoon and did not hear anything about the matter.

I am getting on very well over here and am still not quite a real German. I suppose you see very little of Coates now you are away from Birmingham.

Hoping you are quite well.

Yours sincerely,

R. Le Rossignol.

This informal letter shows that Robert was reasonably aware of what was happening to former Karlsruhe colleagues 'back home' such as Hodsman and Coates. William Arthur Bone was Professor of Coal Gas and Fuel Industries at Leeds University until 1912 when he left to become Head of Department at Imperial College. Hodsman was appointed as lecturer at Bone's old department shortly afterwards, but certainly not as Robert assumed, '*Herr 'Professor*'. Hodsman's previous appointment was in industry in Birmingham where Coates was a lecturer in the University at the time, Robert naturally assuming they would have socialised. Hodsman's '*native town*' was actually York but he had graduated from Leeds University and Robert's comment that

'It must be jolly nice for you to be in your native town ...' may *just* hint of some longing after almost seven years working in Germany. Even so, Robert felt he was *'getting on very well ...'* at the 'Auer', an opinion reflecting his young family, prestigious home, financial stability and professional respect. His transition to *'a real German'* however was clearly still a work in progress. Hodsman was to remain at Leeds despite Robert's letter, as his reply to Haber in the Leeds Special Collections[15] shows. Robert on the other hand was to be bound even more closely to Germany when Emily gave birth to their second son, Peter Walter, in Berlin on the 22 August 1913.[16]

12.5   The Prelude to War

Little Peter Walter was born into troubled times. Germany was unsettled socially, politically and militarily. The 1912 Reichstag elections had resulted in the return of 110 Socialist Democrat deputies,[17] making Chancellor Bethmann-Hollweg's task of liaising between the Reichstag, the autocratic Wilhelm II, and the right-wing military high command almost impossible. To the ruling elite, the German peoples' vote upset the natural 'order of things', but they did not allow this 'red' majority to have any influence on the social, educational and governmental systems, for Germany was not ruled by the Reichstag, but by the Kaiser. The German experience was not isolated. Europe was a continent of social division, simmering resentments, jealousies and rivalries, of national anxiety, fervent nationalism and for some, a thirst for revenge. As a consequence, in the five years up to 1913 expenditure on arms by the major powers had increased by 50% and the foundations of a pan European conflict had been established. Its roots were historic.

Otto von Bismarck, the grumpy, hypochondriac Prime Minister of Prussia and architect of the young Germany had formed his 'nation' from independent states with 'blood and iron' in two devastating wars, against Austria in 1866 and France in 1871. The first brought a collection of German speaking states previously dominated by Austria, into a 'north German federation'. Having created a united assembly in the north, Bismarck determined to achieve the same in the south and so unite all of the German speaking states under Prussia. He therefore 'engineered' a war with the French by attempting to place a German prince on the throne of Spain. The French, fearful of the prospect of a war on two fronts, responded by declaring war against Prussia in July 1870 but the Prussian military machine quickly demolished her forces. There were lasting consequences of this war. Aside from territorial gains— France ceded the coal rich regions of Alsace and Lorraine to Prussia—the southern German states agreed to an alliance with their northern counterparts, resulting in the creation of Bismarck's cherished German nation. But for the French, their territorial loss to Prussia would prove to be an open sore for decades and she was 'itching' for revenge. Neither was Bismarck's work finished with the establishment of his nation. He had to ensure its survival, and this he did by forming alliances which would support Germany in times of war. The first of these was with the old foe Austria-Hungary and then with Russia. But Russia soon withdrew and allied herself with her Slavic cousins in Serbia, whilst Austria-Hungary allied with Romania to impose her will on the same 'troublesome' Slavs who had long been under her rule. Russia's position in the alliance was taken by Italy, but Italy in turn prepared a secret alliance with France in which she agreed to remain neutral if Germany attacked France. Bismarck's manoeuvrings drove other European nations to seek support amongst themselves and this shook Britain out of her position of 'splendid isolation' which she had maintained since the 1870s in an attempt to stay out of European politics. But Germany's aspiration to become a colonial nation—acquiring minor territories in the Pacific and in Africa—directly challenged Britain's global supremacy. German ambitions led to her developing a powerful navy and Britain responded accordingly by allying herself firstly with her European partners, then with Japan to limit Germany's influence in the East, and finally by building a series of immensely powerful battleships, the 'Dreadnaughts'. Between 1904 and 1907 Britain signed 'cordial agreements' with both France and Russia to 'aid one-another in time of war', but a much older treaty, the 'Treaty of London' (1839) also committed Britain to defend Belgian neutrality. Many of the alliances that arose because of the tensions in Europe may well have been just 'posturing' and they were often facilitated by the fact that the Royal families of Europe were interrelated. Whatever the reason, a tangled web of inter-dependencies sprung up in the wake of the newly united German nation.

Even by the turn of the century, the seeds of pending instability in Europe were already sown. Tensions arose between Great Britain and Germany because of the British wars against the Boers in South Africa. Problems continued in 1905 when a dispute between Russia and Japan over the failure to recognise each others interests in Manchuria and Korea led to a humiliating defeat by the Japanese fleet in the naval battle of Tsushima. President Theodore Roosevelt subsequently intervened and facilitated a peace agreement between the two countries but with considerable gains accorded to Japan. The scale of Russia's defeat caused social unrest and contributed to an attempted revolution in 1905. Russia was divided and damaged. The Tsar was determined to find a way to unite the country and what better way than through military conquest? From then on, Russia was 'up for it', and '*it*' could be provided—if necessary—by the simmering cauldron of the Balkans.

Trouble in the Balkans was nothing new. Here, Europe met the East and in doing so it generated a volatile ethnic mix viz, a Christian Slavic majority and a Muslim minority. Between 1912 and 1913 a series of wars between the major nations in the area saw a redistribution of territories and eventually a fragile peace, but nothing had really been settled and tensions remained high. The collection of small nations that had found themselves under Turkish or Austro-Hungarian rule for many years stirred themselves in nationalistic fervour, and while they sought their own individual voice and self-determination, the majority were nevertheless united in identifying themselves as pan-Slavic peoples, with Russia as their chief ally. She in turn was keen to encourage a belief in the Russian people as the Slav's natural protectors, and this was a convenience by means by which Russia could intervene in the Balkans to regain a degree of lost prestige—if she so wished.

Meanwhile, continuing German expansion was 'ratcheting up' tension between herself and Great Britain. Even within Germany tension was high. The desire for concrete political reforms expressed by the people in the 1912 elections was ignored by the conservative ruling classes. Instead, they seized upon demonstrations of national strength as a tactic to rally popular support against a rising tide of socialist sentiment.[18] Chancellor Bethmann-Hollweg came to believe that Germany's only hope of avoiding civil unrest lay in a short, sharp war, or even a European-wide conflict if it resolved Germany's political woes.[19] To this end, and with France and Russia in mind, Germany had her 'Schlieffen Plan', devised by former Army Chief of Staff Alfred von Schlieffen and carefully crafted to deal with a two-front war scenario. Firstly, if provoked, she would conquer France within five weeks on a 'western front' before Russia—whose military frailty against Japan in 1905 had already been noted in Berlin—could effectively mobilise for war on an 'eastern front' (estimated to take *six* weeks). Once France was despatched, Russia would be addressed. By 1913 Europe was a tinderbox of emotion reflected in a network of poorly thought out alliances between powerful nations eager to confront one-another to resolve internal socio-political problems. With tensions high and plans already laid, sooner or later some spark would inflame the tinder and set in motion a mindless mechanical prosecution of the obligations in the treaties. Europe didn't have long to wait for her spark. By little Peter Walter's first birthday the continent would be aflame.

12.6   But in the Meantime …

Even so life went on, and with Robert deepening his understanding of the engineering of glow lamps at the Auer and Bosch progressing the industrialisation of the ammonia synthesis at the BASF, Haber spent July and August 1913 in Karlsbad recovering from 'gallstones and moodiness'.[20] A few months later he had regained his spirits and left Germany to present the *Hurter Memorial Lecture* at the University of Liverpool on Wednesday, 26 November 1913.[14] His talk was eclectic and began by praising the role of the British chemical industry in creating important industrial processes such as the Leblanc soda process, the manufacture of sulphuric acid, of Glaubers salt, bleaching powder, soda, potash, the alkalis, and developing the sound analytical controls which allowed the management of such processes on a large scale. He continued by discussing modern improvements in enamelling techniques using zirconium oxide, progress in gas illumination involving cerium, didymium and lanthanum, together with the use of Thorium in 'Welsbach' (i.e. 'Auer') mantles, and he looked at organic chemistry via the development of synthetic dyes and medicines. Radioactive sources were discussed—again in terms of their medical applications—as was the question of why the blow pipe torch was hotter than the Bunsen burner. This he did with reference to the paper he and Robert Le Rossignol had published at Karlsruhe. (Chap. 8) Finally, Haber demonstrated a novel successor to the 'Davy lamp' viz, the 'firedamp whistle' which he had developed with Dr. Leiser in Germany, and by means of which 'firedamp'—largely methane—could be detected acoustically in a coal mine without the use of a flame[21]—an audible alarm, or 'trill', indicating its presence.

However, neatly buried in-between these concerns Haber addressed the fixation of nitrogen, firstly through the arc process, then via ammonia by-product recovery, and finally from the synthesis of ammonia itself. He discussed the sources of fixed nitrogen in detail, indicating that in Germany by-product ammonia recovery from coke was 'almost universal'. He observed that in the UK the older 'beehive' ovens, which lost the whole of the ammonia, were still not completely replaced and 'up until now', by-product recovery was only employed to a small extent in the USA. In retrospect, considering the tensions in Europe at the time and Bernthsen's talk in New York in 1912, Haber's paper should have aroused further concern that Germany was making considerable progress towards some kind of broad-based fixed nitrogen self-sufficiency—with all that such a progression would imply. Indeed, the following passage from his talk regarding the state of the ammonia synthesis showed that an important component in this progression had already been achieved, built of course upon the earlier work by himself and Robert Le Rossignol—whose contribution he always publicly acknowledged;

> This is the method that was carried through first on a small scale and is now being applied on a large scale … and I cannot emphasise enough the valuable aid of your countryman Robert Le Rossignol who was working with me on this question … The Badische Analin and Soda Company have taken over and further developed our results. Starting from the observation that contaminations in certain cases increase the catalytic activity in such heterogeneous reactions, they have been able to raise the activity of poorer catalysts to that of osmium and uranium. The result was only obtained after a careful study of the influence of small impurities, some of which are useful while others, even when present in the smallest traces, act as poisons. They overcame certain difficulties in the construction of high-pressure furnaces which became apparent by prolonged working. They developed the purification methods for the hydrogen which became necessary with the replacement of our electrolytic gas by the impure hydrogen from coal. It seems the task of managing the process on the largest scale has now been carried through satisfactorily …

In this passage of considerable understatement, he revealed that the 'problem of the catalyst' had been solved, 'certain difficulties' with the high-pressure furnaces had been overcome, and sufficiently pure hydrogen could be produced from Germany's abundantly available coal reserves. Unsurprisingly, no details were disclosed, but his message was clear; in Germany, the technology that he and Le Rossignol had created now formed the beating heart of this most important industrial process.

12.7   Finally, Sarajevo 28 June 1914

Historians have long studied the causes of the first world war. Indeed, it is one of the most studied areas in all of world history and scholars have differed considerably in their interpretation of the event. But there seems little doubt that what triggered the war was the assassinations of the Archduke Franz Ferdinand of Austria and his wife Sophie in the Bosnian capital of Sarajevo. The Archduke—heir apparent to the Austria-Hungary empire but unloved even in his own country—had travelled there as Inspector General of the army to review manoeuvres on the invitation of the Austrian governor of Bosnia, General Oskar Potiorek. Sarajevo was the capital of Bosnia

Herzegovina and her fervent nationalist elements had wanted to unite with Serbia to create an independent state of Greater Serbia or 'Yugoslavia' and rid herself of Austrian rule, so the Archduke was acutely aware of the dangers of his visit, even being *ordered* to go there by his father, the old emperor Franz Joseph. Knowledge of his visit allowed elements of the Serbian Military Intelligence to organise an assassination attempt whose political objective was to break off Austria-Hungary's south-Slav provinces to form the new 'Yugoslavia'. A number of young Bosnian Serb nationalists were chosen, trained as assassins and provided with the necessary bombs, pistols, safe-houses, and agents. While riding in the motorcade through the streets of Sarajevo on 28 June 1914, Franz Ferdinand and his wife Sophie were shot by Gavrilo Princip, just hours after surviving an earlier attempted assassination by another member of the group of dissidents. Princip fired two shots at close range. The first hit the Archduke in the jugular vein, the second inflicted an abdominal wound on Sophie, and both victims were dying while being driven to the Governor's residence for treatment. Franz Ferdinand's last words were '*Sophie, Sophie! Don't die! Live for our children!*' followed by six or seven utterances of '*It is nothing …*'.

Subsequent to the assassination, the obligations embodied in the various treaties fired off a chain reaction of declarations of war. Austria-Hungary, declared war on Serbia on 28 July 1914. Russia, bound by treaty to Serbia, announced mobilisation of her army in her defence. Germany viewed the Russian mobilisation as an act of war against Austria-Hungary, and declared war on Russia on 01 August 1914. France, bound by treaty to Russia, found itself at war against Germany and, by extension, against Austria-Hungary following a German declaration against her on 03 August 1914. Germany quickly executed her 'Schlieffen Plan' and invaded neutral Belgium on 04 August 1914 to reach Paris by the shortest possible route. Britain declared war against Germany the same day—her reason being the 'Treaty of London'. Like France, she was by extension also at war with Austria-Hungary. With Britain's unexpected entry into the war, her colonies and dominions abroad offered military and financial assistance. United States President Woodrow Wilson however declared a policy of absolute neutrality. Japan, honouring her military agreement with Britain, declared war on Germany on 23 August 1914. Two days later Austria-Hungary responded by declaring war on Japan. Finally, Italy, although allied to both Germany and Austria-Hungary, was able to avoid entering the fray by citing a clause in her secret treaty with France whereby she evaded her obligations and declared instead a policy of neutrality.

The events of July-August 1914 of course were to affect the lives of millions of people throughout Europe and the world. Haber and Le Rossignol were no exception. For Fritz Haber his role in the coming war was to eclipse his work on fixation, and re-define his life to future generations. For Clara, her husband's influence on the war was to prove both morally obscene and personally catastrophic. For Robert and Emily too, these were uncertain times. Their commitment to one-another through marriage in 1910 had already isolated them in the eyes of the German state. Under the German citizenship law of 1870[22] and the revised version of 22 July 1913, regulations specified that German women would lose their nationality by marrying someone without German nationality—an embarrassing relic of an era when Germans sought to protect their supposed racial purity. Under the citizenship laws, naturalization was possible but it required living in Germany, being economically independent, and without a criminal record. All of this would have easily applied to Robert but we know that at the time he was still '*not quite a real German*'—a fact soon to be noted by the German state. Emily therefore had lost her legal status as a German, and this by extension applied to John Augustin and little Peter Walter. The Le Rossignols may well have considered leaving Germany prior to the war. But the family's position in terms of citizenship in a country in political and social turmoil—but still technically 'at peace'—had to be balanced against their prospect of significant legitimate financial gain. Europe of course had been simmering for years, but the repercussions of the events at Sarajevo caught everyone by surprise. As late as 28 July 1914, Haber felt confident enough to apply to his Minister of Education for a six-week holiday, with the qualification that;

> If the political situation becomes such that our nation is pulled into a war-like entanglement, then I intend to return from this holiday.[23]

He never had his holiday. War was declared just four days later. The Le Rossignols, completely surprised by the British involvement, were still in Berlin and Emily was no longer a German.

Notes

1. However, at the same time he was comfortable with applications of Planck's quantum theory, the significance of which in chemistry Haber was one of the first to recognise. It formed the basis of most of his Dahlem work. By understanding the application of quantum theory to the photoelectric effect Haber was able to propose a quantum theory of chemical reaction heat. J.E.

Coates, 'The Haber Memorial Lecture', *J. Chem. Soc.*, 1642–1672, (1939), p. 1656.

2. Dietrich Stoltzenberg, *Fritz Haber. Chemist, Noble Laureate, German, Jew.* Chemical Heritage Press, Philadelphia, Pennsylvania, (2004), p. 94.

3. Vaclav Smil, *Enriching the Earth,* MIT Press, (2001), p. 100.

4. Stoltzenberg, op. cit. (note 2), p. 94, his reference no. 44.

5. Nernst's unexpected support of Haber had even more ironic consequences when the Haber-Le Rossignol data was subsequently used in recognising the robust verification of the 'heat theorem' which helped win Nernst the 1920 Nobel Prize—although no mention of either man was made in his acceptance speech. Diana Kormos Barkan, *Walther Nernst and the Transition to Modern Physical Science.* Cambridge University Press, (1999), p. 216. See also Chaps. 15 and 16 regarding Le Rossignol's contributions to other Nobel Prizes.

6. We know from Robert's discussion with Jaenicke in 1959, that 'Haber did the whole of the patent matter …' and so Robert was probably not part of the 'team' that day. Jaenicke's and Robert's letters and the transcript of their conversation are held at the Max Planck Gesellschaft, Archiv der MPG, Va. Abt., Rep. 0005, Fritz Haber. Haber Sammlung von Joh. Jaenicke. Nr. 253 (Jaenicke's letters) and Nr. 1496 (conversation transcript).

7. Stoltzenberg, op. cit. (note 2), p. 95, his reference no. 47.

8. Stoltzenberg, op. cit. (note 2), p. 95, his reference no. 48.

9. From Robert's conversation with Joh. Jaenicke on 16 September 1959. Jaenicke's and Robert's letters from this period and the transcript of their conversation are held at the Max Planck Gesellschaft, Archiv der MPG, Va. Abt., Rep. 0005, Fritz Haber. Haber Sammlung von Joh. Jaenicke. Nr. 253 (Jaenicke's letters) and Nr. 1496 (conversation transcript).

10. Ralph C. Chirnside's sensitive Obituary of Robert appeared in; R. C. Chirnside, 'Robert Le Rossignol, 1884–1976', *Chemistry in Britain,* **13**, 269–271, (1977). Chirnside was a long-time friend and colleague of Robert from their time together at the GEC laboratories. The Obituary was based upon a tape-recorded interview he conducted with Robert on 29 March 1976.

11. Principally Setsuro Tamaru, but also Ch. Ponnaz, A. Maschke, L.W. Oeholm and H.C. Greenwood. Some of these assistants can be seen on the 1908–09 staff group photos at Karlsruhe. See for example Plate 8.6, Chap. 8.

12. Coates, op. cit. (note 1), p. 1656.

13. Stoltzenberg, op. cit. (note 2), p. 93, his reference no. 40. See also Margit Szöllösi-Janze, *Fritz Haber 1868–1934: Eine Biographie.* C. H. Beck, Munich, (1998), p. 866 and her references for publications between 1914 and 1915.

14. Ferdinand Hurter (15 March 1844–12 March 1898) was a Swiss industrial chemist who settled in England. He also carried out research into photography. F. Haber, *Hurter Memorial Lecture,* University of Liverpool,

Wednesday 26 November 1913. Published in *J. of the Society. of Chemical Industry*, **2**, XXXIII, (31 January 1914).

15. Leeds University Library, Special Collections MS 705, *Papers of Henry James Hodsman, (1886–1951)*, with whom Haber had published earlier in 1909 (see references in Chap. 12) Born in York and educated there at Archbishop Holgate's School, Hodsman graduated in chemistry at the University of Leeds in 1906. He took his MSc a year later and then, supported by the award of an 1851 Exhibition, studied applied chemistry at the Technische Hochschule Karlsruhe under Haber and at the Sorbonne in Paris under Henri Le Chatelier. He returned to England and after a short period of employment in industry (at the '*cyanide firm*' in Birmingham), rejoined the University of Leeds as a lecturer in 1912. He eventually became senior lecturer in the Department of Coal, Gas and Fuel Industries. He died on 31 January 1951 shortly before his planned retirement.

16. Curiously however, there was no announcement of Peter Walter's birth in *The Times*. His birth date was found for me by Nicholas Rogers, Archivist at Sidney Sussex College, Cambridge, in response to a request of mine by email. I am grateful to Mr. Rogers. See also Chap. 16. Peter was probably born at 108 Prinzregenten Strasse.

17. 34% of the vote overall but 75% in 'red Berlin' itself.

18. Daniel Charles, *Between Genius and Genocide*, Jonathan Cape, London, (2005), p. 127.

19. The historian Fritz Fischer found evidence in the archives in east Prussia in the 1960s showing that Germany was pro-active in going to war, wanting to secure her eastern and western borders for all time. He rocked the history profession with his first post-war book; *Griff nach der Weltmacht: Die Kriegzielpolitik des kaiserlichen Deutschland 1914–1918* (published in English as *Germany's Aims in the First World War*), in which he argued that Germany had deliberately instigated World War I in an attempt to become a world power. This went counter to the widely held feeling in Germany at the time that she was 'blameless'.

20. Stoltzenberg, op. cit. (note 2), p. 127.

21. In his paper, Haber maintained that the statistics showed that over half the explosions in the mines were due to the flame of the Davy lamp. His link with the mines may well have been inspired by the BASF's dependence on coal for its hydrogen source. But Coates, op. cit. (note 1), suggests that it was at the suggestion of the Emperor. Whichever, the whistle was not a commercial success.

22. Based largely on the model of the Prussian Subjects Law of 1842 which stated that except through later naturalisation, citizenship could only be attained by descent from a Prussian father, or in the case of an illegitimate birth, from a Prussian mother. See also Mathew Stibbe, *British civilian internees in Germany. The Ruhleben camp, 1914–1918*. Manchester University Press,

(2008), p. 28, and Eli Nathans, *The Politics of Citizenship in Germany. Ethnicity, Utility and Nationalism.* Bloomsbury Academic, 2004. ISBN-13 978-1859737767, ISBN-10 1859737765, pp. 169–198.

23. Stoltzenberg, op. cit. (note 2), p. 127, his reference no. 9.

13

Haber and the Chemist's War 1914–18

'Fixation' and gas warfare[1]

'The heathen civilisation of poison gases ... [from] ... the ammonia-breathing Germanic war god' ...

Joseph Roth. The Auto-*da-Fé of the Mind*, 1933.

13.1   Something Wicked This Way Comes ...

Great Britain's declaration of war on the evening of Tuesday 04 August 1914 came as a complete surprise to Germany. Throughout July, His Majesty's Government and its Foreign Minister Sir Edward Grey, attempted to mediate the quarrelsome parties, in turn trying to remain aloof from the dispute. To many this appeared as British 'dithering'. It was only as the war began that the British position suddenly solidified into support for Belgium, predicated upon the arcane, 'Treaty of London', and the fact that a German army just across the channel was far too uncomfortable to countenance. The general feeling at the time was that, had Britain come out clearly on the side of Belgium and France earlier in July, Germany would have instructed Austria-Hungary to settle her differences with Serbia and war could have been avoided. With no such indication, Germany believed Britain would stay out of the war, limiting herself to diplomatic protests. Germany and Austria-Hungary subsequently proceeded under the belief that war would be fought solely against France and Russia. The official report[2] by Sir Edward Goschen, British Ambassador to Berlin, regarding the breaking of diplomatic relations with Germany on 04 August 1914 describes in vivid detail the 'betrayal' Germany felt at Great Britain's position.

© Springer Nature Switzerland AG 2020
D. Sheppard, *Robert Le Rossignol*, Springer Biographies,
https://doi.org/10.1007/978-3-030-29714-5_13

During the afternoon ... I . informed the Secretary of State that unless the Imperial Government could give the assurance by 12 o'clock that night that they would proceed no further with their violation of the Belgian frontier and stop their advance, I had been instructed to demand my passports and inform the Imperial Government that His Majesty's Government would have to take all steps in their power to uphold the neutrality of Belgium ... Herr von Jagow [Foreign Minister] replied that to his great regret he could give no other answer than that which he had given me earlier in the day. I then said that I should like to go and see the Chancellor, as it might be, perhaps, the last time I should have an opportunity of seeing him. He begged me to do so....

I found the Chancellor very agitated. His Excellency at once began a harangue, which lasted for about twenty minutes. He said that the step taken by His Majesty's Government was terrible to a degree; just for a word - 'neutrality', a word which in war time had so often been disregarded ... just for a scrap of paper Great Britain was going to make war on a kindred nation who desired nothing better than to be friends with her.... What we had done was unthinkable; it was like striking a man from behind while he was fighting for his life against two assailants. He held Great Britain responsible for all the terrible events that might happen. I protested strongly against that statement ... it was, so to speak, a matter of 'life and death' for the honour of Great Britain that she should keep her solemn engagement to do her utmost to defend Belgium's neutrality if attacked. The Chancellor said, 'But at what price will that compact have been kept. Has the British Government thought of that?'

After this somewhat painful interview I returned to the embassy ... At about 9.30 p.m. Herr von Zimmermann [Under-Secretary of State] came to see me ... he asked me ... casually whether a demand for passports was equivalent to a declaration of war. I said that such an authority on international law as he was known to be must know ... what was usual in such cases ... Herr Zimmermann said that it was, in fact, a declaration of war ...

13.2   Demise of the Schlieffen Plan ... 'Trench Warfare'

The instant Great Britain declared war on Germany the 'suspect'[3] Schlieffen plan was utterly compromised. The engagement of a third antagonist capable of committing millions of 'boots on the ground' and whose powerful navy was able to blockade German access to her most vital imports was never a consideration in the plan, hence subsequent German fury with Goschen. But the plan had already been executed and throughout August 1914, Germany persisted with her invasion of Belgium. In the meantime, Great Britain sent a 'small' 'Expeditionary Force' of 75,000 regular army men to France, who in turn had began the mobilisation of millions of her own recruits. The Belgian army was only a fraction of the size of the German army, yet it delayed the Germans for nearly a month.[4] Eventually the Germans destroyed the troublesome Belgian forts in Liège, Namur and Antwerp, and when Antwerp was occupied, 50,000 tons[5] of Chile saltpetre was secured from its harbour helping to replace the rapidly dwindling supplies from the import warehouses at Hamburg. At current consumption levels however, such a tonnage would last only until early-to-mid 1915.[6] Still, despite stiff Belgian resistance, the Schlieffen plan was proceeding reasonably well and these reserves should have been adequate. The Germans however failed to anticipate how the delay in Belgium would assist the French mobilisation via her railways. The effectiveness of this moblisation meant that by the time the Germans left Belgium the French and British were waiting for them. The French commander-in-chief, Joseph Joffre, placed the BEF on his left flank, where he believed there would be little fighting. 'Au contraire'. Due to the rapid German advance through Belgium, the small British force took a hiding, but along with the French they provided enough resistance as they fell back towards the Marne river east of Paris to entice the Germans to modify 'Schlieffen'. The Germans turned away from Paris, in an attempt to envelop the retreating allied armies, exposing their right flank and making possible the 'Miracle of the Marne'.[7]

Joffre immediately realised the German armies' tactical error. On 04 September, he made plans to halt the withdrawal and attack the Germans all along the front. In preparation for the attack the German First and Second armies 'wheeled' to face the Allies and in doing so opened up a 30-mile gap— a gap spotted by allied reconnaissance aircraft. The Allies were prompt in exploiting the break in the German lines, pouring in troops between the two armies. On 08 September, the French launched a surprise attack against the German Second Army, serving to further widen the gap. By 09 September, it looked as though both German armies would be totally encircled and destroyed. The German General von Moltke suffered a nervous breakdown upon hearing of the danger and is said to have reported to the Kaiser: '*Your Majesty, we have lost the war!*' His subordinates took over and ordered a general retreat to the Aisne river in northeastern France to regroup. The Germans were pursued by the French and British but halted their retreat after 40 miles at a point north of the river. The German defeat between the 09 and

13 September marked the abandonment of the Schlieffen plan. In the aftermath of the battle which brought about an horrendous loss of life[8] and an enormous expenditure in munitions, both sides 'dug in'; the Germans to halt their retreat and re-group, and the Allies to prevent another lightening attack on France. As a consequence, four years of pitiless 'trench warfare' ensued, a 'gunner's gourmet', for the sniper, the rifleman, the machine gunner, mortar man and the bone-breaking, bunker-busting artillery. These trenches were to eventually stretch 450 miles from the North Sea to the Swiss border, and each day the two sides bombarded one another mercilessly. And at the beginning of the conflict, through the winter of 1914 when if it wasn't raining it was snowing, near the small Belgian town of Messines which lay at the top of a 7-mile ridge running up from Flanders, all this was watched by a 25-year-old German messenger-runner named Adolf Hitler.

The German defeat ended any hopes of a quick victory in the West. As a result, she was forced into a long and costly war on two fronts, especially so as Russia mobilized much more quickly than anticipated and threatened east Prussia. Neither the military nor industry was sufficiently prepared for such a circumstance. Wars at the time were mobile, relatively short-lived affairs, usually lasting a matter of weeks or a few months, in which the consumption of the vital resources of men, munitions and equipment was 'manageable'. There had never been a 'world war' before. Germany therefore, despite her military traditions and after only four or five weeks of conflict, found herself in unknown territory and her military's demands rose to an unforeseen degree. By the beginning of November 1914, the ammunition required by the German armed forces consumed an unprecedented 20,000 metric tonnes of sodium nitrate per month, a level at which—without access to her usual South American sources—her chemical industries were utterly incapable of sustaining.

13.3   Away from the Front

To many at the time it seemed that Germany—and the Allies—had 'sleep-walked' into the first world war, although the British public soon came to believe that the war was necessary. The German General Staff had made no provision at all for a protracted engagement, tacitly assuming that their needs would be served by industry in war as in peace. The general feeling on both sides of the conflict was that … '*It'll all be all over by Christmas!* …'[9] In the event of course it wasn't, and whereas the allies were initially largely immune from issues of resource procurement, the (completely illegal) blockade of

Germany's mercantile shipping by the powerful British navy made her position critical. This scenario however had long been anticipated by some in Germany's scientific community. As early as 1903, Ostwald had raised the issue of the influence of a British naval blockade on Germany's ability to defend herself, and highlighted his 'new' process which oxidised ammonia to produce nitric acid. Early on in the war then, it was not surprising that German scientists, economists and industrialists realised the military predicament and organized to lobby the Minister of War to place industry on a firm war footing—even suggesting that by excluding industry from their preparations, the German General Staff had played into the hands of the enemy. To emphasise their influence, on the morning of 04 October 1914, people across Germany opened their newspapers to find the Manifesto, *'Proclamation to the Civilised World!'* signed by 93 members of the German intelligentsia—including Fritz Haber—reflecting the bitterness of the German government and defending the invasion of Belgium. Principal signatories were Emil Fischer the scientist, and Walther Rathenau the economist, who had already proposed a system of controls to the Minister of War to allow the efficient organisation and distribution of important raw materials barred from import by the British blockade. As a result, Rathenau was appointed as Head of a new 'Board of Wartime Raw Materials' and he quickly appointed a select group of scientists and industrialists to help him—Fritz Haber being appointed head of the Chemistry Department of the Board. They immediately drew up an inventory of raw materials available for the war effort. The result appalled them, suggesting that the war at its current level could be continued for six months at most—foremost amongst their concerns being the stockpile of nitrates from Chile,[10] and this concern had already led to an early, but bizarre, military encounter between the British and German Naval fleets.

13.4 The Battle of Coronel

On 05 October 1914, the day after the proclamation was published, the British intercepted a German radio communication and learned of Vice-Admiral Maximilian Graf Von Spee's plan to prey upon shipping in the crucial trading routes along the west coast of South America. Patrolling in the area at that time was Rear-Admiral Sir Christopher Cradock's South Atlantic Squadron, composed entirely of either obsolete or under-armed vessels, crewed by inexperienced naval reservists. In contrast, Von Spee had a formidable force of five modern vessels which gave them an overwhelming advantage in range, speed and firepower. This advantage was compounded by the fact that the professional crews of these ships had earned accolades (the 'Kaiser's Prize') for their gunnery efficiency prior to the war. On 01 November 1914, off the coast of central Chile near the city of Coronel, and in foul weather, the squadrons engaged[11] and the British lost several vessels including two cruisers and sixteen hundred sailors. Von Spee's casualties amounted to just two wounded. For a short period after the battle, the German navy controlled the shipping on the west coast of South America, to such an extent that insurance on British nitrate ships at the time was refused. A U.S. military expert of the day observed that '*To strike at the source of the Allied nitrate supply was to paralyze the armies in France… the destruction of a nitrate carrier was a greater blow to the Allies than the loss of a battleship …*'.[12]

But German success came at a cost; her superiority was short-lived and her end brutal. Von Spee's squadron had used half of its ammunition, and he had no prospect of replenishing it, for Germany's Pacific territories had already been occupied by the British and allied navies. Von Spee was therefore recommended to return to Germany. British reaction to her defeat at Coronel however was to send a large force to track down and destroy the German cruiser squadron. With no base in the Atlantic or Pacific, Spee it seemed was forced to raid the coal bunkers in the Falkland Islands for fuel to return home. The move was 'anticipated'[13] by the 'Brits', and at the subsequent engagement on 08 December 1914 the German squadron was destroyed. Ten British sailors were killed during the battle and nineteen wounded; none of the British ships were badly damaged, but 2,200 German sailors were lost in the encounter, including Admiral Spee and his two sons. As a result of the action German access to the saltpetre beds of Chile was totally denied. Germany's need for nitrate—already desperate—therefore became acute whilst the Allies quickly doubled the number of their freight vessels to the region.

13.5   'Fixation', Haber, and the Chemist's War

The defeat of Spee's squadron hammered the final nail into Germany's saltpetre coffin, and at the end of 1914 her leaders were forced to form a national 'nitrate commission' to examine ways in which this vital resource could be procured from within her own borders. As head of the Chemistry Department of the Board of Wartime Raw Materials, Fritz Haber was naturally invited to serve, and at the first meeting he immediately declared that he had a 'financial interest' in the nitrogen question. Even before the war had started he was already receiving substantial royalties from the BASF.[14] Just how 'interested' he was however never seemed to be an issue and although the minutes of the meeting recorded that the Commission 'took notice' of his declaration, he was still asked for his continued participation.[15] The Commission obviously realised that there was just one practical form of fixed nitrogen available to Germany viz, ammonia, but there were three sources; by-product recovery from coking plants, the cyanamide process, and the BASF's new direct ammonia synthesis from her modern plant at Ludwigshafen-Oppau. The allied blockade however had an additional affect on the German chemical industry. By also preventing her imports of 'pyrites' (FeS, or Iron(II) Sulphide) —the raw material from which sulphuric acid was synthesized—ammonium sulphate fertilizer could not be produced so easily. The war therefore cut off the German fixed nitrogen industries' traditional markets and they began to suffer seriously declining revenues. As far as the BASF was concerned, without sulphuric acid most of her Ludwigshafen-Oppau plant would be out of production by the end of 1914. The blockade therefore immediately drove the whole industry straight into the arms of the Nitrate Commission and they shifted *en masse* from the production of fertilizer to the production of nitrates for munitions via Ostwald's process. The link between 'fixation' and the ability to conduct war precipitating, in part, the description of the first European conflict as the '*Chemist's War*'.

13.5.1   The Industrialization of Ostwald's Oxidation of Ammonia

Ostwald's process began with the oxidation of ammonia[16] in the presence of a platinum catalyst at ~ 900 °C according to;

$$4NH_3(g) + 5O_2(g) \rightarrow 4NO(g) + 6H_2O(g) \text{ with } \Delta H = -905.2 \text{ kJ mol}^{-1}$$

of ammonia converted, with an almost complete conversion at just 4–10 atmospheres. The step was also strongly exothermic, making it a useful heat source. The nitric oxide (NO) was easily oxidized to nitrogen dioxide (NO$_2$) according to;

$$NO(g) + 1/2 O_2(g) \rightarrow NO_2(g) \ \Delta H = -114 \text{ kJ mol}^{-1}$$

This gas was absorbed by water, yielding (dilute) nitric acid, whilst reducing a portion of it back to nitric oxide which was recycled:

$$3NO_2(g) + H_2O(l) \rightarrow 2HNO_3(aq) + NO(g) \text{ with } \Delta H = -117\,kJ\,mol^{-1}$$

of NO_2 converted. The acid was subsequently brought to the required concentration by distillation. Nitric acid was then used to provide the propellant 'cordite'[17] along with the high explosives 'picric acid' ('tri-nitro phenol', or 2-hydroxy-1,3,5-trinitrobenzene) and 'TNT' ('tri-nitro toluene', or 2-methyl-1,3,5-trinitrobenzene). In addition, by using 'soda' (sodium carbonate) from the Solvay[18] process according to,

$$Na_2CO_3 + 2HNO_3 = 2NaNO_3 + H_2O + CO_2$$

the acid could also be used to produce sodium nitrate for conversion to gunpowder, the whole helping to eliminate Germany's dependence on Chilean nitrate imports by now making *gunpowder from air*.

Ostwald had established the feasibility of the catalytic oxidation of ammonia just after his failed attempt to patent the synthesis of the compound (Chap. 3). On 18 November 1901 he offered the detail to the Fabwerke Hoechst in a somewhat patriotic but perceptive proposal believing that commercialization of the process would obviate the threat of a British naval blockade which would otherwise deny Germany access to Chilean nitrates in time of war (Chap. 7). Ostwald's attempt to patent the method became known in 1903 and the BASF immediately objected to his claim citing numerous independent papers and its own research on high temperature catalytic processes. By 1907 Ostwald had withdrawn his claim and so there were no legal barriers to catalytic oxidation of ammonia, whichever way it was obtained. As usual however, there were technical barriers, but uniquely amongst the German chemical industries, the BASF had already made some progress in the matter, and along with her modern plant and Haber's 'nudging' of her case amongst the bureaucrats of Berlin she was well placed to 'trump' her competitors in the by-product and cyanamide industries—although output from these too was to rise sharply during the war.

As early as March 1914, the BASF had conducted a limited series of laboratory exercises to produce nitric acid from ammonia by oxidation. They had already found a more available and cheaper catalyst—based once again on iron oxides—to replace Ostwald's platinum which was difficult to procure because of the war, and Bosch along with Mittasch therefore knew that the process was feasible—although at the time their yields were insignificant. But the reaction occurred in the gas phase at just a few atmospheres pressure so there were huge volumes to deal with in the 'contact chambers'. The product too—an aqueous solution of nitric acid—was highly corrosive so all the piping, valves, coolers and pumps etc. would have to be fashioned from the relatively new chromium-nickel stainless steel, but both men felt that even without the benefit of a pilot plant, and faced with yet another huge engineering challenge, it would still be possible to switch Oppau's production from ammonium sulphate to nitric acid and hence sodium nitrate.

13.5.2   The 'Masters of War'

In September 1914, Bosch visited the War Ministry who were keen to add the BASF's new ammonia synthesis to the more established processes. Although Bosch found the Ministry utterly ignorant of the chemical requirements of a modern war, he also detected a natural partnership here, a kind of symbiosis; a nation at war desperately needing a product, and a group of companies equally desperately seeking new markets for a product whose old markets had disappeared. This partnership was to herald the rise of the German military-industrial complex[19]—the *Masters of War*—who were to have such an influence on subsequent European history. Bosch however had already learned to be bold where ammonia was concerned, and emphasing the essential bond between the nation and her industries, by the beginning of October 1914 he made what was to become known as the 'saltpetre promise', where the BASF—in an audacious attempt to grab the fixed nitrogen business for itself—offered not only to convert and expand the Oppau plant but also to build a new plant which, after a just few months, could provide 5,000 tonnes of sodium nitrate per month from synthetic ammonia. $60,0000t\ y^{-1}$, equivalent to 9,900t *fixed* N, or 12,000t NH_3.[20]

Under 'normal' circumstances, the funding for the expansion of Germany's nitrate industry would have been expected to borne by the industry itself as an investment. After all it was a commercial venture. By December 1914 as the military's need for munitions became 'ravenous', the BASF initially took the conventional route and agreed to fund the quadrupling of its ammonia production at Oppau to 37,500 tonnes per year,[21] and by May 1915 the first deliveries for military use began. But in early 1915 when Haber on behalf of the nitrate commission pleaded with Ludwigshafen to double that *again* to 80,000 tonnes per year, the BASF 'froze', for this was commercial madness. Even as the flower of Europe was being slaughtered on the fields of France, at some point the war would be over, leaving the BASF with a chemical 'monoculture' and a grotesque overcapacity. Bosch therefore now insisted that the government—who of course precipitated the war—carry the burden and pay for any further expansion. The subsequent negotiations dragged on through the summer and became acrimonious, especially so as Nikodem Caro entered the arena and offered to expand ammonia production using the cyanamide process. But with Haber embedded at the government's tables in Berlin[22] the BASF was kept informed of progress and the developing government consensus was that only the Haber-Bosch process was capable of providing the necessary tonnage of fixed nitrogen. As a consequence of this 'insider' knowledge, the BASF held a firm negotiating line.

By April 1916 the government acceded to Bosch's demands and they provided a loan of 12 million Marks for the new ammonia plant, but by now the 'goalposts' had shifted once again and Bosch had to promise an annual capacity of 38,500t NH_3 (31,700t *fixed* N or 192,500t $NaNO_3$)—over *three* times the original saltpetre 'promise'—simultaneously guaranteeing a reasonable fixed price for the product. The new plant, modeled on Oppau, was eventually built at Leuna in central Germany far away from the increasingly frequent French air attacks[23] that were beginning to plague Oppau as the Allies finally realised the extent of Germany's fixed nitrogen production.

Leuna lay in the heart of Germany's 'lignite' (brown coal) mining region in Saxony-Anhalt near the town of Merseburg. The river Saale lay nearby and provided water whilst a local mine provided gypsum ($CaSO_4 \cdot 2H_2O$) to help replace the sulphuric acid needed for ammonium sulphate fertilizer. Leuna was carefully chosen by Bosch, not only to serve Germany in wartime but also to facilitate a conversion to fertilizer production after the war, and construction began there on 19 May 1916. But by December 1916 the continued pressures of the war ratcheted up the capacity once again and a decision was made to build a plant that quadrupled the previously agreed annual production to 130,000t *fixed* N—entirely vindicating Bosch's insistence on the government taking responsibility for the funding of any expansion. On 27 April 1917 the first converter came 'on line' and produced ammonia a day later. The plant was therefore viable just eleven months after construction began, an amazing feat of engineering considering the strictures of war. Further expansion to 200,000t *fixed* N per year was proposed in July 1918 but this was only achieved post-war in 1923. Soon after, Leuna became the largest chemical plant in the world (Plate 13.1).

Plate 13.1   The sprawling Leuna works looking south in 1920 with its characteristic set of thirteen chimneys for the boiler buildings as the plant now used steam rather than gas to run the synthesis gas compressors. From a painting by Otto Bollhagen. Courtesy of the BASF Unternehmensarchiv (Corporate Archive), Ludwigshafen

13.5.3   Early Allied Understanding of the German Fixed Nitrogen Industry

The exchange of scientists between Germany and the principal Allied nations in the years prior to the war was extensive. Even so, the Allies have often been accused of misreading[24] the state of the German chemical industry in 1914, and by implication underestimating its ability to support her armed forces in their prosecution of the war. But is that a fair assessment? Put another way, should Germany's ability to sustain her fixed nitrogen needs have been obvious to the Allies at the start of the war? As far as 'conventional' fixation was concerned they should have been 'on the ball'. As early as 1915 publications such as 'Manuals of Chemical Technology' (*Manuals of Chemical Technology III, Industrial Nitrogen Compounds and Explosives*. Crosby Lockwood and Son, LONDON, (1915)). devoted a complete issue to a thorough exposition of the European nitrogen compounds and explosives industries—including German production methods, requirements, annual tonnages and issues of patent protection (Plate 13.2).

Allied munitions experts would have realised the extent to which Germany had progressed the well-known by-product and cyanamide processes, but they would have been relaxed about these for both were energy intensive, complex, multi-stage processes and neither was easily capable of a rapid sustained expansion. Even in peace-time they only provided about one half of what

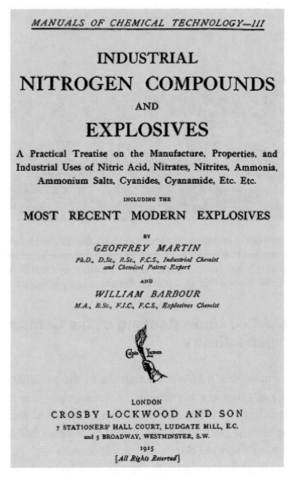

MANUALS OF CHEMICAL TECHNOLOGY—III

INDUSTRIAL
NITROGEN COMPOUNDS
AND
EXPLOSIVES

A Practical Treatise on the Manufacture, Properties, and
Industrial Uses of Nitric Acid, Nitrates, Nitrites, Ammonia,
Ammonium Salts, Cyanides, Cyanamide, Etc. Etc.

INCLUDING THE

MOST RECENT MODERN EXPLOSIVES

BY

GEOFFREY MARTIN
Ph.D., D.Sc., B.Sc., F.C.S., Industrial Chemist
and Chemical Patent Expert

AND

WILLIAM BARBOUR
M.A., B.Sc., F.I.C., F.C.S., Explosives Chemist

LONDON
CROSBY LOCKWOOD AND SON
7 STATIONERS' HALL COURT, LUDGATE HILL, E.C.
and 5 BROADWAY, WESTMINSTER, S.W.
1915
[All Rights Reserved]

Plate 13.2 The title page of the 1915 Manuals of Chemical Technology devoted entirely to the nitrogen industry across Europe

Germany's fixed nitrogen economy needed (Chap. 3). The same would have applied to less conventional but equally well-known processes such as the oxidation of atmospheric nitrogen via the electric arc. Fixation in this manner was only equivalent to about 3% of Chilean imports and less than 5% of 'conventional fixation' (Chap. 3) and with such enormous energy requirements the process was unsustainable in the long term.[25] The potential of these technologies alone therefore should not have fazed the allies.

On the other hand, everyone knew that there was a far more efficient 'new kid on the block', so how aware should the Allies have been regarding Germany's ability to *synthesize* ammonia? Powerful indicators certainly existed even though industrial detail was vague viz, Haber's talk at the Scientific Union meeting in 1910, submission of the BASF patents in both Europe and the USA between 1909–1911, the subsequent battles in the German courts, Bernthsen's talk at the Congress of Applied Chemistry in New York City in 1912, the extensive detail of how to achieve ammonia synthesis revealed by Haber-Le Rossignol in *Z. Elektrochem*, the *Journal of Industrial and Engineering Chemistry*, and *The Times*, Eng. Supplement, 28 May 1913, the Manuals of Chemical Technology in 1915 (pp. 53–55) where the new ammonia process was discussed, and of course Haber's Hurter Memorial Lecture at Liverpool in 1913 which finally announced both the successful industrialization of the process and Germany's ability to produce all the hydrogen she needed from her vast coal reserves. Even so, the German nitrogen industry was clearly not on a war footing at the outbreak of WW1— a common accusation made by the Allies after the war—and their understanding of the evidence was undoubtedly hampered by the BASF's secrecy, diverted by all the tensions in Europe at the time, and confounded by the unexpected ability of German industry to modify and incorporate the new technology so soon after the outbreak of war. It wasn't until 09 May 1915 at the battle of Aubers Ridge that the Allies were confronted with the problem that Germany had immediately recognized after the battle of the Marne. The failure of Aubers Ridge was put down simply to a lack of high explosive shells. *The Times* in May 1915 reported the concerns of the British Commander-in-Chief, Field Marshal Sir John French …'*We had not enough high explosives to lower the enemy's parapets to the ground … The want of an unlimited supply of high explosives was a fatal bar to our success.*' Just as in Germany, the Allies own 'shell crisis' led to the re-organisation of munitions production putting the country firmly on a war footing. Lloyd George was made Minister of Munitions in the newly formed Imperial Munitions Board. A huge new factory was set up at Gretna on the English/Scottish border to produce 'cordite'. An idle factory in Silvertown on the north bank of the Thames in the Borough of Newham was pressed into service to manufacture TNT, and British Commonwealth countries were organised to supply munitions for the remainder of the war. It may not have been coincidence that French air attacks on Oppau began just a few weeks after the failure at Aubers Ridge.

In retrospect, the Allies understanding of Germany's fixed nitrogen capability was predicated on her conventional technologies. They were ignorant regarding the detail of the new technology of ammonia synthesis and of course knew nothing of the industrial genius of Carl Bosch. The stranglehold of the British and Commonwealth navies therefore, would have given them the confidence that the whole affair would be 'over by Christmas'. But the appearance of a new form of warfare, and Germany's ruthless ability to organise and engage her new technology in support of her military machine was totally unforeseen. The Allies too were fortunate, for if German sub-marines had dealt as faithfully with Allied nitrate ships as they desired, or if Chile for any reason had withheld supplies, the Allies would have been forced to give up the struggle themselves[26]—or maybe with the benefit of prevailing winds—do as Germany did, resort to chemical warfare.

13.5.4 The Legacy of Oppau and Leuna

In the absence of Chilean nitrate, Germany initially had to rely on fixed nitrogen contributions from her conventional processes. By-product and cyanamide production struggled to meet the demands of the army in the first two years of the war. Although increasing ten-fold, their contribution then flattened out[27] and there was little to spare for agriculture. Despite the lob-bying of these industries to secure the largest share of the fixed nitrogen business, it was Oppau and Leuna that made the decisive difference, sus-taining Germany's capacity to produce munitions for the remainder of the war. By 1918 these two plants alone produced 60% more than the cyanamide process and surpassed the combined production of by-product ammonia by all of Germany's numerous coking plants.[28] By delaying the Second Reich's inevitable collapse by as much as two years, their existence undoubtedly ensured that the war would not 'be over by Christmas', compounding the human suffering at the front for both sides. But over the war years, with most of the fixed nitrogen now 'feeding' the arms industry and slaughtering mil-lions, a lesser known tragedy also occurred when German farmers, 'starved' of fertilizer, failed to feed their nation. An estimated three quarters of a million innocent German civilians subsequently died of hunger.[29]

The historical importance of Oppau and Leuna is therefore immense, and even for us today the reminders are still seen across Europe; at every national war cemetery, on every roll of honour in every village and town, on every remembrance Sunday, in the faded photographs of all the families who lost the flower of their men-folk in the fields of France, in the tears of the old soldiers who survived the war by as much as 80 years but still sobbed, and through a generation of women on both sides who were never able to find love, each one leaving '*a new page in the book of* [their] *lives, and a chord of* [their] *souls lying prostrate and barren* … '.[30] forever. Oppau and Leuna contributed hugely to the physical, political, economic, and social devastation of Europe, and 'although the conflict began in Europe it spread to the Middle East, to Africa, Asia and beyond. It became the first *world* war and by 1918 four empires were in ruins and four royal dynasties ended. The face of the continent was changed by revolution and by death which it seemed could never claim too many. Tens of millions died, every one mourned by people who loved them, and angels' tears fell on half the world. But here's a funny thing, Austria and Russia, whose quarrel took everyone to the edge were the last to declare war on each other, and when they did … no one really noticed'.[31]

But the provenance of Oppau and Leuna was *so* fragile. Had any link in a chain of otherwise innocuous events been broken, Germany may not have had a synthetic ammonia industry worthy of the name at the start of the war. Had Haber never accepted the Margulies contract, if he had favoured publishing at the lower yield of ammonia of 0.005% in 1905, if Haber had solved the riddle of the Third Law not Nernst, if Le Rossignol had gone to Abegg instead of Haber, if Haber had not met Nernst in Berlin in 1908 and if Mittasch and Kranz had followed Bosch and left the Karlsruhe demonstration in July 1909 things might have been so different. Like Ostwald's claim to be the 'intellectual father' of 'fixation',[32] so too Haber and Le Rossignol became the intellectual fathers of Oppau and Leuna. But their 'true father' was Carl Bosch, and their 'Master' was undoubtedly the German State machine. The infant Haber-Le Rossignol process was conceived in innocence and in hope, but it grew up to change the world in so many unexpected ways.

13.6   Gas Warfare and Haber

Gas! GAS! Quick boys! – An ecstasy of fumbling, fitting the clumsy helmets just in time: But someone still was yelling out and stumbling, and flound'ring like a man in fire or lime …. In all my dreams, before my helpless sight, he plunges at me, guttering, choking, drowning … My friend, you would not tell with such high zest To children ardent for some desperate glory, The old Lie; Dulce et Decorum est Pro patria mori.

Lines from 'Dulce et Decorum est (pro Patria mori)', *It is sweet and proper (to die for one's country)*. Wilfred Owen, 1917.

Haber's contract with the BASF and his membership of the nitrate commission placed him in a unique position. The poacher had turned gamekeeper. Over the years of the war and with an ever-escalating need for fixed nitrogen he was able to use his position to facilitate the BASF's business concerns with Berlin's bureaucracy, which in turn greatly expanded Germany's dependence on 'his' synthetic ammonia. As a result, he became fabulously wealthy.[33] The description of WW1 as the 'chemists war' was of course due in part to the role of 'fixation', and had Haber confined himself to Germany's nitrate problem history might have looked more kindly upon him. After all, a nation would be *expected* to defend herself by conventionally permitted means, and mass slaughter via the progeny of the likes of Oppau and Leuna was then—as today—'de rigueur'. Even early in the war, knowledgeable individuals appreciated the importance of 'Haber's' discovery and marveled at the rapidity of its industrialisation. W. J. Landis, the director of the American Cyanamide Company, wrote in 1915 that '*too much honour cannot be shown to those courageous chemists who have succeeded in placing this process on a commercial working basis …*', and two years later during a debate on nitrogen in the British House of Commons, Sir William Pierce said that the Haber-Bosch process '*… is one of the greatest achievements of the German intellects during the war … It is really a wonderful achievemen.*'[34] But Haber's privileged position on the nitrate commission also gave him a clinical understanding of the perilous position of Germany in the early months of the war, and as a result he immediately placed himself and the resources[35] of the KWI Institute at the service of the nation. It was his work here in support of the war effort that led to him being reviled as the 'father of chemical warfare', for the irritant and toxic gasses investigated, developed and deployed under his direction were the second, highly emotive, and deeply despised component of the 'chemists war'.

This text is not the place to provide a critical analysis of gas warfare in WW1. To dwell upon the controversial aspects of its provenance and questions of international law[36] arising from its application would be inappropriate, for such issues have been fully accounted for elsewhere.[37] But what is inescapable, is that Fritz Haber was intimately and enthusiastically involved in Germany's development and application of the technology. Even so, and in retrospect, his gas warfare had little strategic impact. Towards the end of the war chemical weapons were largely ineffective against well trained and well-equipped troops. Throughout the war fatal casualties were low, around 3%. Only about 2% were permanently invalided and most troops were fit for duty again within six weeks. But these stark statistics hardly convey the horror of being gassed, where men died slowly and in agony, for—unlike 'conventional' wounds—little was available to ease their condition. These observations were in direct contradiction to Haber's early opinion of gas warfare however, which he felt was far more 'humane', in that death would occur rapidly either by the gas itself or through a 'clean kill' as troops were driven out into the open. After all, why should a man have to die in a shell hole slowly bleeding to death over days and calling out for his mother? Gas too had psychological advantages, it was infinitely variable so soldiers had no idea what was heading for them and many substances 'lingered' on the soil thereby confounding a soldier's instincts to fall to the ground for protection. But gas warfare had hardly begun before it played a part in the death of one of Haber's closest colleagues and soon after, that of his wife Clara Immerwahr.

13.6.1 War at the Kaiser Wilhelm Institute

The Institute's initial involvement with the war effort however was innocuous enough. In the early period of mobile warfare—according to Coates[5]—Haber received his first problem from the War Ministry, viz, to discover a substitute for 'toluene' (methyl benzene)—obtained from coal tar—which was no longer available for use as an antifreeze agent in 'benzine' based motor fuel because it was now too important as the precursor for TNT. The institute soon found 'xylene(s)' (di-methylbenzene(s)) and 'solvent naphtha' satisfactory. The sudden and unexpected onset of trench warfare however not only exacerbated Germany's nitrate problem, but also presented her with the need to rapidly re-instate the mobile war by devising some method of driving the Allied forces out of their trenches and into the open where—they felt—their superior numbers and tactical training would prevail. In this respect, the use of 'irritant' substances to render confined positions such as trenches untenable was not a new idea,[38] and none of the combatants regarded irritants to be in conflict with the Hague Treaty of 1899. According to some reports it was the French, who in August 1914, first used grenades containing a few millilitres of 'tear gas' (ethyl bromoacetate or xylyl bromide) on the Western Front. These small quantities however were rapidly dispersed, and the German infantry hardly detected their presence. Soon afterwards, in October 1914, Germany retaliated and deployed fragmentation shells containing a chemical irritant against British troops at Neuve Chapelle with much the same effect. But such warfare was unforeseen by Germany and no preparation had been made in this direction. However, realizing the advantage these substances *might* provide, she called upon her principal chemists to address the problem. At first Nernst was consulted and he suggested the use of the irritant non-toxic powder *o*-dianisidine chlorosulphonate[39]—an intermediate used in the synthetic dye industry—and the lachrymator ω-xylyl bromide[39]—both delivered via normal high explosive and shrapnel shells simultaneously saving on TNT. However, these proved ineffective because the delivery systems dispersed them too widely. Then, towards the end of 1914, Haber was also 'given a share in the work'[39]—some say he 'volunteered'—and asked to improve the effectiveness of the irritants, but not to allow them to persist to the extent that they slow down the subsequent advance of German troops. Haber also had to address the delivery systems[40] which depended upon very scarce steel shells, but experiments in Haber's Institute had hardly begun when they were brought to a tragic end in December 1914 when one of the chemists suffered from a sudden detonation when mixing chemicals.[39] The explosion occurred as a few drops of dichloromethane was added to cacodyl chloride, C_2H_6AsCl,[41] which possibly contained nitrogen chloride as impurity. The cacodyl chloride was sent to the Institute from an industrial quarter and although the purpose of these experiments was to provide an *irritant*, the fact that a substance may also have a *toxic* effect did not seem to exclude consideration of its use.

The chemist involved was the distinguished Otto Sakur, the very same who had worked at Ramsay's laboratory during Robert's time there. (Chap. 1). Sakur was horribly mutilated by the explosion, and had Haber not been called out of the room just moments before, he too might have been involved. But he rushed back into the laboratory and collapsed in shock.[42] Shortly afterwards Clara Haber came running in and, with a mother's instinct to calm a fallen child, opened Sakur's collar so that he may breathe more easily. Sakur however died soon afterwards. At the funeral Haber was inconsolable and seen to be weeping openly, and for the first time Dr. Clara Haber glimpsed the full nature of the work being conducted at her husband's Institute. By all accounts she was appalled (Plate 13.3).

13.6.2 The First Chlorine Gas Attack

Despite the termination of research at Haber's institute, the lack of steel for shells and the problem of dispersion, the German military refused to reject gas warfare as a means of bringing about a quick victory, and additional mechanisms were sought to attain sufficiently high concentrations of gas over a wide front. At the time, Germany had copious amounts of chlorine formed as a by-product from the dye industry and the suggestion was made to utilise this irritant (asphixiant) in a mass discharge from cylinders. We do not know whether Haber suggested the use of chlorine but he certainly wholeheartedly supported its adoption. Accordingly, he was put in charge of the 'gas troops' and the preparations but he soon ran into difficulties with the Generals. Prevailing winds were westerly and therefore in the Allies favour, many

Plate 13.3 Otto Sakur, circa 1913. Courtesy of the Fritz Haber Institut der Max-Planck-Gesellschaft, Berlin-Dahlem

Generals felt that the use of gas was 'unchivalrous', so many would not permit its use 'on their patch' and they all realised that whatever wicked concoction Germany came up with, the Allies would soon develop something equally—if not more—diabolical. Even so, a trial release of cylinder gas was scheduled for the Western Front in March 1915 at Ypres, an area controlled by the reluctant General Deimling.

On 02 April 1915 in preparation for the attack, Haber had an opportunity to test his theory of gas warfare when in a trial well behind the front line, he and an army officer rode too close to the gas cloud and both nearly suffo-cated.[43] Haber was sick for days but he recovered and this incident confirmed all his prejudices … gas warfare really wasn't that bad. Preparations for the attack continued however, and thousands of chlorine cylinders were dug in and covered with earth along a ridge called 'Hill 60'. However, the prevailing winds were particularly unfavourable and the cylinders were recovered and transferred to another location just west of the town of Langemarck. But before many had been removed, the meteorologists called a halt as the winds in the original site suddenly became favourable. General Deimling moved his troops into position at 4 am ready for a morning attack in support of an early gas release, but once again the winds failed. A furious Deimling rode to Haber's headquarters, and not for the first time in the war, the famous scientist received a thorough 'dressing down'. However, as the cylinders were once again being removed to Langemarck, an Allied shell—purely by chance —scored a direct hit, releasing the gas and killing and injuring around fifty soldiers.[44] Presented with first hand evidence of the effectiveness of the new weapon Deimling had a change of heart, and by mid April all the cylinders had been removed and were in position awaiting a northerly wind. On 22 April 1915, late in the afternoon (about 6 pm), the wind arrived from the northeast and the valves were opened releasing an estimated four hundred tons of chlorine gas. As the chlorine drifted slowly towards the French and Canadian lines, another component of the 'chemist's war' came into being. The world subsequently placed the responsibility for chemical warfare squarely on Fitz Haber's shoulders, and despite his justification for its use and the limited casualties it caused, over the passage of time, even down to today, his burden has never been lifted.

The attack however was not as successful as Haber had hoped. Certainly, his poisonous cloud cleared out many of the trenches—especially those occupied by the French/Algerian troops—at the same time opening up a four-mile-wide gap in the Allied lines. The German troops advanced about fifteen minutes after the gas release picking their way between the trenches, through the barbed wire, machine gun nests and over the distorted, still warm bodies of those unable to flee. But the cloud soon dissipated, sufficient support troops to reinforce the breakthrough were unavailable, night was falling and the advance guard had to dig into consolidate the day's work. In later years, Haber became rather bitter regarding the gas release at Ypres. He felt that the German high command should have had more faith in his 'shock of the new'. They should have been more patient and waited for more favourable weather conditions releasing the gas in the morning, and taking full advantage of its novelty by committing enough reserves to sustain the breakthrough. As it was, the battles around Ypres continued for another two weeks and the Germans released chlorine gas four more times. Casualties at Ypres due to gas alone were difficult to quantify but it was estimated that there were around 7000 injured and 350 dead—enough it was felt to effect a *tactical* advantage, but not enough to provide the strategic advantage that would have re-instated a mobile war.[45] Despite some initial enthusiasm for the gas cloud, this remained the opinion of the German high command over the next two years, the Allies however were galvanized and, as Germany had anticipated, they immediately made preparations for both retaliation and protection against gas.

13.6.3 Dahlem Berlin, 01 May 1915

Soon after the first gas attack—somewhere between 24 and 29 April—Haber was recalled to Berlin,[46] and appeared before the Kaiser who, ignoring all military protocol, promoted him from a sergeant in the reserves to a full captain (Hauptmann), a rank aimed at providing Haber and his 'gas troops' with some credibility, and one which he retained for the remainder of the war. Almost 27 years had passed since he had first applied for a commission in the Prussian military (Chap. 2), but finally, the German Jew turned Christian, became an officer and he immediately discarded his sergeant's 'pest controller's outfit' and proudly donned his captain's uniform. Haber's time in Berlin however was short—he had already received orders to travel to the eastern front to organise gas attacks there too—but he found time, probably suitably attired in his new uniform—to hold a reception at his villa in Dahlem on the evening of 01 May 1915. Late that night, under an almost full moon and with Haber sedated and asleep under the influence of his daily allowance of sleeping pills, Clara took her husband's pistol out into the garden, fired a test shot, then turned the weapon on herself. In a final act symbolizing the complete loss of love in her life, she placed a single shot through her heart.

To this day the real reason(s) for Clara Haber's suicide remain unclear. Those friends moved to account for Fritz Haber's life soon after his death mention her demise only briefly or not at all. J. E. Coates[5] for example, whose recollections have frequently been referred to in this work, avoids the incident completely mentioning only that Habers first wife 'died in 1915'; Richard Willstäter's memory too 'skirts' around the event recalling only that when Haber returned home for a short visit, it was the day '… *on which his wife died* …'. Little evidence has emerged over the intervening years from those who have written in detail of Haber's life and Haber himself remained silent, offering no explanation, justification or even a defense against those who suggested his behavior drove her to such despair, for at the time and being 'at the front', death was commonplace in Haber's life. But there seems little doubt that Clara's decision was driven by a heady emotional cocktail of events and that the difficulties between the couple at the time were not minor, but fundamental.

Contemporary accounts however suggest that Clara suffered a persistent, gnawing frustration at foregoing her own career, having to settle for the role of a professor's wife, a mother, housekeeper and cook, roles which she often could not discharge to her own high standards as Fritz' chaotic lifestyle frequently resulted in guests unexpectedly appearing in her home. Dietrich Stolzenberg[47] describes her despair; '*because of her very intelligence, love of truth and perfectionism, she broke herself on the practical aspects of life at the side of a man who lived for his profession and subordinated all else to it.*' Clara's frustration also induced a tendency towards depression which some saw as a family trait,[47] and Fritz's frequent times away precipitated a gradual marital estrangement, amplified grotesquely as the war progressed. Haber it seemed was everywhere, his wife and family had no restraining influence on his enthusiasm for war-related work, and gas warfare in particular which 'appeared to keep him happy'.[48] But Clara saw Fritz' role in gas warfare as a perversion of science, and whether this came up in conversation on the night of 01 May we do not know, but it has been claimed that at some time she 'gave him an ultimatum to break off this activity or she would commit suicide'.[49] In this respect the most compelling evidence for Clara's opposition to her husband's role which suggests it may have played a part in her decision that night comes from Paul Krassa, a friend of the Habers and a distant relative of Clara, who wrote in 1957[50] that '*a few days before her death she visited my wife ... she was in despair over the horrible consequences of gas warfare for which she'd seen the preparations along with the tests on animals ...*'. But there was also the suspicion of an affair, between Fritz and one of his guests that night, viz, Charlotte Nathan, a secretary or business manager at his club —the German 1914 Society—in Berlin. According to one account,[51] Clara came across Fritz and Charlotte in an 'embarrassing situation' which if true, must have pushed Clara's tolerance across all reasonable bounds leaving her feeling deeply betrayed after all her sacrifices. Clara it seems, could see no rainbows in her life, just rain. But amongst all the conjecture what we know for certain is this. On the night of 01–2 May 1915, thirteen-year-old Hermann Haber was awoken by gunshots in the garden of the villa. He rushed outside and found his mother covered in blood and dying. He called for his father and during the time it took for Clara's life to pass away so unnecessarily, father and son were together with her. Unable to secure permission to stay,[52] Fritz Haber left for the front the next day leaving Clara's funeral arrangements to others. She was buried quietly in the cemetery at Dahlem, without the kind of public display of emotion that attended Otto Sakur's passing, and in an instant, young Hermann was left without a father or a mother. Later, on 25 October 1917, Fritz Haber married Charlotte Nathan and they had two more children, Eva Charlotte (b. 21.07.1918, Berlin) and Ludwig Fritz, 'Lutz' (b. 12.07.1920, Berlin). According to some accounts,[53] before her death Clara wrote several farewell letters—none of which survive—but if so her actions on that night seem to have been planned and were not simply an emotional reaction. Clara's suicide therefore may have had the effect she desired; Haber carried the guilt of her death for the rest of his life, a kind of harrowing guilt that only someone so close can ever inflict. Later, during the war, he recalled; '*I hear in my heart the words that the poor woman once said ... I see her head emerging from between orders and telegrams ... and I suffer.*'[54] Even so, as head of the Chemical Warfare Service, in 1916 and 1917 he was awarded—and accepted—the Iron Cross, Class II and I respectively.

13.6.4 Haber's Work Continues at the Institute

Even if Haber's only 'flirtation' with gas warfare had been the organization of the gas attack at Ypres, it would have been enough to condemn him as a 'war criminal' in the eyes of the international community, even though at the time the rules regarding this were 'confused'.[55] But at the time his enthusiasm for the prosecution of chemical warfare was boundless, and his eventual involvement with the technology went way beyond Ypres. At the time, Clara's death certainly made no impact on him, and neither was he concerned with decisions under international law that declared gas warfare as inhumane. His duty as an army officer was to advise and obey those responsible for higher policy.[56] By 'obeying his orders' he saw himself as a patriot, a 'good German' absolved of all the consequences of his actions. His commitment was total and unquestioning, a position he publicly defended in depth after the war.[57] But for all Haber's enthusiasm, some time after Ypres the high command became disenchanted with the cloud attack method because of the lack of a strategic advantage and the potential for 'blow back' which made it difficult to co-ordinate an infantry advance with the gas discharge. Anticipating an Allied response, Haber was therefore asked to concentrate on the problem of gas defence instead. The planning, preparation, and execution of gas delivery systems still occupied his Institute during this period, but over the next six months they also came up with an ingenious respirator that was issued to the troops in October 1915. The development of the respirator was a feat of remarkable organization by Haber, involving not only a clever design in the absence of scarce materials such as rubber, but also the coordinated efforts of a number of huge German companies such as Bayer at Wuppertal/Leverkusen, the 'Auer' in Berlin and the Dräger Works in Lübeck. The respirator Haber helped develop had a close-fitting face piece, held in position with a head harness, incorporating a small lightweight drum or 'cartridge' packed with filters containing absorbent material. During the war the 'Auer' produced the filter inserts, and Bayer the gas-absorbing substances. As the war progressed and different gas agents were encountered, it was possible to develop new filter inserts for the cartridge so that the basic respirator remained almost unchanged throughout the war. But by the spring of 1916 when the Germans had come to regard the gas weapon as of relatively little value, they were forced into a re-appraisal when the French introduced a new gas shell containing 'phosgene' (carbonyl dichloride, $COCl_2$) which, by inducing suffocation when inspired, marked a turning point in chemical warfare. The deployment of 'phosgene' by the French meant that the purpose of gas had changed, no longer was it an 'irritant', it was now meant to kill. As a result, Haber was put in charge of the development of new gas projectiles and through a separate 'Chemical Warfare Service' he became chief advisor to the German army on both offensive and defensive gas warfare. Haber's Institute was regenerated and re-organised on military lines, and the whole of chemical warfare research was conducted there under his direction until the end of the war.

From 1916 onwards, Haber's gas warfare research was concerned with substances that could be used in projectiles and simultaneously counter the growing efficiency of respirators. He introduced the concept of 'Bunteschiessen'[58] or 'variegated shelling' which firstly deployed irritant organic arsenic compounds (the 'maskenbrecher' or 'mask breakers') that penetrated respirator filters and forced men to remove their masks, followed by shelling with poison agents such as 'phosgene'. Of the hundreds of substances the KWI Institute examined under Haber's tutelage, the following were prominent[59]; 'phosgene', trichloromethyl chloroformate ('diphosgene'), chloropicrin, dichlorodimethyl ether, phenyl carbylaminedichloride, ethyldichloroarsine, diphenylchloroarsine, dipenylcyanoarsine, and dichlorodiethyl sulphide or 'mustard gas'. The latter substance—discovered in 1886 by Victor Meyer—was the most effective chemical agent of the war affecting both the skin and the eyes. When introduced by Germany in 1917 it incapacitated large numbers of Allied troops but caused few fatalities and Haber advised the German High Command at the time, that unless they could guarantee that the war would be won within a year, it would be best not deployed, for the Allies would respond in kind and that would bring about Germany's certain defeat.[59] The Allies indeed used 'mustard gas' themselves in 1918 and Germany collapsed soon after, but for both military and economic reasons.[60]

13.7 The End Game

After the armistice was signed in November 1918, the Allies produced a list of 900 alleged war criminals. Haber's name was amongst them. When Germany ratified the Treaty of Versailles in June 1919—which required the surrender of those accused—they demanded his extradition. Convinced by the patriotic nature of his role in the war, Haber did everything he could to remain in Germany. But in late July 1919, after sending his family to Switzerland and giving power of attorney (30 July 1919) to Max Meyer,[61] he put aside his military uniform, grew a beard, and fled to Lucerne. From there on 01 August he followed his family to the Alpine resort of St. Moritz where he acquired Swiss citizenship in an attempt to secure immunity from prosecution.[62] After a few months however, the Allies dropped all their charges against him and he returned to his Institute in Berlin. The war years for Haber were the greatest period of his life. But we shouldn't suppose that he enjoyed war per se. He hated the suffering, the waste of life, the pointlessness of it all, for in his emotional 'ground state' he was kindly, helpful, generous and benevolent. No-one had to 'worship at the altar of Haber's ego'. But it is equally true that the excitation of war transformed him into a ruthless autocrat, a side of him which in peace-time he either suppressed or made manifest as his 'strength of purpose'. He gave his whole being over to the struggle for victory and nothing was allowed to deflect him; neither family, human suffering, nor international law. He brought a vitality to all the positions he held and he always placed Germany first, declaring that, '*Science is for the world … but for your country in time of war!*'[63] For three or four generations past, his family had fought for, or served, Germany and he was proud to have become a Prussian officer and such an influential military man, so when defeat finally came it weighed heavily upon him, as did the personal repercussions of his role in chemical warfare. His already fragile health was permanently damaged by overwork during the war years and when he returned to Berlin to reconstruct his Institute it was as an utterly disillusioned and broken man.

But in the immediate post-war years his spirits rose again as he re-modelled the Institute to take advantage of the scientific experiences that had been gained during the war, applying them now to beneficial ends. Along-side the established areas of physical chemistry, physics and colloid chemistry, he introduced additional sections for organic chemistry, pharmacology, insect pest control[64] and textile chemistry, and there is little doubt that the eventual award of the Nobel Prize for chemistry in 1918 for '… *improving the standards of agriculture and the well-being of mankind ... '*, played its part in his rejuvenation. After the war Haber's Institute became a centre of world class research. Fundamental advances were made in atomic physics, spectroscopy and colloid chemistry. In his own field, he covered new ground in the mechanisms of chemical and photochemical reactions incorporating quantum aspects such as the photoelectric effect, and he continued his life-long studies on flames and combustion, for which he had originally engaged Robert Le Rossignol at Karlsruhe. With the coming of the German hyper-inflation of 1921–1923 Haber once again leapt to the service of his country and spent the next few years re-investigating the 'gold from seawater' problem that had enticed Ramsay in 1905. In an attempt to alleviate Germany's dire financial predicament caused by the heavy war reparations demanded by the Allies, he devised a series of analytical and micro chemical techniques that were novel and pioneering and used much later by others such as in the extraction of bromine from sea-water.[65] His intervention however was a failure, and in 1926 he abandoned the work. But at last he established the truth about gold from the sea, viz, that it occurs at a concentration of about 1/1000th milligram per ton, mainly as the metal, and that samples are very variable, making the whole recovery process economically unviable.

   In the period from 1912 to 1933, Haber's Institute published more that 700 papers. By 1929 he had gathered together more than sixty members—over a half of whom were 'foreigners'—and he remained 'in touch' with each diverse group even though there were heavy external demands on his time. Later life was not kind to Haber, from both a reputational and personal point of view. Towards the end of his life Clara returned more and more to his thoughts. He fell at her feet ... and her tears rained down on him.

Notes

1. There is no doubt that chemistry played a huge part in the war, for the refining of metals, for photography, for dyes for uniforms and the care of the wounded such as with antiseptics, analgesics and the treatment of typhoid and cholera. Morphine was used as a painkiller from 1916 and opium as a painkiller and sedative. However, nitrogen fixation and gas agents were primarily responsible for the description 'the chemists war.' See Michael Freemantle, *Gas! Gas! Quick Boys! How Chemistry changed the First World War*. The History Press, (2012). ISBN-10 0750953756. ISBN-13 9780752466019.
2. *Records of the Great War*, I, Ed. Charles F. Horne, National Alumni (1923).

3. It had no contingency for failure, nor retreat, nor any consideration of improved enemy logistics.

4. Indeed, Belgium almost destroyed the Schlieffen plan single handedly. The railway system across northern France and Belgium was a vital component in the plan, but the Belgians had long prepared to sabotage their own system to slow down the advance if Germany were ever to attack them. If it were not for the large number of failures of the explosives they used, they may well have achieved that. The same railway system also rapidly brought the British Expeditionary Force (BEF) to northern France in time to meet the Germans entering French territory (at Mons). The Germans could not believe they were fighting 'Tommies' only days after the declaration of war! From, Michael Portillo, *Railways of the Great War*, BBC Productions, broadcast 04–08 August 2014.

5. Authors differ here, this author has used a figure quoted by J. E. Coates, in his 1934 appreciation of Haber. (J.E. Coates, 'The Haber Memorial Lecture', *J. Chem. Soc.*, 1642–1672, (1939)). Coates was a contemporary of Haber and his recollections were closer to the time.

6. Dietrich Stoltzenberg, *Fritz Haber. Chemist, Noble Laureate, German, Jew.* Chemical Heritage Press, Philadelphia, Pennsylvania, (2004), p. 130, his reference no. 11.

7. The railway system around Paris had been designed to move troops quickly through the city and to its perimeter. The troops assembled by Joffre were rapidly deployed using this system. From Portillo, op. cit. (note 4).

8. Both the French and the Germans lost around 250,000 men and the BEF about 13,000.

9. A very common saying used by Generals on both sides at the start of the war, Germany relying on her Schlieffen plan and the Allies on the strangulation of Germany's nitrate trade. Lord Kitchener, Secretary of State for War, was unconvinced. He warned the British government that the war would be decided by 'the last million men that Britain could throw into battle'.

10. Barely 14,000 tonnes according to Stoltzenberg, op. cit. (note 6), p. 131.

11. From a British perspective this was a bizarre encounter. Why would Cradock have considered taking on a much superior German squadron? But Cradock had no confirmation that Spee's forces were concentrated, believing that in the vastness of the ocean he might meet one or two enemy craft for which his squadron was a match. The German guns had a superior range and within ten minutes of the first German salvo the result was not in doubt. Geoffrey Bennett, *Naval Battles of the First World War*, Pen & Sword, (2005). ISBN-10 1844153002. ISBN-13 9781844153008.

12. Thomas Hager, *The Alchemy of Air*, Three River Press, New York, (2008), p. 141.

13. German naval experts however were baffled as to why Spee attacked the base —his ships were already fully 'coaled up' and how could the two squadrons have met so coincidentally in so many thousands of square miles of open water? It was generally believed at the time, that Spee was misled by the German admiralty into attacking the Falklands—which acted as a coaling station and wireless relay station for the British—as his intelligence, received from the German wireless station at Valparaiso, reported the port to be free of Royal Navy warships. The probable truth however is that von Spee was lured towards the British battle cruiser squadron by means of a fake signal sent in a German naval code broken by British cryptographers, and although such deception would normally have been against the British Naval Staff's policy, the bizarre nature of Spee's action may well have alerted the Germans to the fact that their code had been broken. There were many in the British Government that suggested that such information should be kept hidden for much more important occasions. Bennett, op. cit. (note 11).

14. He was now to receive 1.5 pfennig for every kilogram of ammonia produced by the BASF, and his royalties had already far exceeded his annual salary.

15. Daniel Charles, *Between Genius and Genocide*, Jonathan Cape, London, (2005), pp. 175–6.

16. However, ammonia is flammable and burns in air to produce water and nitrogen. Because of the lack of motor fuels, this was used in Belgium in the Second World War to fuel buses. From *Inside Science*, BBC Radio Four, 04 September 2014, with Andreas Sella and William David.

17. Cordite' consisted of nitroglycerine (58%), 'gun cotton' or nitrocellulose (38%), and 'Vaseline' (4%). Cordite generates sufficient pressure to propel an artillery shell to its target, but not enough to destroy the barrel of the gun.

18. Potash of course could have been used to produce potassium nitrate directly according to $K_2CO_3 + 2HNO_3 = 2KNO_3 + H_2O + CO_2$ but 'soda' was far more available. Together with the abundant source of Sylvite and the long-standing production of potassium nitrate from sodium nitrate by double decomposition followed by fractional crystallisation, this route was more sustainable and familiar, (Chap. 3).

19. In addition to their contribution to WW1, the ammonia plants were the genesis of the German military-industrial complex which was soon to develop other high-pressure techniques for the production of another essential component of 20th century warfare, viz, oils and fuels, from Germany's vast coal reserves. The original high pressure 'coal to oil' process was created of course by Bergius who worked at Karlsruhe with Haber and Le Rossignol for one semester in 1907 (Chap. 2). The Fischer-Tropsch process however, catalytically hydrogenated carbon monoxide at 200–300 °C and 10–60 atm to create hydrocarbons. Both the carbon monoxide and the hydrogen being obtained from coal. By November 1935, less than three years after Germany's Nazi government came to power and demanded petroleum

independence, four commercial-sized Fischer-Tropsch plants were under construction. These were soon expanded to become nine, and by 1944 4.1 million barrels of oil were being produced annually. 'Liquid Assets', *Chemistry World*, **8**, 5, (May 2011).

20. Stoltzenberg, op. cit. (note 6), p. 131, his reference no. 15.

21. Charles, op. cit. (note 15), p. 176, also (Chap. 11).

22. Indeed in September 1915 the War Ministry asked Haber 'who knows how difficult it is to steer the nitrogen ship', Charles, op. cit. (note 15) p. 177, to mediate on their behalf in its negotiations with the BASF. Haber the 'poacher' had now clearly turned 'gamekeeper', driving the German industrial machine towards synthetic ammonia obviously suited his financial needs.

23. The first of which occurred on 27 May 1915, Oppau being the only German fixed nitrogen plant in range of WW1 aircraft at the time. It also happened to be the largest.

24. See for example, Alan Dronsfield, *Letters, Chemistry World*, **6**, 2, (February 2009), p. 37.

25. There was an alternative view however, especially amongst those who had a 'smattering' of science, such as the wealthy amateur chemist J. A. Le Bel, who became one of William Ramsay's closest 'foreign' friends. On 15 November 15 1915, Le Bel wrote to Ramsay with his concerns that the blockade would not affect the course of the war. '*Germany can manufacture nitric acid by the electric process*', he wrote, '*she can import all the iron ore she needs from Sweden, and if the supply of cotton is cut off she can manufacture explosives from wood pulp …*' See Morris W. Travers, *A Life of Sir William Ramsay K.C.B., F.R.S.*, Edward Arnold London, (1956), p. 289. Needless to say, Germany knew better.

26. An observation from, 'Nitrates from the Air', *The Times*, Thursday 22 May 1920, p. 17.

27. Coates, op. cit. (note 5), p. 1658.

28. Vaclav Smil, *Enriching the Earth*, MIT Press, (2001), p. 105.

29. The German Board of Public Health in December 1918 claimed that 763,000 German civilians died from starvation and disease caused by the blockade up until the end of December 1918. An academic study done in 1928 put the death toll at 424,000. From the Wikipedia entry, 'Blockade of Germany.'

30. Paraphrasing the words of Clara Haber, (Sect. 2.7 *Clara Immerwahr*, Chap. 2), who *so* longed for marriage, seems both poignant and apposite. Little would she have known that her feelings and her words could later have applied to a whole generation of European women. Although some regard this 'lost generation' as a myth. Jeremy Paxman's series, 'The Great War', BBC Television, 2014, claimed 80% of men returned from the war. But in areas where the 'Pals' regiments were raised, the male population was decimated and even amongst

those who returned, many were already married and others were not marriageable due to injury and psychological damage.

31. Adapted from the BBC's, *37 Days to War*, Part 3, broadcast 08 March 2014.
32. Wilhelm Ostwald, *Lebenslinien: Eine Selbstbiographie*. (Lifelines: An autobiography). Volume 2, Chap. 12, Berlin: Klasing, (1926).
33. This aspect is addressed in Chap. 14. Charles, op. cit. (note 15), p. 177, estimates 1,725,000 Marks in 1918 alone, but this is probably an overestimate.
34. Smil, op. cit. (note 28), p. 202, his reference no. 4.
35. Such as they were at the time, for many scientists had been called to the front. Haber himself being promoted to Hauptmann, (Captain) on the express order of the Kaiser. Part of the Institute had even been converted into a crèche.
36. The principal conventions here being the Brussels Declaration of 1874 and subsequent bans by the Hague Conventions of 1899 and 1907. The Treaty of 1899 prohibited the use of projectiles containing asphyxiating or poisonous gasses.
37. Probably the best accounts can be found in; Stoltzenberg, op. cit. (note 6), Margit Szöllösi-Janze, *Fritz Haber 1868–1934: Eine Biographie*. C. H. Beck, Munich, (1998), and Ludwig Fritz Haber, *The Poisonous Cloud. Chemical Warfare in the First World War*, Oxford, (1986).
38. See Stoltzenberg, op. cit. (note 6), p. 132 for a brief review.
39. Coates, op. cit. (note 5), p. 1658.
40. Tear gas was used by the Germans on the Eastern Front in Poland in January 1915 against Russian troops who had 'dug in' near the town of W. Bolimowie, Przedszkol. ('Bolimov'.) In view of the dispersal problem, it was a test-bed for chemical warfare delivery systems and the German forces fired 18,000 gas shells which incapacitated the Russian troops and confused the authorities as to what could have happened. The incident went unreported and never investigated. Huw Strachan, *The First World War*, Part 5/10, BBC, broadcast 18 March 2014. Also published by Penguin Books, (2005). ISBN-10 0143035185, ISBN-13 978-0143035183.
41. 'Cacodyl' is tetramethyldiarsine $C_4H_{12}As_2$, see also Chap. 2 and Bunsen.
42. Charles, op. cit. (note 15), p. 155.
43. Charles, op. cit. (note 15), p. 160.
44. Charles, op. cit. (note 15), p. 161.
45. Stoltzenberg, op. cit. (note 6), p. 138, his reference no. 28.
46. Charles, op. cit. (note 15), p. 165.
47. Stoltzenberg, op. cit. (note 6), p. 175.
48. Stoltzenberg, op. cit. (note 6), p. 140.
49. Stoltzenberg, op. cit. (note 6), p. 176, his reference no. 8.
50. Charles, op. cit. (note 15), p. 166.

51. This story was recounted by Hermann Lütge, a mechanic at Haber's Institute. It is unreliable however as he was not present at the reception himself, but he claimed to have heard of the events from the Haber's household servants. Charles, op. cit. (note 15), p. 166.

52. According the account given by Friedrich (et al.), *One Hundred Years of the Fritz Haber Institute.* B. Friedrich, D. Hoffmann and J. James. http://www.fhi-berlin.mpg.de/mp/friedrich/PDFs/ACh-100-Article.pdf, p. 11. A later account by Friedrich and Hoffmann shows decisively however that Clara's suicide, and the myth that grew up around it, was a much more complex affair than simply gas warfare. See, Bretislav Friedrich, Dieter Hoffmann, 'Clara Immerwahr: A Life in the Shadow of Fritz Haber', in *One Hundred Years of Chemical Warfare: Research, Deployment, Consequences.* Springer Open, (2017), 45–67.

53. Stoltzenberg, op. cit. (note 6), p. 176.

54. From, Chris Bowlby, *Fritz Haber: Jewish Chemist whose work led to Zyklon B,* BBC Radio 4, broadcast Tuesday 12 April 2011.

55. Stoltzenberg, op. cit. (note 6), p. 151.

56. Coates, op. cit. (note 5), p. 1660.

57. For example, F. Haber, *Fünf Vorträge*, Berlin, Julius Springer, (1924). Haber's Lectures on Chemistry and War.

58. The earlier account by Friedrich, (et al.), op. cit. (note 52), p. 12.

59. Coates, op. cit. (note 5), p. 1660.

60. German difficulties were compounded by the entry of the Americans into the war in April 1917, even though the Bolshevik rebellion in Russia freed German troops for the Western Front. By now the troops the were exhausted and demoralized and they lost a number of battles during 1918 largely due to improved Allied technology (tanks and aircraft). The loss of the Balkans deprived Germany of both food and oil, precipitating army and naval mutinies and civilian starvation. In November 1918, after suffering 6,000,000 casualties, Germany signed the Armistice and revolution broke out leading to the fall of Imperial Germany and a republican Government—the 'Weimar' Republic.

61. Szöllössi-Janze. op. cit. (note 37), p. 480.

62. Charles, op. cit. (note 15), p. 189.

63. The recollections of Eva (*née* Haber) Lewis, the 94-year-old daughter of Fritz Haber describing the philosophy of her father in the BBC radio programme, 'Nitrogen', on 04 January 2013, part of the 'The Why Factor' series. He actually said … '*In peace for mankind, in war for my country*', in his resignation letter of 01 October 1933. From the Kaiser Wilhelm Institute, cited in Jens Ulrich Heine: *Verstand & Schicksal. Die Männer der I.G. Farbenindustrie A.G.* Weinheim: VCH Verlagsgesellschaft, (1990), p. 202.

64. Immediately after the war, Haber's Institute developed a cyanide-based compound Zyklon A, based on a substance used during the war to rid flour

mills, granaries and barracks of insect and other pests. But the original substance was odourless and it accidentally killed people as well. So, Zyklon A had a foul-smelling additive that warned people of its use. Zyklon B was a further improvement making it much easier to handle and therefore even safer. However, when the Nazis came to power and began their extermination program, Zyklon B was reformulated to remove the odour then used to exterminate millions of Jews, including some of Haber's own family.

65. Morris Goran, 'The Present-Day Significance of Fritz Haber', *American Scientist*, **35,** 3, 400–403, (July 1947).

14

Robert's War
1914–1918

Hostile crowds in Berlin at the beginning of the war.[4]

Internment at Ruhleben,
release to the 'Auer', then finally home.

P.S. If Le Rossignol is really of much value, the Germans would have let him out long ago ....

UK National Archives, Kew, FO 383-279. A post script added to a note requesting the views of HMG on the proposed employment of British civilians in Germany during WW1. Dated 22 January 1917.

14.1   Berlin, August 1914

Whether or not the Le Rossignols had ever contemplated leaving Berlin[1] after the events in Sarajevo, their decision was made for them by the British declaration of war on Germany; a declaration that immediately confined all British nationals to the German state. The suddenness of the British entry into the war however, meant that Germany at first had no real idea what to do with such people, and it wasn't until mid August 1914 that a vague set of guidelines emerged outlining the responsibilities and the freedom(s)—or otherwise—permitted to those who had now become 'enemy' citizens. In the

© Springer Nature Switzerland AG 2020
D. Sheppard, *Robert Le Rossignol*, Springer Biographies,
https://doi.org/10.1007/978-3-030-29714-5_14

meantime, there was confusion, uncertainty and anxiety amongst the British population in Germany, amplified in the large cities such as Berlin which, in the early part of August 1914, was seething with nationalistic fervour, as huge agitated crowds gathered singing patriotic songs and making clear their dislike of both the French and the British. In the city that the Le Rossignols had come to regard as home, the mood was graphically captured by the respected war correspondent Henry Woodd Nevinson[2] of the *Daily News*,[3] and in these early days, it was a mood that must have concerned Robert, not only for the safety of his young family but for his own safety too. In '*Berlin, a-Tiptoe to War in August 1914*', Nevinson writes[4];

On the evening of July 31[st,] I started for Berlin ... Passing into Germany, we at once met trains full of working men in horse-trucks decked with flowers, and scribbled over with chalk inscriptions: 'Nach Paris,' 'Nach Petersburg,'[5] but none so far 'Nach London.' They were cheering and singing, as people always cheer and sing when war is coming. We were ... six hours late in Berlin ... *[then]* ... for two days I waited and watched. Up and down the wide road of 'Unter den Linden', crowds paced incessantly by day and night, singing the German war songs... Every moment a new rumour whirled through the maddened city. Every hour a new edition of the papers appeared. All day long, and far through the night into the next day, I went backward and forward to the telegraph office, trying to send home all the descriptive news I could ... On the morning of the fatal 4th, I drove to the Schloss, where the Deputies of the Reichstag were gathered to hear the Kaiser's address. Refused permission to enter, I waited outside, and gathered only rumours of the speech that declared the unity of all Germany and all German parties in face of the common peril. A few hours later, in the Reichstag, the Chancellor, Bethmann-Hollweg, announced that ... the neutrality of Belgium had almost certainly already been violated. Then I knew that the long-dreaded moment had come.

In the afternoon I heard that our Ambassador, Sir Edward Goschen, had demanded his papers, and war was declared. I was at the 'Adlon', having been turned out of the 'Bristol' the day before as a dangerous foreigner. While I was dining, I heard the yells of a crowd shouting outside our Embassy in the neighbouring street, and breaking the windows ... Soon the noise came nearer, and in front of the hotel entrance I could distinguish shouts for the English correspondents to be brought out ... Two ... armed police seized me ... and dragged me out, holding an enormous revolver at each ear. 'If you try to run away,' ... 'we will shoot you like a dog!' ... they flung me out into the mob, who savagely set upon me with sticks, fists, and umbrellas ... Holding the revolvers still in uncomfortable proximity to my skull, the police then took me … by taxi to the central police court. There our treatment became more courteous, and after we had made our statements and shown our passports we were dismissed, with a note insuring protection. But as a scrap of paper seemed insufficient insurance against the fury of a mob inflamed, as German, British, French, and all mobs then were by the raging patriotism of war, I demanded to be sent back protected as I had come. So back in a taxi I was sent … protected by only one policeman, who kept his revolver in a more respectful position, and convoyed me to the backdoor of the hotel … I slept as best I could, and next morning I went about the city purchasing a few necessary things. All was quiet, and life seemed going on much as usual but for the excited crowds gathered round the newspaper offices, and the removal of all English and French names from the shops and banks … In the evening, however, I received a kindly invitation from Sir Edward Goschen to come into the Embassy … I gladly went, and was welcomed with amazing courtesy.

Before dawn on August 6th, a string of motors was waiting outside the Embassy, sent by the Kaiser's orders to convey the Ambassador and his staff to a local station … Again, by the courtesy of Sir Edward Goschen, a few of us correspondents were invited to join the staff, and I hardly realized at the time from what a hideous destiny that invitation preserved me. I suppose I should have been kept shut up in Ruhleben or some similar camp for four and a half years … But … our Ambassador saved me, and for twenty-four hours his train carried us all slowly lumbering through North Germany to the Dutch frontier.

On our way we passed … uncounted vans decorated with boughs of trees and crammed with reservists going to the Belgian front. The men had now chalked 'Nach Bruxelles' or 'Nach London' as well as 'Nach Paris' on the vans … At all the larger stations, too, the news of our train's approach had been signalled, and to cheer us on our way all the old men, boys and women of the place had flocked down with any musical instruments they could collect, and, standing thick on the platform, they played for us the German national tunes … Sometimes, to impress their patriotism more distinctly upon us, they brought their instruments close up to the carriage windows, and the shitting tubes of the trombones came right into the carriage. Silent and unmoved, as an Englishman should, sat Sir Edward Goschen, looking steadily in front of him, with hands on his knees, making as though no sight or sound had reached his senses …

14.2   The Early Treatment of British Nationals in Germany[6]

The early opinion of the German state after the outbreak of war was that enemy civilians should eventually be allowed to go home, but not until the army had completed its mobilisation. This opinion however raised a number of issues regarding the German obligations to the well-being of such foreigners in the meantime. If they were of military age—as in the case of Robert Le Rossignol—should they be treated as prisoners of war, permanently prevented from returning home and clothed and fed according to the Hague convention of 1907? If not prisoners of war, what obligations remained regarding their welfare and repatriation? And what to do if some of these people were subsequently found to be spies or saboteurs? By 14 August 1914, some rough guidelines had begun to emerge from the Reich Office of Interior after consultation with the Federal States.[7] Just for now these said, no citizen of an enemy state could leave the district in which they were living without the permission of the local military commander. A curfew on such people could be considered, but the advice urged moderation and there should be a 'mild' supervision of foreigners—if only in consideration of the interests of German citizens held in other countries. These guidelines followed those already issued by the acting military commander for Berlin, General von Kessel, on 10 August 1914 to which Robert was subject, such that; enemy aliens living within the police district of Berlin,

> … as long as they are not suspected of any crimes, and as long as they have a fixed address or a regular job, are to be placed under police supervision … but are otherwise to be left to get on with their lives.[8]

For the British in particular, these early apparently lenient guidelines meant that after the initial German hostility towards enemy nationals had (hopefully) subsided, their position should have been relatively secure and respected. Britons on the whole were well integrated into German society. They were not over represented in certain professions or trades; they were northern Europeans and so did not stand out in looks, mannerisms, culture or habits and frequently spoke German with a high degree of fluency. This was particularly true of the Le Rossignols of course, who were educated, professional and of both German and British blood. Had this early policy remained throughout the war then the 'lot' of the British national may well have been bearable. However, the 'guidelines' were just that, and as public opinion continued to harden against the British, in practice things became much tougher and the guidelines were usually interpreted harshly. A report appearing in the *Evening Post*[9] in New Zealand in 1916 describes the observations of a native New Zealander trapped in Berlin on the outbreak of war. He describes the humiliation of British nationals being ordered to report themselves every day to the police station nearest their home. Men and women alike had to report twice a day; they were restricted to a certain district and compelled to return indoors by eight o'clock at night. Even so, this was still regarded by some neutrals, such as the Americans, as 'considerable liberty'.[10] Later, in autumn 1914 following the failure of the German advance at the Battle of the Marne, things became worse as German policy makers came to favour a mass internment of all British males between seventeen and fifty five, irrespective of their profession, police record, marital status or residence. As a result, Robert Le Rossignol soon became one of the 5,500 Britons that Germany interned over the period from November 1914 to November 1918.[11]

14.2.1 The German Decision for Internment

By late September, early October, a number of Anglo-German agreements were in place that permitted the exchange of women, children, the elderly, doctors and priests etc, but not ordinary men of 'military age', largely because of objections raised by the British War Office and Admiralty. There were accusations for example that the German army was calling up men as old as fifty five, that British nationals of military age were being deported as prisoners of war from occupied Belgium and northern France—none of which was true—but the British subsequently made it known that they would only countenance an exchange of men *over* fifty five as opposed to those of forty five to fifty five who had previously been considered. The Germans in turn complained of the treatment of their citizens in Britain, of the violent anti-German riots in Deptford south London in mid October, of the overcrowding in the Newbury detention camp, and of the arrest of reservists returning home from various parts of the world on neutral ships docking at British controlled ports, or even being taken on the high seas. As early as 17 October 1914 then, and as a result of the bad faith between the two nations which severely taxed the mediating skills of the neutral American diplomats in Berlin and London, the German State seriously considered the internment of British subjects of military age. These considerations of course took into account the reports of apparent 'abuse' of German civilians in the UK, but also accusations of spying and sabotage within Germany by those favouring British interests. The meeting in Berlin between the Prussian Ministry of War, the Reich Office of Interior and the Foreign Office ended without agreement however and the matter was referred to the German General Headquarters at Charleville. But the case for internment they felt was a practical necessity for it would prevent spying and anti-German activity, deprive the British army and navy of manpower and potentially provide an opportunity for the use of important internees as 'bargaining chips'. All of these considerations were reinforced by public opinion which in the early days of the war saw a sustained shift from anti-Russian to anti-English (British) feeling driven by the German press. Only one newspaper (*Vortwärts*) took a stand against internment, the press otherwise being deluged with letters from 'outraged' citizens demanding a stiff response from the German Government to the 'abuses' suffered by Germans in the UK. But public anger seems to have been confined largely to the major cities in Germany. In Hamburg, Stuttgart and Cologne there was palpable dislike of the 'English', and in Berlin police reports for the week ending 02 November 1914 reported an upsurge in Anglophobia, but in the countryside, opinions were less agitated.[12]

But very soon the case for internment became overwhelming. There were fewer than 2,000 British men eligible for arrest in Germany at the outbreak of war, far too few to disturb the German war effort, but along with aspects of pragmatism and public opinion, propaganda too demanded that there was much political capital to be made from their internment. Internment would clearly boost public support for the war and demonstrate Britain's vulnerability to legitimate retaliation for its treatment of German citizens in the UK.[13] The British government's continued obstinacy over the exchange of her own subjects would allow Germany to throw a poor light on her consideration of their human rights and the role of international law. But in an act of blatant propaganda, Germany further decided that she would not proceed directly to internment but be best seen as having to be *forced* into it by issuing an ultimatum to the British government which demanded that unless all German citizens interned in Britain were released by 05 November 1914, British citizens between seventeen and fifty five years old in Germany would be correspondingly interned. On October 31st the German general staff issued instructions to prepare for the mass arrest, and as the British failed to respond to her ultimatum, Germany, 'reluctantly' of course, began her internment of British nationals on the morning of 06 November 1914.

The German action brought to an end a period of profound uncertainty for British nationals such as Robert. Ever since the outbreak of war they had to suffer restricted movement, the surrender of their passports, the 'jingoism' of the German press, rising public opinion against them, and occasional attacks on themselves or their property. Many became reluctant to speak their mother tongue in public places but at least they could return to their families at night. But with the introduction of internment things became worse … much worse.

14.2.2 The Ruhleben Internment Camp, Arrest and Transfer

In some respects, Germany's ultimatum 'backfired' on her, for she was completely unprepared for the consequences of its rejection—especially in terms of housing the British internees. At the meeting on October 17th the case for immediate internment may not have been made but in one aspect the general staff were united, that if internment was introduced then prisoners should *not* be accommodated in or near Berlin. Berlin otherwise would become a 'goldfish bowl' with the whole world and her representatives generating a constant stream of visits. Too much international publicity was bad, for things would inevitably not run as smoothly as international law demanded and the Allies would undoubtedly capitalize on German failings. The Prussian Ministry of War was therefore advised to begin the search for a suitable internment camp for British civilians away from the capital. However, an internment camp already existed in Berlin. It was at the disused Ruhleben[14] racecourse just a few miles from the centre of the city, and it had been used as a place of internment from the very beginning of the war. Acquired by the Ministry of War in September 1914 to house Russian and Polish prisoners from the Berlin area, it also accommodated a handful of Japanese and British citizens who the Germans suspected as 'spies or miscreants'[15] but who more likely drew attention to themselves by challenging the authorities as to the way in which it behoved *them* to order their lives. The authorities therefore ordered their lives for them.[16] When the German ultimatum was rejected and the mass arrests began, the Ministry of War had still not found a suitable camp away from the capital. Because of the rapidity with which internment was 'forced' upon Germany and the sheer numbers of prisoners involved, Ruhleben—despite all the misgivings—became the short-term solution. In the event it was to remain the main camp for British internees throughout the war (Plate 14.1).

Plate 14.1 A sketch of Ruhleben camp circa 1914. *Source* The internet

All across Germany on the morning of 06 November 1914, the internment of British nationals began.[17] Apart from women, children, and men over fifty-five, only doctors, priests, 'lunatics' and those too ill to leave their sick beds were excluded from the new measures. The whole process was completed within three weeks which according to a note issued by the Reich Chancellor[18] was to proceed *'firmly'* … but *'without brutality'*.

Civilians were normally arrested in their own homes and held at local police stations or prisons before being taken (by train) to Berlin. There were conflicting reports of their treatment. A few internees—very few—recalled being treated with some consideration, being allowed to travel to Berlin in comfort, free from unnecessary distress and being permitted to procure provisions on the way. But others recall children coming onto the railway stations to jeer at them or being spat at and reviled for the whole of their journey. Such was the intensity of feeling at the time that even some members of the German Red Cross refused to hand out food and drink, and youngsters selling freshly cooked sausages on the stations along the way often pushed the steaming delicacies in through the internees' carriage windows before withdrawing them then shaking their fists and muttering 'English pig-dogs!'[19] Overall, the common experience of the internee was that despite the nicety of the official guidelines, orders had been issued 'from above' to treat British civilians along the same lines as convicts. Many were driven around their point of 'capture' in Black Marias; they were often forced to carry their luggage through the streets to the railway stations amid the jeers of the local towns-folk and many were kept for long periods without food, observations which were supported by James W. Gerard the (neutral) American Ambassador in Berlin who later wrote[20];

> In the first days of the war it was undoubtedly and unfortunately true that prisoners of war … both at the time of their capture and in transit to the prison camps were often badly treated by the soldiers, guards or civil population … the instances were too numerous, the evidence too overwhelming, to be denied.

Once in the capital, all internees were taken to the *Stadtvogtei* jail[21]—one of the largest criminal prisons in Berlin—which now acted as a clearing house. Robert's address at Prinzregentenstrasse placed him in close proximity to the Ruhleben camp so he was easily arrested on the very first day of internment. Like all internees Robert was allowed to take some personal items to the *Stadtvogtei*; a bed sheet, a pillow case and two light blankets and from there he would have suffered the indignity[22] of being marched publicly through the Berlin streets to the railway station at Alexanderplatz. The Berlin press had made much of the new interment measures and the public turned out in numbers to view the spectacle. In general, it seems that these crowds were well behaved. Some Berliners even offered to help the prisoners with their luggage and others prayed for their early release but there were some reports of spitting and fist shaking. The saddest scene however was that of the wives of prisoners who were previously domiciled in Berlin. Many came to the *Stadtvogtei* to support their husbands and walked alongside them openly weeping and looking utterly distraught.[23]

From Alexanderplatz the prisoners were taken by train to the railway station at Spandau, known as the *Auswanderer Bahnhof* or the 'emigrants' station', where Russian and Polish emigrants to North America were disembarked and disinfected before travelling on to Hamburg or Bremen. At Spandau the internees were lined up in groups of four and marched the short distance to the racecourse at Ruhleben. For many of these men it was to be the last glimpse of the outside world they would see for the next four years. For others, freed in exchange agreements, their stay was mercifully shorter but they all became pawns in a game of international diplomacy,[24] and even though they were to be spared the horrors at the front, Ruhleben was to be an experience they would remember for the rest of their lives. These men, raked in from cities all over Germany, dumped in a swamp and housed in stables, set to work to found a British 'colony' within a few miles of their enemy's capital.[25]

14.2.3 Early Conditions in the Ruhleben Camp

Over the next three weeks prisoners from all parts of Germany streamed into Berlin but Ruhleben camp was hardly prepared for the influx. The place was just a disused trotting race-course. It had three grand-stands, a restaurant known as the 'tea-house', a club-house known as the 'Casino', residential quarters and offices for various functionaries, and eleven stable blocks one of them built of wood and bricks, and the others of brick and concrete. These stables served as barracks for the prisoners, who were housed both in the horse-boxes (some of which were still full of dung when they arrived) and in the lofts above, normally used for storing fodder.[26] The floors of the boxes and the lofts were made of concrete and the first influx of prisoners had to sleep here without an adequate provision of straw being laid. That which was provided was left over from the Russians and not overly clean. Soon however, straw sacks loosely called 'mattresses' were provided and then crude, planked, iron field bedsteads—three of which could be fitted on top of one-another to make 'bunks' to save space. Each man's official allowance of bed-clothing was initially two old horse blankets which stank of the usual, and any other protection from the cold as winter was setting in had to be provided at the prisoners' own expense. As more prisoners arrived there simply weren't enough blankets to go around[27] and each prisoner had to relinquish one of his. Anyone who had brought his own blanket had to surrender both! The horse boxes were about 12 feet square and into each one six men were packed. Those in the lofts that ran the length of the stable blocks fared no better being arranged shoulder to shoulder all around the walls and up and down the middle. In these spaces designed to accommodate 27 horses, over 300 men had to live, eat, sleep, wash, and dry their linen on improvised clothes lines that stretched from beam to beam. They also had to somehow try to get along with one-another.

Clearly the Germans had completely underestimated the number of prisoners they had to accommodate and men openly mocked their much vaunted 'Teutonic organizational prowess'. No arrangements at all had been made to heat the stables, and lighting was grossly inadequate. Such lights as there were hung in the corridors between the boxes and little penetrated the boxes themselves making the prisoners' life during winter one of perpetual twilight. Washing facilities were primitive. Typically, a barrack had just two cold water taps and fifteen earthenware bowls, never-the-less prisoners were expected to parade—washed and dressed—at 6.30 a.m. each day and as the soldiers went around to wake up the sleepers in the morning, they shouted[28] 'Rouse you

English Swine! …' As for those articles necessary for some semblance of civilized life, provision was equally mean. Each man was given a coarse towel which resembled a dish cloth, a tin bowl to hold his food and nothing else. No knives, spoons, forks, plates, cup, nor cake of soap. There was a canteen where these could be purchased but many had no money at all and had to wash without soap, eat with their fingers and drink from the tap. Added to all this there were no mirrors, so shaving was 'hit or miss' and prisoners had to guess whether their hair was parted straight.[29] The 6.30 a.m. parade must have been quite a sight.

The order that exempted the sick from internment too was more honoured by its breach than by its observance. There were suffers of tuberculosis, bronchitis, cancer, rheumatism, infectious skin diseases and all these—together with the crippled and the 'half witted'—were mixed indiscriminately with the able bodied. Even when it came to cleaning out their fouled boxes and scrubbing the floors to make them habitable for themselves, the sick were made to work, but very few were given tools such as shovels or brooms to ease their labours. Exercise for the able bodied too was a problem. At first the racecourse, 1200 m in circumference, was strictly out of bounds; men had to exercise themselves in the stable yards all of which were unpaved. The soil was sandy and loose and as there was no provision for drainage it quickly transformed itself into mud when it rained—and there was much rain in the autumn of 1914. Puddle merged into puddle until they became more like little lakes and through this cold quagmire all the men had to wade up to three times a day, for a distance of almost a quarter of a mile, to fetch food from the kitchens under the grandstand. Each barrack in turn formed up for this 'privilege', often waiting for up to half an hour in the mud to get served. The food consisted of sugarless black coffee or tea in the mornings or sometimes a wretched concoction called 'cocoa'. For dinner there was 'soup' which tasted like cabbage water one day, turnip water another day and rice water on yet another. Sometimes a 'lump' of sausage was thrown in, and a loaf of the infamous[30] standard German military issue 'black bread' (*Kommissbrot*) —made largely with rye, wheat and potatoes—was added every two days.[28,31] The same muddy journey had to be undertaken to reach the latrines which had been dug for them; latrines that the American Ambassador[32] described as 'a danger not only to the camp but to Berlin!' for no disinfectant was ever provided by the authorities. And add to all this the pitiless, piercing winds from Poland and the Pinsk marshes that cut the across the camp each winter.

But out of all the chaos there soon came some order. On 08 November 1914 the prisoners were paraded and told that each barrack had to appoint an interpreter, fluent in both languages, to represent them in any dealings with the authorities. In true British fashion an informal 'election' was held and representatives, the barrack 'captains', were appointed; a process that was repeated as other barracks later filled up. A committee was formed from the barrack representatives; meetings were scheduled, minutes kept, and a camp captain,[33] the 'captain of the captains' was nominated. It was from these small beginnings and the realization that 'unity was strength' that the camp began to organise itself to try and improve the lot of the prisoner.

The first meeting of the committee on 10 November discussed a number of issues; *privileges*, such as permission for the use of a telephone, *latrines*, payment of men to keep them clean, *baggage*, missing baggage to be brought to the camp, *clothing*, especially for destitute prisoners, *supplies*, essentials had to be brought to the camp, and *possible releases*, especially for those caught up in internment that fell outside the age remit or suffered ill health. A principal objective of the committee was to take every opportunity to get as much of the administration of the camp into their own hands and in this respect, they had the backing of the German authorities. Over the next few months the committee appointed postmen, cashiers, firemen and laundrymen and they raised a barrack revenue[34] to pay for buckets and brooms and a small wage for those prepared[35] to keep the barracks clean. They even organised a police force—especially effective at preventing fights and regulating queues—and in Ruhleben the educated prisoners volunteered to teach the ignorant.[10] As time passed prisoners tried to put their own mark on the camp. The alleyways and squares became known by British landmarks such as 'Trafalgar Square' and 'Bond Street' where many 'stores' were in operation such as a tailor's shop, a shoemaker's, and a watchmaker's.[10] All of this reminded them of home, but for those who may have tried to get home by escaping, penalties were harsh, such as indefinite imprisonment in solitary confinement at the *Stadtvogtei* or deportation to the Havelberg camp some 40 miles north west of Berlin where conditions were much worse. Even for relatively minor indiscretions such as drunkenness or smoking in the lofts, there was a three-day confinement in Barrack 11 on 'black bread' and water only (Plates 14.2 and 14.3).[36]

Plate 14.2 'Trafalgar square' with barracks 10 and 11 in the background. The external stairs led to the lofts above the horse boxes. Photograph from Powell, op. cit. (note 6)

Plate 14.3 'Bond Street' alleyway showing the stalls on either side. Photograph from Powell, op. cit. (note 6)

As anticipated, the camp was subject to visits by the neutral powers—especially the Americans—who alongside the Dutch adopted the role of mediating between the Germans and those prisoners and internees of the Allied nations. The first of these was in November 1914 by Mr. Morgan the American Consul General at Hamburg. Disgusted by the sights that he saw he openly spoke before the camp Commandant of the iniquity of 'putting men to sleep in that God-damned hole!'[37] but it was a visit by the American Ambassador himself on 03 March 1915 that coincided with the start of the era of effective reform.

Ambassador James W. Gerard argued that his representatives in Germany should have the right to visit the prison camps on giving reasonable notice—24 h where possible—and that they should be able to converse with the prisoners privately, within sight but out of hearing of the camp officials. In return he agreed that an endeavour should be made to adjust matters complained of with the camp authorities before bringing them to the notice of higher authorities, a 'softly softly' approach that seemed to work well. Gerard later recalled that '… at two periods during my stay in Berlin I spent enough days at the camp to enable every prisoner who had a complaint of any kind to present it personally to me … 'However, his claims may have been somewhat exaggerated as on his first visit, his guided tour was continually interrupted by shouts from prisoners such as 'Let's have some bread not sawdust!' and 'We want bread!'. Many written messages in a similar vein were thrown into his car as he left, despite the best efforts of the guards.[38] On the whole however, the American interventions were positive leading to substantial improvements at the Ruhleben camp. But there were also other factors at play by the spring of 1915, such as the complete breakdown of all negotiations regarding prisoner exchanges, and the increase in the number of German prisoners held in the UK.[39] These additional factors meant that more prisoners would be held for longer than was otherwise anticipated thereby forcing the German and British authorities to improve their conditions. From March 1915 onwards therefore, Ruhleben became reasonably well provisioned in terms of food, clothes, reading materials, board games and sports equipment, all of which was reflected in the number of parcels delivered to the camp; from none in November 1914 and less than 4,000 in December, to over 16,000 during March 1915.[40] By June 1915 prisoners were allowed to use the racetrack for up to seven hours a day giving 'a splendid field for all kinds of games'.[41] But for those dispatched to the camp at the beginning of internment, conditions were undoubtedly at their worst. Two scenes[42] graphically illustrate the change brought about by the various interventions over the four years of the war. In the first, during November 1914 when the German nation was still amply provisioned, a group of cold, hungry prisoners gathered in a shed hiding from the wind. One of them handed around a letter from an educated German lady who rejoiced to hear that 'English' civilian prisoners were to be given only the bare necessities of life because 'that was all they deserved!'. The second scene was from the same month in 1918. Germany's economic and military collapse had begun, the revolution had broken out and men were free to spend time outside the camp. An internee walked out with a small parcel of food under his arm, intending to send it to his wife in Cologne. He was soon approached by a German lady of refined manners and accent. 'Forgive me' she said, 'Of course I oughtn't to speak to you, but I'm so hungry and I know you have food in your camp. Could you—could you—for the love of God and humanity, spare me a few biscuits? For my children they are starving, I should be eternally grateful to you if you would help me' The prisoner did what he could leaving her dissolved in tears, and between these two extremes lies the story of Ruhleben.

14.2.4   22 March 1915

Even though Robert Le Rossignol left us no record of his internment that we can use in any individual biographical sense, there is little doubt that he carried his memories with him to the end of his life.[43] All the evidence suggests that internment was a deeply humiliating and frightening experience, and those such as Robert would not have taken easily to it. These were members of the comfortable middle classes, used to the finer aspects of life, thoroughly integrated into elite German society, often with young families and with no tendency towards any form of political-social disruption or national subversion. Many like Robert who had married German women with whom they had children were often deemed *deutschfreundlich*,[44] but even so he fell through the coarse interment filter of the Reich and found himself in Ruhleben's horse boxes rubbing shoulders with 'fishermen from Hull and Grimsby, black sailors from Africa and the West Indies, Jewish tailors and music hall artists from the east end of London, professional football players and golfers, jockeys from the royal racecourse at Hoppegarten and a host of criminals, conmen, drifters and thugs who happened to be in Berlin or Hamburg when the war broke out'.[45] This bubbling 'melting pot' of humanity babbled a 'kaleidoscope' of languages; Afrikaans, Arabic, French, Hebrew, Italian, Russian, Spanish and Yiddish. Some prisoners even spoke Welsh, and alongside the linguistic cacophony, through the rigours of Ruhleben camp in winter and just a few miles from his wife and children … within the fouled horse boxes Robert must have renewed his acquaintance with an 'old friend', ammonia. Ruhleben turned Roberts's life 'upside down'. His internment coincided with the time when the camp was at its worst. He was present before many of the reforms were enacted, when it was over-crowded and volatile, and during a cold, wet winter—the coldest ever recorded in W. Europe. But suddenly, on the 22 March 1915, after four months as a prisoner, he was released to Berlin 'for the purpose of resuming his former occupation.'[46] Although his release came just three weeks after Ambassador Gerard's first visit, it is Charlotte Haber's memoirs[47] reported by

Szöllössi-Janze[48] that explain the reason for Robert's release viz, 'as a result of the direct intervention by Haber in exchange for the release of a German interned in England.'[49]

At the time, and stimulated by the overcrowding caused by the breakdown of exchange negotiations, German policy began to move slowly in the direction of releasing a small number of 'suitable' internees to their former occupations—if asked to do so by their previous employers.[50] Haber probably took advantage of his senior position within the Reich, together with his connections with Koppel and the 'Auer', to exploit this avenue on behalf of Le Rossignol. In this regard Robert's release would certainly have been supported by Ambassador Gerard who at the time recalled with some concern how even '… men of education … were compelled to sleep and live six in a box stall …' at Ruhleben.[51] Haber's successful intervention at a time when exchange negotiations had faltered, undoubtedly reflected the high regard he had developed for Robert over the previous eight years. It is quite remarkable that he found time to lobby on his behalf and arrange the detail of the exchange when otherwise fully engaged with matters of national importance such as the preparation for the gas attack at Ypres/Langemarck. Even so, Robert was now 'free', but Szöllössi-Janze also records that for the remainder of the war, 'his trace was lost …'.[52] This 'disappearance from the scene' was far from a lack of appreciation for Haber's help however, but a practical expedient. Le Rossignol was widely associated with Haber at the time, but to sustain a relationship with a serving officer who by now had become so prominently involved in the mobilisation of German science to support her war effort would have been impossible to justify to His Majesty's Government (HMG) in the UK after the war. So, on his release Robert returned to the 'Auer' to 'do what he liked' (Chap. 1) which, because the tungsten ductility problem had by now been solved by Coolidge, meant investigating new ways of exhausting lamps. The war therefore broke the working relationship between the two men and Robert stepped 'into Haber's shadow'. To history he has remained there, maybe not out of choice, but certainly due to circumstance.

14.3   Wartime at the 'Auer'

At the 'Auer', the friendship between Robert and Mayer grew. In conversation with Chirnside (Chap. 1), Robert described Mayer as being 'extremely kind to me', so much so that the two men agreed to build and share a house together in Berlin after the war. Over the eighteen months following his release and alongside the sinking of the Lusitania, the evacuation at Gallipoli, the introduction of 'phosgene' on the Western Front and the battles of Verdun, Jutland and the Somme, Robert worked on two processes; for manufacturing, and then exhausting and sealing lamps. Patent applications were prepared in Germany by 24 October 1916 with Robert as Assignor to the GEC company of New York,[53] but a variation on the patent, improving the sealing of the exhausted lamps, was later widely filed throughout Europe and the U.S.A.[54] The distribution of these applications shows that the improvements were internationally significant in terms of the manufacturing of the lamps, leading to improved life-times, economies in effort and raw material, and facilitating automated mass production. Robert's first innovation addressed the efficient exhaustion of lamps[55] which up until then had been achieved by first making an opening in the crown of the bulb, 'welding' a narrow glass exhaustion tube to it which was then sealed off by melting once the lamp had been exhausted. This led to a 'pip' seal on the crown of the bulb,[56] it also introduced a considerable number of steps into the exhaustion process and wasted glass, for these tubes could only be used once or twice before they were discarded. Robert's innovation was to exhaust the lamp from the base of the bulb rather than the crown.

Figures 14.4 and 14.5 show the drawings accompanying Robert's original patent application. 'Figure 1' in Fig. 14.4 illustrates a lamp prepared by Robert's method. The central stem, b, of the lamp supports the filament unit and is flared at its base to facilitate its fusion to the wall of the bulb. The short hooked 'tubulature', a, is the means by which the lamp interior is exhausted and is formed from the glass of the stem alone using the apparatus drawn in both plan and side views in 'Fig. 2' and 'Fig. 3' respectively of Fig. 14.4. To form the tubulature, a stem was firstly secured in the 'chuck' of the device and rotated, simultaneously warming the base with a gas torch. The chuck was then stopped and a narrow blow pipe flame was used to heat up a small circular portion of the stem wall just above the flare until it became 'plastic' and easily deformed. The lever f, pivoted adjacent to the chuck and carrying a hooked, cylindrical, tapered, mandrel on its free end, was then swung into the plastic glass creating the 'tubulature' shown in 'Fig. 2.' which hardened on cooling. Bulbs were placed over the modified stems and fused to the flared base to provide the lamps shown in Fig. 1. The tips of these 'tubulatures' were then broken and the modified lamps placed in the exhaustion/sealing apparatus shown in Fig. 14.5 which Robert described as able to hold a 'plurality' of lamps. Evacuation of the chamber simultaneously evacuated the lamp interior via the tubulatures leading to balanced pressures inside and outside the bulb. This meant that the chamber could then be heated to near

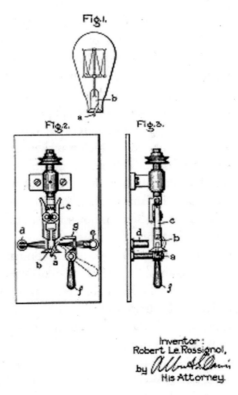

Fig. 14.4 A bulb showing Robert's short hooked 'tubulature', a, and the mandrel used to create it[53]

the softening point of the glass without causing a collapse of the bulb helping to evenly distribute strains and distortion forces. Once the desired level of exhaustion was reached the tubulatures were sealed by fusion[57] using electrical heating elements at the base of the lamps. Sealed and cooled, the lead[58] wires from the filament were soldered to the lamp base, and the base baked onto the bulb. Quite ingenious.

Robert's ideas embodied in these patents eliminated the need for an external exhaustion tube which in turn saved effort, glass, and banished the

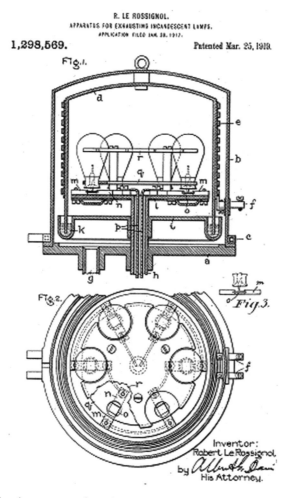

R. LE ROSSIGNOL.
APPARATUS FOR EXHAUSTING INCANDESCENT LAMPS.
APPLICATION FILED JAN. 30, 1917.

1,298,569.   Patented Mar. 25, 1919.

Fig. 14.5 Robert's apparatus for exhausting and sealing the bulbs modified with his 'tubulature'[53]

'pip' seal. The balanced pressures produced within his apparatus were novel, permitting the introduction of procedures that removed defects which otherwise weaken the lamp and which were impossible to apply before. The whole led to a reduction in the number of steps needed to produce the final lamp and facilitated the mass production of a more robust product. By 1919, the whole industry had turned to lamp production much in the fashion described, but Robert's 'tubulature' was replaced by an integrated exhaustion tube running up the length of the stem which remains the situation today—almost 100 years later—although incandescent lamps of course are now being phased out. No doubt these patents were important to Robert in a professional sense for they distributed his name internationally and independently of Haber. But equally they stood as testament to the fact that during the war, when the 'Auer' was actively supporting the German war effort, his time as an employee there was spent 'benignly', working on improvements to the manufacture of the ubiquitous, but humble, domestic 'glow lamp'. Soon, this fact was to assume an unexpected importance when a challenge to Robert's patriotism arose which, if not contested, no amount of successful patents could ever ameliorate.

14.4 'The Case of Robert le Rossignol'. British Foreign Office Notes 1916–17[59]

On 31 October 1916, the American Embassy in Berlin sent the following somewhat vague dispatch to her sister Embassy in London[60];

AMERICAN EMBASSY

BERLIN

The American Embassy at Berlin presents its complements to the American Embassy at London, and has the honor to refer to an enquiry made by Mr. Robert Le Rossignol, a chemist by profession formerly interned in Ruhleben, but now released to Berlin for the purpose of resuming his former occupation, whether any disapproval on the part of the British Government exists, and whether, in case of his eventual return to England, he would subject himself to censure or loss of standing on this account.

Berlin October 31st, 1916.

The detail of Robert's original enquiry has not survived at Kew,[61] but clearly at about the time of his patent applications in late 1916, he suddenly felt moved to seek the approval of His Majesty's Government (HMG) regarding the 'appropriateness' of his wartime work at the 'Auer'. The American Embassy in London soon passed the communication to the British Foreign Office in the following terms[60];

American Embassy,

London

The American Ambassador presents his complements to His Majesty's Secretary of State for Foreign Affairs and has the honour to transmit herewith enclosed a copy of a letter he has received from the Embassy at Berlin, dated the 31st ultimo, regarding the case of one Robert Le Rossignol, who has been released from Ruhleben for the purpose of resuming his former occupation.

November 8th, 1916.

Over the next few months this brief note was to generate a flurry of communications at the highest levels in the British Government, especially between the Prisoners of War Department at the Foreign Office, the Ministry of Munitions, the Home Office and the War Office at Downing Street as the hand written annotations, observations and comments in the records[59] of these departments for 1916 and 1917 shows. However, the initial reaction of HMG to the American communication was hesitant, for there seemed to be no policy in place to deal with such requests and no indication of what Robert's 'former occupation' was. Neither was there any understanding that Robert's release had been arranged by Haber, which—as we will see—may have made his situation more difficult. As a consequence, the following hand written annotation in the Foreign Office (FO) records (Prisoners of War)[62] suggested placing any responsibility for his position firmly on Robert's shoulders.

> Mr. Le Rossignol was an analytical chemist[63] employed by the Osram Electric Lamp Co., and it doubtless suited the interests of the German Gov that he should resume his occupation.
>
> But I suggest that it would be wise not to send a definite reply on the point raised. Mr. Le Rossignol must know for himself how far his work is inimical to British interests and he will have a chance on his return to the UK to explain his proceedings.
>
> ? Reply that on the facts given HMG would prefer not to express an opinion as to the course adopted by Mr. Le Rossignol.

These initial considerations were subsequently passed to the War Office, the Home Office and the Ministry of Munitions on 15 November 1916 for their joint observations regarding the probable nature of Robert's work. The Ministry of Munitions however soon expressed some alarm in their reply to the Prisoners of War Department, alarm driven by Robert's association with Haber, ammonia, and his continued presence in Germany;

Ministry of Munitions of War,

Whitehall Place, LONDON, S.W. 11th December, 1916.

Sir,

I am directed by the Ministry of Munitions to refer to your letter (...) of the 15th. November, transmitting a note from the United States Ambassador relative to the case of Dr Robert Le Rossignol, who has been released from internment at Ruhleben.

I am to inform you that the Minister understands that the gentleman in question, who it is understood has married a German lady, is a Fellow of the Chemical Society and that he is an expert on the processes employed in the production of synthetic ammonia which is urgently required in the production of explosives. He has carried out much work on synthetic ammonia in conjunction with Haber at Karlsruhe, and it is considered probable that this may represent the 'former occupation' which he is resuming. If this were the case, he would doubtless be employed to assist in the production of German military requirements. In these circumstances, Mr. Montague is of the opinion that a reply should be forwarded to the effect that His Majesty's Government would view with strong disapproval the resumption in Germany by Dr Le Rossignol of his former occupation, and he must be prepared to face all the consequences of such disapproval in the event of his eventual return to England.

I have the honour to be, Sir,

Your obedient servant

R. V. Vernon.

To the Controller, (Prisoners of War Department), Foreign Office, S.W.

By 19 December 1916, the Prisoners of War Department—subject to approval by the Home Office and the Army Council—had decided to err on the side of caution and reply to Robert much in the terms suggested by the

Ministry of Munitions. The following letter (from FO 383-211) was sent to both the Home Office and the Council.

Prisoners of War Department,

Downing Street. December 19th 1916

Sir,

With reference to my letter … respecting the release of Mr. Robert Le Rossignol from Ruhleben, I am directed by the Controller of the Prisoners of War Department to transmit herewith a copy of a letter from the Ministry of Munitions on that subject.

I am to state that subject to the concurrence of [*the Home Office and Army Council*] it is proposed to reply to the enquiry by Mr. Le Rossignol in the terms suggested by the Ministry of Munitions.

I am Sir,

your most obedient, humble servant.

G. R. Warner

To:- the Under Secretary of State, the Home Office, the Secretary to the Army Council.

However, early in the new year Robert's case was unexpectedly highlighted by Ambassador Gerard in Berlin who wrote to His majesty's Government via the American Embassy in London outlining the difficulties they faced because of a lack of guidance from London regarding the employment of British prisoners in Germany. In this sense Robert's letter of October 1916 regarding his employment was to help form a policy which until then had received little attention, for the United Kingdom had never fought a war in which such concerns arose. The relevant parts of Gerard's letter (from FO 383-279) appear below;

Embassy of the United States of America

Berlin, January 9th, 1917

Sir:-

I have the honor to refer to a communication of this Embassy of October 31st, in regard to the case of Mr. Robert Le Rossignol, a chemist by profession, who has been released to Berlin for the purpose of resuming his former occupation, and to inform you that a large number of inquiries of the same nature as that made by Mr. Le Rossignol have lately been received by prisoners at Ruhleben. It appears that, where former employers have asked for the services of their former employees a policy is now prevalent on the part of the military authorities to release the latter for the purpose of taking up their former work. A considerable doubt exists on the part of some of these persons as to whether such employment is in accordance with the wishes of the British Government. In such cases, while not wishing to assume any responsibility, I have made a practice of advising the men that where the work is not concerned with the manufacture of the materials of war, it would probably, in view of the physical and mental benefit likely to accrue, not to be unpatriotic to accept the employment offered ...

... I shall be grateful for whatever instruction or suggestion the British Government may wish to make in regard to the whole subject, which is now of considerable importance ... [and also to ascertain whether German civilians detained in England have been engaged voluntarily or involuntarily in outside work.]

I have the honor to be Sir,

Your obedient servant.

James W. Gerard.

Robert's request, Gerard's intervention, together with the fact that German prisoners had long been invited to volunteer for work in the UK, forced the Prisoners of War Department to make its position in the matter quite clear. In a letter dated 26 January 1917 sent to the Secretary of the Army Council, the Ministry of Munitions, the Home Office and the American Embassy and the Dutch Legation in Berlin, the Prisoners of War Department established HMG's position once and for all, and it was a position particularly cogniscent of those such as Le Rossignol who were presumed to have talents useful to the German war machine;

Prisoners of War Department,

Downing Street, January 26th, 1917.

Sir,

I am directed by Lord Newton[64] to transmit to you, herewith, to be laid before the Army Council, a copy of a note from the United States Ambassador with regard to the employment of British civilian prisoners of war in Germany.

Lord Newton is of the opinion that, in view of the fact that enemy civilian prisoners of war are being invited to volunteer for work in this country, His Majesty's Government are not in a position to raise any objection to British civilians volunteering to work in Germany. In so informing Mr. Page,[65] Lord Newton proposes to state that it must be understood that British civilians must not accept any invitation to take part in, or resume a former occupation connected with the production of explosives or other war material...

A similar letter has been sent to the Ministry of Munitions and the Home Office. I am Sir, your most obedient, humble, servant.

Robt. Vansittart

Just a week earlier, the Home Office had finally made its position regarding Robert's employment quite clear in a letter to the Prisoners of War Department dated 18 January 1917, which was largely in agreement with the guidelines issued by Lord Newton;

Home Office, Whitehall.

January 18th, 1917.

Sir,

In reply to your letters of the 15th November ... forwarding papers in the case of Robert Le Rossignol am directed by Secretary Sir George Cave to suggest for the consideration of Mr. Secretary Balfour[66] that as the reply proposed by the Ministry of Munitions is based upon the assumption that Mr. Rossignol's work is connected with munitions of war, a point on which there is no precise information, the answer should rather be that as the H.M. Government do not know the precise nature of Mr. Le Rossignol's occupation, they are unable to express an opinion on the point, but that they would disapprove of him engaging in any occupation connected with the operations of war.

I am, Sir, Your obedient servant,

However, by 12 February 1917 the Foreign Office replied to Robert's inquiry via the Netherlands Minister at Berlin in terms favoured by the Ministry of Munitions rather than the rather 'softer' approaches of Lord Newton and the Home Office. Their reply was 'to the point' and somewhat intimidating;

Memorandum for communication to the Netherlands Minister at Berlin

His Majesty's Government have received a note from the United States Ambassador, transmitting a copy of a note from the United States Embassy at Berlin, dated the 9th January, regarding the case of Robert le Rossignol who has been released from Ruhleben for the purpose of resuming his former occupation.

As it is understood that Mr. Le Rossignol's former occupation is one connected with the operations of war, his Majesty's Government would view with strong disapproval his resumption of that occupation, and in such an event he would have to be prepared to face all the consequences of such disapproval in the event of his return to this country.

FOREIGN OFFICE, February 12th 1917.

The Netherlands Legation at Berlin duly 'transmitted' HMG's opinion to Robert in a letter dated 20 March 1917. Robert's reply to the Legation was immediate. He was clearly distressed that the nature of his work discussed with the American representatives in October had not been properly conveyed to HMG and his patriotism subsequently questioned on the basis of earlier fundamental scientific work he had carried out in time of peace;

Prinzregenten Str. 86,

Berlin-Wilmersdorf.

23. March 1917.

Dear Sir,

I am in receipt of yours of the 20th inst. in which you inform me that the British Foreign Office would view with strong disapproval my resuming my occupation if the work should be in any way connected with the operations of war.

I think there must have been some misunderstanding, as I stated to Mr. Dresel of the American Embassy that my work was neither directly nor indirectly concerned with the operations of war. I am employed in an electric glow lamp factory, my sole work being to improve the methods of manufacture of glow lamps such as are used in every household, I have nothing whatever to do with the construction or manufacture of search-light or similar types of lamps which may be used for the purposes of war. If my occupation had any direct or indirect connection with warfare I would not have pursued the same.

I would be much obliged to you if you would inform the British Foreign Office as to what my occupation is and ask if there is any objection to me pursuing the same, which I state again has nothing to do with the operations of war.

If the British Foreign Office objects to me pursuing my present occupation I will give up the same and take the consequences such as having to leave my family and be reinterned in Ruhleben.

Yours truly, Robert Le Rossignol.

Conditions at the Ruhleben camp had certainly improved since Robert's internment, but the prospect of once again leaving Emily and the two children (now at a different address in Prinzregentenstrasse) for that 'God-damned hole' (as Mr. Morgan had described it), must have filled Robert with dread and so his letter made it perfectly clear that his work at the 'Auer' had not the slightest connection with the war. Robert's reply was sent from the Dutch Legation to HMG with some urgency in a dispatch dated 26 March 1917. By early April the 'difficulty' had seemingly been resolved when a note of clarification on Robert's position was sent from the Foreign Office to Berlin;

Memorandum for communication to the Netherlands Minister (British Section) at Berlin.

With reference to the Note No.B.203 of the 26th ultimo from the Netherlands Legation at Berlin, regarding the case of Mr. Robert Le Rossignol, His Majesty's Government have the honour to state that they have no objection to British civilians volunteering to work on the clear and express understanding that they must not accept any invitation to take part in or to resume a former occupation connected with the production of explosives or other war material, or in any way with the operations of war.

A statement in the above sense was contained in the Foreign Office memorandum of the 20th February last which related to the general question of British civilians in Germany undertaking voluntary labour.

Foreign Office S.W.1,

April 13th 1917.

However, during the summer of 1917 when Robert applied to HMG for an emergency British passport, the letter from the Ministry of Munitions from December 1916 was still in his file. Once again it was raised as an issue but the Foreign Office records of the time show that HMG were now developing a more informed view of Robert's earlier work. Robert's application was sent to HMG from the Netherlands Legation at Berlin via the Hague;

The Netherlands Legation (British Section) at Berlin presents its compliments to the British Legation at the

Hague and has the honour to transmit herewith, for such action as may be considered appropriate, an application form for a British passport duly signed by Robert le Rossignol.

In this connection the Netherlands Legation has the honour to refer to the Memorandum from the British Foreign Office ... dated April 13ᵗʰ 1917, and previous correspondence regarding Mr. le Rossignol.

Berlin, September 27ᵗʰ 1917.

A hand written entry in the FO records (FO 383-312.) dated 16 October 1917 outlined an initial response to Robert's application;

Mr. le Rossignol has not lived regularly in HM Dominions since 1907 and he has a German wife. He is now working in a glow lamp factory and has offered to throw up this work and return to Ruhleben if we object. He is only 33 and only requires the passport for identification. Ask Registrar General if he can confirm statement as to birth ... our correspondence to HO and say we presume no objection.
T of War?... Dept.

In a subsequent letter dated 16 October 1917, the Prisoners of War Department contacted the Registrar General at Somerset House with their request for confirmation of Robert's date and place of birth, and in a letter dated 22 October 1917 referred the request to the Superintendant Registrar, 5 Library Place, Jersey, as records of births in the Channel Island records were not in his custody. In a letter to the Prisoners of War Department dated 29 October 1917, the Registrar was finally able to confirm Robert's statement and forwarded a copy of his birth certificate to the Prisoners of War Department. This certificate remains in FO 383-312.

With Robert's nationality now established, the question of the passport rested on the opinion of HMG regarding his standing in Jersey (where 'his family were well known') and his work with Haber in Germany. A typed note in FO 383-312, subsequently forwarded to the Home Office, the War Office and the Ministry of Munitions between 08-10 November 1917, relates the general feelings regarding Robert's application;

I think we ought to be careful not to be prejudiced in dealing with Le Rossignol's application by the letter of Dec. 11 from the Ministry of Munitions. Le Rossignol's statement as to his present occupation is supported by the description of Le Rossignol in 1914 by his brother Major H. S. Le Rossignol, as 'an analytical chemist employed by the Osram lamp factory'. Le Rossignol was certainly in Berlin at the time of his internment and when released from Ruhleben was released, not to Karlsruhe, where he lived formerly and where he presumably worked with Haber in connection with synthetic ammonia, but to Berlin where he now is. He was not interned in a pro-German barrack.

It seems pretty clear that Le Rossignol's association with Haber was in peace times and it would, I think, be unjust to condemn a British subject for having worked at the study of explosives in peace time in a country where much important work was being formed.

The decision should, I think, turn on any investigations which the Home Office may be able to make in Jersey concerning Le Rossignol, whose family is well known in the Island.

Send a copy of Le Rossignol's application to the H.O. ... and ask that enquiries be made in Jersey and we may be furnished with their opinion as to the issue of a passport which is presumably required only for purposes of identification.

(If Le Rossignol is really loyal in his sentiments it would be a pity to alienate the services of a man who may have considerable knowledge of German processes.)

These sentiments were finally distilled into a letter from Lord Newton on 08 November 1917 to the Home Office, the War Office and the Ministry of Munitions—although no record remains in FO 383-312 regarding the outcome of the investigations in Jersey. One can only assume that they were no barrier to Robert's application and that his passport was duly issued. However, because of his work on synthetic ammonia and its central role in the 'Chemist's war', Robert was still associated with 'explosives' in the eyes of some within HMG, even though in reality he would have had no knowledge of the 'German processes' they referred to. But one question remains, especially so when Robert's approach to establishing the 'appropriateness' of his employment conflicts with what one thinks one would have done in his place, or with what one might think is the more obvious route. Why did it take Robert so long after his release from Ruhleben to seek re-assurance from HMG? The dates on the various letters over the period show that lines of communication were available to pass information between London and Berlin in just a few days. But after his release in March 1915 it was eighteen months before he was to enquire whether 'any disapproval on the part of the British Government existed' regarding his position there.[67] One answer may lie in the confusion surrounding this whole area. HMG clearly had no policy in place at the time that could be used as guidance by officers of the neutral powers to advise released prisoners, and as Ambassador Gerard recalls in his letter, Robert's enquiry was one of many which began to arise at the time— albeit more significant than most. The lack of a policy therefore, probably translated into a lack of obligation on the part of released prisoners to immediately seek approval for their employment, especially so in light of the advice Gerard and his officers were independently dispensing. Or maybe it was because Robert, who after 'stepping into the shadows' and avoiding attention, was beginning to feel some discomfort with the ever-increasing engagement of the 'Auer' with the German war effort, or indeed from his former association with Haber whose involvement in the war was becoming more and more prominent.

But there was another element in play, one which he never revealed to HMG, for Robert was now eligible to receive monies from the private contract he had made with Haber on 01 May 1908.[68] Robert's original engagement with this contract was as an ambitious young man in a time of peace. But now, older, the father of two children, and witness to the terrible suffering the exploitation of this work was causing,[69] within the wider context of returning to the UK at some time, it would not be unreasonable for him to want to discover the British 'establishment's' official view of his position in Germany and whether in any way he had become 'tainted' by association. Whatever Robert's reasons for the delay, his enquiry was to establish HMG's opinion that—despite their misleading association of him with explosives— no discredit should be attached to a British subject for having worked—in peace time—in an area that was now entirely counter to British interests. However, because of his contract with Haber, the First World War was about to make Robert Le Rossignol a very wealthy young man.[70]

14.5 The Haber—Le Rossignol Financial Arrangements

Following Germany's economic and military collapse during 1918, the 'Great War' came to an end with the Armistice of 11 November followed by the Treaty of Versailles on 28 June 1919—exactly five years after the assassination of Franz Ferdinand in Sarajevo. Germany's dependence on synthetic ammonia and Haber's constant lobbying of it's cause at the highest levels led to an increasing escalation of royalty payments to him from the BASF over the period 1914 to 1919. When Robert returned to the UK, he did so with a financial legacy from Haber that began in 1914 and extended to well after the war. Throughout the war, and up until May 1920 when Haber [returned from Switzerland and] finally took control his own legal and financial arrangements,[71] his affairs had been handled by Max Mayer using accounts at the Dresdener Depositenkasse and the Württembergischen Vereinsbank (Chap. 7). Through Mayer, royalty payments from the BASF to Le Rossignol were made because, 'in the circumstances, it was felt inappropriate' that Robert should be paid directly by Haber (Sect. 14.2.4 and Chirnside's conversation with Robert, Chap. 18). When Robert returned to the UK, he authorized Mayer to be his representative in Germany and Mayer continued to organise the payments which Robert's contract with Haber entitled him.

Tracing the movements of monies is a difficult enough task at the best of times, but during a war when *private* transactions between individuals are performed by a third party, it becomes almost impossible. However, through his conversation with Chirnside, (Chap. 18) we know that Robert received a first 'substantial' royalty payment from Mayer in 1914, although we have no idea how much it was. But the work of Szöllössi-Janze[47] on the accounts of Haber and the BASF allows us to construct an 'audit trail' for a nominal period between 1915–1917[72] when ammonia production was reaching a peak and where accurate figures for production and royalty payments are available. Such a period provides a good indicator for Robert's entitlement and in this respect I am entirely indebted to Prof. Dr. Szöllössi-Janze at Munich for confirming my understanding of her figures in a series of private communications in which she was kind enough to answer my questions and clarify my perceptions.[73] Even so, it is impossible to quantify exactly what Robert Le Rossignol received from Haber during this period or indeed over the lifetime of their contract, and neither would it serve any great purpose—other than to possibly rationalise aspects of Robert's later life. But the monies generated by the German fixed nitrogen industry during the war were so overwhelming that curiosity inevitably tempts one to 'have a stab at' Robert's entitlement, and establish just how 'generous', (Chirnside, 1977, Chap. 1) Haber's arrangement with him really was. Throughout this simplistic financial analysis, the influence of issues such as taxation, costs of money exchange/transfer, and the such, can, at best, only be approximated, as Haber's financial matters are notoriously difficult to penetrate. Robert Le Rossignol was still in receipt of monies from Haber after the war but the situation then became even more confused, although we try to address aspects of this in the next chapter. The simplest way to approach our estimate is to firstly establish Haber's *gross* income from the BASF over the period 1915–1917 for Robert's remuneration would have been entirely dependent on this amount.

The Oppau plant went into production on 09 September 1913 producing ammonium sulfate. Its first full year of production therefore was 1914. Together with the sister plant at Leuna, production of synthetic ammonia continued throughout the war and beyond. Payments to Haber can be determined from the royalty levels (Pfennigs kg^{-1} ammonia) that applied at each annual interval. Szöllössi-Janze[74] quotes accurate figures on production and *gross* payments from 1915 to 1919, viz.;

In 1915 Haber received 262,726 Marks @ 1.5 Pfennigs kg^{-1} ammonia from 17,515,054 kg ammonia produced.
In 1916 he received 983,814 Marks @ 1.5 Pfennigs kg^{-1} ammonia from 65,587,621 kg ammonia produced.
In 1917 he received 1,415,643 Marks @ 1.5 Pfennigs kg^{-1} ammonia from 94,376,188 kg ammonia produced.
In 1918 he received 931,070 Marks @ 0.8 Pfennigs kg^{-1} ammonia from 116,383,750 kg ammonia produced.
In 1919 he received 555,464 Marks @ 0.8 Pfennig kg^{-1} ammonia from 69,433,000 kg ammonia produced.

The royalty rate of 1.5 Pfennig kg^{-1} was reduced by Mayer in 1918 to 0.8 Pfennig kg^{-1} to draw attention away from Haber because of the fear that the Allies or the Reich would try and confiscate his wealth, the balance was to be recovered from the BASF later.[75] The data shows that from 1915 to 1917 Haber's gross royalties amounted to 2,662,183 Marks, reducing to ~2,610,000 Marks if we make some adjustment to account for the endowment fund at Karlsruhe (Chap. 7). These earnings represent a colossal amount of money and Germany at the time lacked a unified income tax base, the various Federal States acting independently in this respect. However, the First World War cost Germany 160 billion Marks. Between 1914–1918 only

13.1% of the total costs were financed out of 'recurring revenues' i.e., taxes. Of the remaining costs, 24.8% were paid as 'floating debt' i.e., by printing money, while 62.1% were state bonds signed by German citizens. German fiscal policy was characterized by the 'kid-glove' treatment of company profits and personal income.[76] The war profits taxes that were introduced were inadequate, for not only were they applied too late, but being applicable only to *changes* in income and profit, they exempted by their very nature large profits and incomes insofar as these had been the same as they were in peacetime. Moreover, the manner in which they were levied and collected afforded the profiteers ample opportunity to conceal their gains and Mayer would have made best use of this to minimise Haber's tax liability. In contrast, Great Britain[76] paid 28% of her total costs out of its recurring revenues. The standard UK income tax rate was doubled from $\sim 6\%$ to 12% in the first war budget of 1914, and was then raised progressively, in a fairly linear fashion throughout the war, finally reaching 30% in 1918/1919. Combined with super-tax, this meant that higher incomes were eventually taxed at a rate rather more than 50%.

If Haber's tax liabilities[77] were roughly in step with German recurring revenues over the period, say $\sim 15\%$, then the *net* value of Haber's income from 1915 to 1917 was around 2,218,500 Marks. At a contract agreement with Le Rossignol of 40% this would amount to a gross payment to Robert of 887,400 Marks, which at a nominal (pre-war) exchange rate of ~ 1000 Marks:50 BPS, amounts to $\sim £44,000$. If Robert subsequently paid 'top end' tax in the UK at 30% to 50%, his net income for the period is estimated to be somewhere in the region of £22,000–£30,000. Now there are so many other ways to approach this estimate and each one would employ a similar 'dog's breakfast' of logical and financial analysis, but it is hard to avoid the conclusion that, over the period 1915–1917 alone, Robert's entitlement must have amounted to thousands, if not *tens* of thousands, of pounds. What arrangement(s) Robert made to transfer monies from Germany,[78] and to accommodate the swingeing British income tax rates is of course entirely private. And alongside the effect of later German hyperinflation on his entitlement there also remains a 'fly in the financial ointment'. In conversation with Jaenicke in 1959 (Chap. 18), Robert declared that 'the Badische made hundreds of millions and Haber only got a little bit …'. 'A *little* bit'? Was Haber telling Robert the truth about his royalties? Did Robert receive what he was entitled to? The answer to the first question seems to be an emphatic, *yes*. In conversation with Chirnside in 1976 (Chap. 18), Robert said of Haber, that he was 'a nice man and a kind one … and he played fair by me'. But as for the second question, Robert also told Chirnside that he eventually received only ~10% of what was due to him, a point we try to throw some light on in later chapters. Even so, he also told Chirnside that what he received was 'enough for any man!' and he would certainly return from Germany far wealthier than when he arrived.

14.6   The Nobel Prize for Chemistry, 1918

The period from 1906 to 1918 was clearly an amazing time during Robert Le Rossignol's life. He and Fritz Haber had solved one of the most difficult problems confronting modern chemistry and in doing so they created the foundations for an industrial application that became the most important technological development of the 20th century. Equally importantly maybe, Robert became a husband, a father, and he began to accrue the kind of wealth that should allow him to support, comfort, and indeed cosset his family throughout their lives. At the same time, he had seen suffering, not only in Ruhleben but across Germany because for many Germans, starvation and despair followed military and political collapse. Throughout his long-life Robert was acutely aware of the dichotomy in the human condition, privilege versus poverty. Even as an 'impecunious' young man—as he described himself to Chirnside (Chap. 18)—he was content to agree to divert a pro-portion of his potential earnings to set up the endowment fund at the Hochschule (Chap. 7). Late in life he was to follow his father, using his 'rewards' to establish substantial legacies and 'help those less fortunate'. But as Robert was leaving Germany with Haber's financial legacy, another legacy—the work he did at Karlsruhe on high pressure chemistry—was helping the Royal Swedish Academy of Sciences (RSAS) decide the award of the Nobel Prize for Chemistry for 1918.

During his lifetime, Robert may not have been fully aware of events here. Only in the late 1960s—in Robert's 9th decade—would the nominations and considerations of the Academy be made available to those who were interested enough to enquire of events that had occurred some fifty years earlier.[79] However, when the documentation became available, Haber was already a forgotten man, his name living on in chemistry largely through the 'Haber/Haber-Bosch' processes, and the 'Born-Haber cycle'.[80] There were no substantial objective biographies of Haber until long after Robert's death and Robert had already put this part of his life to rest. But for the Academy at the time, consideration of an award for those involved in the creation and development of the technology of nitrogen 'fixation' proved to be the most controversial episode in its history.

The RSAS is responsible for the selection of the Nobel Laureates in Chemistry from among the candidates recommended to them by the Committee for Chemistry. This committee takes into account nominations made to them by persons—including previous Laureates—invited by them to do so. They are also able to nominate candidates for awards themselves. Laureates are chosen by the full Academy in early October through a majority vote. The decision is final and without appeal. The names of the Laureates are announced in November. During the selection process in 1918, when Haber was a nominee,[81] Haber's son, Ludwig Fritz ('Lutz'), in his book *The Poisonous Cloud*,[82] describes a 'disagreement' amongst the committee members that year. He says;

> In 1918 the Chemistry Prize Committee was not unanimous. One member argued that the Haber-Bosch process was still secret, which was against the principles of the Nobel Foundation. Another member considered that the fixation of atmospheric nitrogen had prolonged the war and that on those grounds Haber should be disqualified. No criticism of chemical warfare was recorded. There was also the question to what extent Haber had been assisted by [his collaborators] Le Rossignol, Mittasch and Bosch. The discussions are a curious mixture of ignorance and irrelevance …

Now one would not wish to deny the love of a son for his father and to care for his place in history, but a more recent examination of the considerations of the Chemistry Committee by Robert Friedman,[83] reveals a much more complex picture than Lutz Haber was willing, or able, to concede at the time; the 'curious mixture …' reflecting tensions in the committee and a tendency for some members to pursue a personal agenda, to play politics, and to engage in favouritism. According to Friedman, the facts are these;

Even before 1918, the Chemistry Committee was already experiencing some disagreement. As early as 1916 the ageing Peter Klason, who had been a committee member from the inception of the Nobel Prizes, had proposed Haber for the prize. Klason had rarely been able to gain any agreement amongst other members regarding his nominations—who were almost invariably industry-orientated chemists. But he was determined to have his way and in his choice of Fritz Haber he was able to play to the committee's undeniable affinity for German chemistry.[84] Klason already had a 'track record' in nominating those who had worked on nitrogen 'fixation', but none of the processes he supported in the past were viable. But just before the war a new 'universal' solution—the 'Haber-Bosch' process—had been announced and he lobbied to reward at least the academic half of this apparent 'partnership'. Haber, he felt, *must* have a prize regardless of the lack of external nominations. Klason received some support from a committee member who called for a joint award to Haber and Bosch, but Klason suggested both nominations should go to Haber. Committee chairman Hammarsten was concerned about the secrecy of the 'Haber-Bosch' process and suggested that Sweden's neutrality might be compromised by awarding a prize for a process that prolonged the war. Over Klason's protests, the Academy sided with Hammarsten and reserved the prize for 1916.

However, Haber returned to the nominations list in 1918, receiving just a single endorsement from a Munich chemist (Wilhelm Prandtl) amongst a meager 11 nominations that year spread across eight candidates. Klason seized his chance again, but by now there was 'blood on Haber's hands', for by 1918 almost half of all German artillery shells fired on the Western Front were filled with 'his' gas. Additionally, with such few nominations Haber was not felt to be representative of the international chemical communities' opinions, although of course circumstances were far from 'normal'. Even so, Klason's report in the spring of 1918 once again singled out Haber. Yes, he argued, the process had been used to produce munitions, but the Americans had used it too, so one was not favouring Germany here. Instead, emphasis should be on the use of the process in agriculture and its 'benefit for mankind'. During the summer a preliminary vote produced three members who backed Haber's nomination, Klason, Åke Ekstrand and H.G. Söderbaum. But because the process was still secret, the motion was opposed and failed. At a full committee meeting on 02 September 1918, disagreement regarding Haber continued. Hammarsten argued that sufficient information to make a sound decision was simply not available to the committee. Klason's argument that an award to Haber would be politically neutral because the process had helped both sides was rightly rejected because in truth it hadn't. And as far as agriculture was concerned the case was hardly proven, the evidence so far suggested all it had done was to prolong the horrors of war. For 'reasons of expediency' therefore, and with no mention of gas warfare, Hammarsten recommended against awarding the 1918 prize to Haber. But his reasoning failed to gain support and other members—all of whom had close ties to German science—voted with Klason to propose Haber. The final vote was seven to Haber, two against and one absentee. The Academy met on 12 November 1918, the day after the sudden signing of the Armistice. Hammarsten as chairman of the Chemistry Committee was entrusted by the Academy to guarantee impartiality and to observe its statutes. He deemed that the time was not yet politically right to award the prize to Haber and his voice had to be respected above the majority vote of his committee. Because of the raw proximity of the war, the Academy decided that none of the year's nominations met the criteria as outlined in the will of Alfred Nobel. Accordingly, no award for Chemistry was made in 1918. However, even though on this occasion Hammarsten had done his duty by the statutes, there was a growing illogical movement towards Haber amongst committee members and the whole episode resurfaced a year later when, in a bizarre but shameful turn of events, the committee abandoned impartiality and its responsibility to evaluate *all* the evidence, and without dissent, but with clear political intent, proposed the Academy award a prize to Haber alone.

On 06 December 1918, almost immediately after the signing of the Armistice, ignorant of events in Sweden but with far more pressing things on his mind, Robert Le Rossignol and family fled Germany.[85]

Notes

1. When war seemed imminent, some British people were able to leave the country, but for the many travelling was difficult because of the huge movement of troops. The Le Rossignol family would have found it difficult to leave.

2. Henry Woodd Nevinson was a committed socialist and respected war journalist covering both the Boer War, the Great War and numerous conflicts and disturbances between and after. His work was characterized by eye-witness accounts, attention to detail and he was scrupulous in gathering and conveying his facts. His work was widely appreciated and his writing often inspired those striving for freedom. A Wikipedia appreciation of his life is available at https://en.wikipedia.org/wiki/Henry_Nevinson.

3. Founded in 1846, by Charles Dickens no less, as a radical alternative to the right wing 'Morning Chronicle'.

4. Adapted from H. W. Nevinson, *Berlin, a-Tiptoe for War in August 1914—A British Reporter in the Enemy Capital.* Nevinson was the 'Daily News' War Correspondent in the early days of the war. The complete article can be found widely on the internet. One such occurrence is at http://outofbattle. blogspot.co.uk/2008/08/4th-august-1914-british-reporter-in.html.

5. To Paris! To Petersburg!

6. My understanding of the treatment of British nationals in Germany at this time is primarily due to the following texts which I have freely drawn upon and recognise. The copyright remains with the authors; Mathew Stibbe, *British civilian internees in Germany. The Ruhleben camp, 1914–1918.* Manchester University Press, (2008). ISBN-978 07190 7085 3, and Joseph Powell and Francis Gribble, *The History of Ruhleben. A Record of British Organization in a Prison Camp in Germany*, W. Collins Sons and Company Ltd., (1919). Digitised for Microsoft Corp. by the Internet Archive in 2007

from the University of Toronto. A copy may be found at http://archive.org/details/historyofruhlebe00poweuoft.

7. Stibbe, op. cit. (note 6), p. 30.
8. Stibbe, op. cit. (note 6), p. 30, his reference no. 36.
9. http://paperspast.natlib.govt.nz.
10. James W. Gerard, *My Four Years in Germany,* BiblioBazaar, (2007). ISBN-10 0554117665, ISBN-13 978-0554117669. Also online at http://net.lib.byu.edu/estu/wwi/memoir/Gerard/4yrs3.htm.
11. Stibbe, op. cit. (note 6), p. 2.
12. Stibbe, op. cit. (note 6), pp. 32–34.
13. Stibbe, op. cit. (note 6), p. 35.
14. Ruhleben lay between the boroughs of Charlottenburg and Spandau.
15. Stibbe, op. cit. (note 6), p. 31.
16. Powell, op. cit. (note 6), p. 1.
17. The necessary information being gathered some weeks earlier. Stibbe, op. cit. (note 6), p. 3.
18. The Reich Office of Interior. Stibbe, op. cit. (note 6), p. 41.
19. *Schweinhunde*! Stibbe, op. cit. (note 6), p. 42.
20. Gerard, op. cit. (note 10), p. 312.
21. Some were taken to another jail, the Plötzensee.
22. *Some* of those domiciled in Berlin, such as Mr. Joseph Powell, later to become a barrack representative and then Captain of the camp, insisted that they should travel by taxi, at their own expense, but they did so with Landsturm men (irregular forces) armed to the teeth sitting in the front! Powell, op. cit. (note 6), p. 5.
23. Stibbe, op. cit. (note 6), p. 43. We have no idea if Emily was involved here, but with two small children—one a babe in arms—it seems unlikely, as the spectacle would have been incomprehensible and too distressing for them.
24. Stibbe, op. cit. (note 6), p. 44.
25. From the Preface, *The History of Ruhleben*, Stibbe, op. cit. (note 6).
26. Powell, op. cit. (note 6), p. 3.
27. The American Embassy, who now became responsible for Allied prisoners in Germany, tried to help here. Gerard, op. cit. (note 10) writes; '*In the beginning of the war the Germans were surprised by the great number of prisoners taken and had made no adequate preparation for their reception. Clothing and blankets were woefully wanting so I immediately bought up what I could in the way of underclothes and blankets at the large department stores of Berlin … and sent these to the camps where the British prisoners were confined.*'
28. 'Ill-treatment of British Prisoners', *The Times*, 22 March 1915, p. 7.
29. Stibbe, op. cit. (note 6), p. 67. See his reference no. 60.
30. Prisoners regarded this as vastly inferior to the white doughy loaves baked and sent to the camp from the British Red Cross in Switzerland and Denmark.

31. 'Life at Ruhleben. Conditions in the Prison Camp', *The Times*, Saturday, 16 January 1915, p. 7.
32. Powell, (note 6), p. 5. Apparently, disinfectant had to purchased by the prisoners themselves.
33. However, Ambassador Gerard, op. cit. (note 10), made an acerbic observation in that; '*The man who finally appeared as head of the camp was an ex-cinematograph proprietor, named* [Joseph] *Powell ... Naturally he was always subject to opposition from many prisoners among whom those of aristocratic tendencies objected to being under the control of one not of the highest caste in Great Britain ...*'.
34. The British relief fund provided four or five marks (four/five shillings) a week for each prisoner channeled via the American Embassy. All who received this money however, had to sign a note to say they would repay the British Chancellor of the Exchequer when in a position to do so! See, op. cit. (note 28).
35. Ambassador Gerard, op. cit. (note 10), expresses some frustration here; '*... I found it impossible to get the British prisoners to perform the ordinary work of cleaning up the camp ... always expected of prisoners themselves ... With funds furnished me from the British Government, the camp captain was compelled to pay a number of the poorer prisoners to perform the work ... an allowance of five marks a week permitting them to purchase little luxuries ...*'.
36. Stibbe, op. cit. (note 6), p. 122.
37. Powell, op. cit. (note 6), p. 18.
38. Stibbe, op. cit. (note 6), p. 68.
39. Stibbe, op. cit. (note 6), p. 111.
40. Harvard Law Library, Cambridge, MA, Ettinghausen Collection, Box 5, File 16.
41. 'Life at Ruhleben, Improvement in the Lot of British Prisoners', *The Times*, Tuesday 22 June 1915.
42. Powell, op. cit. (note 6), pp. 11–12.
43. See Chap. 18 regarding Penn Mead flatlets ... '*they must be modern*'.
44. Although we know that Robert was never held in any form of 'pro-German' barrack. From file FO 383-312, National Archives, Kew. See Sect. 14.4 in this chapter.
45. Adapted from Stibbe, op. cit. (note 6), p. 2.
46. From file FO 383-279, National Archives, Kew. See Sect. 14.4 in this chapter.
47. *Mein Leben mit Fritz Haber, Spiegelungen der Vergangenheit*, (My life with Fritz Haber, reflections of the past). From, Margit Szöllösi-Janze, *Fritz Haber 1868–1934: Eine Biographie*. C. H. Beck, Munich, (1998), her reference no. 123, 'Ch. Haber, Leben, S. 157', p. 747.
48. The explanation appearing in Szöllözi-Janze, op. cit. (note 47), p. 177.

49. At the time of writing, the on-line catalogue of the National Archives at Kew made just one reference to Fritz Haber during the war years viz, '*Reference FO 383-186, Germany: Prisoners, including: Mutual cancellation of paroles given by Lieutenant-General Wylde and Geheimer Regierungsrat Haber ...*' although staff were unable to pinpoint this reference for me within FO 383-186. The detail of Le Rossignol's release therefore remains elusive.

50. A movement already detected by Gerard in one of his letters to His Majesty's Government in 1917. See Sect. 14.4 in this chapter.

51. Gerard, op. cit. (note 10), Chap. IV.

52. An observation made by Charlotte Haber, op. cit. (note 47), and recounted by Szöllözi-Janze, op. cit. (note 47), p. 176.

53. These applications were filed at the U.S. Patent Office on 30 January 1917. The apparatus for exhausting lamps was subsequently patented on 25 March 1919 (U.S. Pat. No. 1,298,569) and that for manufacturing the lamps on 09 September 1919 (U.S. Pat. No. 1,315,783). At the time, it was probably 'inappropriate' for Robert to prepare a patent on behalf of an enemy company. The USA however was a *neutral* country and the next section shows that Robert had a close association with the American Embassy in Berlin.

54. From 18 May 1918 to 19 July 1920 Robert filed improved applications in Germany, Austria, Hungary, Holland, Great Britain, France, Italy, Norway, Sweden, Switzerland, Bohemia, the United States and Denmark, See U.S. patent no 1,430,118 patented 26 September 1922.

55. Lamps were exhausted of course to remove the oxygen which would otherwise 'burn-out' the filament. The use of inert gas fillings had recently appeared on the scene but exhaustion was still the primary protection mechanism.

56. Plate 9.3, Chap. 9. This pip was sometimes ground down adding more effort and expense but improving the 'look' of the bulb.

57. Early forms of the apparatus resulted in less than satisfactory sealing of the tubulatures which were easily ruptured when baking-on the lamp bases. Robert's improvement(s) are described in U.S. Patent No. 1,430,118, patented 26 September 1922.

58. The 'lead' wires are not shown in Robert's diagrams but they protruded from the base of the stem and connect the filament to the external circuit.

59. All the correspondence referred to in Sect. 14.4 of this chapter is held in the National Archives at Kew in files, FO 383-211, FO 383-279 and FO 383-312.

60. National Archives Kew, FO 383-211.

61. Probably because it was not phrased in terms of a *letter*, but rather as a conversation with a 'Mr. Dresel', an American Embassy official in Berlin, the Embassy later communicating Robert's concerns to London.

62. FO 383-211. Prisoners etc., Germany (British civilians), 08 November 1916.

63. A description of Robert's position which had previously been supplied to HMG by Robert's brother Major Herbert Sorel in 1914, probably at the time of Robert's internment, (FO 383-312), and it was assumed that this was the position he was returning to. Herbert Sorel's description of Robert as an 'analytical' chemist may have been more benign in time of war than 'research' chemist which Robert undoubtedly was. Equally, it could have been made from a position of ignorance.

64. In 1916 Lord Newton became Assistant Under-Secretary of State for Foreign Affairs, and was put in charge of two departments at the Foreign Office, one dealing with foreign propaganda and the other with prisoners of war. In October 1916 he was appointed Controller of the newly-established Prisoner of War Department, and in this position, he negotiated the release of thousands of British prisoners of war. From a Wikipedia entry.

65. Walter Hines Page was the American Ambassador in London at the time.

66. Secretary Balfour was no other than Arthur Balfour, the former Prime Minister, subsequently First Lord of the Admiralty in 1915 and then Foreign Secretary from 1916–1919. Robert's case had reached the highest echelons of HMG.

67. This delay may well have extended to being able to inform his family, for even in 1917, *two years* after Robert's release, S. J. Le Rossignol's, *Historical Notes (Local and General) with special reference to the Le Rossignol Family (and its connections in Jersey)*, Trowbridge, (1917), p. 135, records that Robert '… *as a consequence of the European War now in progress, is interned in Germany* …'.

68. Chapter 7.

69. By 1916 Germany's synthetic ammonia production was making an increasingly significant contribution to her ability to continue the war. Indeed, without the technology the war would probably been over by the time Robert wrote to HMG.

70. But should we be 'searching for sin' here? In all honesty, Robert would have become wealthy on the back of nitrogen fixation, war or no-war. If peace had prevailed German hi-pressure fixation would have had an immense advantage over 'traditional' technology and huge amounts of money would have accrued to Haber–Le Rossignol from the BASF's patent royalties from within Germany and the licensing of their process abroad.

71. Szöllözi-Janze, op. cit. (note 47), p. 481.

72. Robert returned to the UK before any payment for 1918 could be made, and payments for the period from September 1913 to December 1913 are not so firmly established. Vaclav Smil, *Enriching the Earth*, MIT Press, (2001), quotes a total output for 1913/14 of 6.8 kt *fixed* N probably at a royalty rate of 1 Pfennig kg^{-1} ammonia according to Szöllössi-Janze, op. cit. (note 47),

pp. 482–483. This in turn translates to 8,257,143 kg ammonia yr^{-1} and royalties of ~82,600 Marks. Readers can factor this figure into the discussion if they wish.

73. Personal communication by email, 01 August 2011. According to Szöllözi-Janze, op. cit. (note 47), p. 482, Haber's financial dealings are so confusing that in the 1980s Adolf-Henning Frucht, grandson of Adolf von Harnack who planned a biography about Haber, commissioned an audit company to examine the files that were left. The results of the investigation were inconclusive and so his financial affairs 'remain unclear, even puzzling'.

74. Szöllözi-Janze, op. cit. (note 47), pp. 481–482.

75. Chapter 15. In May 1920 Haber and the BASF agreed an additional sum to cover these years.

76. Gerd Hardach, *The First World War, 1914–1918. History of the World Economy in the Twentieth Century*, University of California Press, (1977). ISBN-10: 0520030605 and ISBN-13: 978-0520030602.

77. By 1920 Haber complained to the BASF about his tax liability and the erosion of his wealth by the hyper-inflation that was beginning to set in. See Chap. 15. He seemed to be content prior to that.

78. A likely route would have been to the Channel Islands via Switzerland, avoiding UK income tax completely. However, this is pure conjecture on my part. The Le Rossignols however had banking contacts within the family. Edith was connected to the merchant bank 'The Old Bank' (Chap. 1) and Bertie had been employed in the Capital and Counties bank in St Helier prior to the war. There was also considerable banking experience in other parts of the family who were well versed in international commerce. Alfred for example, Augustin's brother, was a merchant in Buenos Aires. But what evidence we have also suggests that at least some of Robert's wealth remained in Germany after the war. See Chap. 15.

79. Nominations for the Nobel Prize in Chemistry was then, as today, by invitation only. The Nobel Committee for Chemistry sends confidential forms to persons who are competent and qualified to nominate. The names of the nominees and other information about the nominations cannot be revealed until 50 years later.

80. The former often taught to illustrate gaseous equilibria and the latter developed by Max Born and Haber in 1919 to correlate the lattice energies of ionic solids to other state-based thermodynamic data.

81. The award of the Nobel Prize to Haber was discussed as early as 1910 however, probably after the meeting of the Scientific Union at Karlsruhe, (Chap. 10) when Friedrich Schmidt-Ott from the Ministry of Education travelled to Sweden to consult with Svante Arrhenius on the appointment of the Director of the Kaiser Wilhelm Institute for Physical and Electro-chemistry. See Dietrich Stoltzenberg, *Fritz Haber. Chemist, Noble*

Laureate, German, Jew. Chemical Heritage Press, Philadelphia, Pennsylvania, (2004), p. 215, his reference no. 2.

82. Ludwig Fritz Haber, *The Poisonous Cloud. Chemical Warfare in the First World War*, Oxford, (1986). ISBN-10: 0198581424, ISBN-13: 978-0198581420. 'Lutz' Haber was Haber's third child from his second marriage to Charlotte Nathan. 'Lutz' settled in the UK and eventually became a respected historian at the University of Surrey. He died in 2004.

83. Robert Marc Friedman, *Neutrality in Twentieth-Century Europe. Intersections of Science, Culture, and Politics after the First World War.* Routledge Studies in Cultural History, (2012). ISBN-10: 0415893771, ISBN-13: 978-0415893770.

84. One initially has some sympathy with Klason and the Chemistry committee here. External nominations for Haber from the Allied nations were naturally non-existent because of the war. In normal circumstances, or had Haber confined his enthusiasm to nitrate production only, one could have expected much more support for what was widely accepted as a process worthy of an award. In proposing Haber, Klason was pursuing a personal agenda by 'normalising' the situation.

85. Szöllözi-Janze, op. cit. (note 47), p. 481. The date provided by Sollozi-Janze differs from that given by Robert in conversation with Chirnside (Chap. 18). Here, Robert says he left 'in November' 1918, but this was probably a mistake by Chirnside. In his three-page letter to UCL in 1961, Robert confirms leaving in December (Chap. 1).

Part III

The GEC Laboratories, the Second World War and Retirement

15

The Post-war Years 1919–1930

Of Airships, unfinished business from Germany, and the new GEC Laboratories.

'O brave new world That has such people in't! …'

Miranda, from 'The Tempest', Act Five, Scene One.

15.1 Post-war Germany

When Robert Le Rossignol fled Germany in December 1918, he left behind a country the international community held responsible for both starting and prolonging the war. Much of Belgium and northern France had borne the brunt of the conflict and were devastated. Soon, a particular bitterness towards Germany arose—especially amongst the French, a bitterness amplified by the realization that Germany had survived the war with much of her homeland and industry intact. Driven by a French imperative to 'de-militarize' her borders, and by an American insistence on occupying German territory, otherwise 'we will have to do this all over again,' the Allies advanced into Germany immediately after the Armistice occupying all her lands south of the Rhine. This occupation included the BASF plants at Oppau and Ludwigshafen which they entered on 06 December 1918.[1] Eventually, through the Treaty of Versailles, the Allies saw Germany as a valuable asset to be exploited as compensation for the war. As a result, they imposed swingeing penalties on her. They forced her to pay billions of Marks in 'reparations'.[2] they ceded some of her sovereign and overseas territories to allied countries, emasculated her armies, denied her the use of tanks and an air force, scuttled her capital ships and submarines (many of which had anchor chains made by Brown Lennox in South Wales), prevented future

© Springer Nature Switzerland AG 2020
D. Sheppard, *Robert Le Rossignol*, Springer Biographies,
https://doi.org/10.1007/978-3-030-29714-5_15

mobility by reducing her railway network, rolling stock and platforms, and seized much of her physical and intellectual industrial property. All of this precipitated a feeling of national humiliation which fuelled a social, political and financial transformation in the country. The immediate post-war years therefore saw chaos in Germany, with the emergence of unusual and transient political systems, a sweeping away of the old certainties with the abdication of Kaiser Wilhelm II, and eventually by August 1919, the replacement of imperial government with a relatively stable, but weak, parliamentary democracy, viz, the 'Weimar' Republic.[3]

The Republic lasted for the next 14 years, but it faced many problems, one of the earliest being the hyper-inflation of June 1921 to January 1924. Hyper-inflation was caused by increases in prices and interest rates, by redenomination of the currency, by 'printing money', by consumer flight from cash to physical assets, and by the rapid expansion of industries that produced those assets. Hyper-inflation for example saw the value of one gold Mark escalate from one paper Mark to one trillion paper Marks. By late 1922, Germany announced that—simply to survive—she had to default on her reparation payments until inflation could be brought under control. Many recipients of the monies were content to give Germany the time she needed to stabilize her economy, but not the Belgians and the bitter French. Consequently, on 11 January 1923 their troops occupied Germany again, this time in the heavily industrialized Ruhr region so that reparations could be made in tangible goods such as coal. In the longer term, hyper-inflation, the resentment that Germans felt at their humiliation and the deepening of social unrest led to the rise of Adolf Hitler and National Socialism which, in 1933, finally ended the Republic. In his book *Mein Kampf*, Hitler makes many references to the German debt and its negative consequences.

Events in Germany during this time were also to affect other European countries. Seizure of the German coal stocks for example flooded reserves, which depressed the international market and led in part to the devastating general strike of 1926 in the UK. Funding the war in the UK came at a severe economic cost. From being the world's largest investor, post-war she became one of the biggest debtors. The value of the pound fell by more than 60% and inflation rose, then peaked in 1920. But the military historian Correlli Barnett[4] argued that 'in objective truth the Great War in no way inflicted crippling economic damage on Britain' but rather, the German naval bombardments of Scarborough, Hartlepool and Whitby on the English east coast in 1914, and the new innovations of aerial bombing, chemical warfare and the U-boat 'crippled the British psychologically …', hitherto being a nation that had regarded herself as 'impregnable'. Now one might have thought that for the Le Rossignols, safe in the quiet backwaters of the UK[5] and financially comfortable enough (potentially at least) to 'ride out' the travails of the UK economy, events in Germany would have had little effect upon them.[6] But they were to cause Robert's financial arrangements to suffer, and of course in the long term the rise of National Socialism in Germany was to touch millions of families across Europe—including the Le Rossignols. On the credit side, one of the 'crippling psychological' effects the war had on some industrialists in Britain was to never again permit themselves to be dependent upon German science, and this quickly led to the creation of an institution which gave Robert employment for the rest of his working life. But on his immediate return to the UK and 'in between jobs', he seems to have found time to indulge himself by filing another patent, this time curiously enough, regarding 'airships'.

15.2 Improvements Relating to Airships

Of all the patents Robert submitted during his professional life, (***Appendix*** F) either independently or on behalf of others, there is one which seems to hold little connection to the 'day job', but never-the-less the problem it addressed still has resonance today. The provisional specification, titled *Improvements relating to Airships*, was submitted in the UK on 07 May 1919 (UK Pat. No. 148,008). The complete specification was submitted by 04 December 1919 and accepted on 29 July the following year.

Now history has not recorded the reason for Robert's interest in airships. Maybe there was a 'boys own' element involved. Could it have been H. G. Wells' novel of 1907, *The War in the Air,* when German aerial forces, consisting of airships and 'Drachenflieger',[7] attempt to seize control of the air before the Americans could build a large-scale aerial Armada? Alternatively, his imagination may have been fired by the German Zeppelin attacks on the UK during the war. The first attack on the UK mainland was in January 1915 but by October 1915, raids began involving 'squadrons' of Zeppelins, as in Wells' book. These lumbering machines were the terror weapons of their day and although filled with hydrogen gas, they were not as easily destroyed as one might have thought, the British taking some years to develop special incendiary bullets to pierce the gas bags and inflame the hydrogen. Certainly, Zeppelins were awesome iconic machines, but ever the practical man, it is more likely that Robert's interest was driven by aspects of safety and cost.[8] Crossing large areas of the globe, such as the Atlantic, in ocean liners was slow. But with the advent of the airship, or 'dirigible', (meaning able to be directed, controlled, steered or 'navigated', thereby distinguishing it from a balloon), speeds of about 80 mph could be achieved—very fast for the early 1900s. Between 1910 and 1914, ~35,000 passengers passed between cities in Germany and other European countries on dirigibles such as the Graf Zeppelin, Although the Hindenburg disaster was many years in the future, the 'Achilles heel' of the early airship was always its 'lifting gas', hydrogen, a choice dictated because of the American embargo on helium at the time. Hydrogen of course is the lightest element and even then, it was relatively cheap to produce. But the lower *flammability* limit of hydrogen in air at atmospheric pressure is just 4%, the upper limit is 75%. The *explosive* limits at atmospheric pressure are 18.3–59%.[9] Robert's innovation therefore was really quite obvious—especially considering his association with Ramsay. Replace hydrogen with a more benign lifting gas viz., helium. Helium is neither flammable, nor a supporter of combustion, but it is expensive, non-renewable, (released helium escapes the earth's atmosphere) and its buoyancy is about 8% less than that of hydrogen—although Robert calculated just over 6% less in his patent. Airships filled with helium therefore are inherently safer but less practical, for they lift smaller loads and travel shorter distances because they can carry less fuel. Helium airships for example would not have had the range to cross the Atlantic with obvious economic consequences. To help overcome these difficulties but at the same time preserving safety and minimizing costs, Robert's patent proposed substituting a proportion of the expensive helium with cheaper hydrogen, but within non-explosive limits.[10] These he suggested were between fifteen and twenty-five percent hydrogen—'perhaps even more'. The total buoyancy he claimed would increase 'from one to two percent' which meant an increase in 'anything from two to ten percent in carrying capacity'.

Now discussions regarding the use of helium/hydrogen mixtures persist to this day, but detailed studies show that no mixture would be safe for use in an airship unless it contained <8.7% hydrogen in admixture with helium[9]—a *much* lower limit than Robert proposed. Even so, Robert's idea never 'took off'. The superior economics of hydrogen eventually led to more than 30 years of passenger travel on German commercial Zeppelins during which time tens of thousands of passengers flew over a million miles on more than 2,000 flights without a single injury. But the inevitable conclusion of the contest, 'economics versus safety', finally came at Lakehurst, New Jersey on 06 May 1937 with the catastrophic explosion of the hydrogen filled 'Hindenburg' bringing an abrupt end to the age of the airship. Nowadays, an ~8% substitution of helium by hydrogen produces little economic advantage, and even if theoretical calculations and experiments show that we can go to 'Le Rossignol' proportions before we have a dangerous mixture, we would still have a hard time convincing the public. Robert's early idea naturally ignores our modern psychological concerns. Are the savings made by using a mixture of hydrogen and helium worth all the trouble and effort it takes to convince everyone that the mixture is safe? It is simpler to avoid the question altogether. Robert's airship patent therefore may not have widely been adopted but he was about to enter a period of his professional life that produced 30 more patents in as many years (*Appendix* F). In early 1919, shortly after his return from Germany, Robert Le Rossignol joined the newly formed GEC Laboratories at Hammersmith in London. Industrial research laboratories were rare in those days but their work was eclectic and admirably suited to his talents. Over the next three decades he rose to a very senior position within GEC and in addition to his profound contribution to the fixation of nitrogen and the engineering of 'glow lamps', he was to influence other 'cutting edge' technologies, ones that laid the foundations of the modern telecommunications world we inhabit today and the success of the Allies in the Second World War.

15.3   *The GEC Laboratories*[11]

The GEC[12] Research Laboratories were formally opened on 27 February 1923 by Sir J. J. Thomson and Lord Cecil, but they had come into being in 1919, and the idea for their formation was at least as early as 1915/16. In 1906 GEC had collaborated with the DGA of Berlin[13] to import Osram filaments from Germany to include in their own light bulbs, the scientific and technical resources however remained in Germany. By 1909 GEC had made a substantial investment in bulbs and completed the Osram Lamp Works at Hammersmith, London, to manufacture high quality Osram bulbs, offering one-third shares to the German and Austrian patent holders as enticement (Plate 15.1).

However, by 1915 with the war in its second year, the large German shareholding in the Lamp Works had provoked a very negative press. Mr. Hugo Hirst, the Chairman and Managing Director of GEC and Chairman of Osram, told a company meeting of GEC that because of the increasing vulnerability that many felt at being dependent on German science, a way had been found that would keep future control of the works in British hands viz, the creation of their own industrial research laboratory to serve all of the Osram Lamp Work's needs. Soon after in January 1916, Mr. Christopher Wilson, the General Manager of Osram together with Mr. Driver the Works

Plate 15.1 The Osram Lamp Works, Hammersmith in 1932. Taken from the Osram Lamp catalogue of the time. 01 March 1932, follow he link; http://www.lamptech.co.uk/Documents/Catalogues/GEC%20-%20Catalogue%20-%201932%20UK.pdf

Manager, visited the National Physical Laboratories to speak to Clifford Paterson, Principal Assistant in charge of Electrotechnics and Photometry. Paterson was approached because he had substantial contact with the Hammersmith Works and he was asked to recommend a man who could organise a research department along the lines envisaged by Hirst. Paterson's original reaction was that he felt GEC were doing the right thing, rather than being reliant on University departments to advise them. He suggested two or three people who might be interested, discussed with the visitors the type of organization they ought to consider, and as industrial research institutions those days were very rare, advised them that to some extent they were embarking on the unknown. But the next day, Wilson asked to see Paterson again having detected that he may not be averse to taking on the job himself. Paterson's initial response however was hesitant; no-one would support the kind of freedoms *he* would want in such an organization; freedom to pick the problems to be investigated, to choose the staff, to set their remuneration, conditions of service and to demand the space and apparatus he felt was necessary to do their job. But he had misjudged his visitors, they agreed wholeheartedly with his vision and asked him to join them as soon as hostilities ceased.

Almost three years elapsed before the idea was considered again, but with the Armistice of November 1918 Paterson approached the Osram management and asked them if they wished to reconsider the project before he and others 'took the plunge' and resigned from the NPL. Amazingly, Wilson told Paterson that far from their resolve weakening, an even larger scheme was envisaged. The German share of the Lamp Works had been purchased by GEC from the Government Trustee of Enemy Property and Osram was to be incorporated into GEC. Paterson therefore was now asked to plan an industrial laboratory to serve the *whole* of the GEC, but beginning with the problems of the Lamp Works. Clifford Paterson became the first Director of the newly formed 'GEC Research Laboratories' in January 1919.

15.3.1 The Laboratories' First Year

The original staff were housed in a wooden building at the Lamp works in Hammersmith, previously used to make 'valves'[14] during the war. The salary book still exists[11] and shows that the Laboratories began with just four members, but in March and May 1919 there was an 'influx' of staff who over time, were to become senior members of the Laboratories. This initial influx included Robert, who was invited by Paterson to join the new venture.[15] After returning from Germany, and in conversation with Chirnside in 1976 (Chap. 18), Robert described himself as 'in the doghouse for a while'. Family, it seems, did not talk about his internment, they were *not* happy with the unwanted attention of HMG in November 1914 and again later in 1916 when his patriotism was questioned. Even so, Mr. Christopher Wilson of the GEC 'sought him out' and, although other offers were available, Robert decided to join the new Laboratories. Wilson subsequently sent him to see Paterson who invited him to become a founder member.

During 1919, twenty-nine people (including six physicists, two chemists, two engineers and a metallurgist) joined the staff, and even at this early stage the Laboratories' interests diversified and began to turn to the design and manufacture of thermionic valves as well as the more familiar filament lamps. In both respects there was much work for the laboratories to tackle—in facilities described as 'very trying for experimental work'—and in a climate of almost total ignorance regarding the application of physics to technology in the UK.[16] Robert's appointment of course was 'a natural' for the Laboratories. He more than most was familiar with the German side of Osram and he brought their current understanding of the technical processes and procedures required in the manufacture and exhaustion of high-quality filament lamps. But the old order was changing. Reports from America during the war years showed that there had been major developments in bulb manufacture—particularly in reducing glass wastage and by the introduction of the 'gettered' filament[17] technique. Additionally, the gas filled tungsten lamp had been introduced in America by Coolidge and Langmuir in 1914, but no tungsten wire had ever been drawn in the UK. There was therefore much metallurgy and physics to be done to understand the process of producing a ductile tungsten[18] and to prevent damaging 'filament sag'. Robert's initial role within the laboratory was concerned with the borderline between the physics, chemistry and engineering of lamps and valves. For example he studied lamp manufacturing processes, he proposed the first laboratory research programme on tungsten, examined the purity of tungsten wire used in vacuum lamps together with techniques of tungsten wire drawing (Chap. 9), he also examined slow leaks as a cause of lamp failure and made progress towards the elimination of impurities in argon used in gas filled lamps.[19] Many of these problems of course were the 'curse' of colleagues in the Lamp works who wanted quick solutions from the Laboratories. These did not always work as hoped and there was initially some tension between the two groups. But laboratory staff consoled themselves by the prospect of their new premises which were planned almost immediately the research laboratories had come into existence. The new laboratories were eventually built at East Lane Wembley, a point we return to later in the Chapter.

Even so, working in temporary premises with newly appointed staff, the range and amount of work achieved in the first year would have done credit to more established organizations. In total, 111 reports were produced during 1919 by people who tackled all sorts of problems in a variety of fields, and who eventually settled down into areas in which they were to become recognised specialists. Even at this early stage in their development, the laboratories began attracting attention from other GEC institutions with a view to funding research and they had their first experience of consultancy in dealing with proposals from outside GEC. They examined proposals for vacuum jackets for increasing the efficiency of gas filled lamps, a buried shell detector, and a fire damp detector. They also addressed other makers products and technology and examined claims in competitor's patents. At the time, Robert for example assessed the domestic ice-making machine of another manufacturer whist a colleague (L. B. W. Jolley) undertook the first work on domestic appliances and examined the design and manufacture of an electric kettle. Robert also helped produce the first report the poor shelf life of batteries, being a topic that was to exercise GEC for some years to come.[19]

Some evidence of Roberts efforts regarding his investigations into the ice making machines still exist today, having been passed down through the family. A single sheet of paper dated 29 November 1919 in Robert's hand lists patent literature of interest for injector refrigerating machines covering the period 1909–1915. Throughout his professional life, Robert it seems studied the scientific literature meticulously, keeping hand-written records of papers and patents of interest in 'filofax' style. One such filofax—addressed '7 St Johns Road'—remains within the family, mainly relating to references for articles written in the 1930s and '40s, and including formulae and topics such as refrigerators, centrifuges and atomic clocks. Several pages covering the period from the late 1940s to the early 1950s are devoted to refrigeration which seems to have been a life-long interest for him, and early refrigeration engineering, comprising compressors, and circulating pumps moving heat transfer agents such as ammonia through heat exchangers, may well have played a part alongside Travers' hydrogen liquefier in the engineering solutions Robert devised for Haber (Plate 15.2).

At the same time as the technical side of the Laboratory was developing, Paterson was developing strong egalitarian issues. Organisation, he felt, should be minimal. Simply appointing someone as a 'leader' was not acceptable. They had to be recognised as such through their scientific ability, experience and their character. In this respect there was just one[20] grade of

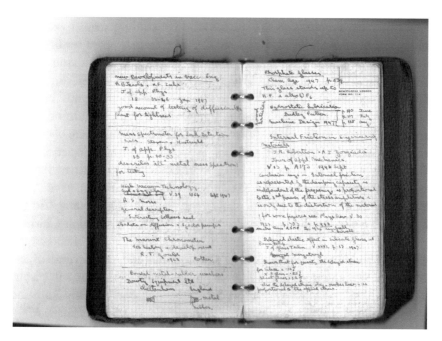

Plate 15.2   Robert's 'Filofax' from the 1940s with detailed notes on papers and patents. Photograph courtesy of the Le Rossignol family

seniority—'Leading Staff'—which *anyone* suitably experienced could attain if they were recognised as such by their peers and therefore 'elected' by them. Paterson also had the wonderful idea of shielding his research staff from organizational issues by delegating mundane laboratory administrative procedures to 'Laboratory Stewards' who therefore dealt with all the day-to-day problems such as procurement of materials, the tidiness and cleanliness of the labs and ongoing repairs and improvements to laboratory facilities.

These ideas of course were unorthodox by 1919s standards, but there was more to come. Paterson held very strong views on punctuality amongst the staff. The day began at *precisely* 9 am and lines were drawn across the attendance register at 3 min past nine and again at six minutes past. Bad punctuality led to the withholding of salary increments however good performance had been during the day and however long staff decided to stay after the normal working hours. Paterson held a punctuality 'fetish' throughout his tenure as Director and one story goes that during the second war when one employee blamed his bad time-keeping on his alarm clock, he procured a new one for him and debited the cost from his salary next month![11] But Paterson didn't begin his tenure by ruling with an iron rod. On the contrary he placed particular importance on personal contact between all levels of staff within the laboratories. He emphasized this philosophy by holding regular parties for staff hosted by himself and his wife at their own home, and a feature of the Laboratories from the earliest time was the range of social activities undertaken. In the heyday of the Laboratories, the 'Research—GEC Social and Athletic Club' as it came to be known, covered an enormous range of recreational activities; cricket, football, netball, tennis, badminton, horse riding, rambling, squash, judo, dancing and ice skating. But there were more cerebral aspects, such as a bridge club, music and dramatic societies. Such activities were an important feature of the laboratories and led to life-long friendships, marriages and even a 'second generation' of laboratory staff.[11] For Robert and Emily this was particularly true, as Robert's nephew Clement Sorel[21] later became a distinguished member of staff, whilst Ralph Chirnside for example, who eventually became Chief Chemist at the Laboratories, remained a friend throughout Robert's professional life and beyond. But as Robert was beginning a new life, events across the world were still playing out, events that had their genesis in Karlsruhe way back in July 1909 but which continued to influence him both financially and professionally.

15.4   Europe and America, November 1918 to July 1924

In the immediate post-war years, Haber's royalties were still tied to the BASF's fortunes, but as foreign competition increased and hyper-inflation 'bit', these were amounting to virtually nothing—especially between 1923 and 1924 when Haber was paid by the BASF in useless paper Marks. There are traces of a later financial settlement between the BASF and Haber for this period but as for Robert's private contract at the time, 40% of nothing would have amounted to nothing. But the period from November 1919 to March 1924 also saw the migration of fixation technology out of a defeated Germany, the award of a Nobel Prize to Haber and then to Nernst, the premature end of Haber's royalty payments, a 'spat' between Robert and Haber over money, and the end of a weak pre-war trend which had occasionally associated Robert's name with Haber's in both a professional and a 'popular' sense. But through it all Robert remained very much in 'Haber's shadow', because events across the world—especially in Germany, the UK, Sweden and America—led to Haber/Haber-Bosch stories dominating the history of fixation. For historians and biographers alike, these momentous events suffocated any memory of Robert's contribution to fixation, so when the story eventually came to be told, he was abandoned.

15.4.1   Events in Germany

Even before Robert left Berlin, the French began their occupation of German territory declaring the occupied region a 'de-militarised' zone. But 'de-militarisation' was not the only Allied objective, and Carl Bosch realised that his factories and their secrets could now be regarded as 'assets', or even worse, destroyed completely in an attempt to prevent Germany ever waging war against her neighbours again. For Haber, the impending occupation of Ludwigshafen meant uncertainty regarding the payment of his future royalties,[22] and so he generously made a new private contract with Robert to make a direct payment to him for the year 1918, but this time from his *own* funds.[23] Presumably, the contract was again at 40% but whether Robert's share of the additional 0.7 Pfennig kg^{-1} due to Haber from the BASF (Chap. 14), was included we simply don't know as, according to Szöllösi-Janze (Chap. 14), Haber's post-war financial matters are utterly impenetrable. In the event, and probably due to the confusion in Haber's life at the time, Robert did not receive the money. A point we return to later (in this chapter, Sect. 15.5).

The French entered Ludwigshafen the same day Robert fled Germany. Two of the three BASF factories therefore came under French control, but not before Bosch had moved much valuable material across the Rhine and shut down the ammonia plant at Oppau. This he maintained was 'due to a lack of coal', but in reality, it was to prevent the Allies understanding how the ammonia synthesis worked. Even so, the French confiscated over a million Marks worth of finished dyes and shipped them to France.[1] The French of course regarded Oppau as a munitions factory and crawled all over the plant trying to 'tease out' her secrets. Bosch knew that patents alone would not explain his industrial process, and with regard to the visit of the French inspectors *The Times* in London on Wednesday 12 November 1919 carried a quote from the BASF in which they maintained that … 'if they, the French, saw the works they would not be able to duplicate them, and even when erected, they could not operate them'. Even so, if the Allies were permitted to study a working plant and disassemble a reactor or two then the BASF's competitive advantage in post war fertilizer production would eventually be lost. Bosch and his staff therefore became totally obstructive believing that in law the Allies had no right to steal German technical secrets.[24] Bosch used every device to avoid restarting the plant, the French inspectors therefore had to content themselves with taking measurements and photographs, gathering samples, climbing ladders, peering into pipes and drawing diagrams.[1] The BASF workers made it even more difficult. Access ladders suddenly disappeared, important gauges were removed or 'de-natured' so that their purpose was disguised, and when inspectors appeared workers would 'down tools', switch off machines they were using to maintain the plant, and stare at them until they left.[1] Oppau stood dumb in defiance, the voices of men and machines falling silent.

But where the French failed the British were determined to succeed and at the end of May 1919 a delegation arrived at Oppau which included representatives from the huge Brunner Mond company—forerunner of ICI—part of a group of 20 members of the association of British Chemical Manufacturers investigating the present position of the German chemical industry. The 'Brits' were amazed at the scale of Oppau but they too met with little success and when they demanded the disassembly of parts of the plant for examination, the BASF threatened to 'mothball' Oppau throwing all the workers on the mercy of the French who would have to deal with the consequences of their unemployment. Just like the French then, the British could only make sketches and notes, storing them temporarily in a railway wagon under an armed guard. But on the morning they left Oppau they found the notes had gone and there was a rather suspicious hole in the floor of the wagon.[1] On their return home, the official report of the inspectors had to be made from memory only. *The Times* on 09 February 1920, published it on page 9 under the title 'Secretive German Chemists'. Undoubtedly, this article was also read by Robert and later he too was to experience just how secretive they still were (in this chapter, Sect. 15.5).

During the French occupation Bosch kept his workers on the payroll even though the loss of production[25] meant huge financial penalties each month but that, he decided, was better than losing the technology. Equally, Bosch knew that he couldn't protect his plants indefinitely. By shutting down Oppau he was simply stalling for time until some decision could be reached regarding her fate. He hoped that the new German government would soon become strong enough to lobby on his behalf and he argued that Ludwigshafen, Oppau and Leuna needed to remain viable in order to help stabilize Germany by providing employment, by feeding the people, and by generating wealth. All his hopes were centred on the peace negotiation at Versailles in the summer of 1919 which he attended as a representative of German industry, but when he arrived at the talks, he soon realised that these were not 'negotiations' but rather a device by means of which a peace settlement would be *imposed* on Germany.

At Versailles Bosch presented his argument to the Allies. He was listened to politely, but ignored. It was clear that the French wanted to shut down the German ammonia plants completely, but what was of more concern to him was the unexpected talk of 'reparations'. The figures put forward were of 'stellar' proportions and ensured that Germany would remain economically crippled for decades. It became clear to Bosch that in order to ensure the protection of his plants he had only one thing to offer. Everyone in the Western world wanted an ammonia plant. The French alone were 'hell bent' on destroying Oppau and Leuna and so Bosch made them an offer they could not refuse, viz, to build a Haber-Bosch plant on French territory.[26] Unsurprisingly, from that point onwards the French attitude towards the German ammonia plants became more relaxed. The deal was signed on the first anniversary of the Armistice and in early 1920 French forces pulled out of the occupied territories. After they left, the reactors were re-started and Oppau began a slow march back to some sort of normality, although on 21 September 1921 Oppau was all but destroyed by a huge explosion when a silo containing 4,500 tonnes of a mixture of ammonium sulfate and ammonium nitrate fertilisers detonated.[1] Even so, Bosch had kept his plants but at what price? From a personal perspective he was rewarded for his negotiating skills and made head of the BASF—a position he both coveted and dreaded because he was happier with a spanner in his hand than a balance sheet or an agenda. From a technical perspective, the French were not to be 'privy' to BASF's commercial secrets and the 'letter' of the Treaty of Versailles made no demand upon Germany to reveal her technology in areas unrelated to war—which the ammonia plants now were. However, it was inevitable that the technology would soon be duplicated in other countries and BASF's domination of the post-war nitrogen fixation market ended. The same applied to dyes. The French flooded the market with confiscated materials and German patents were often stolen or disregarded leading to rising dye production and depressed markets in the Allied countries. All of this was later compounded by German hyper-inflation which meant that even after the settlement with the French, the BASF was still in trouble as a business. In the early 1920s then, Bosch began the process of finding another 'blockbuster' technology. By 1929 he had succeeded, predicated on the belief that the world was running out of oil and the age of the automobile was about to arrive. The BASF's synthetic gasoline ('petrol') process—developed alongside the fundamental research work of Friedrich Bergius—produced high octane fuels via the hydrogenation of 'fluidised' coal which generated something resembling crude oil. Coal was plentiful in Germany but oil was not and the successful development of the process meant that Germany—along with the later development of 'Fischer-Tropsch' (Chap. 13)—eventually became self-sufficient in fuel oils in the same way 'Haber-Bosch' had made her self-sufficient in nitrates. Because of his role before, during and after the war Bosch became a major player on the European chemistry scene, and Germany's additional self-sufficiency at the time naturally 'fueled' fears of another world war. For historians, such momentous events tended to leave little room for Le Rossignol.

15.4.2 Events in the United Kingdom

However, by the 1920s, anyone involved with the fixation of nitrogen in Europe must have been aware of the original fundamental contribution of Robert Le Rossignol, but by then the terms 'Haber' and 'Haber-Bosch' process had become ubiquitous—although in Germany Robert's name did become famous amongst chemists but only because of the 'Le Rossignol' valve (Chap. 18). But, this period probably represents the last time the 20th century ever recorded Robert's contribution to fixation, certainly until the Haber biographies began appearing from the late 1980s onwards. Only then was his memory revived, and only in the rare case was there a glimpse of an appreciation of his importance to Haber and of his contribution to fixation, primarily through the work of Szöllözi-Janze and Stoltzenberg. But in the immediate post-war years probably the last 'popular' reference in the UK to Robert's involvement with 'fixation' accompanied reports of the decision of HMG to finally construct a 'nitrogen factory' on British soil.

Late in 1919 the Ministry of Munitions in the UK published a report by the Nitrogen Products Committee which addressed various aspects of 'the nitrogen problem' both at home and in the Empire. An interim report had been produced some two years earlier and recommended the construction of a large factory at Billingham-on-Tees in County Durham (near Middlesborough). Late in 1917 land was purchased, roads laid down, some buildings erected and materials ordered, but the work was never pressed forward with any vigour and ceased completely at the conclusion of hostilities. The new report looked in detail at all aspects of the supply of fixed nitrogen and recommended that HMG build a synthetic ammonia plant at Billingham to produce ammonia/nitric acid on a commercial scale using the 'Haber-Bosch' process. The Government accepted the report's findings but decided that progress was best made by private enterprise and advertisements for contracts and tenders appeared in *The Times* and other 'nationals' on Wednesday, 19 November 1919.

Subsequently, a syndicate consisting of Messrs. Brunner Mond and Explosive Trades Limited[27] made a successful bid which was again reported in *The Times* on 22 April 1920. The same article rationalised the need for a 'nitrogen factory' in terms of a 'national insurance' to guard against British merchant shipping being prevented from conducting the saltpetre trade by German submarines, *or* the possibility of a future Chilean Government withholding supplies of nitrate. The article described the process to be adopted at Billingham ' … being that generally known by the name of the German chemist Haber', but went on to mention that 'an Englishman, Le Rossignol, was associated with him in its inception'. But there was no conviction or qualification concerning Robert's involvement and this was a common thread in post-war reports concerning the 'Haber' process, universally accepted as the 'fountain' of synthetic ammonia without which—the article went on to say—'[Germany] would not have dared to make war and certainly could not have continued to fight as long as she did'. A simplistic view of the attitude of the British Press in those raw post-war years would conclude that they were reluctant to acknowledge the involvement of a British national in what was seen as a magnificent but ultimately cruel German innovation. And whereas there must be some element of truth here, it is equally compelling to learn that pre-war reports of Robert's involvement—when the 'Haber factor' should have been a positive—were just as downbeat so that on the whole, the Press were probably also influenced by ignorance, the immense respect for German science[28] at the time and Robert's reluctance[29] to engage the public with *his* story. It is also certainly true that during this time, the term 'Haber-Le Rossignol' process never really achieved any popular currency. In fact, at the time this author cannot trace its use at all.

Involvement of an Englishman or not, the 'Haber process' was *German* and HMG were determined to 'clear the decks' to ensure that Billingham progressed. Adequate safeguards were adopted to prevent the new company (Synthetic Ammonia and Nitrates Ltd) passing out of British control. In keeping with the psychological effects of the war Directors were to be British born, measures were devised to protect against interference by the Germans in the working of the process in the UK and the government placed all enemy patents bearing on the process at the disposal of the new company on a royalty basis. The payments under this arrangement were to go to the Custodian of Enemy property for account under the reparation clauses of the peace treaty.[30] Everything the HMG knew about Oppau was given to the new company[31] but progress was enhanced spectacularly during 1920 with the arrival in London of two opportunist German engineers who approached Brunner Mond with armfuls of drawings, claiming to know all about Haber-Bosch at Oppau, and willing to sell what they knew.[32] The legality of such an approach of course was dubious and the Company considered its position, but eventually they decided to buy the (evidently stolen) material which turned out to be all that was promised. By 1923 Billingham became operational and began moving towards its initial target of producing 60,000 tonnes of ammonium sulphate per year.[31]

The success of Billingham led to a lucrative market in German ammonia engineers and America soon followed Billingham's path. Between November 1919 and August 1924, *The Times* of London published over 200 articles referring to 'nitrogen' in one context or another—usually with regard to the nitrate industry—illustrating the intense public interest in the subject. The same period saw any number of articles referring to the 'Haber' or 'Haber-Bosch' process but now no mention of the 'Englishman'. By the end of the decade Billingham had grown to rival the technical magnificence of Oppau and Leuna. So much so that its modernity and the symbol of regeneration it represented is believed to have inspired Aldous Huxley to pen his *Brave New World*, often regarded as one of the best novels of all time. But what must Robert have thought of the appearance of a 'nitrogen factory' on UK soil? The fleeting association of his name alongside Haber's in the article announcing the 'go ahead' for Billingham would not have escaped his attention for *The Times* was well read by the Le Rossignols, being the organ by means of which the family had long recorded 'milestone' events.[33] And even though during the war HMG felt Robert to be 'a man who may have considerable knowledge of German processes ....', (Chap. 14) there is no biographical evidence available to suggest he had the slightest input to Billingham or indeed if he ever visited the place. But one 'snippet' suggests that over the next five decades Robert took a close interest in the maturing of his 'prodigal', for as Ralph Chirnside recalled in his 1977 obituary of Le Rossignol '... he derived great satisfaction from the contribution that the Haber-Le Rossignol process had made to the feeding problems of the world and particularly those in the third world ... only a few weeks before his death we discussed the energy balances in the ammonia process as now carried out at Billingham ...' and surely, if Robert still followed events at Billingham at the age of 92, it can only reflect his lifelong interest. It is also touching that his friend afforded—the by now elderly gentleman—the dignity of describing the ultimate foundation of Billingham as the 'Haber-Le Rossignol' process, a lead which surely, we chemists should now follow. But of the immediate post-war events that briefly mentioned Robert's contribution to fixation, what happened in Sweden in late 1919 eclipsed them all.

15.4.3 Sweden, November 1919–21

Robert's post-war 'popular' exposure was both momentary and shallow, but following the announcement by the Royal Swedish Academy of Sciences on 13 November 1919 of the award of the Nobel Prize in Chemistry for 1918 to Fritz Haber, his contribution to nitrogen 'fixation' was recognised at the highest possible professional and academic levels—but not rewarded. However, this recognition came at a price, for it was to be acknowledged by a man whose status as a named war criminal had only just been revoked and by way of an Academy whose decision was to fall into disrepute, for the same committee that had considered the Chemistry award the previous year now reversed its decision, and without dissent, awarded the prize to Haber alone with scant regard for any of its earlier concerns. According to Robert Friedman in *Neutrality in Twentieth-Century Europe*,[34] the facts behind the Academy's 'volte face' are these.

During 1919 the chemistry committee had regrouped to support Haber. Klason and Ekstrand sent their own letters of nomination, and with the war now over, Hammarsten, who by all accounts had nothing but 'contempt for the English', broke with his previous position, gave voice to his German sympathies, and actively helped overcome the problems that had previously hindered Haber's candidature. In short, with absolutely no mandate from international nominators, the committee shamefully set out to unilaterally engineer support for Haber alone. The committee, for example, did not object to members evaluating those whom they proposed, the claim that nominators are partial whilst committee members are objective seems to have been ignored or conveniently forgotten. Had they waited a year or two things would have been clearer. The Americans had failed to build nitrogen factories, the process did not benefit both sides, what was the full extent of Le Rossignol's or Bosch's involvement in the creation of this technology, were others worthy of a share in the prize? Committee members agreed to a collective 'tunnel vision', their 'fixation' on Haber was overwhelming. At the preliminary meeting in May 1919 all members agreed to support Haber. Some small reservations were raised but his candidacy sailed through the committee. In October the Academy supported the committee's decision but 50 members—anticipating the furore—subsequently sent a petition to academies and journals abroad pleading that German science not be ostracized because of their decision, and on November 13th 1919 three Germans, including Haber, were honoured. Friedman[34] describes it perfectly. He says,

> Committee members did not simply assess the scientific merits impartially and then turn the matter over to the Academy to decide whether such an award entailed moral and ethical issues. They played the role of partial advocates; they produced flawed evaluations. The Academy too seem to have desired to 'cut loose' in an atmosphere of utter pessimism and dejection; to cast caution to the wind and embrace those with whom its members wanted to stand in solidarity.

This remarkable 'cutting loose' by the Academy appears to have been driven by a belief that in the confusion that embraced Germany at the time, no political party there offered any hope for culture and learning. On the contrary, 'a gaudy, materialistic, commercial middle class sprang up amongst wartime profiteering' and Swedish scientists therefore turned to comfort their German colleagues. 'Surrounded by economic ruin, shamed by humiliating defeat, buffeted by revolution and cast as international pariahs', German academics needed encouragement to heal the wounds of war, and by deliberately ignoring the principles of the Academy, Sweden could 'stand in solidarity' with them to help them do so. Sweden it seemed, had no intention of turning her back on Germany.[34] In this sense the Academy stood behind the view that although the imperial government of Germany had fallen, her elite culture remained, and it was this culture that was honoured by their awards in 1919. This view however proved both naïve and overly optimistic because international reaction was swift and damning. Friedman,[34] in *Neutrality in 20th Century Europe,* sums up the position;

> *Of all the actions during and immediately following the war, none so clearly revealed the process of bias as the efforts in 1918 and 1919 of the Nobel committee for chemistry to award an undivided prize to Haber. Committee members themselves elevated Haber as a contender for the prize at this time. They abandoned established principles and routines when evaluating his worthiness for a prize. They could have waited until information as yet held secret was made available; they could even have waited for signs of support from nominators other than themselves for Haber alone, or for dividing a prize with his collaborators. They acted with clear partial interests. They then insisted in public statements that they acted without bias; indeed, that they had no choice, that the statutes require that they reward solely on pure scientific merit. Some may have convinced themselves that this was indeed the case; others knew full well that in the name of protecting the image of the prize they needed to fabricate a lie.*

This searing indictment, based on current and contemporaneous evidence,[34] means that during his lifetime Robert probably knew nothing of the Machiavellian maneuvers behind the decision to award an undivided prize to Haber as the deliberations of the Nobel committees are not made public until 50 years has elapsed (Chap. 14). In terms of collaboration in the *creation* of the technology of high pressure nitrogen 'fixation'—for which the award was given—there was only one contender for recognition; every paper Haber published on the technical realization of the ammonia synthesis was with Le Rossignol, every public lecture given by Haber praised his contribution, and each of the two critical patents (high pressure and circulation) bore Le Rossignol's name as well as his own, and if ever we were to fabricate a set of circumstances that militated against the award of a prize in spite of overwhelming evidence to the contrary, one need look no further than the case of Robert Le Rossignol for inspiration. Yes there were elements of Robert's candidature that may not have helped the committee had they behaved objectively and seriously considered dividing the prize; the BASF's ruthless prosecution of commercial secrecy prevented the two men publishing as quickly and as widely they wished; Robert's move to Berlin from Karlsruhe broke their working relationship allowing Haber alone to dominate the dissemination of their work; the 'hijacking' of fixation by the military was an obstacle—but not one of Robert's making; Haber's public praise fell silent because of the war, and at the same time by moving into the shadows of Haber's life to avoid any association with a man so closely associated with the German war machine, Robert became isolated and anonymous. But it was the duty of the chemistry committee to look beyond such aspects and examine the body of work the two men produced in partnership and in peace-time. Had they consulted widely they would have learned that the only input the BASF had to the pilot high pressure fixation process was funding, and learned too of Haber's private contract with Le Rossignol which would have been another critical indicator of the depth of his dependence on the young man. Under normal circumstances could the committee have expected nominations for Le Rossignol? He was not without potential support, from such as Donnan,[35] Trouton and maybe even Ramsay had he not died in 1916. In this respect however, Robert would not have promoted his own candidature; that would have been undignified. Being a product of Victorian England he 'knew his place'. All the circumstances of commercial secrecy, the war, Haber's dominant personality and the high regard in which German science was held at the time were stacked against Robert. But with regard to this young man the Committee totally abrogated their responsibility to the integrity of the most prestigious award in science, and this above all was the killer obstacle to Robert's candidature. The committee employed 'quota think'. Having acknowledged the British by an award to Charles Glover Barkla (the reserved 1917 Physics prize) the Academy felt no need to make further gestures to the victors. Hammarsten's 'contempt' of the English meant that Le Rossignol had scant consideration. By 1920, 'protest' nominations for Bosch had already begun to arrive. A strong sense of injustice eventually forced the committee to admit it was in error and they later awarded Bosch a share in the 1930 Chemistry prize for his industrialisation of the process. But for Robert there could be no such retrospective. The award for the *discovery* of the technology had already been given to Haber alone. He was left unrecognized. But had he been recognised, and had he been comfortable with the Academy's reason for the award, given the bitterness of the time, could he have stood alongside Haber to accept his prize? Only one person could ever answer that question.[36]

But the announcement of Haber's award had hardly been made before Karl Hjalmar M. Branting,[37] leader of the Swedish Social Democrats, Finance Minister and soon to be both Prime Minister and a Nobel Laureate himself, criticised the Swedish Academy in the Democrats' newspaper, *Socialdemokratin*,[38] equating Haber with 'barbarism'. By celebrating Haber and German chemistry the Academy had destroyed its position as a neutral institution he claimed. He called for vigorous protest from within Sweden and asked whether a comparable British or French chemist who had worked directly for the war effort would have been awarded a prize 'while sores are still dripping of blood … but because Haber is a Prussian professor we find it acceptable!', hinting in turn at an uncomfortable collusion between Sweden and Germany. British newspapers hardly covered the story of the awards in terms of a leading article, possibly out of contempt for the whole affair. Branting's criticism however—'pure politics' according to the RSAS—was reported briefly in *The Times* 'Imperial and Foreign News Items' on Thursday 30 November 1919, where it was claimed that he 'deplored that an international distinction should fall to one who during the war devoted all his powers to inventing new means of destruction …'. But British disinterest belies the outcry amongst the various international communities which spilled over into the 'quality' press on both sides of the Atlantic.

The next few weeks saw displeasure with the Swedish Academy grow and grow, and it made the front page of the *New York Times* on 26 January 1920 where severe criticisms—attributed to 'French sources'—prompted an immediate defence from the First Secretary of the Swedish Legation in Washington D.C.—published in the same newspaper on 28 January. Secretary Hammarskjöld was 'by order of the Minister from Sweden, to set forth the following *facts*', viz, that the award of the Prize was for the synthesis of ammonia from its elements, that the invention was of the greatest value to the world at large leading to the production of cheap nitric fertilisers, that the detail of the invention was made available [*equally*] to all before the 'Great War' by Bernthsen (in 'Philadelphia') in 1912, that the 'Haber plants' in Germany were erected with a view to the production of agricultural fertilizer alone, and that in order to generate explosives, ammonia has to be converted to nitric acid which was *not* part of the process described by the Award. Secretary Hammarskjöld finally made it clear that during the war Sweden had a strict export ban on 'all sorts of war materials', a point made to deflect a growing feeling against Sweden because it was felt she had accommodated German requests for certain material(s) during the war.

Hammarskjöld's 'clinical' justification of the award was entirely in keeping with the fabrication the Academy wished to propagate, but now the Swedish Government was also implicated in the ruse. His intervention however, drew a skeptical response. In this example—once again from the *New York Times*[39] and written by the eminent American colloid chemist Jerome Alexander[40]— the author takes the First Secretary to task (Plate 15.3).

Alexander accepted Hammarskjöld's view that the perfection of the commercial synthesis of ammonia from hydrogen and atmospheric nitrogen

NOBEL AWARD TO HABER
Source of resentment felt in Allied
countries.
To the Editor of the New York Times

In today's TIMES, there appears a
letter from Mr. Hammarskjold, First
Secretary of the Swedish Legation,
containing some statements of fact
regarding the award of the Nobel Prize
for Chemistry to Professor Haber.
While most of these statements are
correct, not all of the facts are stated
and there are besides some erroneous
conclusions drawn. I would comment
upon Mr. Hammarskjold's numbered
statements as follows;

Plate 15.3 Beginning of Alexander's letter to the New York Times. Reproduced by the author

warranted the award of a Nobel Prize, and that the 'Haber process' would ultimately be of great value to the world. But he refused to accept the 'half-truth'[41] that Germany's primary reason for the erection of her nitrogen plants was the production of agricultural fertilizer, believing instead that ammonia was always meant to be the first step in the production of nitric acid for munitions. He cited Bernthsen's talk at the 8th International Congress of Applied Chemistry in NY City (*not* 'Philadelphia') in 1912 (Chap. 12) where, contrary to Hammarskjöld's views, technical detail was scarce and *not* 'known to all nations before the great war …' but where it became clear that 'Germany could conduct a war even if the British Navy cut off the Chile supply'. Alexander's letter crystallized the Allies bitterness towards Germany in that it was universally accepted at the time that she had long organized her industries against the probability of war—a view repeated many times e.g., in *The Times* on 22 April 1920, However, history has shown that many of Hammarskjöld's statements of 'fact' were indeed just that. The 'Haber process' was indeed originally conceived as a solution to the world's need for cheap nitric fertiliser. The Germans had *not* prepared for war by building nitrate factories. An industrial process to convert ammonia to nitric acid was never part of the initial brief for Oppau, but came later as Germany's engagement with the war—initially confident and typically Teutonic—soon gave way to grave concern as her nitrate deficiency became acute. But what Hammarskjöld's 'facts' did not reveal was the political maneuverings of the

RSAS. Alexander however, like many in the scientific community realised that 'something was up' and he was not prepared to have his prejudices compromised by their version of the truth and so he turned his fire on Sweden herself believing (correctly as it happened) that a Swedish bias towards Germany—exemplified for him by the behavior of certain Swedish citizens—may well lay at the heart of the Award. He pressed home this opinion in the final point of his letter[42]; (Plate 15.4).

These exchanges illustrate the remarkable contemporary feelings against all things Swedish and 'Haber', but despite fierce international objection the award ceremony went ahead, and on 01 June 1920, the 1918 Nobel Prize for Chemistry was accepted by a delighted[43] Fritz Haber 'for the synthesis of ammonia from its elements'. Conveniently in light of the controversy, but officially owing to the sudden death of Crown Princess Margaret,[44] no member of the Swedish royal family was present when Dr Å. G. Ekstrand (intimately involved in engineering Haber's prize himself) gave his presentation speech,[45] which, in keeping with Hammarskjöld's earlier 'sanitized' justification of the award, made no mention of the war, nor the manipulation of the ammonia plants, nor gas warfare, and was in keeping with Nobel's original intention for the prizes which were to be 'distributed annually … to

(5) Although Mr. Hammarskjold disclaims knowledge of the manufacture of gas masks in Sweden it is possible Germany got wood or charcoal from Sweden for gas mask purposes, just as she got iron ore. No criticism attaches to Sweden for this, and her fear of Russia and proximity to Germany across the Baltic (a German lake) ready explain her attitude to her powerful neighbour. The pro-German activities of certain Swedes and Swedish-Americans, and especially the abuse of Swedish diplomatic privileges by Germans such as Count Luxberg of "spurlos versenkt" fame, have naturally created amongst the allied people an atmosphere of suspicion against Sweden; and since Professor Haber is understood to be one of those who advised and helped developed gas warfare, it is easy to understand hoe many believe that the award of the Nobel Prize to him is at this time is ill advised.

JEROME ALEXANDER.
Ridgefield, Conn., Jan. 28, 1920.

Plate 15.4 Continuing Alexander's letter to the New York Times. Reproduced by the author

those who during the preceding year have conferred the greatest benefit on mankind.' Ekstrand concluded his ... speech thus;

> *Geheimrat Professor Haber. This country's Academy of Sciences has awarded you the 1918 Nobel Prize for Chemistry in recognition of your great services in the solution of the problem of directly combining atmospheric nitrogen with hydrogen. A solution to this problem has been repeatedly attempted before, but you were the first to provide the industrial solution and thus to create an exceedingly important means of improving the standards of agriculture and the well-being of mankind. We congratulate you on this triumph in the service of your country and the whole of humanity. Please, accept now your prize from the President of the Nobel Foundation.*

Haber's acceptance speech[46] followed the next day and similarly confined itself to the historical development of the 'Haber process' and its industrialisation by the BASF. The tone of his speech was in keeping with the 'humanitarian' reason for his award and this was summed up in two final sentences, firstly;

> *... in combination with other method of fixation ... they relieve us of future worries caused by the exhaustion of the saltpetre deposits that has threatened us for these 20 years ...'*

and then, as Sir William Crooks had exhorted all those years before (Chap. 2);

> *... the chemical industry comes to the aid of the farmer who, in the good earth, changes stones into bread ...*

But as Haber had done at every opportunity in the past, he made time to recognise Robert Le Rossignol's contribution to his award from the Nobel platform, such as with;

> *During the course of these investigations, together with my young friend and co-worker Robert Le Rossignol, whose work I would like to mention here with particular sincerity and gratitude ... and*
>
> *The construction and operation (carried out in collaboration with Robert Le Rossignol) of a small-scale plant ...*

Now, in the history of the most coveted prize in science, such recognition—immensely important to those who had contributed to a prize but were never awarded it—has often been denied them by egotistical Laureates. The cases of Rosalind Franklin, Oswald Avery and Lise Meitner come to mind immediately. As scientific discoveries are almost invariably 'collegiate', *and* because the prize bestows a kind of Wagnerian 'heroic genius' upon the winner, the exaltation of the individual has often led to a profound resentment dividing former colleagues. In this case however, Haber's unconditional endorsement of Robert's work was probably genuine, heartfelt and deserved, and Robert in turn treasured their friendship until Haber's death in 1934. In the same speech Haber also generously acknowledged Nernst's contribution to his prize. At the ceremony, all the German Laureates from 1914 to 1919 were present who, because of the war were previously unable to travel to Stockholm. Only one other Laureate was present, the English physicist Charles Glover Barkla. The German Laureates however had a wonderful time reveling in the celebrations laid on by their Swedish hosts.[47] But discontent at the Swedish Academy's decision rumbled on. Two Frenchmen awarded prizes that summer (for Medicine and Economics[48]) refused to accept them. Some such as Theodore Richards declared that they would not shake the hands of German Scientists at Nobel ceremonies or anywhere else until sincere remorse was forthcoming for signing the 'Manifesto of the 93', and in such a climate, given the bitterness that infected the world after the war, even Haber's public endorsement of Robert's work was still not enough to permit the accolade 'Haber-Le Rossignol process' to ever gain any popular or professional currency, and replace the—by then—ubiquitous, '*Haber* process'.[49] Later life was not kind to Robert and in comparison this lack of universal recognition seems unimportant. However, although no resentment ever lay in his heart, it may have been a life-long regret for Robert that, unlike 'Haber', the name 'Le Rossignol' never became synonymous with nitrogen fixation. But Robert was to suffer a Nobel 'double whammy', for along with the Award, Haber also received 138,198[50] Swedish Kronor equivalent to around a quarter of a million pounds today which, because of ratcheting German inflation, he kept in a Swedish bank.[51] It would therefore have been perfectly possible for Haber's 'endorsement' of Le Rossignol to translate into a royalty payment for 1918, in accordance with their additional private contract (Sect. 15.4.1) made only eighteen months earlier. In the short term however, no payment was made.[52]

There were however some compensations for Robert as significant professional recognition followed the Nobel Award to his friend. The American Chemical Society Member listing for May 1921 shows that Robert was admitted to this prestigious organization in 1920,[53] and less than a year after Haber's award, he was elected a Fellow[54] of the Institute of Chemistry of

Great Britain and Ireland, (later of course, the Royal Institute/Society of Chemistry). Additionally, in November 1921 the 1920 Nobel Prize for Chemistry was awarded to Walther Nernst, a prize that the work of Haber and Le Rossignol was widely acknowledged as helping justify. By 1921 Nernst had gathered 90 nominations in Physics and Chemistry.[55] Like Haber, Nernst's involvement in the war also brought criticism as yet *another* German scientist received the highest award. But the tenor of doubt seemed to fade away from the nominations received after 1916. The successful synthesis of artificial ammonia supported by the experimentation and the theoretical calculations performed by Haber and Le Rossignol using Nernst's approximation formula (Chaps. 5 and 6), was considered by many at the time to be a robust verification of the heat theorem.[56] But in the records of the Nobel committee, this theorem alone was not the main reason for his award —indeed the draft of the award citation was modified at the very last minute to delete the remark 'with special emphasis on his heat theorem'[57]—possibly because of its use in the history of ammonia synthesis and the subsequent controversy surrounding that technology. Rather it was offered[58] to Nernst by the committee in a wider less contentious sense, 'as recognition of the exceptional merit of your work on thermochemistry', and as van't Hoff, Ostwald and Arrhenius had already received the prize, with this latest award all four founding fathers of physical chemistry had been recognised and indeed Haber and Le Rossignol had surely played their part. Nernst made his acceptance speech on 12 December 1921[59] and whatever the opinion of the Nobel committee he made it clear from the outset that it would 'best fulfill my obligation to give a lecture relating to my prize-winning publications, if I discussed my heat theorem …'. Even so, there was no mention of Haber and Le Rossignol in his speech and like Haber, Nernst also kept his money (134,100 SEK[50]) in a Swedish bank.[57]

15.5   More Matters Financial

After the Nobel awards, and especially over the years from 1923–1924, Haber's financial arrangements with the BASF became even more complex. Accurately tracing the amounts involved and any subsequent payment(s) to Robert becomes impossible as there is only fragmentary evidence available. However, during these years when Haber was paid in essentially worthless paper Marks (Sect. 15.4), Szöllössi-Janze[60] describes traces of a further financial settlement between himself and the BASF, but nobody knows of its exact conditions. Some talk about a very generous settlement to protect

Haber from the effects of hyper-inflation, but nobody gives an exact amount. About the same time, there was also a change in the German patent law (09 July 1923) which should also have benefitted Haber but he seems to have been unaware of it. This change extended the duration of a patent from 15 to 18 years meaning that the 'circulation' and 'high pressure' patents—which would have expired in 1923/24 respectively—now continued through to 1926/27. He should therefore have received substantial royalties from the BASF in hard currency for the years 1924–1926/27, but in the event he did not. It seems that the BASF had 'conned' him.[61] Friends advised him to go to court, but by now his health was already too weak and he let the matter rest. Even so, putting the loss of the additional revenue aside, Haber's 1920 'compensation' for the reduction Mayer had arranged to protect Haber's finances from scrutiny by the Allies,[52] (and Chap. 14) the Nobel Prize money and a possibly 'generous' settlement with the BASF for the hyper-inflation years, should have put enough money 'in the pot' to honour his financial obligation to Robert for 1918, possibly even through to the end of the original patent protection.

Whether Haber made Robert aware of the extent of these settlements we don't know. But what we *do* know is that the payment for 1918 does not seem to have been made and that around the time Haber was receiving his Nobel Prize in Stockholm, Robert traveled to Berlin to check on his 'affairs'. In a letter from St Johns Road written to 'Bertie' dated 22 July 1920 which remains in the family, we gain some insight into Roberts financial arrangements, which appear to have been a little insecure. The salient points are;

Dear Bertie,

… I have a suggestion which I would like to make and which I hope you will agree to. Unfortunately, I have to put a condition on it for the present but I am sure you will understand. Years ago, you suggested jokingly that when the ammonia turned trumps, I should give you a car. I would like to do that now; you said when I was last in Jersey you would sell me your car for £600. I wish to buy it, for which I enclose a cheque for £600 and I wish you to accept the car from me as a present, I hope you will like the car, as far as I know the last owner looked after it very well. The provision is that if I do not get some of my money in a year, you sell the car for me, if it fetches more that £600 you are to take the difference.

Now, I would like to tell you what happened in Berlin. I found my affairs in good condition but it was a good thing that I went as several matters had to be seen to. Whilst I was there, I had £1000 from my money which was due since peace transferred to London, thus the above offer. In Cologne I went to Lloyds and saw the manager, but was very 'disappointed' with their terms. They give no interest on deposit accounts which I think takes the bun, so I am not going to transfer any money to them. I may have my securities sent there later on ... I enclose a cheque for £650 as I owe you £50 ...

Your loving brother, Robert.

The letter also goes on to reveal some other interesting aspects of Robert's life at the time viz., the arrival of Mrs. Walter and Flora (Emily's mother and sister) from an impoverished Germany to St Johns Road the previous day, both looking 'very thin—especially Mrs. Walter, the installation of electric heating in the house to save on labour as they have no servants yet, the theft of their linen from their previous home whilst in transit, the safe arrival of their furniture, and the hope that their silver will arrive soon. The 'nonsense' regarding the car too seems to have been just a ruse to transfer a substantial amount from his ammonia royalties to Bertie, a harbinger of Robert's subsequent generosity regarding his 'rewards'. But clearly, at least some of Robert's money 'due since peace time' remained in Germany even though normal bank transfer arrangements between Germany and the UK seem to have been re-established after the war. But this money, it seems, did not include the separate contract Haber made with Robert to cover 1918 and this little 'difficulty' it seems rumbled on, and over the next few years we have addition evidence of Robert's return to Germany to meet Haber.

In conversation with Johannes Jaenicke in 1959, (Chap. 18) Robert mentions a 'business' trip to Berlin in '1923 or 1924' when he firstly met Haber at the Kaiser Wilhelm Institute, moving from there by rail[62] to Dresden to meet his friend Max Mayer. Although Robert reveals no purpose to the meetings—other than renewing old acquaintances—the suspicion must be that 'matters financial' may well have been discussed, with Haber probably bemoaning the fragility of his finances.[52] However, what evidence we have suggests that shortly after his trip Robert became increasingly frustrated by the lack of movement regarding his payment for 1918, and he placed a complaint before the joint German–English Arbitration Board. This seems to have taken place in 1924, but the details are vague. One fragmentary source comes from the German representative at the Board, referring to the lawsuit, 'Le Rossignol *vs* Haber' dated 01 March 1924.[63] Another hint is in a memo of the BASF patent department regarding Haber's letters to Bosch of 15 and 18 March 1926,[63] but the evidence is ill-formed and so it is difficult to reconstruct any reliable detail for the case. However, what seems firmly established is that Robert later withdrew his action.[63] Robert's legal action may have been a final 'nudge' to his friend. Certainly, his lawsuit seems to have been in place by March 1924, some time before he was due to attend the first World Power Conference[64] in London opening on 30 June 1924 where 1700 experts from 40 countries were to meet to discuss energy provision. France and Belgium were opposed to the participation of Germany but the UK and the USA were in favour, and Robert knew that a German delegation was invited, bringing Haber as a member. Haber arrived in London in late June accompanied by his wife Charlotte, and Robert may have seen this as an ideal opportunity to both re-acquaint himself and to resolve their stumbling financial relationship. In short, the lawsuit may have been a ploy to apply a pecuniary 'pressure push' on Haber.

During the conference, the Germans and the Austrians were marginalized by the Allies and made to sit amongst the 'neutral' states. Sir Ernest Rutherford, President of the Royal Society, neither spoke to Haber, nor shook hands with him. He extended the same discourtesy to Charlotte. Very few delegates made any effort to meet with Haber, but Robert was an exception, shaking hands with him in a very public manner.[65] History has not recorded what passed between these two at the time but Haber remained in London for the next two months and some time after the conference Robert withdrew his legal action. The conference closed on 11 July 1924 and was regarded by Germany as a success, a step on her road to international redemption. The World power conference of 1924 was the last occasion where it was *publicly* recorded that the two men met. However, in conversation with Jaenicke in Beaconsfield in September 1959, (Chap. 18) Robert describes a further meeting with Haber and Mayer in the health-resort of Wurzach, Baden-Württemberg near Karlsruhe, in 1926. Haber at the time was on the Supervisory Board (the *Aufsichtsrat*) of the BASF[61] and Robert's observation, described by Jaenicke's clumsy English, that;

> *Years afterwards, when Haber was Aufsichtsrat in the Badische, they showed us the work. But they were so tight; Haber was so angry he didn't see anything …*

Might well refer to this meeting, whilst in another letter to Johannes Jaenicke in August 1959 (Chap. 18) Robert recalls last seeing his 'dear old Professor and friend, Haber, in 1933(?)' although no such meeting is mentioned in Jaenicke's notes of their subsequent conversation in September 1959. What passed between Haber and Robert at these meetings however, has not been recorded.

And so what do we make of it all? Szöllössi-Janze[66] correctly maintains that Robert Le Rossignol was in receipt of monies from Haber not only during, but also after, the Great War. We know too that Robert's first substantial payment was in 1914 (Sect. 14.5) and according to Chirnside's transcript (Chap. 1) his final payment was in 1924, a payment of two million Marks— nominally £20,000, in keeping with our calculations for 1915–17 (Sect. 14.5 again)—but worthless because of German hyperinflation. Robert's meeting with Haber in London and his complaint before the Arbitration Board probably precipitated this final payment. But how much money did he receive from Haber over the life-times of the 'circulation' and 'high pressure' patents? We know he received only ~10% of this entitlement because of the War and hyperinflation. Even so he felt it was 'enough for any man' (Sect. 14.5) but we cannot apply any precision to these transactions. And in any case what right do we have to know? These monies were certainly 'generous' as Chirnside's obituary of Le Rossignol describes them, (Chap. 1) but they were also part of a private business arrangement freely entered into by the two men enabling them work together to create a technology that the Nobel committee saw as an 'exceedingly important means of improving the standards of agriculture and the well-being of mankind'—the only level at which Robert ever engaged with it. Robert's royalty payments ended by the mid 1920s and although one could argue that his life was never really free of Germany, events here began to have less of an influence. Accordingly, this seems a good time in our story to let the financial issue rest for a while as no further progress can be made here in view of the complexity of Haber's finances. Significant though Robert's 'rewards' may have been, there is a more important question to ask now; not 'how much money'?, but rather 'what became of it'? A point we address later in the book. The 1920s were a time when Robert got on with building the rest of his life, a professional life at least undertaken wholly at the new GEC Laboratories at Wembley.

15.6 The GEC Laboratories and Developments Over the Decade

During 1920, 26 more people including two chemists, two engineers and a patent agent[67] had joined the staff providing a scientific 'critical mass' that urgently needed more suitable accommodation. Although the laboratories were originally housed in the Osram Lamp Works at Hammersmith, planning for the new buildings started almost immediately. Luton was the preferred site initially, but the proximity of Wembley to central London became a deciding factor.[68] The design of the new building was 'fluid', Patterson asking for three extensions even before the buildings existed, but by April 1920 the plans had stabilized, a site at East lane Wembley was decided upon and competitive tenders opened. The contract was placed by May at a total building cost of £120,000 and construction began in October 1920. Staff took possession of the building at the beginning of 1922 but the official opening by Sir J. J. Thompson and Lord Robert Cecil wasn't until 27 February 1923.

The original design was to have a type of building that could expand in as many directions as possible because no-one could anticipate developments. This indeed proved a wise decision and extensions to the original building continued for many years, the basic structure however was one of long corridors off which opened workshops and laboratories. This architecture also served one of the most enduring and enjoyable social events of the Laboratories, viz, the annual children's party, the first of which was held on 10 January 1923.[69] The laboratories and workshops were cleared on Friday evenings and Saturday mornings to provide large open spaces for the children. By Saturday afternoon decorations were up together with elaborate sideshows —often with a technical or scientific basis. Over the years, children (and parents!) reveled in model railways, radio-controlled cars, pony rides, a circus, a 'cockshy'[70] at ('spent') lamps out of the Life Test Labs, shadow plays and silhouettes and the magic of the 'Wizard of _OS_ram'! Children from local orphanages or children's homes were also invited; it was time of innocence and so refreshing after the travails of the Great War. Robert's children, Peter and John were just 9 and 11 years old at the time and probably joined in all the fun for many years to come.

Although these social events were a credit to the GEC its '_reason d'être_' of course was research, and whereas the Hammersmith factory was originally largely concerned with lamp technology, all this changed in October 1919 when Osram combined with the Marconi Company to set up the jointly owned Marconi-Osram Valve (M-OV) works.[71] Valve research had been conducted at Hammersmith during the war, but it wasn't a major effort of the company until the creation of M-OV. A gradual division of forces therefore began at GEC as valve production took place at Hammersmith, but considerable research was undertaken at Wembley. At Wembley therefore, one group continued towards lamp technology, but as radio communication was becoming a new and exciting field another group, whose founding members included Robert Le Rossignol and A. C. Bartlett, developed the 'valve team',[72] charged with improving both the manufacture and performance of these devices, especially for transmitting purposes. In terms of radio communication lamp technology was still of considerable value, especially for small receiving valves, but the development of large high powered[73] transmitting valves presented major new interdisciplinary problems; problems which crossed the boundaries of mathematics, chemistry, physics, and electrical engineering, at the same time generating new disciplines such as statistical testing and materials science. Making sense of Robert's efforts here demands at least some understanding of 'thermionic' technology and interested readers are referred to **Appendix** E for a very basic grounding, but undoubtedly the eclectic nature of the discipline of thermionic technology was what motivated Robert to move from lamps to valves. This progression can be seen by the nature of the patents he filed during the decade, sometimes with others. Four patents were filed on behalf of the GEC between 30 October 1924 and 17 November 1926. viz, 'Improvements in electric discharge tubes', 'Improvements in thermionic valves', 'Improvements in or relating to the construction of electric discharge tubes', and 'Thermionic Valve'. (see **Appendix** F) By the mid 1930s, the GEC valve team had developed some of the most powerful radio transmitting valves in the world, developments with which Robert was closely associated and which he chronicled in a substantial paper in 1936.[74] We develop this aspect of his life more fully in the next chapter. However, when the Wembley valve team was formed, transmitting valves were then quite different.

15.6.1 Improvements in Early Transmitting Valves

At the end of the first world war, the largest commercially available wireless transmitting valves, the so-called 'glass' valves, were capable of dealing with an input of just 1 kilowatt (kW). Bearing in mind that the efficiency of a valve was seldom more that 70% and often no more than 30%, it was necessary to parallel a great number of them to achieve a high-power amplification. But with regard to improvements in amplification, the glass valve suffered from a number of deficiencies.

The anode dissipation[75] of a typical glass transmitting valve at the end of the war was around 300 watts, and a major problem confronting the development of more powerful valves was the requirement that the temperature of the glass should not exceed ~ 200 °C, otherwise gas would be evolved causing the vacuum to deteriorate and grid to lose its control over the valve's characteristics. Therefore, to build a more powerful valve with a much larger anode dissipation, the glass bulb had to increase in surface area to allow radiation to sustain the temperature constraint. This quickly led to the realization that such valves would require enormous, impractical, glass bulbs. 'Hard' glasses such as 'Pyrex' were becoming available to replace 'traditional' lead glass, and these could withstand much higher temperatures without gassing. However, this advantage was countered by the fact that hard glasses were not so transparent to heat as lead glass.

Higher power also meant larger anodes, otherwise the excess (wasted) heat would evaporate the (usually nickel) metal too rapidly causing blackening of the bulb and raising the glass temperature above the safe limit. Later glass valves employed molybdenum as anode which withstood higher temperatures without undue evaporation and increased dissipation from ~ 300 to (a still modest) 600 W. Neither was the grid absolved from concern. Too great a rise in temperature caused it to act as a filament giving rise itself to 'primary' electrons which once more destroy control of the valve.

Another important part of the valve was the filament, the intended source of the 'primary' electrons and a component that influenced both the efficiency and the lifetime of the valve. Filaments, transmitting or otherwise, were usually made of tungsten, a metal with which Robert was familiar both from his time at the 'Auer' and from his earliest days at the GEC. A heated tungsten filament under vacuum has definite rates of electron emission (per unit surface area) and of metal evaporation at any given temperature. If a longer life was required for a given rate of emission, the total surface area of the filament had to increase; this makes the filament larger taking it longer to evaporate and at the same time increasing its emissivity. Equally, larger filaments consume more power and this had to balanced against the longer life and reliability. Experience showed that filaments usually 'burnt out' when evaporation has reduced its diameter by 10%. But as other factors are always present to influence the life of the valve, it was pointless to increase the filament life beyond a certain value. For the GEC, lifetimes between 3000 and 6000 h for glass transmitting valves were considered 'economic', but these lifetimes were not particularly long, and if we were to parallel a large number of such valves there was an unavoidable risk of failure and interruption of transmission service.

In terms of high-power radio frequency (RF) transmission therefore, the decade quickly established an obvious fact for the GEC; an entirely different type of valve had to be developed, one which would permit much higher anode dissipation. To achieve this, a radical re-design was necessary, a design in which the anode would have to be part of the glass envelope of the valve so that it could be immersed in water, or in oil, or blown by air to facilitate a very high surface heat dissipation. Thus, in 1923, the 'cooled anode transmitter' or 'CAT' valve was conceived, but only after a way of making reliable joints between glass envelopes and copper anodes was discovered.[76] Even so, this magical 'art' was practiced only by the most highly skilled and highly paid glass blowers, very few of whom were available to the vacuum industry. There was a clear 'bottleneck to the progression of cooled anode technology'. The CAT valves that subsequently appeared were of two types,[77] the 'demountable' and the 'sealed-off', the difference being that the demountable valve could be taken down and dis-assembled to replace burned out filaments or other components, whilst the sealed valve, just like a spent light bulb, would have to be completely replaced once the filament had deteriorated. Replacing a filament in a demountable would cost just a few pennies and could be completed in a few seconds or minutes, but in reality, the tube was put out of working condition probably for some hours. Time has to be allowed for the valve to cool before opening and releasing the vacuum, otherwise oxidation could occur. The process of opening the tube takes time too—depending on the complexity of the valve—and once the valve has been re-furbished, the airtight seals and gaskets have to be re-instated and the valve continuously pumped down to maintain the vacuum. The primary advantage of the demountable was that it could be used at very high-power levels because burned-out filaments could be easily replaced. The primary disadvantage comes in a situation when a service via the valve cannot be interrupted, as in broadcasting. This, coupled with an industry dislike of the inclusion of a delicate vacuum system[78] which has to be continuously operated, made the sealed-off valve the main contender for continuous high-power broadcasting systems. This was the valve the GEC concentrated their efforts on, a position fostered by developments in the USA where sealed-off tubes equally as good as their demountable counterparts were being produced. Nevertheless, high power demountable valves were made in the UK, principally by the Metropolitan-Vickers Company of Manchester and used in the Post Office radio station at Rugby. Demountables were also used within the GEC Laboratories to study valve cathodes,[79] and the valve was later to be the first choice in probably the most critical technology of the second world war (Chap. 17). During the decade, the work of the GEC valve group overcame the many technical deficiencies of the old glass valves to create a new generation of sealed-off transmitting valves which became world leaders during the 1930s. Robert Le Rossignol—along with others—made a fundamental contribution to the development of large radio transmitting valves. Indeed, by 1944, (the now *Sir*) Clifford Patterson said in a lecture reviewing the laboratories contributions[80]; 'from the time he started working on the CAT valves, I do not believe Le Rossignol ever looked at another lamp problem! …' By the mid 1930s, driven by the demands of radio and television broadcasting, short wave wireless telegraphy, and the rapid long-distance communication required by both ships and aircraft, commercial transmitting valves were available which were comfortable with inputs up to 500 kW … and the world was transformed.

15.7   And Finally …the Decade Comes to a Close

During the decade, Robert established himself as an important member the GEC Research Staff but his life was not all work. The 1920s was also a time to attend to family after all the uncertainties and worries of life in Germany. Living at 7 St John's Road, Harrow, in the Parish of the Church of St John the Baptist, Greenhill, it was a time for the family to relax a little more. Emily it seems settled comfortably into the social life of her new country being found for example at the annual Eton–Harrow cricket matches, whilst for the boys, John Augustin and Peter Walter, their education became a priority which Robert and Emily addressed—for Peter at least[81]—by means of Brighton College. Founded in 1845, today the College is one of England's top independent schools and pupils there often secure places at the UK's most esteemed universities, particularly 'Oxbridge'. For Robert however, the choice of Brighton College may have had an additional provenance. From 1892 to 1895 the Rev. Robert Halley Chambers was headmaster at Brighton, but he had previously been headmaster to Robert's brothers, Austen and Herbert, at Victoria College (1881–1892) prior to G. S. Farnell taking the appointment, (Chap. 1). Brighton therefore may have been well known to the family and Peter became a pupil there during the late 1920s and early 1930s.

   The decade therefore ended well for Robert, but Germany entered his life once more; this time in terms of an invitation to contribute to a special issue, a 'festschrift', of the German Journal *Die Naturwissenschaft* (Chap. 8) on the occasion of Haber's 60th birthday in December 1928. For Haber, by now a portly, cigar smoking character, the previous few years had been turbulent. An increasingly difficult marriage to Charlotte—much of it his own making—had led to divorce the previous year. After the divorce Haber's sister Else, whose husband had recently died, moved into his Dahlem villa and became devoted to him. Charlotte and the children, Eva and Lutz, spent their time at their new home in central Berlin, funded by Haber as part of their separation agreement.

Haber also wanted to support Hermann, his son by Clara, who had moved with his family to France. But failed investments in South America, taxes, and the economic crisis (the 'Wall street Crash') of 1929 consumed more of his wealth than he could emotionally bear. Coupled with the rise of Hitler's National Socialist Party Haber became a somber man. In a letter to his friend Richard Willstätter some years later on the occasion of Hitler becoming Chancellor, he made his feelings about his life at this time clear[82];

> *I battle with diminishing energy against my four enemies; sleeplessness, the financial claims of my ex-wife, worry about the future and the feeling that I have made serious mistakes in my life …*

For Haber, his divorce was a great shame. He avoided contact with colleagues and when his sixtieth birthday arrived, he took little pleasure in the cards, the congratulations and the planting of a Linden tree in his honour in the courtyard of the Institute. He avoided the whole affair by travelling with Herman and his wife Marga to Egypt (Plate 15.5).

For Robert however, the invitation to contribute to *Die Naturwissenschaften* was significant, and for two reasons. Firstly, it was an unusual recognition of a 'foreigner' by the German scientific elite, confirming Robert's seminal role in the fixation of nitrogen. Here, his article was in auspicious company, other contributors including Haber's friend and Nobel Laureate Richard Willstätter for example. But secondly, 20 years after the event and free of the shackles of commercial secrecy it was an opportunity to reveal the true story of the manufacture of synthetic ammonia. In the event, the paper, written in German, was an historical disappointment, divulging little that we did not already know and, naturally I suppose, being largely a homage to Haber. Robert might have considered doing what the Chemistry Committee of the RSAS should have done in 1919 by revealing his role in the engineering of the high-pressure apparatus in Karlsruhe in 1909. But equally, it would have been difficult to upstage the old man on his birthday.

The next decade saw the fall of the Weimar Republic, the dispossession of all Haber held dear by the newly elected National Socialists, a profound tragedy for the Le Rossignol family, the development of the awesome CAT14 valve by the GEC, the most important pre-war experiment ever, and then the outbreak of a second world war. Once again there were troubled times ahead.

Plate 15.5 The linden tree planted at the KWI on the occasion of Haber's 60th birthday. Courtesy of Archiv der Max-Planck-Gesellschaft, Berlin-Dahlem

Notes

1. Thomas Hager, *The Alchemy of Air*, Three River Press, New York, (2008), p. 178.
2. The Allies demanded that payments be made in gold Marks or foreign currency and not in the paper Mark. They initially demanded 132 billion Marks, far more than the German gold and foreign exchanges could possibly raise. The last (reduced) installment was made in October 2010.
3. So-called because it was named after the city where the constitutional assembly took place. It was the seat of German culture, both Goethe and Schiller were born there.
4. Corelli Barnett, *The Collapse of British Power*, Pan, (2002), p. 424 and p. 426. ISBN-10 0391034391.
5. The patent(s) described in Sect. 15.2 of this chapter record Robert's address from 05 May 1919 to November 1919 as 2 Caesarea Place, St Helier, Jersey.

The Le Rossignol family home. However, it was probable that during much of this time he lived at 7 St John's Road, Harrow (Chap. 1).

6. Emily of course must have worried about her family left in Germany. Chapter 18 shows that they were still in contact with one-another at the time she died in 1975.

7. The Drachenflieger ('flying dragon') was an iconic canvas covered experimental aircraft developed in Austria-Hungary in 1901. Although taxi-ing trials were successful, this first heavier-than-air flying machine was woefully underpowered by its automobile engine. It crashed and was wrecked. Wells seemed to ignore this little difficulty! Nowadays, the German word 'Drachenflieger' refers to a hang-glider.

8. There was much debate at the time regarding the susceptibility of airships to explosion. In the new (Dutch) Boerner airship, reported in *The Times* on 30 March 1922, the hydrogen gasbag was surrounded by nitrogen used as a diluent and asphixiant.

9. Coward H. F., *Limits of Flammability of Gases and Vapors*, Washington D.C., Universities of North Texas Digital Library at http://digital.library.unt.edu/ark:/67531/metadc12662/. See also http://en.wikipedia.org/wiki/Hydrogen_safety and http://hq.msdsonline.com.

10. When helium is added to hydrogen, it dissipates some of the heat of combustion so that more hydrogen is needed to sustain the reaction—a 'dilution' effect. However, in the patent, Robert provided no mathematical or scientific justification for his permitted 'non-explosive' proportions of hydrogen. Robert's patent however seems to be the first time this (now) common suggestion was recorded—at least in the UK.

11. Robert Clayton and Joan Algar, *The GEC Research Laboratories, 1919–1984*, IEE History of Technology, Series 10, Peter Peregrinus Ltd., in association with the Science Museum London (1989). ISBN-0 86341 146 0, https://doi.org/10.1049/pbht010e. This title was later acquired by the Institution of Engineering and Technology, (30 June 1989). I am entirely indebted to these authors for my understanding of the GEC laboratories and Robert's early work there.

12. The General Electric Company of *England*, not to be confused with GEC of New York. Much of what appears in Sect. 15.3 of this chapter is from Clayton and Algar, op. cit. (note 11), pp. 1–3 and http://gracesguide.co.uk/Osram.

13. The DGA/Osram of course was at Friedrichshain, Berlin, where Robert had been working since September 1909, (Chap. 9).

14. Thermionic 'valves' were invented at the beginning of the 20th century. These powerful electronic components formed the driving force behind the development of all aspects of electronic engineering and were used widely up until the 1960s when they were eclipsed by transistors and integrated circuits. They could be used for example as a binary device, i.e. an 'on-off' switch

(hence the name 'valve'), as a 'rectifier' converting AC to pulsed DC, and as an amplifier, especially to amplify signals in radio transmitters and receivers. Even today there are still applications that are best served by 'valves'. For example, in high-power short-wave transmitters, some TV transmitters, and a special sort of valve is used in microwave transmitters for radar, and the microwave cooker in a kitchen. Also, ask any electric guitarist about the sound quality of a 'valve' amplifier.

15. Robert joined the GEC Laboratories in his mid thirties. A recently published salary comparison (*Chemistry World*, December 2012, **9**, 12, p. 78) suggests that the *median* salary for chemists of this age at the time was about 3 times the UK *mean* salary. Robert therefore would have been well remunerated in comparative terms. For the whole of his employment at GEC the Le Rossignols took up residence at 7 St Johns Road, Harrow, a property which had been 'in the family' at least since Austen Clement's time in London (Chap. 1).

16. Clayton and Algar, op. cit. (note 11), p. 5.

17. A 'getter' is a deposit of 'sacrificial' reactive material that is placed inside a lamp bulb or thermionic 'valve', for the purpose of completing and maintaining the vacuum. When gas molecules strike the getter material, they combine with it chemically or by adsorption. Thus the 'getter' removes small amounts of impurity gas from the evacuated space improving filament/emitter life. The getter material was strung below the filament and connected to the lead wires in the same way. When current first passes, the getter immediately vapourises and is deposited on the inside of the bulb as a small silvery metallic film. Older readers may recall this. From a Wikipedia entry.

18. Tungsten is the most refractory metal in the periodic table and therefore an obvious choice for a filament, but the metal is brittle and therefore difficult to draw into a wire. Coolidge's solution to this was to produce high purity tungsten which was subsequently converted into an ingot by reduction pressing and sintering. Mechanical working through a series of 'swages' (a forging process in which the dimensions of an item are altered using a die or dies, into which the item is forced) did the rest. The tungsten wires were then as strong as steel. From, Andrea Sella, 'Classic Kit, Coolidge's X-ray tube', *RSC Chemistry World*, **10**, 2, (February 2013), p. 62.

19. Clayton and Algar, op. cit. (note 11), p. 7.

20. Two scientific staff grades overall, Staff and Leading Staff, a situation which was to remain until at least 1950. Clayton and Algar, op. cit. (note 11), p. 19.

21. Indeed, C. S. Le Rossignol built the first 'all British' ruby laser at Wembley in the early 1960s using in-house Verneuil 'flame fusion' growth crystals and xenon flash tubes made by the lamp group. Clayton and Algar, op. cit. (note 11), p. 263. Clement Sorel was born to Robert's brother Herbert and his wife

Mai on 15 August 1922 at a nursing home in London (*The Times*, Births, 19 August 1922).

22. Of course Max Meyer had already been taking action to 'hide' Haber's money, (Chap. 14). The appearance of the Allies at Ludwigshafen and the subsequent inclusion of Haber's name on a list of 'war criminals' meant that his royalties might be confiscated as the Allies took financial control of the BASF's plants. The new contract was probably agreed just before Robert left Germany and can only be seen as an honourable act by Haber in favour of a respected colleague.

23. Confirmed for me by Prof. Szöllössi-Janze in her email of 01 August 2011. See also Chap. 14.

24. But as a munitions plant, Bosch was mis-informed. Hansard 07 April 1925 quotes Article 172 of the Versailles Treaty, which provided that 'the German Government should disclose to the Allied and Associated Powers the nature and mode of manufacture of all such chemicals as are here in question'.

25. Production figures at the time, upon which Haber's royalties were based, reflect the influence of the occupation. (See Szöllözi-Janze's figures in Chap. 14). In 1918, 116,383,750 kg ammonia were produced. In 1919 however, the figure was just 69,433,000 kg, roughly 60% of the previous year and largely due to Leuna which continued to work 'normally'. If Robert received any payments during this time they fell accordingly.

26. The plant, at La Grande Paroisse in the Seine-et-Marne department in the Île-de-France region in north-central France would be built to BASF's specifications, it would produce up to 100 tonnes ammonia a day and have access to BASF's technological improvements for the next fifteen years. In return, the French would allow Oppau and Leuna to continue production, and they would pay the BASF five million Francs together with a small royalty on every tonne of ammonia produced. Hager, op. cit. (note 1).

27. The latter responsible for producing the nitric acid and explosives.

28. Before the WW1, German was undoubtedly the language of science. Only after the war was it sidelined by English in a deliberate act which tried to disengage German science from the mainstream.

29. For whatever reason, Robert never seemed to 'publicly' acknowledge his role in nitrogen fixation. Was it a natural reluctance, a fear of upstaging Haber, or maybe not wanting to attract attention to his private contract?

30. 'Synthetic Ammonia and Nitrates Limited', *The Times*, 05 June 1920, p. 21.

31. Reported in, *Hansard*, on 07 April 1925. (also op. cit. (note 24)).

32. Hager, op. cit. (note 1), p. 206.

33. Births, deaths and marriages within the family were almost invariably announced in London in *The Times*.

34. Robert Marc Friedman, *Neutrality in Twentieth-Century Europe. Intersections of Science, Culture, and Politics after the First World War*. Routledge Studies in Cultural History, (2012), references, 1–60, pp. 111–114.

35. Donnan of course was Robert's tutor at UCL, but during WW1 he was a consultant to the Ministry of Munitions and worked with the renowned chemical engineer K. B. Quinan on plants for the fixation of nitrogen. For this work, Donnan later received the CBE (1920) and he would have fully appreciated the importance of Robert's work.

36. There was a precedent here however; Pierre Curie (and others) demanding that his wife Marie too be nominated for the 1903 Physics prize alongside himself. Strings were pulled and she was awarded jointly with him. However, there is record of Haber doing the same for Le Rossignol.

37. Branting was leader of the Swedish Social Democrats 1907–1925, Minister of Finance at the time of the announcement of the award, and Prime Minister in 1920, 1921–1923 and 1924–1925. Branting was jointly awarded the Nobel Peace Prize in 1921 for his contributions to the League of Nations. His displeasure therefore would have carried influence both in Sweden and beyond.

38. The issue of 17 November 1919, from Friedman, op. cit. (note 34).

39. Published in 'Letters' on 03 February 1920.

40. Jerome Alexander (1880–1953) was a chemical engineer and consulting chemist whose work in colloid chemistry made fundamental contributions to the subject. He was also a prolific essayist, poet and communicator exchanging letters with chemists and physicists—including many Nobel Prize winners—regarding their contributions to colloid chemistry. His opinion(s) were highly valued and like Branting, his observations regarding Haber's prize would have been noted by the international community.

41. Although he accepted that Germany's people and armed forces had to be fed.

42. In the summer of 1917, Count Luxburg sent secret dispatches to Berlin through the Swedish Legation via Stockholm, which were made public by United States Secretary of State Robert Lansing. These dispatches urged that certain Argentine ships should be sunk 'without a trace' (*spurlos versenkt*). The publication of the documents resulted in the dismissal of Count Luxburg from Argentina, and the entry of Argentina into the war. Luxburg was also Minister to Uruguay, and on his dismissal from Argentina, he asked for a passport to Montevideo instead of to Berlin.

43. In a letter to his friend Richard Willstätter, Haber found it inconceivable that Stockholm would consider 'Germans nominated by Germans'—being aware of course of the reluctance of Allied chemists to nominate him. Only Wilhelm Prandtl (Chap. 14) had supported him in 1918 and he knew nothing of the Committee's maneuverings on his behalf. He was convinced that his legacy hadn't really amounted to much, being an academic 'rock hopper', jumping from one thing to another. See Daniel Charles, *Between Genius and Genocide*, Jonathan Cape, London, (2005), p. 195.

44. Due to an infection following a mastoid operation.

45. "The Nobel Prize in Chemistry 1918". Nobelprize.org. 03 July 2012 http://www.nobelprize.org/nobel_prizes/chemistry/laureates/1918/press.html.

46. http://www.nobelprize.org/nobel_prizes/chemistry/laureates/1918/haber-lecture.pdf.

47. Dietrich Stoltzenberg, *Fritz Haber. Chemist, Noble Laureate, German, Jew.* Chemical Heritage Press, Philadelphia, Pennsylvania, (2004), pp. 215–216.

48. Daniel Charles, *Between Genius and Genocide*, Jonathan Cape, London, (2005), p. 196.

49. The designation Haber-*Le Rossignol* of course ran completely counter to Germany's established reluctance to share her innovations with 'foreigners'.

50. https://www.nobelprize.org/nobel_prizes/about/amounts/.

51. Diana Kormos Barkan, *Walther Nernst and the Transition to Modern Physical Science.* Cambridge University Press, 1999, p. 217. Haber's loyalty to Germany was clearly not unconditional!

52. But Haber was becoming increasingly concerned about his finances. Just prior to receiving his award, at the beginning of May 1920, he made two requests to the BASF, Szöllössi-Janze, op. cit. (note 60), pp. 482–483. The first of these asked for an eleven-month extension of his payments from the end of the 'circulation' patent on 12 October 1923 until the end of the 'high pressure' patent on 19 September 1924, (Chap. 8), the length of patent protection at the time being 15 years. This adjustment essentially compensated him for the loss of royalties for years 1918 and 1919 which Meyer had arranged to protect Haber's finances from scrutiny by the Allies (Chap. 14). According to Szöllözi-Janze this adjustment brought Haber's royalties to about 1,600,000 Marks for 1918 and 950,000 Marks for 1919, an adjustment to which the BASF agreed (but a little less than the extra 0.7 Pf kg^{-1} would have generated). The second request brought him back to the idea of having a fixed payment from the BASF, making him independent of the companies profits over the remaining years of the patents (1920–23/4). In uncertain times Haber was looking for assurance in his finances, justifying his request by citing his increasing tax burden, his insecurity because of the Margulies' legal action, and his financial obligations toward Le Rossignol. In the event the BASF refused the second request on the basis that none of these concerns had anything to do with the original contract conditions. The BASF were also even less obliged to concessions given the foreign currency Haber was to about receive from the Nobel Prize.

53. They also report that the last member listing for him was in July 1947. Thus, his ACS membership tenure should be stated as 1921–1947. From ACS, Office of Society Services help@acs.org by email Thursday, 19 February, 2009.

54. Recorded in the Open University's Biographical Database of the British Chemical Community, 1880–1970, http://www.open.ac.uk/ou5/Arts/chemists/ and confirmed for me by the RSC at Burlington House

whose records show that it was awarded some time between 22 April 1921 and 28 May 1921. The same records show that Robert was a retired Fellow at his death in 1976. Robert would have paid 5 guineas to transfer from Associateship to Fellowship.

55. Kormos Barkan, op. cit. (note 51), p. 216.
56. See for example Kormos Barkan, op. cit. (note 51), and also *One Hundred Years of the Fritz Haber Institute*. B. Friedrich, D. Hoffmann and J. James, p. 8. This paper can be found at; http://www.fhi-berlin.mpg.de/mp/friedrich/PDFs/ACh-100-Article.pdf.
57. Kormos Barkan, op. cit. (note 51), p. 217.
58. The presentation speech along these lines was made on 10 December 1921 by Professor Gerard de Geer, President, RSAS.
59. http://nobelprize.org/nobel_prizes/chemistry/laureates/1920/nernst-lecture.html.
60. Margit Szöllösi-Janze, *Fritz Haber 1868–1934: Eine Biographie*. C. H. Beck, Munich, (1998), p. 486.
61. Carl Bosch suffered some remorse regarding this 'oversight', he therefore had Haber elected to BASF's supervisory body (the *Aufsichtsrat*) from June 1925 until May 1932 in a final act of atonement. See Szöllössi-Janze, op. cit. (note 60), p. 486.
62. Robert recalls in conversation with Jaenicke (Chap. 18), that at this time of German hyperinflation, the cost of the rail journey was just tuppence!
63. Szöllössi-Janze, op. cit. (note 60), p. 812, her note no. 17 and personal clarifications for me by email.
64. The World Power Conference still exists today as the World Energy Council.
65. Along with George Donnan Robert's former tutor at UCL. Donnan, Harold Hartley and William Pope were Britain's WW1 'gas warriors' (Charles, op. cit. (note 43), p. 227) and former adversaries of Haber. Post war however, the three made strenuous efforts to bring Haber back 'into the fold' especially after Haber had been ousted by the Nazi regime in 1933. See Chap. 16.
66. Szöllössi-Janze, op. cit. (note 60), personal communication by email.
67. Clayton and Algar, op. cit. (note 11), p. 46. By the middle of 1920, a Patent Department existed at Hammersmith but responsibility for all GEC patents was soon centralized at the Laboratories. (Clayton and Algar, op. cit. (note 11), p. 401). The central patent office kept a patent index and a library, and dealt with matters of licenses, amendments and patent applications. It also kept a watching brief on possible infringements of existing GEC patents. By 1956 more than 4000 patents had been filed and for Robert at least, this seems to be the main area in which he published.
68. If Luton had been chosen, Patterson would have demanded season tickets for staff to attend scientific meetings in London such as the World Power Conference. For Robert living in Harrow, Wembley happened to be particularly convenient.

69. Clayton and Algar, op. cit. (note 11), p. 43.

70. A target to be thrown at, as in a 'coconut shy'. Glass shards? … What would 'health and safety' have to say today!

71. Registered on 20 October 1919. See, W. J. Baker, *History of The Marconi Company* 1874–1965, Routledge, (September 1970). ISBN-10: 0415146240, ISBN-13: 978-0415146241.

72. As a result of the formation of the valve group, Robert, Bartlett and other group members became research staff of the Marconi-Osram Valve Co. Ltd. at the G.E.C. Research Laboratories. A.C. Bartlett was a highly respected theoretical physicist and went on to author a number of patents with Robert.

73. In this context, 'high power' meant anything from 1 kW to many hundreds of kW.

74. R. Le Rossignol, E. W. Hall, *The Development of Large Radio Transmitting Valves,* GEC Journal, **7**, pp. 176–190, (1936).

75. The electrical efficiency of a typical glass valve was at best, about 70%. With 1 kW input (power), 300 watts were 'wasted'. This energy manifests itself at the anode plate as heat which has to be dissipated. The greater the power and the poorer the valve efficiency, the greater is the wasted energy and the need for efficient heat dissipation.

76. The fundamental techniques being first described by Houskeeper in the paper; W. G. Houskeeper, *The Art of Sealing Base metals through Glass*, Trans. A.I.E.E., **42**, pp. 870–876, (June 1923).

77. I am indebted for my understanding here to I.E. Mouromtseff et al., 'A review of Demountable *vs.* Sealed-off Power tubes', *Proceedings of the I.R.E.*, pp. 653–664, (November 1944).

78. The original demountable systems employed mercury condensation pumps with liquid air traps. This had improved by 1929 when it was found that the distillation of lubricating oil in a molecular still produced a fraction that was about a thousand times less volatile than mercury. If such a liquid could be used as the working fluid in a condensation pump, a vacuum of the order 10^{-6} mm (good enough for vacuum tubes) was obtainable without the need for refrigerants. Later systems therefore employed a two-stage oil diffusion pump backed up by a rotary pump. But many still felt that this was not an acceptable part of the equipment of a transmitting radio station. From C.R. Burch and Dr C. Sykes, 'Continuously Evacuated Radio Transmitting Valves', *Nature* **135**, pp. 262–263, (16 February 1935).

79. Clayton and Algar, op. cit. (note 11), p. 117.

80. Recounted in Chirnside's Obituary, (Chap. 1) probably from the lecture; '*A confidential history*' given by Paterson on the 21st anniversary of the opening of the 'new' laboratories in February 1944, the only speech he ever made regarding the nature of the laboratory's work. See also Robert Clayton & Joan Algar (Editors), *A Scientist War. The War Diary of Sir Clifford Paterson,*

1939–45, Peter Peregrinus Ltd., in association with the Science Museum, (1991), p. 457.

81. John Augustin's education remains a mystery. The couple were wealthy but there is no record of John at Eton, Harrow, Victoria College nor Brighton College. Neither does John's military record (Chap. 17) contain any detail of his schooling,

82. Charles, op. cit. (note 43), p. 216.

16

The GEC Laboratories 1930–1939

Of CATs, Cambridge and the BBC.

'Nation shall speak peace unto nation.'

The motto on the coat of arms of the British Broadcasting Company,
adopted in 1927.

16.1 The 1931 Nobel Prize in Chemistry

In September 1930, the RSAS' Chemistry Committee sent out Nobel Prize nomination forms around the world; to selected professors at universities, to Nobel Laureates in physics and chemistry, to members of the Royal Swedish Academy of Sciences and others. These forms, submitted by 01 February 1931, were screened by committee members and a list of preliminary candidates for the 1931 prize was assembled. From March to May 1931, specially appointed experts were asked for their assessment of the candidates' work, and by September the Committee submitted its recommendations to the Academy. Early in October, Academy members selected the Nobel Laureate(s) in chemistry by a majority vote. Their decision was final and without appeal. The names of the Nobel Laureates were then announced, and thus—strictly according to the rules and with due recognition of its previous failings—the Academy awarded the 1931 Nobel Prize in Chemistry[1] jointly to Carl Bosch and Friedrich Bergius 'in recognition of their contributions to

© Springer Nature Switzerland AG 2020
D. Sheppard, *Robert Le Rossignol*, Springer Biographies,
https://doi.org/10.1007/978-3-030-29714-5_16

the invention and development of chemical high-pressure methods.' At the award ceremony on 10 December 1931, the debt these men owed to the pioneering work 'of Haber' at Karlsruhe was mentioned during the speech by Professor W. Palmær, Member of the Nobel Committee for Chemistry[2];

> When Haber, in 1908, approached Germany's then largest concern in the field of the chemical industry, namely the Badische Anilin und Sodafabrik … in order to try to interest them in producing ammonia by the direct combination of its components, the gases nitrogen and hydrogen, he was able to point out that he had succeeded in finding two substances … which acted as powerful … catalysts … to such an extent that, with the help of pressure, a practical exploitation … was now conceivable as a means of producing ammonia … from the nitrogen in the air. In this case … pressure promotes the change to a very great extent … By taking this step, Haber had given rise to a new method … for which he has already been rewarded by the Nobel Prize for Chemistry.

At the banquet that evening, neither Bosch nor Bergius mentioned Haber in their speeches, but both did so later in their subsequent Nobel lectures.[3] With the award of these prizes, the Academy felt that all the founding fathers of high pressure chemistry had now been recognised, and that it had also moved somewhat to atone for the way it had conducted itself over the universally discredited award to Haber in 1919. For Robert too these awards signified a closure. Over the years of course, and especially in Germany according to Jaenicke (Chap. 18), his name had become famous, not for fixation, but for the 'Rossignol-ventil' (Rossignol-valve), a piece of 'classic kit' familiar to generations of chemists even though—like the author of these lines—they were not always familiar with its provenance. But during Robert's life, no-one ever bothered to do the 'joined-up thinking' and connect the valve to the birth of high pressure chemistry. So, despite his profound contribution, with the recognition of Bosch and Bergius he later told Jaenicke (Chap. 8) that he had come to accept that, 'my name will not be remembered for big things.' And 'big things' were certainly being addressed here, for when Palmær summed up his speech he said;

> … two problems of the utmost significance to humanity have been resolved by the chemical high-pressure methods … nitrogen [has been] made available to mankind in inexhaustible quantities, in a form suitable for agriculture …[and] … by the injection of hydrogen under pressure, pit coal, brown coal, and other carbon-bearing materials can be processed to liquid fuels which are considered indispensable in modern life for the propulsion of ships and vehicles.[2]

At the time, the high-pressure synthesis of ammonia had become ubiquitous. In 1930, the giant BASF plant at Leuna alone saw the production of 250,000 tons of benzine[4] from the local brown coal. Along with these awards therefore, came a recognition that Germany, should she choose to do so, could once again become independently capable not only of feeding and arming herself, but also of providing all the fuel she needed—for whatever purpose. 1933 saw the election of Adolf Hitler as German Chancellor, an individual embittered by Germany's humiliation after the war and who saw the Jews as responsible for the failures in his life. Hitler offered the German people a different path to the one they had followed under the Weimar Republic, and with the high-pressure chemical industry as a foundation he began to re-build German self-esteem. But for Robert Le Rossignol, now forgotten to all but a few, the 1930s saw his attention drawn to another valve, an entirely different 'Rossignol-ventil', the 'CAT' valve.

16.2   The Development of High-Power Radio Transmitting Valves

Chapter 15 showed that to achieve an increase in power to possibly many hundreds of kilowatts, a radical re-design of the glass transmitting valve was necessary. During the 1920s, the work of the valve group at the GEC overcame the many technical deficiencies of the old valve to create a new generation that were to become world leaders for a decade and more. Much of what follows here is taken from Robert's own description of the problems the group met, and the solutions they provided, which appeared in his paper for the GEC Journal in 1936, viz., *The Development of Large Radio Transmitting Valves*, GEC Journal, 7, pp. 176–190, (1936). The technical detail of these developments is largely omitted, but what readers should be aware of here is the broad range of scientific, engineering and practical skills needed to construct what were at the time, 'state of the art' electronic devices. The team was gradually expanded over the years and eventually surmounted the many problems of CAT valve construction until their techniques could be passed on to the workshop craftsmen. These high-power transmitter valves were most prominent in the 1930s and eventually ranged in size from a cricket ball to a beer barrel, and they could weigh over half a hundredweight (25 kilos).

16.2.1 CAT Valve Construction

Robert's involvement with the CAT can be traced back to as early as 1922 when in November of that year public broadcasting in the UK began with the creation of the British Broadcasting 'Company'.[5] The first daily broadcasts, (radio only) included news, politics and music. A GEC laboratory report of May that year entitled, 'Valves for broadcasting sets',[6] signaled the laboratories' engagement with the new medium. In that year, experiments were started on large water cooled valves[7] and Robert's initial involvement addressed the various glass-to-metal joints that they depended on—one of the most important being the integration of the cylindrical anode plate with the glass envelope. In this way an outer jacket containing coolant could surround the envelope and come into direct contact with it—heat at the anode being dissipated through the glass to the circulating coolant. Robert was later joined by his friend and fellow chemist Ralph Chirnside, and together they attended to the problems with coolants, but between 1922-3 Robert, together with C. A. Morton, were the first to successfully 'glass' pieces of nickel-iron (anode) tube to the envelope, the first specimen of which was christened by them, 'the Great Seal of England!'[8]

Much work had already been done in America regarding metal-glass bonding by William G. Houskeeper (no 'e') of the Western Electric Company, which he patented in the USA in 1919.[9] Houskeeper was able to make robust glass-to-glass joints between hard and soft glasses, and to hermetically seal metal conductors into vacuum tubes by simultaneously welding glass to the opposite sides of an intermediate metal disc, usually of copper. The disk was very thin (sometimes thicker, but with very thin chamfered or 'feathered' edges) and the distortion caused by the differential cooling was accommodated by the flexibility of the disc (or rather the disc's edge). In this way the glass expansion predominates and the disk is *forced* to adopt the thermal expansion coefficient of the glass. Subsequently in 1923, the 'art' of hermetically sealing the anode tube to the valve envelope was described by Houskeeper in a second publication, (Chap. 15), but for many the process remained a mystery, achievable only by the most talented of glass blowers. Progress could only be made with the new cooled anode valves therefore if this technique could be mastered and devolved to the laboratory craftsmen who had to build the valves. It was here that Robert made a fundamental contribution by using a variation on an already known process. Alloys of nickel and iron could be made which had approximately the same coefficients of expansion as the glass envelope and which could be joined directly to the glass. Usually the alloy was thinly copper plated at the joint as glass adheres readily to copper, and this was facilitated by coating the copper with borax which allowed the glass to 'wet' the metal. As with Houskeeper's technique, the nickel alloy tube was usually turned thin at the edge (tapered or 'feathered') to compensate for any small difference in expansion during manufacture, and during the heating/cooling that occurred when the transmitter was turned on and off. This second method, pioneered by Robert Le Rossignol, was the one adopted by the M-OV in the manufacture of all its cooled anode valves, viz, transmitters-'CAT's, modulators-'CAM's and rectifiers-'CAR's and so early in the development of cooled anode technology, the engineering of probably the most critical component had been mastered. For the craftsmen in the workshops however, his process must have been magical when seen for the first time. For Robert tells us in his 1936 paper—in typically modest fashion—that;

> ... the making of a large glass-to-metal joint usually inspires awe in the uninitiated, but as a matter of fact, once the technique has been grasped there is no great difficulty in making such joints ...

The manufacture and operation of these valves however involved far more than just the perfection of Robert's anode joint. Much electronic engineering and theory was involved; as in the design and materials of the filament; in minimizing inter-electrode capacity, the inductance and 'flash arcs'; attending to the 'magnification' of the valve which depended upon the filament length, its diameter, the radius of the grid and the anode/grid voltages, none of which are remotely suitable for treatment here. The valve group also had to design special apparatus for the workshop craftsmen to build the valves. The anode of course is opaque and so it was impossible to see whether the electrodes were sealed centrally in the anode tube. A 'glass lathe' was therefore designed on which the electrode assembly (filament and grid) could be mounted, centred, and then inserted into the anode tube to be sealed in. The lathe had two heads to achieve this (one for the anode tube, the other for the electrode assembly) in accurate axial alignment with one another. The heads, which could be rotated at equal speeds, replaced the hands of the glass-blower who would otherwise be unable to cope with the size and weights involved. Mounted on the lathe was an array of blow pipes which the glass-blower could adjust and move along the lathe bed to construct the valve. Once constructed, the valves could only become functional by 'pumping out' to obtain the necessary high vacuum. This was done on the 'pump table'. For the older glass valves the pump tables could accommodate just one valve, the pumping process taking about a day to complete. The M-OV therefore had banks of pump tables each with an attendant operator. For the much larger cooled anode valves the pumping process could take as long as a week and tables were subsequently redesigned to hold a plurality of valves. Photographs of these 'modern' lathes and pumping tables appeared in Robert's paper of 1936.

16.2.2 CAT Valve Cooling[10]

Cooling systems too had to be addressed by the valve group, another multi-disciplinary area involving chemists and mechanical engineers. It was here that Robert and Ralph Chirnside (Chap. 18)—who was given an opportunity to take on the responsibilities of the principal chemical analyst, W. Singleton, on his death in 1930—collaborated in the evaluation, procurement, preparation and analysis of coolants for valve types used all over the world, Chirnside eventually becoming GEC's Chief Chemist in the mid 1950s. CAT valves consisted of an integrated glass envelope and cylindrical anode enclosing the grid and filament, the leads to which were taken to outlet points in a further glass cylinder attached to the envelope of the anode and forming a continuation of it. The valve was mounted perfectly vertically in an outer jacket, so that the envelope was contained within it but separated from the walls of the jacket by a space through which the cooling liquid could circulate. The complete unit was then mounted vertically on insulators, (using 'spirit levels' on the jacket) because the anode and its jacket were at a potential of thousands of volts above the frame of the transmitter, the filament, and the grid circuits. For water cooled systems, water was taken to and from the space between the anode envelope and its jacket by means of long lengths of rubber hose. The purpose of these was to provide a sufficiently long column of water between the anode and earth to provide a resistance of such magnitude that it prevented excessive leakage of anode current to earth. Therefore, with valves of this type it was essential to have water which was as pure as possible; free from matter suspended or in solution which might otherwise conduct and/or deposit a layer of 'scale' on the envelope inhibiting the dissipation of heat. Distilled water was therefore generally used, being continuously circulated through the jacket and then through an external cooling radiator before being returned. Variations on the water-cooling method were also designed where the water was allowed to boil away rather than circulate. These were only feasible when large disposable quantities of quite pure water were available.

Water has a very high thermal capacity so water cooling systems were relatively compact. In some of the early valves of the CAT type however, oil instead of water was adopted to retain thermal capacity but avoid the problem of high frequency losses in the insulation of the water systems. In practice however, the valve group found that efficient cooling only took place when incipient boiling occurred at the anode. With ordinary transformer oils, this temperature was so high as to cause the glass to 'gas'. However, with oils of lower boiling point such as 'kerosene' (140–160 °C), the vacuum remained unimpaired. But with all oils, 'sludging' had to be guarded against. This problem was well known to electrical engineers—especially when unsuitable oils were used in oil cooled transformers. Here a sludge formed on the copper conductors—especially if oxygen was present—and with CAT valves, this sludge inhibited dissipation at the anode, raising its temperature and 'cracking' the oil just as in a refinery.

Other CAT valves developed by the M-OV during the period—generally those of about the same wattages as glass valves—were cooled by air, either by convection or by a forced draught from large air blowers. These were a development of the *metal*-envelope valve made under the trade name 'CATkin' which in turn was a development of the water-cooled valve. Such valves offered certain advantages over ordinary glass valves; they could operate at shorter wavelengths, and because of their much lower operating temperature the anode had an enormous 'clean-up' property which was able to absorb large quantities of gas to maintain the vacuum. And whereas we have concerned ourselves largely with anode cooling here, it's important to realise that the filament leads, which carried a heavy current in the more powerful valves, also had to be cooled. This was also generally achieved by water or by air.

16.2.3   CAT Valve Modulation

In addition to his contributions to the construction and cooling of the various CAT valve(s), Robert also turned his attention to improved methods for modulating high frequency transmitters via the CAM valve. In this respect he devised the method of 'series-modulation'[11] which he subsequently patented on behalf of GEC in 1933.[12] CAT valves of course were used in the final stages of amplification of RF signals. These signals consisted of two parts; the carrier wave (say 1500 m wavelength for a long-wave transmission), upon which was imposed a *modulating* signal which varies the properties of the carrier in accordance with the information to be transmitted, which for radio broadcasting was the speech signal. At the receiver, filtering out the carrier leaves the speech signal which we then hear through our speaker. By the early 1930s the two established methods of modulating the carrier in radio transmitters were widely accepted to be wasteful of power. Robert's 'series-modulation' provided a more efficient method of modulating; the transmitter being fed with current varying according to the speech current received from the microphone(s), and radiating *power* correspondingly. The method was especially applicable therefore when large amounts of power were to be modulated and provided a much higher quality signal which in turn vastly improved reception for the listener. Series-modulation was therefore ideal for the new 'super-broadcasting' stations and although an alternative system, 'phase modulation', had been developed by the Société Française Radio-Electrique, GEC-Marconi did not consider it to be superior to, or had any economic advantages, over series modulation—even though Radio-Luxembourg and the French state broadcasting stations were already equipped with this system. Robert's series-modulation was subsequently widely adopted in telecommunications.

16.2.4 CAT Valves and British Broadcasting

The first water cooled transmitting valve made by the company, naturally called the 'CAT1', was similar to an earlier American design. Each valve could accept 10 kW of input power when used as an RF amplifier and these valves were employed at the Rugby radio station of the G.P.O. in Warwickshire (alongside demountables, Chap. 15), at the Marconi wireless station in Caernarvon North Wales, (where 48 old glass valves in parallel were replaced by 10 new CAT1 valves), and at the new BBC long wave station at Daventry in Northamptonshire in 1925. With an overall output of 25 kW, at the time Daventry was the most powerful transmitter in the world. Later in 1934 (Sect. 16.5), the new larger Droitwich transmitter adopted the frequency and the famous Daventry '5XX' call sign, but Daventry was retained, firstly as a reserve for Droitwich, and later for transmitting coded messages across Europe during the second world war. The availability and the raw power of the transmitter during the 1930s also saw Daventry host probably the most important pre-war experiment ever in the UK, an experiment that laid the foundations of a technology that more than any other contributed to the successful outcome of the 1939–45 war, viz, the 'Chain Home' (CH) and 'Chain Home Low' (CHL) 'radar' stations and the airborne radars.[13]

Three types of Marconi water cooled valves were used at Daventry, five CAT1s as the R.F. transmitting valves, eight CAM1s as the modulator, and eight CAR2s used as HT rectifiers for the other valves. All needed a minimum of 2 gallons of water per minute for cooling. As a point of historical interest, in 1949 the 'Old Gentleman' as it was then known was scrapped and the building cleared for installation of the new BBC medium wave 'Third Programme' transmitter. Three of the water-cooled valves, one of each of the types that were used, were kept on display in the entrance hall of the building. In 1992, when all transmissions finally ceased at Daventry, the 3 valves were put on display in the main transmitter hall for the closedown day, but they disappeared. They were subsequently found at the BBC Television Centre in London as it was being cleared for the move to Salford in 2012. The valves are currently at the Town Council Museum in Daventry and they are probably the earliest examples of Marconi water cooled valves to survive. They are also precious examples of Robert's contribution to this important technology.[14]

During the mid-to-late 1920s a series of more powerful CAT valves were developed by the valve group viz., the (oil cooled) CAT2 through to the CAT9, each driven largely by the growing needs of the world's broadcasting companies. By 1930–31 'super power' broadcasting stations began to be considered and a major advance was made with the CAT10 valve which could handle an input of 100 kW. By the end of the decade, the M-OV were world leaders in CAT technology. A large cooled anode transmitting valve (the CAT12) operated the transatlantic telephone line,[15] and even larger more powerful valves, the CAT14 and CAT14C (500 kW), were being designed for use by the BBC, each modulated by Robert's 'series-modulation' method. These valves were water cooled transmitting triodes (30–40 gallons min^{-1}) with tungsten filaments suitable for use as high frequency amplifiers operating at up to 10Mc sec^{-1}. The CAT14 and CAT14C weighed 71 lb (32.2 kg) and differed only in that the CAT14C had air cooled valve filament leads whilst the CAT14 had water cooled leads. These valves were originally built in the GEC Laboratories—little room being available at Hammersmith—and for many years during the 1930s and 1940s they were the most powerful transmitting valves in the world. A mark of their success being that valves of the CAT14 type were to remain in service in the UK until 1957 (Plate 16.1).

When using CAT valves in a broadcasting station, automatic arrangements were provided to close down the transmitter should the temperature rise too much or the coolant flow cease. These arrangements were influenced by the general atmospheric temperatures, and hot weather could sometimes cause a station to go 'off air'. Large valves of the water-cooled variety were also

Plate 16.1 The GEC CAT14 valve in its water-cooled jacket. The two filament leads appear at the top. Photograph reproduced with permission,[6] published by the Institution of Engineering and Technology (1989) https://doi.org/10.1049/PBHT010E10

expensive items, costing many hundreds of pounds. They were also fragile and precautions were taken to prevent damage during handling. Marconi therefore designed special trolleys to move the valves from the valve store to the transmitter hall and then lift them into position, not only to prevent damage but also to avoid undue delay in the event of a valve failure. To make doubly sure, spare valves were already kept mounted in position. The life of these valves was only measured in thousands of hours, they were therefore always run under optimal conditions and constantly monitored.[16] As an example of the dependability of CAT valves, in 1932 the BBC's transmissions amounted to 58,163 h. Breakdowns amounted to 0.023%, just half a second per hour.[10] The design of a broadcasting station was therefore complex, a 'many splendid thing', but by September 1934 when BBC Droitwich began broadcasting with the debut of the awesome CAT14, GEC and Marconi had perfected a technology which they subsequently exported by designing, and then installing their equipment, in broadcasting stations all over the world.

Now one can hardly credit Robert Le Rossignol alone with the perfection of the cooled anode valve, but equally it is undeniable that once again his 'fingerprints were all over the technology', a technology that helped lay the foundations of the world of telecommunication we live in today. Resolving the problems of CAT valve construction, its attendant modulation and cooling system(s) were to be Robert's greatest technical contribution to the Wembley Laboratories but his knowledge and experience, crystalised for us in his 1936 paper, was to see him influencing developments in a range of other valves in coming years. The techniques he developed, particularly the glass-to-metal seals, were equally applicable to smaller valves. These became increasingly important to the Laboratories as the decade progressed and they in turn were to meet the requirements for power valves in radar and navigation in the 1939–45 war. But by the early 1930s, with Robert now in his mid-to-late 40 s and firmly established at Wembley, he was not only progressing technical innovation but also beginning to develop in a more managerial sense. By the end of the decade he was to take charge of the production of valves for the war effort, the kind of role well known to his 'dear old professor and friend, Haber' (Chap. 18) who he may have met again in London in 1933 (Sect. 16.3). But whereas Robert's professional career was blossoming, Haber, by now a tired old man, battling illness, financial worries, the dispossession of all he held dear, and suffering the tragedy of the German Jew,[17] was entering his final years. The relationship between these two friends, broken by the first world war but re-formed during the 1920s, was about to be broken again, but this time forever.

16.3 Fritz Jacob Haber—Dispossession and Disillusion

When Robert wrote his appreciation of Haber's life for the *Die Naturwissenschaften* 'festschrift' in December 1928, (Chap. 8) Haber's Institute was a centre of world class research. But although Haber had regained much of his energy and enthusiasm after the physical and emotional devastation of the Great War, the years from 1928 to 1933 were difficult ones for him; the shame of the failure of his marriage to Charlotte, his financial worries, his continued poor health and his worries about the future for Germany all weighed heavily. The Great depression of August 1929 in the

USA led to the 'Wall Street Crash' in October and a world wide economic down-turn. In Germany by 1931, more than 4 million workers were unemployed. Inclined to despair by a feeling of uselessness, Haber was to write to ministers warning of a collapse 'even worse than that of 1918',[18] but the feeble Weimar Republic was unable to take the necessary 'root and branch' reforms that those such as Haber proposed. The German people, attracted by the dark charisma of Adolf Hitler, made the National Socialists the single largest party in the elections of 1932. The elections of January 1933 saw Hitler appointed as Germany's Chancellor on a wave of nationalism. He set out to crush all opposition. Day by day he claimed more and more powers and by March 1933 the democratic Weimar Republic collapsed when the German parliament gave Hitler's government absolute power. Germany was ready to accept a führer, a führer whose bitterness towards Jewish people was used to construct flag-ship, and undeniably popular, political policies.

For Haber this new order was completely at odds with his world. By April alarm bells were sounding as it soon became clear that now 'Jewishness' was a matter of ancestry not religion; conversion was no indemnity. Nazi leaders organised a boycott of Jewish businesses; in Prussia all Jewish judges were required to voluntarily request leave of absence; a law was passed that demanded Jews be removed from the civil service within a year unless they had served in the Great War. This law applied to every professor at a German University and nearly every scientist at the Kaiser Wilhelm Institute. Haber was stunned by the immediacy and the ruthlessness of it all. Even so he could have remained in office because of his 'distinguished' war service. But, in the absence of Einstein, who was travelling at the time and who made it clear he would not be returning to Germany, Haber became the most prominent Jewish scientist at the Institute. Because of its prestige and the large proportion of Jewish scientists there (about 25%) the Nazis targeted it for special repression. They ordered the Institute to 'change its make-up' by dismissing some scientists, and Haber, as director, had to make the choice(s). Should he 'dispose' of those of lower rank in order to protect the top scientists? Or was it better to remove the more prominent for they would find new employment more readily? In an act of humanity, he chose to remove the Institutes two top scientists Herbert Freundlich and Michael Polanyi who had already secured alternative employment. But Haber had enough of the 'new order' and on 30 April 1933 he tendered his own letter of resignation to the Prussian Minister of Culture, effective on 01 October 1933.

Over the next few months he remained the Institute's Director, busying himself with the welfare of his young scientists and trying to arrange employment for them abroad. But of his own future he seemed paralysed, wanting to leave Germany but not knowing how to achieve it. He was too weak to research or teach full time but not ready to retire. Like some ageing modern rock star, Haber decided it was better 'to burn out than rust'. But who would want him? During the summer of 1933 he tried to find out, travelling to Holland, France and England in a sad attempt to 'sell himself' and procure an honorary position at a scientific institution. He found none, but whilst in London he may have met with Robert Le Rossignol one more time, Robert recalling in his letter to Johannes Jaenicke in August 1959 (Chap. 18) that 'I last saw him in 1933? ...' but the question mark suggests Robert's recollection was vague. What is more substantially established however is Haber's meeting with the Zionist leader Chaim Weizmann at London's Russell Hotel in July 1933 when Weizmann asked if Haber would consider moving to Palestine to help develop scientific institutions there.[19] Soon after, fascinated by Weizmann's offer but having made no commitment, Haber returned to Berlin. But even as he was being pressured to leave his villa in Dahlem he still could not make up his mind what to do, and so on 03 August, he went on his travels once more, this time to meet Hermann in Paris, then to attend a scientific conference in Spain, fully intending to return to Berlin in three weeks time.

In Paris the offer that Haber had been longing for finally arrived. Harold Hartley, Frederick Donnan and Sir William Pope—Britain's WW1 'gas warriors'[20]—had convinced Cambridge University to offer Haber an honorary, unpaid position there, with little teaching or research. From a room in the Avenue George V,[21] Haber composed a letter, in German, to Sir William. Pope's offer, he explained, gave him all he hoped for, to remain active in science and possibly acquire British nationality both for himself and his children. Haber explained that;

> ... the most important goals in my life are that I not die as a German citizen and that I not bequeath to my children and grandchildren the civil rights of second-class citizenship as German law now demands.[22]

Haber however was concerned that he achieve an 'honourable separation from Germany'. Such permission was important because if he settled in the UK without it, Germany could demand an 'emigration tax' (the 'Reichsfluchtsteuer') for 'unpatriotic desertion'[23] which would decimate his finances. Haber's monies however were held in Switzerland and (also being a Swiss citizen) he could have ignored the tax, but that in turn would brand him a (dishonourable) tax-evader, such was his emotional attachment to Germany that was something he could not countenance. Confused as to how the 'honourable separation' might be achieved but offering Pope a possible solution,[24] he finally added; 'as soon as I have this permission, I will accept your generous invitation …'. After posting this conditional letter of acceptance, he left Paris for the meeting of the International Chemical Union at Santander in Spain.

At the conference, which was marked only for its mediocrity, Haber's health worsened. His presentation—well prepared but poorly attended—was a shambles, with Haber often having to reach for his nitroglycerine spray as he gasped for breath. In late August with his heart problems worsening, he left Spain for Berlin to try and facilitate his 'separation', but just before entering Germany, as the train reached the Swiss city of Basel, he suffered an attack of total exhaustion and had to rest a while. In Basel, he heard again from Chaim Weizmann, and having recovered somewhat, but against the advice of his doctors, he travelled to Zermatt to discuss the Palestine position once more. This time, Weizmann recalls, Haber accepted the idea with enthusiasm,[25] then left again for Germany. Homeward bound, he suffered another physical collapse which his doctor, Rudolph Stern, thought was probably a heart attack. Whatever, he never made it into Germany and was taken to a sanatorium in the Swiss town of Mammern where he was joined by his sister Else. As the effective date of his resignation grew closer, he remained in Switzerland leaving Else to return briefly to Berlin to clear the Villa at Dahlem. But on 21 September 1933—only a few days before the effective date of his resignation— Haber wrote to the Deutsches Museum in Munich with a remarkable offer which expressed his desire that, after his departure, Germany retain something 'physical' regarding the historical significance of the work conducted by himself and Le Rossignol at Karlsruhe. This letter, provided to the author by Dr Stefan Wolff at the Museum and annotated by the author, is reproduced below;

To the German Museum in Munich.

Years ago, [1922] upon request, I sent my laboratory apparatus to the Deutsches Museum. [The apparatus] was essentially the same as the small apparatus built for the laboratory, which I had used together with my younger friend Le Rossignol in Karlsruhe. I then developed this apparatus into the larger model that BASF took over and gloriously designed into a large-scale industrial process. I have described the Karlsruhe piece, with the permission of the company, in detail in the Journal of Electrochemistry, (Chap. 8) but [the original apparatus] was not included in the transition to Berlin-Dahlem, and it has been lost. During the 22 years in Dahlem, the work of the institute's workshop has always been more urgent than the reproduction of Karlsruhe's larger, more advanced model. Now, however, [with] my farewell at the end of this month, I have had … reason to reproduce the larger Karlsruhe apparatus. If the Deutsches Museum wishes to accept it as a gift, the master mechanic of my institute, [Hermann] Lütge, will effect the transfer to Munich on my behalf and [ensure its] proper installation there, without any costs for the Deutsches Museum. It only requires his immediate assignment to the task, so that the [transfer] can [be] done this month. during my time of service.

Respectfully F. Haber.

The museum inventory records show that the apparatus Haber originally sent to the museum in 1922 was that produced by the BASF in 1910 and shown in Plates (11.1) and (11.2), (Chap. 11). This apparatus had been used at Karlsruhe for teaching/demonstration purposes. But as the original Haber-Le Rossignol apparatus had been 'lost' after Haber's departure from Karlsruhe, he instructed his master mechanic to produce a faithful re-construction for posterity. The apparatus held at the Deutsches Museum is therefore a replica, but its mere existence reflects Haber's concern that he be remembered by the German people, above all, for the ammonia synthesis (Plate 16.2).

Haber's wanderings across Europe were taking their toll on him both physically and mentally. For his own peace of mind, he had to decide between Palestine and Cambridge. With no prospect of achieving an 'honourable

Plate 16.2 Hermann Lütge's re-construction of the original Heber-Le Rossignol apparatus at the Deutsches Museum, Munich. Photograph courtesy of the Deutsches Museum Archives, CD_68715, with special thanks to Dr Stefan Wolff

separation' if he went to the Zionists and fearful of the long journey there, he settled on Cambridge but still with reservations regarding his finances. At the end of October 1933, Haber and Else travelled to Paris to meet with Hermann and from there on 01 November 1933, Haber wrote[21] to Sir William Pope (in English);

Dear Sir William,

Since I had the invitation from you to come to Cambridge, I have passed a time which has been no pleasant one for me. Finally, this Saturday I will be able to come over to England using the Golden Arrow via Calais and Dover. My sister Frau Dr Freyhan, will come with me. We will spend the time from Sat 4 Nov evening till probably Monday afternoon in London and be at Cambridge no later than seventh November ... Perhaps you will allow me to discuss the matter of the public announcement of my immigration with you before any thing is published about the matter ... We stay in London at the Russell hotel, Russell Square ...

On 07 November 1933, Haber and Else took up residence at the University Arms hotel in Cambridge.

Over the next two months or so Haber wrote frequently to Pope regarding his position at the University.[21] He also wrote to Carl Bosch, the only member of the Board at IG Farben, (which had by now merged with the BASF) who had shown him any sympathy on his resignation in April and who had also sent him a card on his 65th birthday.[26] Haber pleaded with Bosch to use his influence in Germany to allow him to live out his 'remaining years of pitifully diminished health and strength in peace and decency ...'. But Bosch had already done what he could, even arguing with Hitler that persecution of the Jews would only harm German science. Haber's letters went unanswered. Bosch was also in league with the new 'Masters of War' in that the IG Farben had signed a contract with the Government to supply large quantities of 'gasoline' from coal, the Government contract making the process financially viable. Eventually, this gasoline was to fuel Hitler's coming 'European tour'.

The stress of these few months, coupled with the damp British climate, saw Haber become even weaker and once again he decided to travel to relieve his symptoms, this time to another sanatorium in Switzerland. Asking his friend and doctor Rudolph Stern to accompany him, the two agreed to meet in Basel. On 25 January Haber wrote a final letter to Sir William, and four days later he and Else arrived, exhausted, in Basel. As well as Stern, Hermann and Marga also stayed with the couple at the Hotel Euler and they were shocked by how much worse Haber had become. Walking had long been a problem but now even talking for a few minutes was to bring on severe chest pains. Stern asked Haber to rest and then examined the old man, at the same time re-assuring him of the regenerative properties of the milder climate. Soon however, Haber rose and began discussing his options for the future. But that night, Haber called Stern to his bedside immediately after retiring; he was gasping for breath as his heart failed catastrophically. Slipping into unconsciousness, Stern tried all he knew to revive him but despite his best efforts Haber passed away. His remains were cremated in Basel on 01 February 1934.

A few weeks before his death, Haber had given instruction to Hermann; if the political climate allowed it, he wanted his ashes to rest alongside Clara in Dahlem, if not, Hermann was to find a suitable resting place and, in time, arrange for Clara to brought there too. Hermann chose the latter, and found a cemetery in Basel close to the German border. Haber's ashes were buried there on 01 February 1934. The funeral of one of the 'mightiest men in Germany'[27] was attended by just four people, Rudolph Stern, Hermann, Else, and Richard Willstätter, who had travelled from Munich to say a few words at the graveside. Haber too had wanted a few words on his gravestone; 'In war and in peace, as long as it was granted him, a servant of his homeland'. His grave was as close to Germany as Haber was to get, but Hermann decided not to include the epitaph. Later, Clara's remains were brought to Basel as Haber wished, and today both Fritz and Clara lie together, their memorial stone simply reading;

```
Fritz Haber

1868–1934

Clara Haber

Geb. Jmmerwahr 1870–1915
```

Everyone dies, but not everyone 'lives'. And behind the austerity of their memorial lies a story of two people who experienced all aspects of their humanity; of love and neglect, triumph and tragedy, war and peace, and life and death. Together now forever.

Fritz Haber's story has dominated the history of the discovery of 'fixation'. The circumstances of his life, his powerful personality, Jaenicke's 'critical mass' of biographical evidence (Chap. 18) and the romantic Victorian notion of the 'gentleman scientist' bound by honour and patriotism, (Chap. 10), has proved attractive to historians. In terms of honour, Haber fulfilled the expectation made of him. His sincere recognition of Le Rossignol for the work he did, the fact that he 'played fair' by him in their financial arrangement, (Chap. 14) and intervened to get him released from Ruhleben testifies to this side of his character. It is reflected too in the support he provided for his children and his second wife Charlotte after their divorce, and in his resignation from the Kaiser Wilhelm Institute when ordered by the Nazis to remove his Jewish scientists. And in 1933 when leaving for a position at Cambridge, he sought an 'honourable separation' from Germany to avoid the 'Reichsfluchtsteuer' for 'unpatriotic desertion' which would not only decimate his finances. but brand him a tax evader.[23] Such was his emotional attachment to Germany, it was a legacy he could not countenance and illustrates the 'tension' he felt at being torn between his obligations on the one hand and his pursuit of money as reward on the other.

His patriotism too was unquestionable. In his resignation letter from the KWI he said of WW1, 'Science is for the world … but for your country in time of war!'. But we should not suppose that he enjoyed war per se. He hated the suffering, the waste of life, the pointlessness of it all, for in his emotional 'ground state' he was kindly, helpful, and benevolent. But it is equally true that the excitation of war transformed him into a ruthless autocrat, a side of him which in peace-time he either suppressed or made manifest as his 'strength of purpose'. He gave his whole being over to the struggle for victory and nothing was allowed to deflect him: neither family, human suffering, nor considerations of international law. He brought a vitality to all the positions he held and he always placed Germany first, For three or four generations past, his family had fought for, or served, Germany and he was proud to have become a Prussian officer and such an influential military man, so when defeat finally came it weighed heavily upon him as did the personal repercussions of his role in chemical warfare.

In the wider terms of his science, Haber was characterized by versatility. There was hardly an important branch of contemporary physical chemistry to which he did not make a fundamental contribution. But despite all his talents, his eclectic and fundamental contributions to chemistry, his energy and his devotion to duty, he is remembered by many today not by the epitaph he composed for himself: 'He served his country in war and peace, as long as it was granted him', (Stoltzenberg, Chap. 2) but in a way he would have never have wanted. 'As a Jew … the 'father of chemical warfare'.'

But in the history of science, Haber's papers of 1905–1908 with van Oordt and Le Rossignol stand alongside those of Einstein's 1905 *Annus Mirabilis*, not of course in terms their imagination and scientific insight, but in their subsequent impact on humanity. Einstein's explanation of the photo-electric effect, his evidence for the existence of atoms, and his monumental $E = mc^2$, took decades to influence the lives of ordinary folk, the vast majority of whom even today remain utterly ignorant of the ways in which Einstein has marched into their lives. But within a few years of the publication of %*NH₃* @ *equil.* = 0.012%, (Chap. 2) the most influential technology of the 20th century was born, carving out the Europe—and indeed the world—we live in today. The first half of the century may have belonged to physics, the second half to biology, but the equation that defined the century is, $N_{2(g)} + 3H_{2(g)} \leftrightarrows 2NH_{3(g)}$, and it belongs to chemistry. It is for this, above all else that has passed, that we should remember Fritz Haber.

16.4   11 Carlyle Road Cambridge, May 1934

Fritz Haber died a German-Swiss citizen, and his estate probably avoided the Reichsfluchtsteuer.[23] Neither were his young children to inherit the rights of a second class citizenship, for soon both 'Lutz' and Eva became British citizens.[28] Haber's passing was recorded briefly in *The Times* on Saturday 03 February 1934, and then more fully in an Obituary by Professor H. E. Armstrong, FRS, on the following Tuesday, who begged leave to 'pay unreserved tribute to the service rendered to all nations by our gifted German colleague …' Each article recorded Haber's honourable exit from Germany, his recent stay at Cambridge, and the importance 'to civilization' of his fixation work ('with Bosch'). There is little doubt that these articles were read by Robert and the family, but (quite understandably) neither mentioned 'the Englishman, Le Rossignol …' as had sometimes been the case in the British Press both before and after the war; for these were eulogies for just *one* man. Robert's prediction that 'my name will not be remembered for big things' however, was being fulfilled, but for those who knew the real story, the public praise of Haber only served to emphasise how important a piece of chemistry went on at Karlsruhe between 1906–1909. One 'in-the-know' of course was young Peter Walter, now an undergraduate at Cambridge during Haber's stay there, reading chemistry as part of his Natural Sciences degree. Although Cambridge was littered with famous men, Peter must have been immensely proud of his father, or 'Daddy' as Robert was affectionately known. One cannot help feeling too that, in time, Robert may have thought of meeting Haber once more at Cambridge, possibly accompanied by Peter, but in the event we know that it never happened, Robert later recalling in a letter to Jaenicke (Chap. 18); 'I very much regret that I did not see him during his stay in Cambridge …'.

Plate 16.3   A group photograph of pupils at Walpole House Brighton College, 1931–1932. Peter Walter is seated, second row, and second from the right. Courtesy of Mrs. J. M. Heater, Hon. Archivist Brighton College.[29]

There is only so much one can do for one's children, and the pressure of the imminent commission of the Droitwich broadcasting station probably meant that Robert 'shelved' any meeting for a later date. After all, Haber was to become a permanent fixture at Cambridge, so it could wait. It was a shame, for meeting Haber would surely have been a privilege for Peter but he was already living quite a privileged life. From 1929 (aged 16) until 1932 (aged 19) Peter was educated at Brighton College[29] under the Rev. William R. Dawson, headmaster there from 1906 until 1933 but previously at Corby Grammar School and King's School Grantham (Plate 16.3).

Under Dawson's tutelage Peter did well, winning a place at Sidney Sussex College Cambridge to study the Natural Sciences Tripos,[30] being admitted to the first year of the 'Prelims' in October 1932. It appears that Peter, probably like his father, was a very quiet, studious young man, but not 'abnormally' so, because he entered fully into the life of the college and made many friends with whom he was particularly close. In his first year he took accommodation close to Sidney at (P9) Garden Court, and as he was fond of rowing, he joined the College Boat Club rowing in the third crew. Peter wrote home regularly to Robert and Emily with cheerful letters reflecting a settled time at Cambridge—even attending the dinner of the 'Old Brightonians' at Cambridge in 1933.[29] Peter's first year at Cambridge was un-eventful, but when the summer examinations came along—as they always do—Peter, together with a number of others that year, was awarded a 'Special Examination'. Today such an examination would be awarded to students with particular educational needs and Cambridge recognises a number of difficulties faced by students. Candidates may be granted additional time if they have a condition or disability which would prevent them from completing the examination in the specified time; they may apply to defer course-work if they have very good reasons; they may be permitted to bring food and drink into the examination room if they have a medical condition that requires them to consume such at regular intervals; if candidates experience excessive fatigue or pain as a consequence of a medical condition they may be granted supervised rest breaks; if they are visually impaired, question papers may be enlarged or produced in an alternative format, and if candidates are unable to write because of a temporary or permanent disability or injury they can sit their examination by dictating to an amanuensis. A similar set of disadvantages was recognised in the 1930s although it is not certain what Peter's special educational needs were. The first year would have been quite demanding requiring the study of a wide range of subjects such as chemistry, physics, biology, earth sciences, mathematics etc.,—not entirely dissimilar with Roberts matriculation examination at UCL—but special needs or not, Peter passed the examination(s) and proceeded to the second year 'Prelims'.

For his second year, Peter moved to 11 Carlyle Road, taking rooms with Ms Louisa J. Peters. A compact mid terrace Victorian house, number 11 is in a quiet road, close to the river Cam, and just a short walk to all the main Colleges of the University. Peter lived here when Haber was 'in town' and one could just imagine the discussions with his friends as news of Haber at the University Arms hotel raced around the campus (Plate 16.4).

However, the second year, as is often the case for students, became more demanding, as the depth of Peter's understanding of the various subjects was now more seriously questioned. Responsibly, Peter realised that he had to devote more time to his studies and to this end he gave up his rowing—family members today recalling that it was Robert who made this 'suggestion' to Peter. But unbeknown to his friends and tutors, Peter was struggling to cope. Having been awarded a special examination in the first year, the University had already expressed a willingness to support him, but for some reason he never sought further help.[31] Inevitably, the second-year examinations came around and on the morning of Friday 25 May 1934 Peter sat for the chemistry paper. Two or three-days earlier Peter had told his friend Mr. M. H. Oddy—more as a joke than anything else—that he may not be 'up'

Plate 16.4 11 Carlyle Road today, second house from the left. Courtesy of Google Earth Pro, street view, 2019

next year as he might not pass his examinations. However, on Friday afternoon the two took tea together and worked through some past papers. That evening in Carlyle Road, Peter confided in his landlady's sister about the morning's chemistry examination and the daunting six hours he would have to spend in the examination room next day. When Ms Peters also asked about his chemistry examination, he was downbeat, saying that he'd done 'much as expected'. But, Peter was quite relaxed and on retiring that night he said he would like to be called at seven thirty in the morning for breakfast at eight, and no doubt with a heavy day ahead for the young man, Ms Peters would have planned a substantial, comforting repast.

The following morning Ms Peters began her preparations, after which she went upstairs to call Peter. She knocked politely on the door …. 'Seven thirty Sir' … but no answer. She knocked again, several times, but still no response. This was odd. The door was unlocked and she took the unusual decision to open it and step inside the room. But as she did so she felt some resistance as a bath towel lay on the floor across the bottom of the door. Inside the room a window was wide open and there was the faint but unmistakable smell … of gas. Peter lay perfectly still on the bed, dressed in his pyjamas with the bed clothes over his body and a bed rug over his head. There was a gas bracket in the room and Mrs. Peters' first thoughts must have been that there had been a terrible accident … but moving closer she saw a long piece of rubber tubing. One end was attached to the gas bracket, tied on by a handkerchief, the other end lay next to Peter in the bed. This was no accident. The gas mantle and the burner had been carefully removed from the bracket and placed on the dressing table; the rubber tubing had been deliberately attached and brought to the bed, and on the dressing table was a sealed envelope addressed to Robert. Alone and in distress, Peter Walter Le Rossignol took his life in the early hours of 26 May 1934, and considering the relationship between his father and Fritz Haber, he chose a path of utter irony … gas poisoning. The story of the 'fathers of fixation' had suffered a second suicide, but by leaving a window wide open Peter had ensured that the excess gas would escape and hopefully not endanger anyone else in the house. His actions were clearly well planned for later it emerged that Peter had purchased the rubber tubing rather than 'borrowing' it from one of the laboratories which might have raised some suspicion as gas poisoning was often a preferred route for those considering suicide in those days. Equally, by leaving the bedroom door unlocked, was this a cry for help that went tragically wrong?

P. C. Palmer and Dr. Youngman were called to Carlyle Road, and by midday the news was conveyed to Robert and Emily. The same evening saw a brief report on page 5 of the *Cambridge Daily News* under the headline; '**UNDERGRADUATE GASSED Cambridge Landlady's Discovery**'. Establishing the cause of death as coal gas poisoning, there was no need for a post-mortem examination, but an inquest was set for the following Monday, 28 May, under Deputy Borough Coroner Mr. W. R. Wallis, at the Mill Road Institution in Cambridge. The inquest sat without a jury and witnesses were Ms Peters, P. C. Palmer and Inspector Bellamy for the Police, Dr. Youngman, Mr. Oddy and Peter's tutor at Sidney the Rev. B.T.D. Smith, Mr. Metters from the Cambridge Gas Company, and Robert. By the morning of the inquest, Peter's death had been reported nationally in both the *Daily Mirror* ('**UNDERGRADUATE'S FATE**') and *The Times* ('**UNDERGRADUATE FOUND DEAD**') but as yet with little detail. During the inquest the Coroner read the contents of Peter's letter to the court. Described later in some of the press as a 'pathetic last note', it began; 'Dear Mummy and Daddy', and continued, 'I cannot go on like this any longer. Now I have made a mess of the chemistry paper. I am a failure'. If he went on, the letter continued, he would not only be a disgrace to himself but also a drag on his friends, who had all been so good to him. Referring to Robert and Emily, he said that he could never thank them enough for what they had done for him, but somehow, he had felt 'this coming for years' as there was something in his mental make-up which made it impossible for him to make the best of his life. The letter concluded 'Please try to forgive me … My love to you both, Peter.'

Witnesses gave evidence; The Rev. Smith explained that Peter had never appeared to himself or his supervisor as in any way depressed, and when asked by the Coroner; 'He never intimated to you that he felt himself a failure?' Smith replied, 'No, never.' Mr. Oddy explained that Peter had never made any remark suggesting he was tired of life. Medical evidence offered by Dr. Youngman suggested that death had occurred three to four hours before he was called. The scene of death was described by Ms Peters and P. C. Palmer. What evidence remains suggests that Robert had little input to the proceedings but he was called to confirm the body as that of his son, and that the letter was in Peter's handwriting. Robert also explained that Peter had never given reason to consider he had contemplated anything against his life and he could find no reason why he should do so, unless he had been overworking, which may have 'affected his brain'. At the end of the inquest Mr. Wallis returned a verdict of 'Suicide by coal gas poisoning during temporary insanity'. He summed up as follows;

> It appears that the deceased had been unduly worried as to the results of his papers. He apparently took it to heart. As indicated in the letter which I read, he thought he was a failure. From the evidence we have had I think he was altogether wrong. He had been of a very cheerful disposition, as the evidence stated by various witnesses shows. As so often happens, you do a paper and you think that you have not done at all well and he got depressed on that account. It is to be regretted that he apparently purchased that tubing with the full intention of committing this act … there was no doubt that he was in a very difficult state of mind but had he thought a little, he would not have committed the act which has brought us all here today.

The Coroner expressed his sympathy with Robert; Inspector Bellamy associated himself with this, as did Mr. Metters of the Gas Company. The inquest was then closed.[32]

Peter's body was returned home and on Tuesday 29 May his remains were buried in consecrated ground at Harrow cemetery, the only interment that day. The morning was quiet and warm with gentle winds, cloudy, but occasionally the sun broke through. The service was conducted by the Rev. Henry Wolferstan Beck, Vicar of the family's Parish Church of St. John's Greenhill. In time, the family arranged a full traditional headstone and kerb for Peter's grave with the simple inscription;

In Loving Memory of Peter Walter Le Rossignol, died May 26th 1934, aged 20.[33]

In England and Wales at the time of Peter's death, attempted suicide, was seen as an immoral, criminal offence against God and the Crown. It was stigmatized and punishable for those who survived and any report of suicide attracted the attention of the Press. On the morning the family buried Peter, the popular national newspaper the *Daily Mirror* carried an account of the inquest. It was reported alongside several pictures of the Daily Mirror's Golf Tournament and the antics of Miss Jean Batten, the New Zealand 'Air Heroine' on her record-breaking flight to Australia; a harbinger of today's insensitive press (Plate 16.5).

UNDERGRADUATE'S
DESPAIR

"I Am a Failure" Letter
Before Examination

FOUND GASSED

Why he committed suicide by gas poisoning was explained in a letter left by Peter Walter le Rossignol, an undergraduate of Sidney Sussex College, Cambridge University, which was read at the inquest at Cambridge last evening.

He was to have sat next day for the Second-Year Preliminary examination in Natural Sciences.

The letter was addressed to his parents, who live at Harrow, Middlesex. It read: —

"Dear Mummy and Daddy—I can't go on like this any longer. Now I have made a mess of the chemistry paper. I am a failure.

"If I go on I will not only be a disgrace to myself, but a drag on my friends, who have been so good to me; but it is much harder to treat you like this, who have given me everything.

P. W. le Rossignol.

"There is something in my mental make-up that makes it impossible for me to make the best of my life."

"Overworking" Theory

Mr. Robert le Rossignol, of St. John's-road, Harrow, a research chemist, said that his son's letters were always cheerful. He thought overworking had affected his brain.

Other evidence showed that Rossignol told friends that he might not be up at Cambridge next year as he might not pass the examination.

He had also talked to his landlady's sister of the examination and of the six hours he would have in the examination room next day.

Recording a verdict of Suicide during temporary insanity, the coroner said that the evidence given showed that the young man was not a failure.

Plate 16.5   The report of Peter's death in the Daily Mirror, 29 May 1934

By Friday 01 June, a more measured account of the inquest was published in the *Cambridge Independent,* an account which we have relied on heavily here. Even so it carried the gaudy headline;

'EXAMINATION LEADS TO SUICIDE.

Cambridge Student Who Gassed Himself

I AM A FAILURE—Pathetic Last Note

By the standards of the day, the reports of Peter's death went 'viral'. For such a private family, the public exposure of their grief and the nature of their son's death must have been so difficult to bear. For his friends at Cambridge, Peter's Obituary appeared in the Sidney Sussex 'Annual' just before the new Autumn term that year. Short and uninformative, its very reticence tells of a life unfulfilled. It read (Plate 16.6);

Clara's death haunted Fritz Haber for the rest of his life. He knew he had made serious mistakes and none more so than the neglect of his first wife. For Robert and Emily too, an inevitable feeling of guilt must have overwhelmed them when they read in Peter's letter, 'Now I have made a mess of the chemistry paper. I am a failure …' For some reason Peter had felt his despair 'coming for years …' Was he too proud to accept another special examination? Had he chosen to study science simply because of his father? Did Haber's presence in Cambridge, his father's friendship with the 'great man', and the expectation of his friends finally inflict a pressure on him to achieve a standing in chemistry which was beyond his capabilities? Whatever the turmoil in the young man's mind, by 'bottling it up' he left the same legacy of guilt to Robert and Emily as Clara left to Haber; a guilt that never eases, one that comes in the 'black dog' hours of the night, when you hold something of theirs or pass their room. But it's at its worst when you realise that time and time again you lost the chance to hug and to hold, to let them know that no-matter what happened everything would be alright, because with

PETER WALTER LE ROSSIGNOL

Peter Walter le Rossignol died in Cambridge on May 25th, 1934 at the age of twenty and in his second year of residence. He came to us from Brighton College and was reading for the Natural Sciences Tripos. He was a member of the Boat Club, and his death was a great shock to his friends.

Plate 16.6 The entry in the Sidney Sussex Annual October 1934. Courtesy of Sidney Sussex and Mr. Nicholas Rogers

hind-sight, by 'putting two and two together', you really should have seen their despair building … but you didn't. Years later, Richard Feynman who suffered the death of his lovely young wife from 'TB', learned to accommodate science within the important things of life[34];

> *[Tell your son to …] stop trying to fill your head with science - for to fill your heart with love is enough.*

Robert and Emily now not only carried the most unnatural burden of a parent, the death of their child, but guilt and stigma. Maybe they learned something from the way Augustin and Edith coped with the death of Austen (Chap. 1). Even so, our current understanding of child bereavement is such that Robert and Emily's life would have changed forever. Men and women grieve differently. Men try to rationalise their life by continuing to do all the 'normal' things. But mothers are always crying inside. Even years later, a phrase, a mannerism, a look … reminds them of what has passed bringing silent floods of tears which for fathers, although understood, is both sudden and unexpected. Women more than men look for support amongst family and friends to talk about their bereavement. But it is not so simple. Friends we have had for years are often uncomfortable with the depth of our grief, 'crossing the road' to avoid contact for fear of saying the wrong thing, and for the Le Rossignols this would have been compounded by the manner of their son's death. There was no need for a public announcement of Peter's passing, the Press had already attended to that, and without a post mortem examination, within the context of the times, his burial was simple and swift. Along with the irony of Peter's death by gas poisoning there was a final irony; an examination system which his father had survived at the turn of the century was still the 'norm' and it became a last impossible hurdle in Peter's young and fragile life. But Peter's death may have left Robert and Emily another cruel burden. When Emily died in 1975, she had been almost completely crippled by arthritis for over forty years, a virtual recluse. Throughout this time Robert devoted himself to her care. According to Chirnside's Obituary of Robert (Chap. 18), Emily's condition appeared about the time Peter died. Today, the link between emotional trauma and physical illness is well established.

16.5   September 1934, a Professional Pinnacle —BBC Droitwich

It is difficult to understand how one gets life back on track after such a tragedy. The social inclusion between all levels of staff at Wembley under Paterson may have helped both Robert and Emily to come through this difficult time. Even so, as the years passed, Chirnside's Obituary of Robert (Chap. 18), records that Robert rarely spoke of Peter, maintaining instead a quiet dignity which was simply 'his way'. Like many men, he may too have eased the pain by throwing himself into his work, and at the time there happened to be much work to do with the imminent commissioning of the new giant BBC super-broadcasting station at Droitwich which had the M-OV valve group's awesome CAT14s at its beating heart. Other than the coming war years, Droitwich probably represented a pinnacle in the professional careers of many of the original Marconi valve group. In turn, Droitwich was a milestone for the BBC, and indeed for the broadcasting world. Today, radio and television broadcasts are ubiquitous; available 24/7, digital, high quality, terrestrial and satellite. But in 1934, Droitwich made broadcasting history and became the very model of a modern transmitting station whose technology was exported world wide. Through his work on the development of large radio transmitting valves, Robert had helped to make a significant contribution to what was at the time another 'cutting edge' technology. Without which super broadcasting and even radar may not have been achievable.

16.5.1   Droitwich, Construction and Opening[35]

The development of broadcasting in the UK over the previous ten years had naturally mirrored transmitter technology. By 1925 there were over 20 medium wave stations in operation centred in large towns and cities all over the country. Today, these would be regarded as local radio stations. With the creation of the British Broadcasting *Corporation* in 1927[5] the broadcasting structure gradually developed into a regional service, the low power stations being replaced by 'high-power' transmitters serving seven regions of the British Isles. The new stations—transmitting on two wavelengths— gave listeners the choice of a regional or national programme. By 1930 for example, the Midland region was served by the two transmitters situated at Daventry; the regional midland programme on medium-wave and the national programme on long-wave. At just 25 kW, Daventry 5XX long-wave was unable to cover the whole of the British Isles as originally intended and it was necessary to provide several fill-in transmitters for the more distant regions. It was therefore decided that a single new 'super high-power' long-wave transmitter (taking 5XX as its call sign) should be built radiating the national programme on long-wave from a site where it would serve most of the highly populated areas of the UK. In this respect the seven regional areas; London, the Midlands, South Wales and South West England, Yorkshire and Lancashire, Glasgow, Edinburgh and Belfast were all to receive their national broadcast from this new transmitter. Areas outside these regions being served by the existing services. The 'centre-of-gravity' of this population map was Birmingham but technical reasons made it difficult to locate a high-powered transmitter in such a built-up area. An alternative site had to be found to the south of Birmingham and the short-list included Barnt Green and Wychbold, Droitwich Spa. In 1932, after the BBC conducted a series of transmission tests with a mobile transmitter and discounted the existence of underground brine streams, Droitwich was selected. The purchase of a 24-acre site was completed in February 1933 and in addition to a national service by means of the long wave transmitter (at an initial 200 kW), a regional service for the midlands was to be provided by means of a medium wave transmitter (50 kW).

The later CAT valves were designed specifically for 'super high-power' broadcasting. Working closely with the BBC engineers, construction of the valves and the transmitter began at Wembley under the direction of the valve group. The system of modulation chosen by Marconi was Robert's 'series modulation' and it was expected to provide faithful reproduction of sound broadcasts between 30 and 10,000 cycles. But the original (long wave) transmitter output power was planned to be 200 kilowatts and this caused concern in certain quarters. Some said Droitwich would be the 'loudest voice in the world' and cause interference with other stations on an unprecedented scale. As a compromise it was eventually agreed to operate at a lower radiated power. Even so, some curious things happened.

By April 1933 the approach road to the new transmitter site was started; by September the two 700-foot steel lattice aerial masts were nearing completion; by November the transmitter building was well advanced; by July 1934 testing of the transmitter began and an opening date for the station was set for 06 September 1934. In August, listeners who happened to tune to the long-wave band of their radios (or 'wireless sets') in the early hours of the morning would hear an announcement 'This is the long-wave National transmitter at Droitwich testing' followed by a programme of 'gramophone records'. Reports on the reception from all over the country confirmed the high-quality signal that had been the promise of the new Marconi series-modulation system. A trade boom was already apparent and GEC/Marconi received orders to build similar transmitters for Finland, Sweden and Rumania. At home, the Radio Exhibition at Olympia, 'Radiolympia', coincided with the testing period and manufacturers were able to display their new receivers with 'Droitwich' marked on the tuning dials alongside the more exotic Moscow, Luxembourg, Warsaw and Radio Paris etc. Concern was expressed that some of the older radio sets would have such poor selectivity that their owners would be unable to tune to stations other than the strong Droitwich signal. Some of the radio journals began to publish articles on the construction of simple technical fixes to help overcome this problem, although manufacturers hoped for a boom in sales of new sets of the 'super-heterodyne' type with their superior selectivity.

On Thursday 06 September 1934, a party of 150 pressmen from all over the country were taken from Droitwich railway station to the transmitting station by Midland Red bus. Together with B.B.C engineers, Regional Directors and representatives of overseas broadcasting undertakings, they formed the audience for the opening ceremony. At 3.45 pm 5XX went on air replacing the transmission from Daventry with a radiated power of 150 kW. The audience heard a programme from London by the B.B.C. Orchestra, conducted by Aylmer Buest. It began with the overture from 'The Merry-makers' by Eric Coates followed by three English dances in the old style. The newsmen and guests were then given an extensive tour of the station and equipment. Locally, the 08 September edition of the Bromsgrove, Droitwich and Redditch *Weekly Messenger* gave a full-page report with pictures of the new installation, whilst nationally, on the morning of the opening, *The Times* devoted two columns to the event together with a full radio programme for the day. For the first month after the opening, Droitwich replaced Daventry only during the mornings and evenings. The official changeover took place on 07 October 1934 with Daventry closing down after nine years of transmitting the National Programme, Droitwich taking over the full service. The change was chosen to coincide with the end of British Summer Time and, to quote the Bromsgrove Droitwich and Redditch Weekly Messenger, 'the return of normal time, that harbinger of the long winter evenings, when the wireless is especially welcome in the home' In those early days the programme began with the Daily Service at 10.15 am and ended at midnight with dance music. Such was the interest shown in the new revolutionary high-power transmitter that technical visitors from all over the world came to inspect it. In time, some 'teething troubles' naturally emerged at Droitwich long wave. By June 1935, a report by the research department identified electrolysis of the water in the cooling systems leading to the oxygen product dissolving in the water and causing excessive aeration. Longer lengths of rubber hose or smaller diameter hose were recommended as a solution, both of which reduced leakage of current by increasing resistance. Some criticism the series-modulation arose concerning the grid and filament circuits at the high-tension end of the circuit, but as the visitor book shows, interest in Droitwich continued throughout the world.

Over the next few years however, an interesting phenomenon that the local householders became used to, especially those living adjacent to the transmitter site, was the sound of programmes mysteriously emanating from objects such as electric cookers and iron ranges, such was the power of the new transmitter. This was due to long lengths of house wiring and large metal objects acting as receivers and picking up energy from the powerful transmitting aerials. This gave rise to an audible, but rather 'tinny', signal. One lady claimed to be able to hear a programme every time she touched the poker on the firebars of her metal fireplace! In later years, an attraction that appealed to visitors, especially the children, was the pond. It was part of the cooling system for the large transmitting valves, holding 300,000 gallons of water containing thousands of goldfish to keep the water free from algae. In time, the fish learned that when the metal guard rail was tapped, food was about to appear. The canteen kept a supply of stale bread and visitors were able to command shoals of fish to appear in response to the signal. It was against regulations to remove fish from the reservoir but over the years many a local ornamental garden pond was stocked from this source! Even so, one cannot help feeling that the pond should ideally have contained some CATfish(!) (Plate 16.7).

Plate 16.7   The valve cooling pump crypt at Droitwich in 1934. Distilled water driven by these pumps passed through the jackets of the CAT14s and out to the external radiators in a closed-circuit system which were cooled by spraying water on them from the pond containing the goldfish. Photograph from Le Rossignol's paper, The Development of Large Radio Transmitting Valves, GEC Journal, 7, pp. 176–190, (1936)

16.6   Wembley Through the 1930s and the Outbreak of War

The beginning of the 1930s coincided with the industrial depression in the States and Europe and the Laboratories were encouraged to be as financially prudent as possible. In the event however, the steady growth which began in 1919 was to continue until the outbreak of the second world war in 1939. During the decade, progress was made in research groups across most areas. The lamp group with which Robert had been initially involved developed the first high pressure mercury lamps, a practical and commercial low pressure sodium lamp, a mains voltage tubular florescent lamp and the illumination-group matched these developments with practical applications of the mercury lamps for street lighting and the use of fluorescent lamps in big stores—and indeed in the laboratories themselves. Throughout the 1930s the transmitting and receiving valve groups continued their development of triodes and multigrid valves, developments which culminated in the CAT14 series. But the techniques involved were equally applicable to small valves, and this work, especially relating to the metal-to-glass seals, was to be important in meeting the requirements of various valves for radar, navigation etc. during the war.

Robert's remarkably eclectic contribution to the work of the Laboratories' valve-group can be seen from the fifteen patents submitted by him on behalf of GEC-Marconi over the decade, the majority of which were in his name alone and submitted after Peter's death (see also ***Appendix*** *D*). Thus, we had '*improvements in or relating to …*';

> methods for modulating high frequency transmitters (1932), electric discharge devices (1933), electric arc converters (1934), thermionic valves (1935), switches for converting electric currents (1936), electric discharge devices (1936), thermionic valves and other electric discharge devices (1936), electric discharge devices with indirectly heated cathodes (1936), electric discharge devices with external cooled anodes (1936), demountable electric discharge devices (1937), thermionic valves (1937), large thermionic valves (1937), air-cooled thermionic valves (1938), thermionic amplifiers (1938) and, the cooling of external metal parts of electric discharge devices (1939).

The decade also saw the first appearance of the *GEC Journal* (1930), in which much of the Laboratories' work was published, but apart from his 1936 paper on large transmitting valves Robert seemed to confine his publication efforts to patent applications.[36]

Towards the end of the decade the Laboratories were becoming more and more involved in preparations for national defence as tensions rose once again across Europe. Hitler's election as Chancellor in 1933, his abolition of democracy and his seizing of absolute power in Germany, led almost immediately to a massive German re-armament campaign. Many political scientists at the time predicted the possibility of a second 'Great War', a view firmly established when in 1935 Hitler rejected the Versailles Treaty, accelerated his re-armament campaign and introduced conscription. In August of the same year, concerned about developments in Europe, America declared herself neutral. Over the next few years a variety of territorial occupations occurred across Europe and Treaties were signed as nations tried once again to protect themselves against the catastrophe of another conflict. When the Spanish Civil War broke out in 1936, both Germany and Italy supported the Nationalists against the Republicans and used the event to test new weapons and methods of warfare. The war ended in 1939 and in January of the same year Hitler ordered a major covert expansion of German Naval power to challenge British supremacy. On 01 September 1939, the inevitable occurred and Germany invaded Poland on the pretence that Poland had violated German territory. On 03 September Britain, along with France and much of the British Commonwealth, declared war on Germany. At the Laboratories, in anticipation of another war, the illumination-group had already carried out a study on visibility from the air in relation to a potential national 'blackout', and the development for GEC's own purposes, of metal-to-glass seals in valves with which Robert Le Rossignol was intimately concerned, led to the earliest involvement of any industrial organization in the Armed Services programme to develop valves for radar. The pre-war work of the television and cathode-ray groups also allowed them to rapidly transfer to military work on radar.

Immediately on the outbreak of war, Lord Hirst sent for Paterson and told him to place the Laboratories at the disposal of the Government and not to permit commercial considerations or industrial rivalry to stand in the way of the war effort. Nor did he wish the Laboratories to make any profit on research done for the Services.[37] This instruction would have been entirely in keeping with Paterson's own views, and for all those working at the Laboratories this was how they conducted themselves over the war years. Work was undertaken because staff had the best expertise to achieve a solution and not because it would place GEC in some advantageous position after the war. Sometimes, solutions provided by the Laboratories were taken to production by other companies and so GEC 'lost out', even so Paterson had no time or respect for anyone in any position whom he believed to be furthering their personal agenda or playing corporate politics. For almost every member of staff the years from 1939–45 were devoted, without exception, to projects aimed at winning the war. In this respect the major contribution of the Laboratories was their work on valves and cathode ray tubes, particularly for navigation and radar.[38] There was a tremendous expansion of facilities for the pre-production[39] and testing of a variety of valves developed specifically for wartime. Over 1000 newcomers joined the staff but the Laboratories relied heavily on their senior staff and Robert Le Rossignol's intimate knowledge of the construction, manufacture, operation and characteristics of thermionic devices naturally made him a key player in 'Wembley at war'. Like Haber, Robert found himself at the service of his country. But in the event, like thousands up and down the country, his dedication was to be cruelly rewarded.

Notes

1. Reported in a single sentence without comment in 'News in Brief', *The Times*, Friday, 13 November 1931.
2. Professor Palmær's speech is available at http://nobelprize.org/nobel_prizes/chemistry/laureates/1931/press.html.
3. CARL BOSCH
   The development of the chemical high-pressure method during the establishment of the new ammonia industry. Nobel Lecture, May 21, 1932.
   http://www.nobelprize.org/nobel_prizes/chemistry/laureates/1931/bosch-lecture.html.
   'Partly as a result of the Lecture given by Haber here in Stockholm in 1920 has been learned that in 1908, he approached the Badische Anilin und Sodafabrik with the suggestion to attempt the technical synthesis of ammonia from hydrogen and nitrogen under high pressure. In the course of his studies on the position of the equilibrium in the system ammonia ↔ hydrogen + nitrogen he ascertained that the foreseeable shifts in equilibrium for high pressures actually occur in favour of the ammonia, but—and this was a particularly encouraging and for us decisive fact—he also found that osmium and uranium markedly accelerated the attainment of the equilibrium and brought it about at a far lower temperature than the catalysts exclusively used hitherto, i.e. manganese, iron, platinum, etc.'
   FRIEDRICH BERGIUS
   Chemical reactions under high pressure. Nobel Lecture, May 21, 1932.
   http://www.nobelprize.org/nobel_prizes/chemistry/laureates/1931/bergius-lecture.html.
   'In 1908 and 1909 I was given an opportunity in the laboratories of Nernst and Haber to witness the use of the high-pressure methods in investigations into the ammonia equilibrium and ammonia synthesis, and I tried my hand, in these

laboratories, at that time, at syntheses by high-pressure techniques, with the then imperfect apparatus, and with little success.'

4. The figure is from Palmaer's speech. 'Benzine'—not to be confused with *benzene*—is a colorless, flammable liquid, normally obtained from petroleum distillation. It is a mixture of hydrocarbons (chiefly pentane, hexane and heptane), much like gasoline and naphtha. Benzine was used as a solvent in the manufacture of products such as drugs, dyes, and paints, and because the vapors of benzine, when mixed with air, are highly explosive, it was used as a motor fuel.

5. On 01 January 1927 the British Broadcasting Company was dissolved and the British Broadcasting *'Corporation'* (BBC) was constituted under a Royal Charter.

6. Clayton and Algar, *The GEC Research Laboratories, 1919–1984*, IEE History of Technology, Series 10, Peter Peregrinus Ltd., in association with the Science Museum London. (1989), p. 47.

7. However, a satisfactory commercial article had firstly been produced in America by the General Electric Company.

8. Clayton and Algar, op. cit. (note 6), p. 119.

9. *'Combined Metal and Glass Structure and Method of Forming Same'*, W. G. Houskeeper, US Patent Office Serial No., 210,396. Patented 04 February 1919. Robert refers to Houskeeper in his 1936 paper, but he mis-spells his name as Housekeeper. A very common mistake.

10. In addition to Robert's paper, some of what appears here is based on an article by M. M. Goddard B.E(ng) in the *Sidney Morning Herald* on Wednesday 22 February 1933 under the title, *'Transmitting valves, removing surplus heat'.*

11. For any poor unfortunate who wishes to know; the name derives from the fact that the main modulating valves in the high-frequency stages were connected in such a way that the voltage on the anodes of these valves varied in accordance with the low-frequency input to the grids thus varying the high-tension supply to the anodes of the high frequency valves, with which they are 'in series'. Now go back to bed.

12. *An Improved method for modulating high frequency transmitters.* UK Patent Office, no. 393,397, accepted 08 June 1933.

13. 'Radar'—a British invention, not so much in concept but rather in terms of a practical realisation—was originally called RDF (Radio Direction Finding), but the American term RAdio Direction And Ranging soon took preference. During the 1930s a government scientist, Robert Watson Watt, was developing an experiment to monitor the height of the ionosphere, reflections from which are used to 'bounce' radio signals around the world. The experiment was important because at the time, Britain was searching for a reliable method of radio transmission across her Empire. Watson Watts' idea was to transmit a very short but powerful RF burst and measure the time it

took to return from the ionosphere as an 'echo'. To this end he successfully used a new invention—the cathode ray tube—and he published his results in 1933 in a report called *The Use of the Cathode Ray Oscilloscope in Radio Research*. But while monitoring the ionosphere, he noticed that occasionally, he received other echoes, which appeared to come from aircraft flying in the area of the experiment. When war seemed imminent in the late 1930s, this experimental curiosity took on a whole new meaning as a mechanism for detecting aircraft still at some distance from the UK mainland *i.e.* an 'early warning' system.

To prove the practical capability of this idea, Watson Watt required the use of a very high-powered transmitter. His original idea was to 'pulse' the signal in high energy radio bursts, but only *continuous* transmitters existed, and still one of the most powerful in the world was that 'idling' at BBC Daventry. A test was scheduled for the morning of Tuesday, 26 February 1935. Watson Watts' equipment was installed in a large caravan in the meadow around the transmitter, and an RAF bomber, a Handley-Page Heyford [K6902], was procured and flown back and forth along a route between the two transmitter masts. The tests went perfectly. The bomber was easily picked up and tracked over 9 miles—a much greater ranges than had been expected. Soon, from a laboratory on the East coast at Bawdsey Manor, new equipment was installed which was able to detect aircraft from over 50 miles or so out to sea. From this were developed the 'CH' (Chain Home) radar stations that covered the entire East coast before the war and the whole of the British Isles by 1940. They were not good at finding very low flying aircraft, but could find the higher aircraft up to 120 miles away. Later, the 'CHL' (Chain Home Low) stations were added to fill in the lower altitudes. These were the monsters of radar, with huge towers (350ft or more) transmitting pulses of radio energy of hundreds of kilowatts using demountable cooled anode transmitter valves in the power stages and 75ft high 'curtain' antennae as detectors. The technology was minaturised at the GEC by the beginning of the war and their valves formed the basis of various airborne radars.

14. Much of the information regarding the transmitters in Chap. 16 is taken from the website maintained by retired BBC engineers at; http://bbceng.info. I fully acknowledge their contribution here. Copyright remains with the web authors.

15. Clayton and Algar, op. cit. (note 6), p. 119.

16. By 1957 the CAT14C was still in use at Droitwich, four in the transmitter and nine as spares. The cost of each then was £650 and the average lifetime was 11,000 h, op. cit. (note 6).

17. Albert Einstein famously called Haber's life, '*The tragedy of the German Jew, the tragedy of unrequited love.*'

18. Daniel Charles, *Between Genius and Genocide*, Jonathan Cape, London, (2005), p. 219.

19. Weizmann found the famous chemist, 'broken, muddled, moving about in a mental and moral vacuum …', Charles, op. cit. (note 18), p. 226.

20. Gas Warriors. General Harold Hartley was head of Gas Warfare on the British side. William Pope, Chair of Chemistry at Cambridge, was knighted (K.B.E) in 1919 for developing mustard gas for the Chemical War Service. Frederick Donnan of course was Robert's tutor at UCL. These three men had a remarkable affinity with Haber's views on chemical warfare which they too thought was more humane than conventional weapons.

21. '*A Genius for Friendship*', pp 1–16, from *Fractured Biographies* (German Monitor), Editor Ian Wallace, (c) Editions Radopi B. V., Amsterdam—New York, NY (2003). ISBN-10: 904200956X and ISBN-13: 978-9042009561. Containing ten letters from Haber to Sir William Pope from Paris and Cambridge over the period August 1933 to January 1934. The letters are also held as part of the Haber-Sammelung at the MPG, Berlin. These letters are a moving testimony of Haber's state of mind at the time.

22. From the letters in note 21 quoted in Charles, op. cit. (note 18), p. 228.

23. The dreaded '*Reichsfluchtsteuer*', the 'Reich Flight Tax' to prevent the flight of capital from the Weimar Republic.

24. The letters in note 21 show that Haber suggested a form of 'ceremonial honour' that the British Ambassador in Berlin might present to the German Foreign Office, which as an act of international courtesy, they might feel obliged to support.

25. Charles, op. cit. (note 18), p. 232.

26. Charles, op. cit. (note 18), p. 234.

27. At the meeting with Weizmann in Zermatt, Haber had apparently announced that; "*I was one of the mightiest men in Germany, but at the end of my life I find myself a bankrupt*". Charles, op. cit. (note 18), p. 231.

28. Charlotte brought the children to London in 1934 after Haber's death. 'Lutz' became a Reader in Economic History at the University of Surrey whilst Eva married and settled in Bath. Lutz died in 2004, Eva died in 2017.

29. I am completely indebted to Mrs. Joyce Heater (Hon. Archivist) at Brighton for my understanding of Peter's time there. She tells me; Peter Walter first appears in a list of new boys at Brighton in May 1929 when he would have been almost sixteen. The school has no indication of his previous education but he would hardly have come from a 'prep' school at that age. This raises the possibility of health problems and perhaps private tuition, but the school has no evidence of either. He *may* have attended Brighton College 'prep' school at some stage but it was independently owned at the time and records are incomplete. Peter was a boarder at Walpole House, the building still exists but it closed as such in 1934 due to the depression. Peter was a Walpole House Prefect there in January 1932 and a school Prefect from May 1932. This was a great privilege and he is shown wearing the school tie in Plate (16.2). He was also Platoon Commander in the OTC and achieved

Certificate 'A' in the School Certificate Examinations in summer 1932. On leaving Brighton, it was usual to include the University destination of the leaver in the 'school Valete'. There is no such reference in Peter's entry. This *might* be linked to his health; we can only assume his admission to Sidney Sussex was confirmed after he left Brighton.

30. The Natural Sciences Tripos is the framework within which most of the science is taught at Cambridge. It is taught by a number of departments designed to allow a highly flexible curriculum and includes a range of physical and biological sciences, together with the history and philosophy of science. Much of what appears here is due to Mr. Nicholas Rogers, Archivist at Sidney Sussex (by email) to whom I am grateful.

31. Sidney Sussex records Peter's award of a special examination in the first year but not the in the second. Courtesy of Nicholas Rogers, archivist at Sidney, (by email).

32. The original inquest papers have not survived and this account is based on a report published in the Cambridge Independent on Friday, 01 June 1934. I am indebted to Mr. Josh Acton of the Central Library, Cambridge for taking the time to provide me with a copy.

33. I am indebted to the Cemeteries Department of Harrow Council for their help in obtaining the inscription for me.

34. Richard P. Feynman to the mother of Marcus Chown, Feynman's student, who was trying to get his mother to understand more about his life in science. Quoted in *The fantastic Mr. Feynman*, last shown on BBC2, Tuesday 20 August 2013.

35. The source material for this section and the next was taken from the BBC engineering web site, http://www.bbceng.info, op. cit. (note 14), constructed and maintained by ex-BBC engineers. I am particularly indebted to the reminisces of Mr. J. F. Phillips in his article '*Droitwich calling*' which I have relied upon heavily and modified. Readers may also wish to see a short (edited, ten minute) film made in 1934 by the celebrated documentary maker, John Grierson, showing the construction and operation of Droitwich. In, *Droitwich. The world's most modern long wave transmitter*, amongst many other things referred to in this section, we see the mobile transmitter lorry, the CAT14 valves being moved on their trolleys and installed in the new transmitter, together with the external radiators for cooling the valves. Readers are warmly recommended to look at this film if possible. Visit the bbceng.info web site or 'You Tube'.

36. Clayton and Algar, op. cit. (note 6). Appendix, *Some Publications by Staff of the Laboratories'*, contains just one paper by Robert; R. Le Rossignol, E. W. Hall, *The Development of Large Radio Transmitting Valves*, GEC Journal, 7, pp. 176–190, (1936).

37. Clayton and Algar, op. cit. (note 6), p. 21.

38. Clayton and Algar, op. cit. (note 6), p. 51.

39. The needs of the Services were so urgent that for many programmes there was no time to transfer a design to a valve factory, at least until early demands had been met 'pre-production' units were therefore set up in the Laboratories. Clayton and Algar, op. cit. (note 6), p. 124.

17

The GEC Laboratories 1939–1949

Wembley at war, John Augustin, 'St. Helier'
and retirement.

… There is one point in which I failed this morning. I do not think I made clear to Megaw, Gossling, Le Rossignol and company our appreciation of the wonderful contribution they have made to our armoury by bringing these valves into being ….

Letter from Capt. Willet, H. M. Signal School, to Clifford Paterson, 18 August 1941, concerning the NT98, NT100, and E1248 valves.[1]

17.1 The Wembley Laboratories During the War

The major contribution the Wembley Laboratories made to the war effort was undoubtedly the range of valves developed and produced for the services, particularly microwave triodes and the 'magnetrons'.[2] With the outbreak of war, all the research groups engaged on television, and transmitting or receiving valves turned their attention to devices for radar, navigation and communications, either in research and development or—as in Robert's case —taking charge of valve pre-production groups and the associated monitoring, testing and quality control activities. Throughout the war Robert Le Rossignol was the senior man on the Laboratories' staff concerned with the design of valves,[3] and in all probability there was hardly a new valve developed during this time that did not benefit from his understanding and advice.

Few, if any of the UK research laboratories—industrial or otherwise— could match the breadth of activity and the multidisciplinary approach built up by the GEC in its two decades of existence, embracing as it did applications in mathematics, physics, chemistry, materials science and metallurgy,

© Springer Nature Switzerland AG 2020
D. Sheppard, *Robert Le Rossignol*, Springer Biographies,
https://doi.org/10.1007/978-3-030-29714-5_17

together with electrical and mechanical engineering. They were therefore able to undertake a wide range of tasks in support of the war effort. Even so, although deeply involved with issues of radar throughout the war, the GEC were not *initially* involved with the terrestrial 'Chain Home' (Chap. 16) and 'Chain Home Low' systems.[4] These transmitters were manufactured by Metropolitan-Vickers, the receivers by A. C. Cossor Ltd., and the 'curtain' antennae array by Marconi. The transmitter was designed by Dr. J. M. Dodds[5] of 'MetroVick' using his 'de-mountable', water cooled tetrode valves (Chap. 15). Whitehall had very little faith in the British radio community in those days[6] and although GEC were one of the few they believed to be capable of work of a fundamental nature, the contracts went to 'MetroVick' and Cossor simply because Watson Watt was familiar with their research staff, and Dodds' demountable valves were the only ones that could give the high pulsed output power necessary (about 300 kW, later rising to 750 kW–1 MW) at the wavelengths required for this form of radar. But for the development of mobile radar in military applications, viz., artillery gun laying, airborne and naval radars, the principles involved in 'CH' and 'CHL' had to be drastically miniaturised and air-*cooled* anode valves naturally had to replace their water-cooled cousins. It was here that GEC staff were far more intimately involved.[7]

During the late 1930s, major developments in the Laboratories led to a range of small 'sealed-off' oscillator valves which were immediately recognised by those developing mobile radar systems as suitable for use as powerful compact microwave transmitters. The first of these was made in 1939 and became known as the E1046[8] 'micropup', later as the 'VT90'. Peak pulse powers of as much as 200 kW were obtained from pairs of valves of this kind.[9] The breakthrough in construction of these valves came when the Laboratories realised that Houskeeper's technique (Chap. 16) of connecting two glass tubes together via a thin copper disc would make an ideal RF connection to internal valve electrodes and this was adopted by GEC, firstly in the 'micropup', and then in a series of increasingly more powerful 'co-axial' valves[8] such as the 'milli-micropup' (Plate 17.1).

The Laboratory also concerned itself with modulator valves for radar, and small rugged valves for proximity fuses[10] in explosive shells, but developments by other groups also attracted their attention—especially the 'cavity' magnetron—'invented' by John Randall and Henry Boot at Birmingham University.[11] This became an essential concern for the Laboratories and a 'game changer' during the war. The Randall-Boot magnetron was a vacuum tube with a solid cylindrical copper anode containing six 'cavities'

Plate 17.1   A GEC 'micropup' E1046 (VT90) microwave triode valve used widely in early airborne radar, gun laying radar and naval gunfire control. Note the 'co-axial' glass tubes; the filament leads enter the tube at the left; the grid lead enters the tube at the right. In the centre is an air-cooled copper anode. Photograph courtesy of Mr. Allan Wyatt and the National Valve Museum, UK. http://www.r-type.org/

surrounding an axial thermionic cathode placed between the poles of an electromagnet. Their device was water cooled and its vacuum tube was continuously pumped, but by 21 February 1940 they had used it to generate 400 kW power on a short 'centimetric' wavelength of about 10 cm. Quite simply, this performance was revolutionary.

Work on magnetrons had long been conducted at Wembley—ever since 1932 in fact—but this new small device generated astonishing levels of microwave energy ideally suited as the source of power in mobile, high resolution radar. Randall and Boot's cavity magnetron however only became a working reality through the resources and experience of the GEC, and in April 1940, in great secrecy, the pair established contact with E. C. S. Megaw at Wembley to help develop the device into a practical concern. By July 1940, Megaw had produced a powerful pulsed signal (10 kW, again on a centimetric wavelength of 10 cm) from a 'sealed-off' version of their magnetron, with a permanent magnet, an air cooled copper anode, a new type of oxide coated cathode recently discovered by Henry Gutton in France, and gold ('Housekeeper') discs which allowed compact end plates to be attached to the magnetron to make the final seal.[12] With Megaw's improvements, the Randall-Boot cavity magnetron was transformed into a unit suitable for aircraft, capable of being manufactured in quantity for operational use. This development more than any other allowed an unrestricted growth in mobile microwave radar to take place over the war years.[13] The GEC device was eventually introduced into ground, airborne and naval radars as the 'E1189' (NT98) and its essential features are retained in radars today—even in your microwave oven (Plate 17.2)!

Because of these developments, it was inevitable that the Laboratories would become intimately involved with the main airborne radar systems used

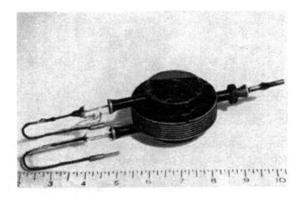

Plate 17.2 *The 'E1189'—Megaw's version of the Randall-Boot magnetron—with its large cylindrical air-cooled copper anode and 'sealed-off' valves. After the war Megaw was awarded the MBE and elected a Fellow of the Royal Society. Photograph provided courtesy of the Canadian Association of Physicists,*[14] *from an article by Paul A. Redhead: 'The invention of the cavity magnetron and its introduction into Canada and the U.S. A.', Physics in Canada, (Nov/Dec. 2001), pp. 321–328*

during the war viz., Aircraft Interception (AI) radar, Air-to-Surface-Vessel (ASV) radar and the airborne ground scanning navigational radar system, 'H2S', this involvement being particularly in terms of electronic circuits, microwave components, antennae, display units, together with flight testing. Also, a scanner developed at Wembley was the first to be used in the ASV and H2S systems.[15] Other aspects of the Laboratories' work during the war was unknown to all but those who worked on them and to senior members of staff. These involved developing electronic-countermeasures,[16] equipment to confound German bombing beams, 'jamming' enemy ground and airborne radars, and developing search receivers to detect various enemy transmissions. Unsurprisingly, little is known of these activities even today[15] but as a senior member of staff one can only assume that, along with others, Robert's advice in these matters would have been sought and valued.

17.2 Airborne Radars; the Context of Robert's War

During August 1936 Watson-Watt, by then the Superintendent of the Air Ministry Research Establishment at Bawdsey, formed an Airborne Group under Dr Edward George 'Taffy' Bowen to develop AI and ASV radar systems. The 'CH' station at Bawdsey had a transmitter that weighed several

tons, the receiving antennae were 75 ft high strung between 350 ft transmitting masts, Bowen's task was to miniaturize the whole system and produce
a transmitter that weighed just 200 lbs with a receiving antenna that could fit
inside an aircraft's nose or belly 'blister'. The first trials of a complete system
installed in a Heyford bomber (transmitter, receiver and display operating at
6.7 m and ~ 45 MHz[17]), were conducted by Bowen in March 1937. By
August of that year, the wavelength, and hence the size of the antenna had
been reduced and installed in an Avro Anson for flight trials at Martlesham
Heath. This setup demonstrated ranges of a few miles[18] and formed the basis
for all the wartime airborne radars. Subsequently, the first-hand built
examples of AI Mk.I (now 1.5 m, 200 MHz) were installed in six Blenheim
fighters of No. 25 Squadron at Northolt in August 1939,[19] whilst similar
examples of ASV Mk.I—using 'micropups' as a power source[20]—were fitted
to a very small number of Coastal Command aircraft. When the war broke
out in September 1939 therefore, Britain had working versions of both terrestrial and airborne radars.

After the success of RAF fighter command in the 'Battle of Britain' in the
summer of 1940, it was quickly anticipated that the Luftwaffe would resort to
night bombing. The battle of course was fought in daylight, 'CH' and 'CHL'
directing intercepts to the general area of the enemy aircraft which were then
engaged by sight. At night, or in adverse weather conditions, a 'blind'
interception mechanism was urgently needed by the RAF and the need for
the widespread adoption of Bowen's AI system became acute. 'Taffy' Bowen
worked closely with the GEC during the war—Paterson describing him as '*a
fine chap ...*' in his war diary entry of 16 January 1940—and whereas the
'micropup' did stirling work, when the new E1189 magnetrons began coming
through the workshops at Wembley in July 1940,[21] they were immediately
deployed for AI radar. One was also flown to the USA on the 'Tizard mission'[22] in late September 1940 and demonstrated there to the amazement of
the Americans. Such was its importance, James P. Baxter, the official US
historian of science developments in the second world war, described it later
thus;

*When the members of the Tizard Mission brought one [E1189] to America in
1940, they carried the most valuable cargo ever brought to our shores. It
sparked the whole development of microwave radar and constituted the
most important item in reverse lend-lease ....[14]*

If only for the development and manufacture of the E1189 then, Wembley
staff were at the forefront of the scientific war against Germany. Naturally, its

existence was kept secret but—much to the annoyance of the Americans—its wider deployment was delayed in the belief that as soon as it was used the enemy would be able to adopt the technique both in radar and in counter measures.[23] The delay however did not apply to airborne interception—these aircraft were not expected to leave UK air-space and so any lost equipment could be recovered.[23] But for ASV, a vital tool in detecting and attacking surfaced U-boats charging their batteries at night when travelling to and from their bases along the Bay of Biscay, the operational delay of about 9 months[23] cost the Allies an estimated 1.5–2 million tons of merchant shipping in the North Atlantic.[23] The eventual failure of the U-boat campaign in May 1943[23] was undoubtedly due to the deployment of the magnetron along with the work of the code-breakers at Bletchley Park.[23] The delay also affected RAF Bomber Command—although they already had the electronic bombing aids GEE and OBOE that extended their capabilities as far as the Ruhr—but the eventual deployment of the magnetron in continental bombing raids via the 'H2S' radar system—largely again at the insistence of the Americans—led to a huge improvement in Allied bombing accuracy. 'H2S'—a name probably derived from the term '*Home Sweet Home*'—was the first ever 'blind' airborne, ground scanning radar system. It was designed to pinpoint targets for night-time and all-weather bombing to improve the accuracy of RAF bombing over Germany. Although air-crews invariably claimed their bombs had dropped 'on-target', the photographic interpreters showed that fewer than half the bombs dropped at night fell within five miles of their target; photographic reconnaissance also revealed that only one in three of the bombers reaching the vicinity of the target, bombed within 1 mile of it and of the total sorties, the bombs were scattered over an area of 41 sq miles around the assigned target.[23] H2S allowed an attacking aircraft to map the ground it was flying over and find features such as towns, lakes and rivers. The images were crude at first but improved steadily after its deployment in 1943. The system was installed in relatively few 'Pathfinder' bombers such as the de Havilland Mosquito. These used 'H2S' to accurately identify targets and illuminate them with incendiary flares; for the main force of 'follow-up' bombers, targeting their objective was then—in comparison—'a piece of cake' (Plate 17.3).

However, the crash of a British aircraft near Rotterdam in February 1943 allowed the Germans to recover the magnetron—its 'self destruction' mechanism having failed on impact. But the Germans were already well aware of the technology, so that nothing had really been gained by delaying its deployment operationally. However the German failure to employ the technique themselves, their inability to understand the extent to which the

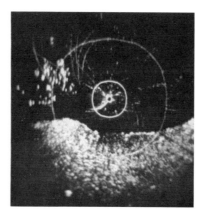

Plate 17.3   This remarkable photograph was taken from the display of an American H2S radar during the liberation of France, 06 June 1944. It shows part of the Normandy beach-head a few minutes before the British and Canadian landing craft reached the Juno and Gold beaches. The town of Caen is the bright spot at '7 o'clock'[24]

Allies had already done so, and the almost non-existent relationship between their scientists and the German Operational Command, gave the Allies a huge advantage.[23] In contrast, the British scientific community were completely integrated with the industrial war effort and it is against this whole background that we have to rationalise Robert's war.

17.3   Robert's War 1939–1945

Robert Le Rossignol—unlike his brother Herbert—seems to have had no inclination whatsoever towards a military life, and by the time the second world war broke out he was already fifty-five years old and beyond conscription. What remains in the public domain of Robert's war is due almost entirely to the Diary of Clifford Paterson,[1] which has been transcribed by Robert Clayton and Joan Algar and published by Peter Peregrinus Ltd. Paterson makes over 50 references to Le Rossignol[25] over the war years reflecting his position as a senior member of the Laboratories, and through these it is possible to construct a chronology of Robert's war and the work he undertook. In turn, some of these references can be expanded to provide a more complete understanding of the importance of Robert's involvement within the areas the Government asked the Laboratories to address. Even so, the diary only provides us with a 'course-grained' picture of Robert during wartime.

There is no evidence to suggest that Paterson kept a diary prior to the war, but he was moved to do so from the day Germany invaded Poland on 01 September 1939, until victory in Europe was celebrated on the 08–09 May 1945. The diary is a remarkable historical document. It reflects the progress of the war, the work undertaken by the Laboratories, the planning and organization of the British scientific war effort, the administration and housekeeping of the Laboratories and most importantly, the roles and sometimes the personal problems of many of the staff. The diary demonstrates the thorough integration of the British operational war effort with her scientific community; it records visits to Wembley from senior military, government Ministers, officials of the various Government departments, together with visits by Laboratory staff to Company departments, factories and abroad—usually to the USA. It is clear from Paterson's diary that Robert's contribution was largely one of organization; of meetings, of travelling, recording, production monitoring, quality control, accounting, and of advice as a senior member of staff. His contribution during this period however, seems far less in terms of technical innovation. Just the one patent (Appendix D) during the war years, applied for on 20 October 1943, viz., 'Improvements in gas-tight joints between metals and ceramics', probably in relation to difficulties KLG were having in rendering their aircraft spark plugs 'airtight'.[26] But it is equally clear the he was a very close colleague of Paterson's, involved at the highest levels of decision-making within the Laboratories, 'privy' to the most sensitive information and liaising with many of the principal scientific, military and governmental characters involved in the British war effort.

17.3.1 Paterson's Diary Sept 1939–May 1941

Plate 17.4 shows the very first page of Paterson's war diary; and so at the very beginning of the war, through experience built up over two decades in the valve group at Wembley, Robert, ('RLeR'), was placed in charge of the pre-production of the 'micropup', probably one of the most important of the Laboratories' valves considering the urgent need for AI and ASV. By October 1940, under Robert's guidance, the Laboratory was producing around 300 of these valves a week. His role here was not just technical but also one of recording every detail of the production process; costs, waste, supply and so on. Pre-production of valves and developments for the Services were to complicate the Laboratories' previously simple accounting system.

WAR DIARY

C.C.PATERSON.

September
1st to 4th
1939.

Considered and arranged many staff transferences to work of more urgent importance.

Ryde and Harris	to Klystrons.
Laycock	to receiving valves.
H.G.Jenkins & Bowtell	to CRT screen efficiencies.
Lewer	to radio work Coventry.

Started organisation under RLeR of 1046 valve production at Laboratories.

September
4th to 8th.

Intensification of activity on valve and radio work for Bawdsey and Signal School (ZZ).

Despatch by car of special CRT's to S.Coast for Fogg and Atkinson's application - drivers Waldram and Stevenson.

Heavy alloy bullets made for certain trials.

Negotiations with Mr.Sims for making Elmet Metal at Wembley.

September
11th.

Discussed with Stevens their plans for ARP lighting particularly in regard to IES-BSI activity. Considerable activity on subject at Wembley.

September
12th.

CVD meeting with Signal School. It appears that with Air Ministry moved 'N and Admiralty W the Research Laboratories is the most convenient fixed point for future liaison between all concerned. It was agreed that meetings must be held in future more frequently, i.e. a full meeting 2 monthly and intermediate meeting sectionalised.

September
13th.

Attended meeting of National Register - dearth of Radio men.

Meeting with all the tonnage in C.S.Wright's office to decide programme of special research to be undertaken at Birmingham by Oliphant with treasury grant.

Watson Watt /

Plate 17.4   The first page of Paterson's war diary recording the appointment of Robert, 'RLeR', to 'micropup' pre-production at the Laboratory. Page courtesy of the Institution of Engineering and Technology[1]

In the past they had just charged other GEC units for research work. Now Robert and others had to charge for goods made and supplied. This aspect however was not one which came naturally to him, compounded by the fact that in the early days Robert had no clerical support, having to record every detail of the mundanity of the process himself. These early months of the war also saw the Laboratories make contact on a regular basis with those at the Air Ministry, the Admiralty etc., together with leading scientists such as Watson Watt, Bowen and Sir Henry Tizard—himself a physical chemist 'by trade' and who, as Chairman of the Aeronautical Research Committee during most of the war, enthusiastically championed the development of 'RDF'. Indeed, by November 1939 Tizard visited the Laboratories to convince himself that all the resources of industry were being focused on radio and RDF, Paterson recording on 06 November 1939 that Tizard was 'sure that this is a 'radio war', and radio will win or lose it'.

Shortly after Tizard's visit, on 15 November Robert and a colleague visited E. K. Cole Ltd. ('EKCO'), at their factory at Southend–on-Sea, to advise them on the mounting and cooling of the E1046 for 'Bowen's job'—clearly an involvement with AI at a 'hands-on' level. At the time EKCO was one of the largest manufactures of radios in the UK, with a wide range of iconic 'Bakelite' cabinets that readers 'of a certain age' might remember. They began the production of thermionic valves in the 1930s and were involved in the early days of RDF/radar. In 1938 the Government held secret discussions with EKCO regarding help with radar research and production methods, and by 1939 EKCO were getting the night fighter radar AI Mk II[27] into production—probably still at Southend, although by 1940 radar work had moved to a 'shadow factory' in Malmesbury, Wiltshire away from the bombing. Shortly after Tizard's visit, and at the request of Watson Watt, the Wembley Laboratories were to take on the 'R and D' for AI, 'breaking away if possible, from orthodoxy ... 'if they could get an expectation of better results therefrom. Watson Watt therefore encouraged progress by thinking 'outside the box'.[28]

By the end of 1939 however there were mixed fortunes in the war; German submarines dealt a psychological blow by sinking HMS Royal Oak 'at home' in Scapa Flow, in November magnetic mines sank 60,000 tons of shipping on the English East coast in just one week, but in December at the Battle of River Plate, a 'hunting group' of British cruisers seriously damaged the German pocket battleship the Graf Spee and forced her into Montevideo harbour for repairs where she was scuttled and taken out of the war. At the Laboratories however, the year ended with a good deal of illness and absence, Paterson recording 'I hope the New Year will start off with a good push.'

17.3.1.1   1940

January 1940 saw bacon, butter and sugar rationing introduced in the UK, this was followed by meat, tea, jam, biscuits, breakfast cereals, cheese, eggs, lard, milk, canned and dried fruit. Alongside this austerity however, *The Times* on 14 February 1940 recorded an 'Anonymous' gift of £1000 to the Red Cross and St John Fund by a 'Mr. and Mrs. Le Rossignol'. Also, at this time AI work at Wembley was intensifying, Paterson and senior members visiting Northolt aerodrome to consider the possibility of installing the radar in a number of aircraft—a task which Paterson acknowledged, 'does not become easier as we examine it'. For Robert however, January brought a welcome diversion from the tedium of recording E1046 pre-production, as his experience in making seals between various ceramics (glass) and metals led to him investigating a problem from the KLG company who were experiencing difficulties making their aircraft spark plugs gastight. This work also probably led to his only patent of the war. However, there was a growing concern regarding AI, viz., the need for 'centimetric' wavelengths—Espley suggesting as low as 5 cm may be necessary. In February, a development of the 'micropup', the 'milli-micropup' (E1190), was effective at 25 cm and by April this has been reduced to 20 cm at 1 kW peak pulse, but progress was slow and the sickness at the laboratories continued with staff 'dropping like flies' from colds, fevers, 'flu and measles'. The war too took a turn for the worse as Germany invaded Norway, Denmark, Holland, Luxembourg, Belgium and then France. Chamberlain resigned and Churchill formed a Government of National Unity, but amidst this turmoil, on 01 May Paterson's Diary records; 'Randall from Birmingham came ... to get us to start quickly with replicas of three large magnetrons which Megaw thinks highly of ... ', an entry of little significance at the time but in hindsight it marked a turning point of the radio war. But as the bad news continued, Paterson decided to hold a service in the Library on Sunday 24 May for those who cared to join him in the National and Empire Day service of prayer. It was the first time a religious service had ever been held at the Laboratories, and Paterson observed ... 'we may only have a very small attendance'. Prayers it seemed went unanswered, British troops were evacuated from Dunkirk and German troops entered Paris. By 17 June Paterson records; 'the French have caved in ... so we are to fight the war alone ...'

But from April, when Randall and Boot first approached Megaw, progress on the magnetrons had been spectacular, and towards the end of June two types had been progressed; Megaw's sealed-off, air-cooled and Randall's water-cooled. The importance of the magnetron was now clear to everyone, operating as it did at the desired 10 cm wavelength and at 2 kW, rising—in Megaw's case—to 4 or 5 kW, and then to 10 kW by early July when it went into pre-production at the Laboratories. About the same time, just as the Battle of Britain was beginning, Robert's stewardship of the micropup saw 300 'good valves' per week emerging from pre-production and he was already making preparations for the E1046B, the first enlargement of the micropup, with a target of number of 1500 to make. By July 20th Paterson wrote, 'AI dominates everything …' but with good reason, because the effectiveness of the radar was finally demonstrated on the night of 22–23 July 1940, when a British Blenheim night-fighter shot down a German Dornier bomber off Bognor Regis using the first production version of Bowen's AI, the Mk III.[29] This attack enthused everyone, and by early August all those involved in AI at Wembley got their wares into the 'shop window' for the visit of Sir George Lee and Sir Frank Smith; Director of Communications Development and Chair of the Technical Defence Committee respectively. During the visit, Robert 'showcased' the magnetrons and milli-micropups for the visitors in the main workshop of the Laboratories, and on 06 August Lord Hirst was escorted around the facilities with Paterson, Dudding and Robert again 'doing the honours'. There was naturally intense government pressure on Wembley to progress AI after these visits, but even so, with the Battle of Britain now at its height, Paterson was trying to arrange two weeks holiday for everyone, recording on 10 August, ' … it is really important'(!).

After the Battle of Britain, the anticipated bombing attacks—'the blitz'—began on 23 August by both day and night. London was hit heavily, and a seven-hour raid on the night of 26–27 August made many staff that morning 'good for nothing'. During August the raids continued, often with little damage to the Laboratories but with Paterson concerned for those members of staff whose homes and families suffered, the first stick of bombs falling on North Wembley on 20 September narrowly missing the Lamp Works. London of course was heavily bombed, but neither did Robert's Harrow escape German attention, because of the nearby Northolt aerodrome. From 07 October 1940 to 06 June 1941, 337 high explosive bombs fell in Harrow district[30] and the whole gamut of this new kind of war was experienced here. Every day children would go out looking for shrapnel, and the competition was quite keen to find the biggest bit, water tanks were kept on the streets and incendiaries or unexploded devices were thrown into them, 'vacuees' were eventually sent to safer places across the country, bomb disposal men appeared on the streets often sitting on the devices they disarmed as they were taken away on their lorries, there was a wailing of sirens, Anderson and

Morrison shelters abounded, queues at local shops for rations, dogfights in the air above the town and aircraft taking off and landing continuously at Northrop—children often 'counting them all out' and (sometimes) counting 'them all back in'. Disturbingly, rumours abounded about how local land marks such as Harrow-on-the-Hill church were used by the Luftwaffe as 'markers'.[31] 7 St. John's Road however survived and 'Harrovians', with infinitely more conviction than some others, could certainly 'look the 'East-Enders in the eye'.

By 24 September incendiary 'roof watching' duties were organised at the Laboratories, with Paterson and other senior staff taking their turn, often experiencing 'more excitement than … expected'. In late September Paterson and Robert attended the Air Ministry with a view to a new contract for 'Signals', and although the Diary makes no mention of the Tizard mission during September 1940, by 17 October 1 Paterson records that at a meeting 'in the Club' after the mission, Tizard reported 'laudatory remarks regarding our valves from experts in the USA'. Mid-to-late October saw a new edition of the milli-micropup achieve 30 kW at 25 cm, Robert and a colleague travelled to the Government's Telecommunications Research Establishment (TRE) at Swanage, where they had 'none too easy a time', (see below) and in November with heavy air raids continuing on London, Bristol and Coventry, working times at the Laboratories were 're-jigged'; work now started at 7.30 am, and with Paterson recording that ' … we don't seem to have mastered the night raider yet', staff had to be well-away before dark. But even on Boxing Day 1940, commitment was such that the Laboratories had a 'full house'. However, apart from Roosevelt being elected for a third term in America suggesting some continuity in the relationship between the two countries, the passing of 1940 gave little to lament.

17.3.1.2   January–May 1941

The new year however began on an optimistic note following the capture of both Tobruk and Benghazi in late January and early February. At the Laboratories, these early months were marked largely by 'teething troubles' with the new magnetrons, whose reported 'idiosyncrasies' began to grip all concerned, the very same that caused Robert to have 'none too easy a time' at Swanage the previous October. These alleged anomalies prompted Paterson to record in early January that they had by now 'stirred up a regular panic, and half of Westminster appears to be boiling!'. Paterson received a number of worried telephone calls demanding 'the truth' about the devices, but he dismissed them, recording that it was just the kind of 'dreadful happening' which often arises from a new device, and placed the blame squarely on non-expert handling at TRE Swanage. Even so, Megaw was charged with re-examining the performance of the device—concluding that its disconti-nuities were small—before being dispatched to Swanage to pour 'oil on the waters'. By the end of the month however, Paterson was firmly of the opinion that; 'The real trouble is that when we sent our first raw sample of 10 cm magnetron to Swanage, they chose to regard it as a thoroughly investigated and established valve, instead of an article whose main development lay ahead of it'. By early February, Paterson felt the need to bring concerned parties to Wembley to re-assure them of progress, and Air Commodore Leedham was 'lunched' and then shown around the Laboratories by both Paterson and Robert. However, the general feeling of 'cold feet' at Westminster and else-where was not improved when news arrived from the USA that Bell Labs had improved the efficiency of Wembley's magnetron from 10 to 50% *and* achieved an increase in power of some 300–400%! When actual samples of these valves arrived from the USA in early February—by 'The Ferry'—Megaw immediately tested them and established that they were 'just the same as ours'. With this evidence, and reports now coming in that consistent service was being achieved from the magnetrons, Paterson observed; 'I fancy the cold feet have been warming up somewhat … [although] … Swanage are too self-satisfied to learn the lesson'. Even so, as the months passed and little was now heard of the alleged vagaries, Bell Labs still claimed 40–60% effi-ciencies for their devices, ('a complete enigma …' according to Paterson) whilst difficulties with Swanage rumbled on causing Paterson to consider charges of 'amateur irresponsibility' against them and remarking ' … what a lot of trouble these people cause by doing things the wrong way.' At this time, staff at Wembley increased their visits to various establishments, catching up on the 'uniformed' use of these valves before it was too late. But there is little doubt that all these 'wranglings' also contributed to the delay in the eventual deployment of the device.

By late February, more and more members of the families of the staff chose to join in the valve-making at Wembley, Mrs. Megaw and Miss Dudding being mentioned in particular. This it seems was in response to the difficulty of keeping a 'stable staff of girls' for valve-making, Paterson denouncing them (or their parents) as 'giving no consideration to the importance of sticking to their jobs in the national interest'. Because of the increasing production demand, Paterson's wife had already begun working there in November 1940 and when Mrs. Megaw joined, she started as a 'pumper'. The employment of the ladies was understandable from the point of view of their availability, dexterity, and their ability—unlike most men—not to get bored with the mundane repetition of some of the tasks. Paterson's hope was that even more wives of staff would be willing to engage with the work, because getting enough women trained to do the most difficult operations was problematic and by now there were very big 'outfalls' in some pre-production lines at Wembley. No mention is ever made of Emily in the Diary, even so, the Le Rossignol family like thousands of others, were thoroughly immersed in the national war effort, being bombed and rationed, disrupted and inconvenienced, deprived of the comfort of a normal life, fearful of the turn of events, but trying hard to make a contribution, which for Robert at least meant his stewardship of the 'micropup' which by April 1941 had achieved a cumulative total of 10,000 valves.

As for the war, the 11 April 1941 saw the 'Lease-Lend' Bill[22] signed by President Roosevelt, an event which the Diary records as 'bringing the USA definitely out of her neutrality' and a source of elation in the UK. Any optimism however, was tempered by concerns over the invasion of Yugoslavia and Greece by Germany, the latter presenting a threat to the critical British interests of Egypt and Suez. For some families in the country these events may have seemed remote, but for others such as Emily and Robert the summer of 1941 saw their immersion in the conflict become deeper and far more intimate.

17.3.2 Detroit Michigan, June 1941

On 07 June 1941, a young man entered the USA at the border crossing control in Detroit, Michigan.[32] He was 5′ 11′ tall, with dark brown hair, a fresh complexion and blue eyes. He was unmarried, accompanied by 'no-one', and destined for the Graham Aviation Company at Americus, Georgia. The young man was British, but he was born in Berlin and his home was at St John's Road, Harrow. Robert and Emily's son John Augustin, by now an RAF volunteer, had arrived in the USA from Canada, a member of the very first draft (SE-42-A) of the 'Arnold Scheme'.[33] The Arnold Scheme was created under 'lend-lease' by Roosevelt and U.S. Major General Henry ('Hap') Arnold, to use spare USAAF capacity to train British pilots. Since the skies above Britain were a war zone, the Air Council realised that it had to pursue aircrew training outside the UK, initially in Canada, but subsequently through the Scheme in what was, at first, a neutral United States. The Scheme lasted from June 1941 to March 1943 by which time over 4000 pilots from cohorts SE-42-A through SE-42-K, graduated. In 1940 the 'Battle of Britain' had been played out for all to see in the skies above the cities of southern England. It made the RAF fighter pilot a national hero and inspired thousands of young men to apply for service via the Royal Air Force Volunteer Reserve (RAFVR).[34] Once accepted for aircrew training, (pilot, navigator or wireless operator), these civilian volunteers took an oath of allegiance (the 'attestation') then normally returned to their jobs until 'called-up' for duty. During this time, they could wear a silver lapel badge to signify their status. John Augustin was one such young man and after his 'call-up' he joined the Arnold Scheme immediately after the signing of the 'lease-lend' Bill.[35]

Having 'reported for duty', transporting thousands of these volunteers 'round-trip' for training in Canada or the USA was not a simple process. In 1941 the waters of the North Atlantic were very dangerous, and in an effort to avoid German reconnaissance aircraft, the battle cruisers and the feared U-boats, troop transport ships (HMT) embarked from western sea ports, mostly Glasgow and Liverpool, but occasionally Avonmouth and Milford Haven. Some of these ships—often converted ocean liners—zigzagged their way across alone, others travelled in small groups or convoys with armed escorts, and the anxiety of training aside, this alone must have concentrated the mind. But most of the volunteers had never travelled by sea before and anxiety was compounded by sea-sickness which almost everyone suffered, the lower decks of these transports often awash with vomit. Afraid of being rejected for aircrew training, many tried to hide their indisposition and large numbers of them hardly ate anything on a journey which normally lasted about ten days. It was thus, John Augustin, along with the other 548 members of SE-42-A, crossed the perilous North Atlantic on the 27,000-ton former Cunard White Star Liner HMT *Britannic*.[33]

Disembarking on 29 May 1941 at Halifax, Nova Scotia—just two days after the Royal Navy sank the *Bismarck* in the northern Atlantic—the men were greeted by a freak winter-like storm with gale force winds driving snow flurries across the docks and bringing the temperature down to the low thirties Fahrenheit. Because of sea-sickness many were very weak, and the cold caused some to faint, which in turn led to rumours that people in Britain were starving. Even so, the men were soon out of Halifax, boarding (un-heated) train cars for a further grueling 30-h journey to RCAF Manning Depot, Toronto. But the misery continued. Surrounded for the first time in months by an abundance of food, cigarettes, chocolate and drinks, some-one had forgotten to issue the men with Canadian money, and they had to wait until Toronto to receive an advance on their service pay. There was however a brief stop at Truro, Nova Scotia, where locals boarded the train and distributed sandwiches, milk and fresh fruit juice, and even though the journey from Truro to Toronto passed through some of the most stunning scenery in North America, these volunteers were generally unimpressed with the whole experience. SE-42-A reached Toronto on 31 May and things began to 'look up'. They were greeted by a welcoming band and within an hour of their arrival they had been fed, showered and were in their bunks. During that first week of June 1941, they were fingerprinted, photographed and issued with temporary American visas. They also received many invitations to dinner or supper from locals, to stay for the weekend, or take a ride in their cars. As representatives of the RAF—victors in the Battle of Britain—the Canadian people treated them like royalty, welcome members of an admired culture. Preparing to enter the (neutral) United states, uniforms were discarded and the men donned their service-issue gray flannel or pin-stripe worsted suits. These were generally of the 'one-size-fits-all' variety; some men were swallowed up by them, others burst out all over. Many closet comedians came-out that day and for some, although their suits were complemented by black shoes, socks, blue shirts and black ties from their regular uniforms, a sartorial dignity in-keeping with the adulation of their Canadian hosts was difficult to achieve. Boarding cars of the Canadian Pacific Railroad on Friday 06 June, the men travelled to the United States by two different routes. Those bound for the Primary Training School at Camden, South Carolina, travelled via Black Rock and Buffalo, then across the states of New York and Pennsylvania to Washington D.C., and finally south. Those bound for Primary schools in Alabama (at Tuscaloosa), or Georgia (at Americus and Albany) and Florida (at Lakeland and Arcadia) went via Detroit, Cincinnati and Chattanooga. At Chattanooga the Alabama contingent changed trains for Tuscaloosa, the remainder continued on their journey. At about 8.30 am on Sunday 08 June the train carrying John Augustin rolled to a halt at Union Station in Atlanta. Those allocated to training schools in Americus and Albany disembarked and boarded a train of the Central of Georgia line heading out of Atlanta to Americus and Albany via Macon. That Sunday afternoon, fifty-three creased and crumpled men of SE-42-A disembarked for the final time at the railway station in Americus. Training school staff and citizens of the small town provided vehicles and transported them some nine miles past bountiful peach orchards on red clay or hard packed gravel roads to their barracks at 'Souther Field',[36] home of the Graham Aviation Co. where Charles Lindberg made his first solo flight in May 1923. After showering, shaving and making themselves presentable, supper was served in the dining hall and if they thought the Canadian welcome was warm, the locals assured them that, 'they ain't seen nothing yet!'[33]

Plate 17.5   SE-42-A, from the Americus 'Times Recorder' 12 June 1941. John Augustin is somewhere in this group, but the quality of the photo is poor. Photograph courtesy of Sumter County Archivist Alan Anderson

Over the next few days, khaki uniforms and cotton twill overalls were issued helping overcome an image that many felt of being like 'poor cousins' of the Americans. Off-duty however, trainees were expected to wear 'civvies', and most put their service-issue aside and spent their meager pay on lighter clothing more suited to the hot and humid Georgian summer. The presence of these British 'Tommies' in Americus was big news and reporters and photographers from the Americus *Times-Recorder*—the local afternoon daily newspaper—ensured that articles appeared in the paper for weeks. The issue of Thursday 12 June 1941 in particular, devoted the front page to a picture of the 53 member SE-42-A wearing their new uniforms and their 'colonial' sun helmets.[37] The *Recorder* describes these young men as 'all dressed up in British uniforms and in military formation' (Plate 17.5).

Wherever these young men went 'off-duty' in the next ten weeks the wives, mothers and daughters of Americus showered them with their famous 'Southern hospitality'. They had iced tea, Southern fried chicken, corn on the cob, creamed corn, fried okra, fresh tomatoes, Spanish rice, peach cobbler and pecan pie. But there was more; fresh melons, peaches, cantaloupes, strawberries, water melons, and in the town of Americus itself they found ice-cream, milkshakes, banana splits and sundaes.[38] All of this was in stark contrast to the austerity in the UK and some trainees naturally felt guilty, but at least they could give their families the precious assurance that, after dodging the U-boats and the *Bismarck*, for the next six months they were all 'safe'. However, amongst many trainees in the South, there was soon a

Plate 17.6 The understandable cause of chronic constipation amongst some trainees! Photo courtesy of The History Press[33]

problem, … chronic constipation … because at some barracks e.g. *Albany*, although the toilet blocks contained the usual apparatus of pans and cisterns, these were all in a row, without cubicles, or even partitions! Americans were obviously a sociable bunch, but British modesty found this arrangement very difficult to use (Plate 17.6)!

But the volunteers were not here for the hospitality nor the social life; they were about to make a substantial investment in themselves, one which they un-conditionally accepted could be lost in the blink of an eye. The emotional tangle of anxiety and hope however was punctured by excitement, because with hard work and no small measure of luck, some of these young men knew they would become fighter pilots, the 'darlings' of the RAF, the *crème de la crème*, the pride of family and friends. Many wrote home with their excitement, describing their journey from the UK, the barracks, their progress in learning to fly, the hospitality, the discipline, the weather and inevitably the local girls—especially those at Georgia's South Western College! Some too may have had a private agenda to address, to prove themselves or to achieve on behalf of others. Maybe John Augustin brought with him the frustration of his brother's life, but whatever cross each man had to bear, waiting to swallow them all were the open cockpits of Souther Field's Stearman biplanes (Plate 17.7).

Over the next ten weeks, the trainee pilots learned to become confident in the Stearman, beginning with landings and take-offs, stalls, power-on and power-off spins, 'soloing' in around three weeks and culminating in a basic mastery of the aircraft with chandelles, pylon eights, lazy eights, snap rolls, slow rolls and Immelman turns. During their stay at Souther Field there were

Plate 17.7   1941 (Boeing) PT-17 Stearman biplanes, the leading American trainer of WW II, at Souther Field maintenance hangar. Photograph courtesy of Wikimedia Commons, https://www.commons.wikimedia.org

no fatalities, but 2 of the 53 were 'held over', 1 was eliminated for 'non-flying' (a euphemism for desertion) and 20 were 'washed out' for flying deficiency,[39] but 30—including John Augustin—graduated successfully.[40] On 15 August 1941, these trainees moved on to the Basic Flying School at Cochran Field, Macon, Georgia, and joined 114 others who had graduated from Tuscaloosa, Albany, Lakeland and Arcadia. Training at Cochran lasted until 03 November, when a further 138 successful graduates moved on to Advanced Flying Schools at Craig, Napier and Maxwell Field in Alabama (single engine), and Turner and Moody in Georgia (twin engine).[41]—RAF Disclosures[35] showing that John Augustin did his advanced training at Maxwell. During the second week of December, after the bombing of Pearl harbour and the declaration of war by both Britain and the USA, John Augustin and the final 119 members of SE-42-A were posted to Florida for the fixed gunnery training course. These young men established an outstanding record, the instructors being impressed by both their accuracy and aggression, and on the 01 January 1942 the 30-year-old airman—by now Sgt., 1288222, Le Rossignol J.A.—became Pilot Officer 120555, the first commissioned rank in the RAF.[42] But more importantly ... he had his 'wings'.

17.3.3 Paterson's Diary May 1941–April 1943

With John Augustin 'safe' and learning his trade in the Southern States, Robert's war moved along. From June to December 1941 his main concerns involved the transfer of the micropup pre-production from the Laboratories to the full production facilities in four new factories managed by M-OV, the protocol for such processes being agreed in early June. However, by the end of the month Sir Frank Smith contacted Paterson warning of a shortage of the valve and Robert along with Dudding agreed to concentrate on the 'VT90' at the expense of less critical units. By September new contracts saw more pre-production groups set up in the Laboratories by clearing out space formally used in peace time, but all of this tedious, administrative effort was beginning to wear Robert down, Paterson recording in his diary of 25 September 1941;

> *The financial check on these contracts involve a good deal of work and have to be kept well up to date … Le Rossignol's check, which he keeps on the costs of valve pre-production, are a weariness to his soul. I shall have to try to get him some intelligent clerical help.*

Eventually, such help appeared in the form of a 'Miss Lock' who worked alongside Robert gathering, analyzing and reporting on the costs of valve pre-production at the Laboratories for the remainder of the war.

By October, the problem of getting 'girl labour' in sufficient numbers persisted—the loss of just one or two leaving the laboratories 'badly landed'— Paterson even writing to Sir Frank Smith asking for a letter from 'influential quarters' emphasizing the importance of their work in an attempt to retain their female staff. For Robert, the remaining months of 1941 concerned both magnetron and VT90 production, the latter becoming much healthier by late October with M-OV alone now producing over 500 a week and Paterson describing the valve as 'a sound egg for both manufacture and operation …'. On Christmas Eve 1941, Paterson received a letter from the Chief Superintendent of TRE Swanage, reporting the first combat engagement of an aircraft fitted with the new magnetron based AIS[43] radar, and although the outcome of the engagement was unknown, the radar operated perfectly. TRE also added exactly the kind of qualification that Paterson was looking for by praising the work of Laboratory staff[44];

> *... although it must seem to your people that we get the fun at the end, I would like to assure you that we always link your people with ours in referring to this common effort...*

Such praise it was hoped would galvanize the newcomers and the 'slackers' at Wembley to become more dedicated to their work. And dedication to the national cause was certainly needed because on Christmas Day 1941, iconic Hong Kong surrendered to the Japanese.

1942 began with more depressing news from the Far East. By March, Manila, Burma, Singapore and Java had all fallen, but to make matters worse, at home soap was now added to the rationed list! Neither was the war going too well in the European theatre with heavy and unexpected losses for RAF bombers which some attributed to German radar stations along the French coast. The exact nature and purpose of the equipment at these stations however was a mystery but pressure eventually led to an airborne raid on one at Bruneval in northern France in late February 1942 bringing back elements of the equipment for examination at GEC. Paterson's Diary of 17 March describes little of technical interest in one of the German receiving valves recovered from Bruneval, but subsequent work identified the equipment as 'Würzburg radar', the principal gun-laying radar used by both the Luftwaffe and the German army. Deconstruction work eventually established that the radar was robustly impervious to 'jamming' by conventional methods which in turn led to the deployment of the new British 'chaffing' countermeasure system, '*Window*', which 'blinded' enemy radar. However, at the time Paterson once again seemed concerned with 'slacking' at the laboratories, the entry of March 17th also declaring;

> *We are organizing a campaign against slacking in the Labs – wasting time and the like. There are a lot of new people who have no tradition and who don't do a full day's work. We must get this mended ...*

Exactly who 'we' were is not clear, but by 23 March *every* member of the Laboratory received the memo, 'WASTING TIME', a move Paterson later justified in an entry of 02 April because '*a number of ... younger people (and some older) have not learnt the way to work hard.*'[45] (Plate 17.8).

Whatever the outcome of 'WASTING TIME', late March saw the start of the Laboratories' engagement with H2S, a difficult problem that Paterson was keen for his staff to '*get their teeth into*'. Late April saw Robert present yet more accounts to Paterson who, in recognition of a life in science, was elected F.R.S in early May. At the same time, night fighter interception techniques

23rd March 1942.

(A copy of this is being sent to every
member of the laboratory staff).

<u>WASTING TIME</u>.

There is a large section - probably half - of the people
in the Laboratory who work hard and pull their weight for the
whole of the daily working hours. They are in to time and start
up work at once; they are determined to get as much into the day
as possible. At the end of the day they take the minimum possible
time for clearing up.

But a very considerable number of people fall short of
this. They are not working "all out" as the country requires of
them and as the importance and urgency of the Laboratory work
justifies.

People stroll about the place instead of proceeding
briskly.

There are women and girls who, even if they arrive in
time, slack about until ¼ or ½ past eight. This is obvious in
some of the experimental laboratories but is not confined to them.

Youths are seen talking and hanging about their work when
men are risking life and may be dying for lack of what it is their
job to supply.

Girls seem to think that attention to their hair and
personal appearance will justify long absence in cloakrooms during
working hours. Their value to the country depends on the amount
they do here - not the hours they spend here.

Others/

-2-

Others, both men and women, forget that it is up to
everyone to help to maintain a cheerful atmosphere and to keep
going in a brisk and buoyant manner - remembering that we are
all working for the cause, and for the men and women whose lives
and safety are in our hands, rather than for ourselves or the G.E.C.

The fighting spirit is an ungrudging one. It goes
"all out", with unbounded loyalty and support to the teams we
are with and their leaders.

In factories, slackness is watched and checked. Here
we rely on people themselves maintaining a high standard of
personal conscientiousness. If some individuals do not do so,
we cannot allow the morale of the place to suffer by their example,
and other means which we are reluctant to take will be necessary
to bring these people into line.

At all costs these Laboratories must pull their fullest
weight in the important work which has been entrusted to them.

This notice is for you who are now reading it - you
know to which of the two classes of worker you belong - so please
face the truth and make sure you are on the right side.

M.W. Paterson

BMO.

Plate 17.8   Paterson's memo to <u>all</u> members of staff which went very much against his
egalitarian founding principles for the Laboratories. Page courtesy of the Institution of
Engineering and Technology[1]

were augmented by another development at the Laboratories, the 'Turbinlite', a powerful searchlight placed in the nose of interceptors to illuminate enemy aircraft which could then be engaged by fighters or from the ground by conventional means. At first quite successful, it was later abandoned due to indifference by senior operations officers, and improvements in airborne radar. And airborne radar interception generally seemed to be working well, Paterson recording in May, '*The Hun seem shy of coming over ...*', but all of this was at a cost to those developing the technology, a crash at TRE in early June costing the lives of three of their most able men. Mid July saw Robert and Gossling 'entertaining' a delegation from the Massachusetts Institute of Technology who were both interested and impressed by the work at the Laboratories, and Robert presented yet *another* set of accounts to Paterson— but this time together with his new 'intelligent' assistant, Miss Lock. Paterson's comment regarding '*the Hun*' too was 'spot on', Watson Watt having delivered a speech in the Machine Shop of GEC's Radio and Television Works in Coventry in early August explaining that over the last few weeks the number of German night raiders brought down by their equipment was greater than the quantity brought down by all the other defensive devices put together, adding exactly what Paterson needed to further exhort his 'slackers' at Wembley, viz.,

> *... what the workers are doing is to make life just a little easier for the ace night fighters who have to fly in a black sky in the middle of the night looking for black-painted aircraft ...*

August also saw a 'heavy push' on H2S with the Laboratories engaged mainly via the supply of magnetrons, a larger (waveguide) version of which—developed by W. E. Willshaw—was beginning to replace the resonant cavity type. This 3 cm magnetron, although radiating half as much energy as the older type, did so at a much-reduced spread of frequencies about the required value and was expected to be more effective. Both Robert and Willshaw briefed Paterson regarding the theoretical aspects of this new device and its implications for production. Early September saw Robert leading consultations regarding the heavy-duty cathodes for these magnetrons but by October both Robert and Megaw had something of a 'set-to' with the Admiralty Signals School regarding the life of these cathodes—such 'spats' being quite a common feature of the Laboratories' relationships with the military and manufacturers in the early stages of new developments. With 1942 drawing to a close, Willshaw made improvements to the new magnetron (more power) and both he and Robert visited Oxford in November to update people there on the modifications. In December, Robert attended to increasing demands for valves using the CV90 (micropup) techniques—GR gun laying radar based on the 25 cm micropup performing particularly successfully—and by late December H2S was introduced to Bomber Command.

1943 began spectacularly for the Laboratories when three staff members—including E. C. S. Megaw (MBE)—were recognised in the King's New Year Honours List. This success was reinforced by a visit to Wembley on Saturday 16 January from Sir Stafford Cripps, then Minister for Aircraft Production, who used the Laboratories public address system to speak to *all* the staff, once again reminding them just how appreciative the Government was of their work;

> ... *there is no doubt that you, who are working in the laboratories doing development and research work, are pitting your brains against theirs [Germans and Italians], and we can be proud that British development, at least in this scientific warfare, has always been able to hold the lead, and it is your work which has given us this lead, and it is your work that is going to keep us in the lead to the day of victory* ...

Clearly, Paterson was now harvesting the kind of recognition that would shame and motivate his 'slackers', and the praise continued with a letter from the 3rd Sea Lord—Controller of the Navy—on 01 February for producing, in quantity, ' ... *several types of especially important valves used in Radio Direction Finding equipment* ...'[46] Paterson's letter to Sir Frank Smith was paying dividends, Miss Locke certainly taking up the baton with several 'mentions' in the Diary regarding her efforts in sorting out the 'various vicissitudes' of the valve pre-production accounts with Robert, who in turn was alongside Paterson when the two men met G. M. Wells in March—who Paterson describes as '*of the extreme Left*'—to discuss the Laboratories' structure, their ideals for its future and the impact on his Trade Union, the Association of Scientific Workers. It was apparently a very friendly 'chat' but backed by the enormous amount of praise the Laboratories had received recently, Wells and his colleagues were left in no doubt that those staff who failed to engage fully with the efforts of the Laboratories were clearly in breach of their national obligations. By early September that year however, the explosion of new staff at the Laboratories—largely beyond Paterson's control—led to his opinion of the Union(s) hardening[47];

> *Truly things are going well – far better than our materialistic and individually grasping tendencies deserve. Trade Unions lead the way and*

everyone else follows the Gospel of 'Get all and give as little as you can'.
But there are thousands who have not bowed the knee to these Gods and
they are the people who are saving the situation ...

The Unions and some workers therefore received 'short shrift' from Paterson and little mention was made of either in the Diary for the remainder of the war.

17.3.4   April–September 1943, RAF Cranwell to AHQ Malta

By March 1943, it seems Robert had been 'lifted out' of the daily 'grind' of valve pre-production to the Valve Pre-production Supervisory Committee which met every Monday morning at 08.45. Each particular group of valves e.g., triodes, rectifiers, magnetrons, and processes e.g., testing, packing, inspecting etc., was subsequently allocated a pre-production leader, an assistant(s) and a supervisor(s) thereby spreading the burden. This kind of detail from Paterson's Diary illustrates that—as for many others of his age— Robert's war work on the 'home' front was essential but often prescriptive and frankly uninspiring. At the 'real' front, the flag was carried by the young, and Robert's eldest son John Augustin joined thousands of others in playing a dangerous part here. After promotion from Sgt. to Pilot Officer in January 1942, John Augustin spent the next ten months on the 'special' duties list. Although his service record[35] sheds no light on what these duties were, during this time he would not have been attached to a 'normal' dedicated RAF squadron but rather been gaining experience in a number of different areas. Consequently, he received further promotion[48] on 01 October 1942 to Flying Officer (F/O) and after a few weeks de-briefing travelled to an Advanced Flying Unit at RAF Little Rissington in Gloucester where he remained until April 1943. F/O Le Rossignol was then posted to No. 51 Operational Training Unit (OTU) at RAF Cranwell, Lincolnshire—a unit once led by the iconic (dam buster) Guy Gibson—to train as an elite night-fighter pilot (probably) flying 'Beaufighters', and to learn to use the latest AI radar. At Cranwell John Augustin teamed up with his navigator Sgt. Lesley Green (1556380) of Darlington, a former Darlington Grammar School lad.[49] The rare photographs below show the pilots and navigators at 51 OTU in April 1943.

After Cranwell, John Augustin and Leslie Green transferred to No. 60 OTU at RAF High Ercall, NE of Shrewsbury to train as 'Intruders', flying the revolutionary de Havilland Mosquito. John Augustin must have been a skilful airman. The Intruder operation was a specialist exercise requiring keen eyesight, cool nerves, patience, and the ability to 'hide in the skies' to seize a chance that would often only last for seconds.[50] The tactics taught at 60 OTU had been perfected over occupied France, Belgium and Holland during the previous year and were used over Germany and Poland as aircraft with greater range became available. They were also deployed in North Africa, Italy and southern France in support of the desert and Italian campaigns. But wherever it was flown, the Intruder operation was risky because it involved a lone aircraft over enemy airfields at night often attacking bombers at their most vulnerable viz., landing or taking-off. The risk was even greater when single engined aircraft were used. These pilots not only had to be a 'one-man band', flying/navigating/observing, but if badly damaged or if their engine failed there was simply nowhere to go. When twin engined aircraft such as the Mosquito were introduced the situation improved because the remaining engine was capable of bringing the aircraft home, and a second crew-man shared the burden of his pilot.

'Intruding' also took a number of forms. In one form, ground observers would report enemy bombers reaching the English coast. A few aircraft would then set off singly and head for the continent, guessing from the direction of the bombers which airfields were being used and laying in wait for them to return. When the bombers came back, they were at their most vulnerable: low on fuel, ammunition expended, crews tired and unsuspecting. Runway lights were briefly flicked on and off and this, together with the slow speed of the bombers as they descended, helped the Intruder pilot locate his target. Bombers were regarded as 'dead meat' at the point where the undercarriage was about to meet the runway and this was the perfect time for an Intruder to attack. Another form of Intruder operation took the initiative, flying to occupied territory early in the night and looking for bombers as they took off. This was much riskier; the German crews were alert and the ammunition racks were stocked. Yet it had the advantage that, if successful, the sortie not only destroyed the aircraft but also its bomb load. A third form came into play when the Intruder couldn't find any aircraft to hit. Rather than return with the rubber sheaths still over the protruding cannon barrels and the bomb rack intact, they resorted to 'beat 'em up', the most obvious targets being trains, although anything military was fair game. Even with these tactics, the chances of finding German aircraft were low. Luftwaffe pilots often returned to a different airfield from the one on which they had taken off and they had many from which to choose. Night after night, many Intruder pilots would not even see an enemy aircraft, let alone engage or destroy one. Some pilots survived the whole war without ever engaging the enemy. Even so, Intruder aircraft were invariably alone, flying over enemy territory close to well-defended airfields playing a *very* dangerous game of 'cat and mouse' and whichever way it was conducted, 'Intruding' was a dangerous, furtive operation, striking an opponent 'in the back' when they least expected it and in all theatres of the war the Germans regarded the brand as 'toxic'. They detested it.

It was within this context that John Augustin and Leslie Green began their Intruder training at High Ercall, and two difficult techniques in particular had to be mastered. One was flying at high speed and low level, 'hedge-hopping' to the target, hugging the ground, becoming lost in the 'ground clutter' of German radar. The other was returning home and landing safely—but at night. Both these techniques were practiced over and over again. Even so, John Augustin had an advantage over the earlier Intruder 'pioneers', he and his navigator were being trained to fly the twin engined de Havilland Mosquito fighter bomber FB Mk II, a versatile aircraft bristling with innovation, luxuriating in two hugely powerful V12 Rolls Royce Merlin engines and carrying the AI Mark VIII and H2S radars developed by his father's GEC at Wembley. The Mosquito was the one Allied aircraft that could be just about anything; a fighter, a fighter bomber, a ground or shipping attack platform, a high-altitude photo reconnaissance aircraft or a long-range Intruder. It could beat a Spitfire for speed in level flight, 'dog-fight' carrying a 2000-pound bomb load, and with extra 'slipper tanks' it could fly from the UK to the Mediterranean non-stop, refuel, carry out Intruder missions in eastern Germany and fly back to its base in the 'Med'. The pilots of 51 OTU would be trained to fly these aircraft to their target at 300 mph just 50 to 100 ft above the ground—and at night. Once they were close by, they would learn to cut speed to just above idle, and slowly make a 'race track' holding pattern 5–10 miles away from the German base. Having hit a target, the pilot would learn to 'firewall' his throttles, and get out of the area flying home at 100 ft and 410 mph, the Mosquito's flat-out, sea level top speed. 'Intruding' and coming home again, could take up to six hours, and at no point would the crew be able to relax, except maybe when they approached their home base. But at night, being relaxed was not always the best condition to be when attempting to land on an unlit runway. For all these reasons, Intruder pilots were the 'best of the best'. But the two-man Mosquito crew worked as a team, and although the radio and the monitoring of engine settings—normally a navigator's domain—was left to the pilot, Mosquito navigation was a very demanding task, requiring a position fix every two minutes. On main force bombers, checks were only needed every five minutes or so but the speed of the Mosquito meant that the slightest error could put the aircraft far off course. The tiny navigators table and angle poise lamp with dim bulb in the cockpit made the job even more difficult. Also, when over enemy territory, he had also to constantly check the H2S screen to identify coastlines, lakes, etc., as navigation landmarks. Like their pilots, Intruder navigators too were at 'the top of their tree'.

The de Havilland Mosquito was generally loved by its crew; its speed and versatility were unsurpassed and its tough construction allowed it to take punishment but still remain airworthy. Octogenarian 'Bud' Badley, a Mosquito pilot with 23 Squadron during the war, described returning from one mission thus[51];

> … *One time I came back … flying a lot of holes held together by an aeroplane!*

But magnificent as the Mosquito was, it had some vices which the pilots would have been made well aware. These aircraft relied on pure speed to function, therefore the engines tended to operate 'flat out' for longer periods than normal, and engine failures or 'cuts' were a problem. Without full power the 'Mossie' could be a bitch, unflyable, a 'brick with wings'.[51] Its light plywood construction and powerful engines gave it a frightening acceleration which made it prone to 'swing' on take-off, especially if the engines were not balanced. But if there was an engine *failure* on take-off, the torque of the remaining engine at full throttle would roll the aircraft onto its back driving it into the ground. There was no recovery procedure here, even with full rudder and aileron correction. Unofficially, the best thing to do was to 'cut' the remaining engine and attempt a straight-ahead crash-landing. But crews were aware of no-one having done this successfully and a relatively high number came to grief in this way. Equally, the landing speed was high (110 mph) and a one-engined Mosquito landing was just as dangerous, although some, like the legendary night-fighter ace Wing Commander Bransome Arthur '*Branse*' Burbridge, did so on several occasions. To add to the pilot's concerns, some evasive and 'zero-g' maneuvers could cause the carburetor(s) to fail and the engine(s) quit. These aircraft, built at factories at Hatfield and Leavesden in Hertfordshire, could never be tested individually, that had to be done 'live' on operation and some aircraft developed so many vices that aircrew refused to fly them. But even for the best aircraft, an engine 'cut' was always a possibility, and dealing with the subsequent 'asymmetric flying' was impossible to prepare for, it had to be learned 'on the day'—or to make matters even worse for an Intruder, 'on the night'. The Mosquito crash log[52] shows that engine failure contributed to a significant proportion ($\sim 11\%$) of Mosquito crashes and for the crews, many were fatal. The 'cut' was the *bête noire* for the Mosquito pilot.

F/O Le Rossignol and Sgt. Green completed their training at High Ercall on the 16th July 1943 and were subsequently posted to 23 Squadron at Area Head Quarters (AHQ) Luqa in Malta, arriving there from the UK on 22 August 1943.[53] For both men, after a period of training that began in June 1941, this was probably their first operational posting. Luqa was an important base for the British, fighting from there for naval control of the Mediterranean, for ground control of North Africa and now providing air cover and Intruder operations for the imminent invasion of southern Italy. 23 Squadron were a seasoned outfit, based in Malta since December 1942, but they also had an impeccable pedigree—Douglas Bader being a member when he lost his legs in a crash in December 1931—and John Augustin and Sgt. Green must have felt quite at home because they met up with other members of 51 OTU viz., F/O Rapson and navigator Sgt. White who had arrived there a few days earlier on 13 August 13. (Plate 17.9) At the time of John Augustin's arrival however, Luqa was in transition. The allied forces gathering in Sicily and North Africa for the invasion of Italy caused the Axis forces to withdraw their bombers beyond the 'economic' range of the Mosquito, it was decided therefore that the Squadron operate from a forward base in Sicily. It was impractical to move the whole Squadron and so an advanced party of 30 ground crew were sent to Sigonella on the east side of the island to prepare the landing ground. From 05 of September 1943, aircraft were flown from Malta to Sigonella each evening, re-fueled, and after operations, returning to Malta at dawn.[54]

Prior to the invasion, John Augustin flew two sorties from Malta. The first was on 30 August 30 in Mosquito FB VI HJ737, an uneventful four-hour patrol over Taranto Bay, 'up' at 18.40 and 'down' at 22.30. The second sortie was on 01 September now in HX814, again an uneventful patrol over Taranto Bay but this time 'up' at 18.30 and 'down' at 22.20. On 03 September the main allied invasion force landed at Salerno near Naples in Operation *Avalanche*, while two supporting operations took place in Calabria (Operation *Baytown*) and Taranto (Operation *Slapstick*). On Sunday 05 September, the first five aircraft of 23 Squadron—including Le Rossignol and Green in HX814—left Luqa for the new advanced base at Sigonella, arriving at dusk and able to operate even though the weather that night was poor. John Augustin flew an Intruder operation in the region of Monte Corvino just north of the landing beaches. 'Up' from Sigonella at 20.20, the two men flew an uneventful sortie for several hours ... and then it happened. Somewhere near Naples an engine 'cut'.

Plate 17.9   The night-fighter pilots (top) and navigators (below) training at 51 OTU in April 1943. Flying Officer John Augustin is seen seated, second from the right, and his navigator Sgt. Leslie Green is standing in the back row on the extreme left. Photographs courtesy of Mrs. Lorraine Eastcott

The Operational Record Books (ORBs)[53] of 23 Squadron at the time are littered with references to engine 'problems' or failures, so these two men must have been trained to cope with their situation. On the return journey to Sigonella they would have tried to separate the emotion of the moment from a clinical analysis of what to do and what not to do. But landing on an unfamiliar airfield, at night, during wartime, in poor weather, with just one engine would require them to create and customize a maneuver to get themselves down safely, a maneuver that they could never have practiced before. It was certainly possible, because only days earlier some Squadron members had achieved just that. On 16 August, F/O Keeling also experienced engine failure over Naples, but brought his aircraft safely back to Sicily where it landed at Termini. On 13 August Sgt. Dawson suffered engine failure 200 ft over Foggia but managed to land his aircraft at Palermo with the aid of lights from two Jeeps on the runway. But some pilots were less fortunate. The same night Keeling was over Naples, Sgt. Farelley was patrolling Taranto Bay and was later heard to call that he was returning on one engine. In an attempt to land at Cassibile the Mosquito overshot and was burned out. His navigator was killed instantly. Farelley died later in hospital. Those pilots who managed to land successfully at night after an engine 'cut' would have passed their experiences on to the others, and everyone hoped to learn from the mistakes of those who failed the maneuver, but the ORBs show that F/O Le Rossignol and Sgt. Green managed to 'nurse' HX814 back to Sigonella, and at one minute past midnight (local time), 06 September 1943, both men and machine were 'down'.

That night Sigonella was hardly equipped to help any of its airmen. Lines of communication and organization were poor compounding all the other problems, the ORB entry for 05 September 1943 reading sarcastically;

> A glorious air of indecision hangs over Sigonella landing ground where the detachment find themselves controlled and managed by no fewer than five units – Administered by 232 Wing, controlled by 326 Wing … on the aerodrome, controlled in operation by T.A.F. supplied by T.B.F. and finally owned by Malta. However, in spite of these and other difficulties, five aircraft arrived at dusk and were able to operate although weather conditions were poor.

The other aircraft that took-off at Sigonella that night carried out reconnaissance and Intruder operations without incident and returned safely. But for John Augustin and his navigator their landing at midnight was catastrophic and both men were killed. In that instant, an investment which began at Americus in the summer of 1941 reached maturity, after just three operational flights. John Augustin of course had already landed successfully at Sigonella, but with both engines and in the 'remains of the day'. Unlike many other fatal incidents, the ORBs contain little detail for this particular crash. Maybe it was pilot error, or possibly a second engine cut, the aircraft could have hit the runway too hard or maybe overshot. But the detail has not been recorded, obscured by the 'fog of war', and all that we know is that on the morning of 07 September, both men were buried in the British war cemetery at Catania, a lonely funeral attended only by F/O Lewis, on behalf of the Squadron's Commanding Officer. And as with the death of Peter Walter, there was a final irony here. John Augustin did not die from enemy action, but rather from a failure to execute what was essentially a 'blind landing'. 'Blind' interception and scanning techniques such as AI, ASV and H2S had of course been long pioneered by GEC and what was clearly needed for the

Intruder was a blind landing device that would help the pilot bring the aircraft down in complete darkness. Such a device was already under trial at the time, Paterson's entry of 19 July 1943 reading;

> *Espley's blind landing experiments (a GEC development) seem to make him sanguine – and we propose to make demonstration equipment if it promises well. The easing of pressure on the Compact Range Equipment may give him a chance to get on with it …*

But for all the Intruders who came to grief in the same way it was just too late.

News of John Augustin's death came to Robert and Emily a few days after their son's burial. On Friday 10 September 1943, *The Times*, carried the following notice from the family;

'ON ACTIVE SERVICE'

Le Rossignol. – In Sept., 1943, killed in action, Middle East, Flying Officer John Augustin Le Rossignol, R.A.F., beloved and only surviving son of Mr. and Mrs. Robert Le Rossignol of 7, St John's Road, Harrow, Middlesex.

In the fullness of time, John Augustin's grave was marked by a headstone erected by the Commonwealth War Graves Commission to commemorate his life 'in perpetuity'. It reads;

FLYING OFFICER
J. A. LE ROSSIGNOL
PILOT
ROYAL AIR FORCE
5TH SEPTEMBER 1943 AGE 31

Robert and Emily later added a personal inscription, just two words more than that for Peter Walter (Plate 17.10);

After the war, the Le Rossignol's Parish church at St John's Greenhill erected a memorial to those from the Parish who fell in WWII and John Augustin's 'sacrifice' is also recorded there.

For Robert and Emily, the loss of a second son must have left them devastated. But one hardly needs a war to create such tragedies. The lessons of our industrial past teach us this. On a single day in our mining communities for example, many a mother lost her husband, sons and her brothers. These poor families were far less capable than the Le Rossignols, but the depth of grief for the one is the same as for the many, and their notice in *The Times* announced that it was to the 'many' that Robert and Emily now belonged.

IN *PROUD AND* LOVING MEMORY

Plate 17.10   F/O John Augustin's headstone at Catania, Sicily. Photograph courtesy of Steve Rogers and the War Graves Photographic Project. http://www.twgpp.org

John Augustin died doing his duty. His death 'normalised' the family, and brought them into a fold of thousands of others up and down the country who suffered in the same way. In this sense the life of John Augustin could eventually be 'celebrated', and in time it also provided a way for Robert and Emily to remember the life of Peter Walter and to help ease the pain and the manner of his passing. But for this couple, Dylan Thomas' lines from 1945 would have had little resonance[55];

> ... *After the first death, there is no other.*

17.3.5   Paterson's *Diary, September 1943–May 1945*

Between April and August 1943, Paterson's Diary makes little reference to Robert, but on 01 September at a meeting of the Coordination of Valve Development group, his entry records a proposed visit to Canada and the USA by both Robert and another colleague L. C. Jesty. But by 20 September events in Sicily became known to all at GEC;

> ... *News came through to Le Rossignol ten days ago that his son had been killed flying (fighter pilot) in the near East. A sad blow which we all feel keenly. This may affect his projected visit to the USA ....*

The full detail of John Augustin's death was not yet understood, but with this sad event, Paterson was obviously prepared—on compassionate grounds—for Robert to postpone, or maybe even cancel, his visit to the 'States'. But in the event, the trip went ahead and on 20 October 1943 Robert left for North America by sea, at a time when German U-boats had been largely withdrawn from the Atlantic[23] and the *Bismarck*'s sister ship the *Tirpitz* put out of action from an attack by British 'midget' submarines at her base in northern Norway. Jesty left earlier by air and was already in the USA. Certainly, travelling by sea was now relatively safe, and for obvious reasons air travel may not have been appealing to Robert. But a sea journey took about ten days and he was to be away from Emily for almost two months at a time when her emotions must still have been very raw. But this was war, and as a senior staff member a national duty had to be discharged, and just as the opening of Droitwich helped Robert to normalize his life after the death of Peter Walter, this duty probably allowed him the same opportunity. While Robert was away, the heavy obligation of attending to their son's affairs fell to Emily. After all, Robert may neither have reached the States nor returned safely, and these duties have to be discharged in order to draw a line in life. John Augustin died just a few days before his 32nd birthday, intestate and a bachelor but, unlike his brother, old enough to have accumulated an 'estate'. Because of Robert's absence, 'Letters of Administration' were granted by the Principal Probate Registry in the name of *Emily Le Rossignol*[6] on 15 November 1943, she being 'one of the persons entitled to share in the estate of the ... said intestate'. Their son's gross estate amounted to £832-17s-10d.

By Friday 10 December Robert had returned to the UK and 'rang up' to let everyone know he would be 'in' the following Monday. A few days earlier Jesty had given a talk on his experiences with CRTs, Paterson noting on 08 December that;

> *He showed the flag well and did credit to his Lab. and country. His reception there showing clearly what friends we have in the USA ....*

The Monday saw Robert return to the Laboratories, walking straight into a fairly severe 'flu' epidemic that had laid low 15–20% of staff. On Wednesday afternoon a meeting of the CVD group was 'enlivened' as Robert delivered an account of American and Canadian expertise with magnetrons. He reported that he had the warmest reception everywhere—particularly from the Schenectady (New York) folk under Coolidge whom he had first met on his honeymoon in 1910 (Chap. 10)—but he brought back an observation that their manpower *per* valve and their general wastage was higher than at

Wembley. As 1943 drew to a close, momentum in the war had clearly swung towards the Allies. After the invasion of Italy, the Italians surrendered, the Russians took Smolensk and then Kiev and the Fifth Army entered Naples. On Boxing Day, the German heavy warship, the *Scharnhorst* was sunk. December also saw a welcome development at the GEC Board meeting on the 9th where post-war plans for the company were discussed for the very first time. The 'beginning of the end' it seems, was now in sight.

1944 began almost as spectacularly as 1943 for the Laboratories when Mr. McLeod, the popular foreman of the metal shop, was recognised in the King's New Year Honours List by the award of a BEM—delighting all the staff at Wembley. The same could not be said however of an 'Atlantic Conference' over cable on the evening of 27 January regarding 'micropup' and CRT valves, Robert describing the event to Paterson as a 'slow and dull affair'. But just a few weeks later, the Laboratories had an outstanding week of self indulgence celebrating the 21st Anniversary of the opening on 27 February 1923 with celebrations organised by the Social Club Committee—entirely independent of 'management'. The program consisted of evening events each day of the week and began with a religious service on the Sunday afternoon then a formal recognition of, and presentation to, the 34 original members of staff who were present at the opening in 1923. These members gathered in the Library for a photograph, Plate 17.11, Paterson is seated centre, front row, with Robert immediately to his right. The implication is obvious.

The principal events that week were a dance on the Monday night, a lecture entitled '*A Confidential History*' given by Paterson on the Tuesday, and a 'social' on the Saturday. Incidentally, Paterson's lecture was the only one he ever gave regarding the war work of the Laboratories and in it he made a formal recognition of Robert's work on thermionic devices over the years (Chap. 15). Paterson also records the effect these celebrations had on the general *esprit de corps* of the Laboratories, noting that it could 'hardly be exaggerated'. The general 'feel good' factor of the Laboratories was also reinforced by events at the front over the next few months. In March, Monte Cassino was destroyed by Allied bombers, then taken by the Allies in May. Rome was entered by the Fifth army on 04 June and two days later came the Normandy landings, 'D-day', Paterson's observations that day being expressed in just two sentences in the diary;

> The invasion has come at last – a cool blustery W to NW wind. One would not have thought it possible in such weather.

Plate 17.11   The 34 'founding fathers in February 1944, a photograph taken in the Library of GEC Wembley. There are actually 36 people in the photograph, the two additional are probably original clerical staff at the time. Photograph courtesy of the Institution of Engineering and Technology[1]

The Diary makes little reference to Robert over the remaining months of 1944. What entries exist are largely concerned with general 'housekeeping' and keeping things running smoothly. To some extent the pressure was now 'off' the Laboratories because D-Day was a successful event which everyone recognised must surely lead to Germany's eventual and inevitable defeat. And so it was; by December the Allies had entered Cherbourg and Caen, landed in southern France, entered Paris, captured Antwerp, liberated Brussels and Calais, and crossed the German frontier at Trier. The final German heavy warship the *Tirpitz* was sunk on 12 November, and as Germany made a last desperate effort to break out at the 'Battle of the Bulge' in Ardennes during December, Robert and Miss Lock brought in their six months valve pre-production returns (Apr-Sept) to Paterson. At noon on 22 December, the Laboratories closed for four days over Christmas. Paterson's final entry for the year was a simple personal observation viz.,

Saturday – this ends the year which has been wonderful in achieving what seemed to be impossibilities. People are tired through the demands which

have been made on them – nervous, emotional as well as physical. It is in their home life that people have felt the controls and restrictions, the privations and the burden of work …

And, anticipating '… the organised quarrelling which is to start up politically after the war …' another year passed.

1945 began with a seventh decoration for the Laboratories when A. G. Pearce, then leader of magnetron pre-production, received a 'well deserved' MBE in the New Year Honours, but January saw little developments of note, except for the first approximation by Paterson of the 'after-war' laboratory organization. By the end of the month he and Oliver Humphreys had produced a second approximation of some 900 staff and annual costs of about £325,000. The remaining few months of the war Diary are littered with ideas for the engagement of the Laboratories in peace time viz., police radio, cinema equipment in discussions with Sir Alexander Korda and the Rank organisation, Jessops entered the fray as did the Kodak company together with the 'knotty problem' of the junior technical assistants as these were dead-end occupations if these fellows were not going in for research. Each advance of the Allies was also recorded but during these months only a rare mention of Robert, now in his sixties and just a few years from retirement.

In an entry on 07 May 7th 1945, at 5.15 pm, the war diary anticipates the final victory in Europe;

The weekend has brought news of surrenders and successes and the end was obviously near. Today we hear that cease fire has been ordered by Hitler's successor to all forces – submarines are to cease operations and return to base. The great announcement of VE day has been expected momentarily but has not yet come through. If it comes through this evening we shall close for Tuesday and Wednesday.

followed on 08 and 09 May with Paterson's economic, unemotional, 'matter of fact' closing remarks;

Public holidays and rejoicing. We have now passed into the post-war era, this five and a half year diary comes to an end. At our pre-production meeting, Pearce announced that we had made 295,138 valves.[57]

And so it was that Wembley's war was finally over.

17.4   The Immediate Post-war Years

As the Laboratories moved into the post-war era there could be little accommodation for Robert, at least in the long term. Sixty-five in April 1949, he would have been increasingly concerned with plans for retirement as well as the future direction of the Laboratories, and there were some final duties to attend to, particularly following the unexpected death of Clifford Paterson. Paterson had been knighted in 1946, but he died just a few years later in 1948[58] soon after returning from a trip to Australia. His position as Director was taken by Oliver Humphries, leader of the Heat Group. But the Board of GEC were not prepared to hand the reins of scientific direction to one man alone in succession to Sir Clifford. Humphreys was therefore 'supported' by a Scientific Panel—known within the Laboratories as the 'Seven Pillars of Wisdom'—who were to 'discuss' policy and programmes with him.[59] There appears to be no strict record of this body or its membership but it is almost certain that Robert served on the Panel for the remainder of his time at Wembley. At the same time, free from wartime duties, Robert was able to indulge himself in his science once more and from 1947 to 1949 he submitted three final patents (Appendix D). But retirement inevitably came and it was as a senior 'management' man, the principal member of staff concerned with valve design throughout the war, and author of 24 patents on behalf of the MO-V and GEC, that Robert finally stepped-down from Wembley on his 65th birthday, 27 April 1949 and into what was to be a long retirement.

7 St. John's Road Harrow had been in the Le Rossignol family for at least 50 years. It was home when Robert studied at UCL, and the matrimonial home for the thirty years he worked at GEC Wembley. Both his boys grew up there and despite the tragedies of their deaths there must have been some happy lingering memories. But for their retirement Robert and Emily decided to move on and they bought a property not far from the retirement home of Sir William Ramsay in Hazlemere, High Wycombe. On 11 September 1951, the *London Gazette*, announced a number of new land registrations.

H.M. LAND REGISTRY.

The following land is about to be registered. Any objections should be addressed to " H.M. Land Registry, Lincoln's Inn Fields, London, W.C.2," before the 25th day of September, 1951.

and amongst the many 'freehold' registrations that day was;

> "St. Helier", Penn Road, Beaconsfield, Bucks,
> by R. & A. E. H. Le Rossignal, both of 7, St.
> Johns Road, Harrow, Middlesex.

'St. Helier' became the new address for 67 Penn Road, Beaconsfield, a substantial property built in the 1920s around the time the railway came, and set in half an acre of private grounds in leafy suburbia beside the road leading from Beaconsfield to the ancient and pretty village of Penn. The house as Robert and Emily knew it no longer exists, having been re-developed in 2012 —largely on the original 'footprint'. But the photograph in Plate 17.12 shows the property in the mid-1960s and as the couple knew it.

In some ways Robert's retirement mirrored that of Sir William Ramsay. He and his wife had always been town dwellers and in all the letters Ramsay wrote prior to retirement he never mentioned a garden. But in a letter to Emil

Plate 17.12 The house that was Robert and Emily's retirement home. Photographs courtesy of the le Rossignol family

Fischer, 06 May 1914 he describes their new home in the countryside about 45 miles from London[60];

> *We have about two acres of ground, partly garden, partly meadow and partly beech wood … my laboratory is nearly ready and it looks nice … My wife loves the garden …*

Robert and Emily too were town dwellers, but unlike Ramsay whose retirement was to be all too short,[61] they were to spend some 25 years at *St. Helier* in comfort and peace after a life battered by tragedy and two world wars. Here, gardeners were employed to keep the grounds in order, maids helped with the 'domestics', Robert built himself a small workshop—in a shed in the garden—and the couple engaged with the local community, especially it seems with Holy Trinity Church, Penn. But as the post-war years brought a new life, there were also echoes of Germany as Haber's son Hermann, whom the couple knew from both Karlsruhe and Berlin, wrote another tragic chapter in the story of 'Fixation'.

At the end of his life, Fritz Haber was close to his eldest son even though Fritz had forced him into a career in chemistry when the young man clearly wanted to study law. Hermann had little interest in the science and over the years he struggled to hold down a job. Desperately unhappy, he drifted from employer to employer and from country to country. He worked in Czechoslovakia, the United States and finally in France after a period in the French Foreign Legion.[62] The beginning of the second world war saw him at the Société des Produits Chimiques. He was unable to obtain French citizenship however, and when the Germans invaded France, he and his family fled to the Dordogne[63] to avoid internment. In constant danger as Jews, they left for Marseilles and from there travelled to the Caribbean on a French ship. Interned by the British for several months, Hermann applied for a visa to allow the family to enter the United States. Their case was supported by Einstein, and the whole family eventually arrived in New York City as 'friendly aliens'. Hermann then worked as an independent patent attorney and the family lived in Long Island. As the war progressed, Hermann and Marga received full residency permits which allowed them to become financially more comfortable. But soon after the war, Marga developed leukemia and died. After her death, Hermann became increasingly isolated and lonely. He took to drinking heavily, became depressed, and towards the end of November 1946, like his mother, he too committed suicide.[64] The 'fathers of fixation' both lost sons in this way, but unlike Robert, Haber did not have to bear witness his son's agony. And tragedy continued to stalk the family.

A few years after Hermann's death, Clara, the eldest of his three daughters also took her life, the third family member in as many generations to do so and the fourth in the story of 'fixation'. Clara, also known as 'Claire', was born in Haber's house in Berlin on 21 December 1928, Marga and Hermann living there as Haber tried to find Hermann a job.[65] According to Kate Mitchells story, 'The Forbidden Zone', presented at the Salzburg Festival, 30 July 2014 and described in *The Times* Article 'The Secret Story of the Poison Gas Family Suicides' the same day, a researcher uncovered Claire's birth certificate and unraveled the story of Haber's granddaughter. Claire—also a chemist—died a horrible death in 1949 gasping for air after swallowing cyanide in a Chicago toilet opposite the American Institute where she had been working on counter-agents to poison gases. The story goes, at least as Mitchell tells it, that in her bag was Haber's Iron Cross (Chap. 13). Claire too it seemed, struggled to cope with the cruel legacy of her family.

Were Robert and Emily aware of these events at the time? We don't know, but they probably heard of them from Johannes Jaenicke when he visited *St. Helier* in 1959 (Chap. 18)—if indeed news had not already filtered through. Over the quarter century of their retirement together, this private couple were to engage in an amazing, but humble, demonstration of their humanity undoubtedly predicated on their life experiences and a desire to help people less fortunate than themselves. And it's always part of the human condition that even when people bear what seems to be more than their fair share of tragedy in life, out of that tragedy often comes triumph.

Notes

1. Robert Clayton & Joan Algar (Editors), *A Scientist War. The War Diary of Sir Clifford Paterson, 1939–45*, Peter Peregrinus Ltd., in association with the Science Museum, (1991). ISBN 0 86341 218 1, https://doi.org/10.1049/PBHT014E. This title was later acquired by the Institution of Engineering and Technology, (30 January 1991).
2. During the war, the Laboratories produced 303,848 valves of 40 different types, Clayton and Algar, op. cit. (note 1), p. 574. The 'magnetron' was a device that generated microwave energy—as its name implies—from the interaction between a _magnet_ic field and elect_rons_. The 'cavity magnetron' developed by the Laboratories was a particularly high-powered 'sealed-off' vacuum tube that generated microwaves in such a way.
3. Clayton and Algar, op. cit. (note 1), p. 588.
4. However, Clifford Paterson's war diary for Sept 1st 1939 op. cit. (note 1), has the entry … '*Intensification of activity on valve and radio work for Bawdsey and Signal School*'. Ironically however, a very readable account of CH and CHL

was published in the GEC Journal in 1985, viz., B. T. Neale, 'CH—The First Operational Radar', *GEC Journal*, **3**, 2, 73–83, (1985).

5. Together with Mr. L. A. H. Bedford at Cossor.

6. Even extending to wartime. Indeed, at a meeting of the National Register on 13 September 1939, Paterson records in his diary the presence of ' … *a dearth of radio men …* '. See Plate 17.4.

7. An excellent history of the development of mobile radar written by 'one who was there'—including the role of the GEC is, E. G. Bowen, *Radar Days*, Taylor and Francis, (Jan 1998). ISBN-10 075030586X. ISBN-13 978-0750305860.

8. The 'E' (*e*xperimental) designation was used in the laboratories during development and pre-production of a valve. Other designations usually followed when in full production.

9. Robert Clayton and Joan Algar, *The GEC Research Laboratories, 1919–1984*, IEE History of Technology, Series 10, Peter Peregrinus Ltd., in association with the Science Museum London. (1989), p. 124.

10. A 'proximity fuse' detonates an explosive device automatically at a predetermined distance from the target. Different proximity fuses are designed for different targets, such as aircraft, missiles, ships and ground forces. They provide a more sophisticated trigger mechanism than the common contact fuse. Along with radar, the proximity fuse is considered one of the most important technological military innovations of World War II. (From Wikipedia).

11. There is no doubt that Randall and Boot invented a cavity magnetron, but they were not the first to do so. They were preceded by the work of Samuel in the USA, Aleksereff and Malearoff in Russia, and by work in Japan. But Randall and Boot were almost certainly unaware of this prior work. however, their design, with the anode being part of the vacuum envelope and the output coupling loop inside a cavity, made the high-power cavity magnetron possible. This design was the basis of all further microwave magnetrons. The development was to revolutionise radar and was the key element in making microwave radar possible.

12. Bowen, op. cit. (note 7), pp. 147–148.

13. Clayton and Algar, op. cit. (note 9), p. 129.

14. The Canadian Association of Physicists represents the Canadian physics community pursuing initiatives that enhance the vitality of physics and the contribution of physicists in Canada.

15. Clayton and Algar, op. cit. (note 9), p. 52.

16. These included WINDOW, (foil strips dropped from RAF aircraft to swamp German radar with false signals), MANDREL, (to jam and swamp German ground radar such as Freya and Wurzburg), BOOZER, (equipment that warned bomber crews when they was being tracked by German radar), MONICA, (fitted to bombers to give warning of approaching German

fighters), TINSEL, (radio transmissions from the bombers to drown out radio communications between German fighters and their controllers), CIGAR, (German-speaking RAF operators assigned to transmit false and confusing directions to German fighters) and SERRATE, (airborne radar fitted to RAF night-fighters on Intruder operations to enable them to track and attack German night-fighters).

17. The speed of light (electromagnetic radiation) is 299,792,458 m s^{-1}. A wavelength of 6.7 m means there are $\sim$45 million complete cycles sec^{-1}, i.e. $\sim$45 MHz.

18. The target range to achieve at the beginning of 1939 was about 4 miles. Bowen, op. cit. (note 7), p. 65.

19. Bowen, op. cit. (note 7), p. 81.

20. Bowen, op. cit. (note 7), p. 89.

21. Bowen, op. cit. (note 7), p. 147.

22. The 'Tizard mission' was an enticement to the USA to provide material help to the UK and the Commonwealth to continue the war 'on behalf of all civilization'. It occurred before Roosevelt signed the 'Lease-Lend' Bill in March 1941. The information provided by the British delegation contained some of the greatest scientific advances made during the war: Radar (in particular the greatly improved cavity magnetron) and design for the proximity fuse, details of Frank Whittle's jet engine and the Frisch-Peierls memorandum, which described the feasibility of an atomic bomb. Though these may be considered the most significant, many other items were also transported, including designs for rockets, superchargers, gyroscopic gun sights, submarine detection devices, self-sealing fuel tanks and plastic explosives (Wikipedia). The Tizard mission, was instrumental in installing the first airborne radars in US aircraft. The greatest achievement of the mission was to pass on the secret of the resonant magnetron to the US only a few months after its invention at Birmingham University. This was the device that eventually brought about a revolution in Allied radar, putting it far ahead of the corresponding German technology for the remainder of the war. Roosevelt's 'lend-lease' was a program under which the United States supplied Great Britain, the USSR, Republic of China, Free France, and other Allied nations with materiel between 1941 and August 1945. 'Reverse lease-lend' refers to the reciprocal relationship where the beneficiaries of lease-lend provided the USA with important material they would otherwise not have had access to.

23. Sir Bernard Lovell FRS, 'The cavity magnetron in World War II: Was the secrecy justified?' *Notes and Records, Royal Society. Lond*, **58**(3), 283–294, (2004).

24. *Radar*, Issue No. 3, 30 June 1944, a US Army Publication. From http://apss. org.uk/projects/APSS_projects/radar/H2S/images/norman01s.jpg.

Photograph courtesy of Bob Thompson of the Aviation Preservation Society of Scotland. (APSS).

25. Along with Robert, the main characters at the Laboratories during the war—at least in terms of the number of times they were referenced in the Diary—were; Dr. D. R. P. Dudding, Dr. D. C. Espley, I. C. Gamage, M. R. Gavin, B. S. Gossling, O. W. Humphreys, G. C. Marris, Dr. E. C. S. Megaw, Dr. A. H. Railing, J. W. Ryde, J. M. Waldram, Dr. T. Walmsley, W. E. Willshaw and G. T. Winch. Robert therefore belonged to an elite group.

26. 15 January 1940, Clayton and Algar, *op. cit.* (note 1), p. 22.

27. AI Mk II turned out to be an operational failure however.

28. By 1942 with Wembley's help, EKCO entered full production of air interceptor radar AI Mark VIII—probably at Malmesbury—and it stayed in production there for the rest of the war (Source Grace's Guide, https://www.gracesguide.co.uk/).

29. From the article, Ian White, 'A Short History of Air Intercept Radar and the British Night Fighter. 1936–1945. Part One, (https://www.yumpu.com).

30. http://www.bombsight.org/explore/greater-london/harrow/.

31. 'A Child's Memories of World War 2 in Harrow', *WW2 People's War, BBC*. http://www.bbc.co.uk/history/ww2peopleswar/stories/87/a2051687.shtml

32. Recorded in the immigration records; Detroit Border Crossings and Passenger and Crew Lists, 1905–1957. Courtesy of Ancestry.co.uk.

33. Gilbert S. Guinn, *The Arnold Scheme*, The History Press, (2007), ISBN 978-1-59629-042, to which I am completely indebted for my understanding and which I fully acknowledge here. The copyright remains with the author. As we write, one of the original members of SE-42-E (Sept 1941), 92-year-old Dennis Payne, returned to Americus, his visit reported in the *Times Recorder* of 02 June 2011 (http://www.americustimesrecorder.com).

34. The RAFVR was the principal recruitment body for the RAF during the war. It held a 'pool' of men who could be called upon to serve 'as and when'.

35. John Augustin enlisted in the RAFVR as an aircraftsman (A.C.2, the lowest RAF rank, a 'private') at the Uxbridge recruitment centre on 14 September 1940. He was 'called up' on 25 January 1941 and transferred to an Initial Training Wing becoming leading aircraftman (LAC) on 05 April 1941. Subsequently assigned to the Arnold Scheme, he moved to a Personnel Transit Centre on 30 May 1941 and from there travelled to the USA. Courtesy of RAF Disclosures, RAF Cranwell, SLEAFORD. (11 November 2013).

But at twenty-nine when he joined the scheme, his age suggests that he had already been a member of the RAFVR for some time. Indeed, the RAFVR Officers website records him at the time as Sgt. Le Rossignol, 1288222. See also footnote, *London Gazette*, 19 May 1942. For the RAFVR visit, http://www.unithistories.com/officers/RAFVR_officers_L01.html/.

36. Named after Major Henry Souther a consulting engineer on many American aviation projects. Today Souther Field is the 'Jimmy Carter Regional Airport'.

37. Guinn, op. cit. (note 33), p. 111.

38. Guinn, op. cit. (note 33), p. 112.

39. This quite high rejection figure was due to a number of factors. SE-42-A were the first cohort, there was a lack of mechanical awareness, cadets were idolized which flattered some and diverted their attention as did the food and climate, many fell asleep in the classroom.

40. Guinn, op. cit. (note 33), p. 71 and p. 118.

41. Guinn, op. cit. (note 33), p. 387.

42. The announcement of John's promotion was made in the *London Gazette* 19 May 1942, p. 2182, 'Sgts. to be Plt. Offs'.

43. AIS, Airborne Interception Shortwave, i.e., 10 cm or less

44. By now the number of Laboratory staff had expanded significantly over peace-time levels, but many had probably not been exposed to peace-time levels of scrutiny regarding their suitability as employees.

45. At the outbreak of the war the Laboratory's staff consisted of some 200 'hand picked' graduate research workers supported by about 350 other dedicated staff. Total numbers however were to rise to over 1500 of whom about 700 were engaged on preproduction work, Clayton and Algar, op. cit. (note 1) p. 584. It is clearly this new intake—'*a large section-probably half*—that Paterson is targeting here.

46. Clayton and Algar, op. cit. (note 1), p. 331.

47. Clayton and Algar, op. cit. (note 1), p. 402.

48. Reported in the *London Gazette*, 20 November 1942, 'Plt Offs to be Flg. Offs.', p. 5039.

49. http://www.rafcommands.com/forum/showthread.php?11962-Details-for-these-Casualties-ex-pupils-of-Darlington-Grammar-School/.

50. An excellent account of the Intruder is by Roger Darlington, *Night Hawk*, William Kimber and Co. Ltd., (Nov 1985), ISBN-10 0718305744, ISBN-13 978-0718305741. To which I am indebted for much of my understanding here.

51. From '*The Plane That Saved Britain*', broadcast on UK Channel Four, 21 July 2013.

52. David J. Smith, *The de Havilland Mosquito Crash Log*. Midland Publishing; 2nd Revised Edition (Dec 1980). ISBN-10 0904597334, ISBN-13 978-0904597332.

53. I am indebted to the web site, oldrafrecords.com which has digitized the Operational Record Books (ORB) for 23 Squadron which include details of John Augustin's service in the Middle East theatre.

54. Summary of operations, September 1943, No. 23 (F) Squadron, National Archives AIR27/Sqn 23 ORB and op. cit. (note 53), ORB 23 Sqn, Summary of Events, 31 August 1943. Appendix A.

55. Dylan Thomas, *A Refusal to Mourn the Death, by Fire, of a Child in London.* Collected Poems, 1934–1953. Phoenix Paperbacks, (1998) ISBN—978 0 7538 1066 8.

56. UK Documents, Bunning Way, Islington, LONDON. N7 9UP.

57. Pearce's estimate was too low. The actual figure was recorded later as 303,848. Clayton and Algar, op. cit. (note 1), p. 574. The figure of course only represents the GEC's efforts. Many other factories produced valves for the war effort as well.

58. At the relatively young age of seventy.

59. Clayton and Algar, op. cit. (note 9), pp. 23–25.

60. Morris W. Travers, *A Life of Sir William Ramsay K.C.B., F.R.S.,* Edward Arnold London, (1956), p. 281.

61. Ramsay died in Hazlemere, High Wycombe on 23 July 1916 after months suffering from nasal cancer.

62. Daniel Charles, *Between Genius and Genocide,* Jonathan Cape, London, (2005), p. 252.

63. Dietrich Stoltzenberg, *Fritz Haber. Chemist, Noble Laureate, German, Jew.* Chemical Heritage Press, Philadelphia, Pennsylvania, (2004), p. 182.

64. Stoltzenberg, op. cit. (note 63), p. 182, his ref. 20, in conversation with 'Lutz' in Bath, 11 October 1989. Also, Vaclav Smil, *Enriching the Earth,* MIT Press, (2001), p. 327, post-script 24.

65. Stoltzenberg, op. cit. (note 63), p. 180.

18

Penn, Buckinghamshire 1949–1976

The retirement years, 'coming-out', and remarkable benevolence.

Forget your personal tragedy.
We are all bitched from the start ....
but when you get the damned hurt, use it - don't cheat with it.

Ernest Hemingway, (adapted) from a letter to F. Scott Fitzgerald, 28 May 1934.
Ernest Hemingway Selected Letters, 1917–1961.

18.1 Penn, Buckinghamshire

Robert and Emily's move to their new home between Beaconsfield and the village of Penn, represented a move to an entirely different way of life. Like Berlin and Karlsruhe, St John's Road Harrow had meant 'big city living'. Here, an eclectic population had easy access to central London; shops, services and transport were all 'on the doorstep'. Not so Penn Road. The pace of life here was much slower, and more traditional values prevailed. This was an area populated by the wealthy, the famous, those from the old landed families, the farmers, the clergy and—because of its proximity to London—the well-heeled city commuters and those with their hands on the levers of power. There were contrasts either side of *St Helier* however. If one required some semblance of city life, then just to the south lay Beaconsfield. Nudging the old town with its coaching inns, small shops, red brick houses and a provenance stretching back to the 1100 s, the larger 'new' town of Beaconsfield arose at the turn of the 20th century when the railway arrived. A busy little place today, things would have been more subdued when Robert and Emily came; the second world war had not long ended, the Korean war had just begun, and for most people there was still the austerity of rationing. A little further away too lay Amersham with its convenient underground connection to central London,

© Springer Nature Switzerland AG 2020
D. Sheppard, *Robert Le Rossignol*, Springer Biographies,
https://doi.org/10.1007/978-3-030-29714-5_18

but just a mile or so to the north of *St Helier* lay Penn village, arguably unchanged in centuries; historic, peaceful, traditional, 'conservative', pretty, and undeniably 'chocolate-box' England.

Penn Road in Robert and Emily's time would have been considerably less developed than today. A drive along the few miles of leafy lane nowadays finds it packed with fine homes often with spacious gardens, tennis courts and swimming pools. But the characterization of the area in Robert and Emily's time is best described in terms of those who chose to live there. A contemporary was Enid Blyton, who from 1938 lived at 'Green Hedges' just off Penn Road, a name chosen for her in a competition by the children who read her books. Another children's author, Alison Uttley, of 'Little Grey Rabbit' fame, moved to the area during the second world war. The two ladies however did not get on; Uttley describing Blyton as 'a vulgar curled woman',[1] an observation which in its turn provides a characterization of many of the 'locals' who were articulate, who spoke their mind, and were always careful to preserve that which they held dear. The poet Robert Frost came to Beaconsfield in 1912 and remained there until his death in 1963. In the 1950 and '60s the popular singer and film actor Stanley Holloway and his family lived in Penn village, as later did the eminent philosopher Sir Karl Popper, and the novelist Elizabeth Taylor lived here from the 1940s. But the area was also littered with the echoes of historic residents such as Edmund Burke, the philosophical founder of modern conservatism and Benjamin Disraeli, Earl of Beaconsfield. Some maintained that the Quaker William Penn (of 'Pennsylvania' fame) regarded Penn as his ancestral home, and over the centuries, many of the landed families of the area had close connections with the various 'Royals' of their time receiving 'grace and favour' in terms of property and privilege which persists amongst them to this day. Robert and Emily spent a long retirement at *St Helier* from which we can only assume that here, they were content. They made friends in both Beaconsfield and Penn, but Penn village seems to have been held in special affection by them.

The name 'Penn', which applied both to the village and to the wider Parish, is probably an Anglicization of the ancient Brythonic[2] word 'pen', meaning 'head' or 'top', the village standing on a promontory of the Chiltern Hills some 3 miles NW of Beaconsfield. The eleventh century church of Holy Trinity[3] dominates the tiny village, and from its tower—or so residents would have it—a dozen counties can be seen. Much of the village today is just as Robert and Emily would have known it. Now designated a conservation area, the oldest building is Holy Trinity[3] with its late 12th Century nave, 14th Century Tower and 15th Century 'Penn Doom', a 12-foot-wide painting of the Last Judgment on oak panels and one of only five surviving wooden 'tympanums' left in the country. The Church also contains reminders of the influence of the landed families of the area. In the Lady Chapel, the Tudor and Stuart brasses of the Penn's can be seen, owners of large parts of the Parish and the latest representatives of a family originating in the Welsh borders. Throughout the church there are wall monuments dedicated to the 18th and 19th Century Curzons and Howes who followed on from the last of the Penns through marriage. Six grandchildren of William Penn the Quaker are buried in a family vault under the centre of the nave. In the churchyard the earliest marked grave is that of William Penn, the local Lord of the Manor who died in 1693 but no relation to the Quaker Penns although throughout the 17th an 18th centuries both families mistakenly thought they were related[4] (Plate 18.1).

Quiet and peaceful today, over the years the church has seen its share of colourful vicars. During the 11th Century the incumbent was murdered with an axe, and in 1549 the vicar of Penn was jailed at Aylesbury by his churchwardens for 'uttering certain opprobrious words'. But the genealogical connection between the Quaker Penns and the Penns of the parish—or rather the absence of it—suddenly became a contentious issue at the beginning of the 20th Century because of the rivalry between the Vicar of Penn and the Vicar of neighboring Tylers Green.[5] This difficulty arose as the Vicars competed for the ear of their patron, the fourth Earl Howe. On the death of his father, the new Earl chose the long-serving Vicar of Tylers Green to

Plate 18.1   A photograph of Holy Trinity in February 2012 taken by the author

officiate at the funeral. Seeing this as a snub, the newly appointed Vicar of Penn 'gate-crashed' the funeral, and his colleague asked the local policeman to take him in charge for causing a disturbance. The incident caused 'bad blood' between the two men and discontent erupted again in 1913 when the Vicar of Tylers Green accused the Vicar of Penn of serious libel in the form of 'anonymous' letters and postcards sent to himself and other notaries signed by such as a 'well-wisher' or a 'disgusted parishioner'. In these, the Vicar of Tylers Green was accused of having a 'drink problem', that he had used money destined for the Church for his own use, that he was a liar, a thief, a 'drunken little devil' who conducted services whilst 'under the influence' and who played cards after Sunday services. The case came to court in the Bucks Assizes before a special jury and the representative of the Vicar of Tylers Green gave examples of the bizarre behaviour of the Vicar of Penn over the years; one was forcing himself into the funeral and another that he had gone to America to collect monies for the restoration of Holy Trinity under false pretences, claiming that Penn village was connected with William Penn the Quaker. There was however, no known record of such a connection and the Vicar of Tylers Green had written to the court pointing this out. After much argument, an amused judge concluded; 'I do not know how many Penns there are in Penn but there is certainly a pen in Penn Parish which has penned letters that might have been un-penned.' The Vicar of Penn was found guilty and ordered to pay £400, an enormous sum of money in those days.

Hardly a village, in the sense that in the 1950s Penn had no square, nor Post Office nor shop, what it did have throughout Robert and Emily's time there however was another colourful vicar, the fifty-fifth at Penn, the Rev. Oscar Muspratt, who was to 'resurrect' the contentious issue of the link between the Penns of Penn and the Quakers. Up until Muspratt's intervention, the existence—or absence—of a genealogical link between the families was largely their private affair. But Muspratt's vigorous efforts to connect Penn village Bucks, to Pennsylvania USA via the family link aroused fierce opposition amongst 'the locals' and, as we shall explain later, his maneuverings drew the Le Rossignols into the controversy. But why would Robert—neither particularly devout nor, at his age, an active parishioner—get involved with Muspratt's contentious campaign? Well, there may have been a number of reasons for that, but these two apparently quite different men had more in common than one would have at first thought.

18.2   The Rev. Oscar Muspratt. (1906–2000)

Muspratt's father Frederic, himself once a 'man of the cloth', took his family from the UK to Jersey in a home-made gypsy-style caravan[6] at the turn of the last century to escapee the 'tedious restraints of Church life'.[7] Like Robert, Muspratt was born there (in 1906) and for a time he too lived in St Helier—next to the local Vicarage. But his upbringing was far less privileged than Robert's. His father had no qualms about teaching his young son the finer points of a rugged outdoor life—the hard way. He and his elder sister helped on their father's home-made 30ft fishing boat, subsequently lugging a heavy basket door-to-door and selling their freshly caught lobster to make ends meet. But besides the accident of birth, both men also suffered the loss of an admired elder brother early in life, Oscar being just eleven when his brother 'Freddie'—himself just nineteen—was killed in action at the battle of the Somme. But the Muspratt's had family in Australia, and in 1920 they de-camped Jersey when another older brother, Eric, took over an aunt's pineapple plantation in Queensland. Once again there was hard work for Oscar but he moved on to a succession of other tough jobs; running a dairy farm, harvesting wheat in New South Wales, then becoming a docker in Melbourne. And although up to this point his education had been both sparse and intermittent, he had by now managed to save some money and was able to attend university at Melbourne where, like Robert, he studied chemistry. But during the summer holidays when he again took a job wheat harvesting in NSW, an accident badly damaged his left leg and made him reflect on his life. This led him to change his course to theology, eventually being ordained in Australia in 1929 and thereby abandoning over a hundred years of family scientific tradition.[8]

Muspratt began his ministry as a 23-year-old curate in Caulfield, a suburb of Melbourne, moving from there to Panton Hill, a Bush Parish covering some 500 square miles. But in 1936 he returned to England to study theology at Cambridge under Regis Professor of Divinity, Rev. Dr. C. E. Raven, where he met Margaret Hooton of St Pauls Road, Cambridge, who became his wife in 1937, the couple spending that year in Switzerland at Adelboden and Grindelwald near Interlaken where Muspratt was a young chaplain. Returning to Australia he served in Fitzroy, the worst slum area in Melbourne where he helped organise a scheme through which many of the drunks and 'down and outs' could get help from the state by arranging a permanent address for themselves via the Salvation Army—a scheme which lasted some time and of which he was always proud. On the outbreak of the Second

World War he and Margaret braved the U-Boats and returned to the UK where he became an Army Chaplain on 27 March 1941,[9] subsequently serving at El Alemain, volunteering for service in Malta at the time of the siege, then in Sicily where he tended the wounded in a field hospital in 1943, and finally taking part in the Normandy landings. Muspratt knew Malta well and he was in Sicily immediately prior to the invasion of Italy and at the time Robert's eldest son John Augustin was killed there. There is little doubt too that he would have been familiar with Catania military cemetery. His active war service ended in late 1944 when he became Chaplain for Buckinghamshire. Soon after, on Xmas Eve 1944, Muspratt, became vicar for Penn retiring on 31 December 1989 as the second longest serving vicar at Holy Trinity. This 'potted history' of Muspratt shows him to be a dynamic, adventurous man, a 'mover and shaper', and he certainly made his mark during his tenancy of Holy Trinity. But he courted controversy, for Muspratt was a 'Marmite' man, generating fierce loyalty from some but utter exasperation from others. Two examples illustrate how his decisions could exercise parishioners.

Shortly after Robert and Emily moved to *St Helier*, Muspratt became caught up in the story of the last woman to be executed in Britain. Ruth Ellis, a nightclub hostess born in north Wales, was hung in Holloway prison after an affair with racing driver and Penn-man David Blakely went horribly wrong. Their tempestuous relationship was spent in Penn, and in Bucks pubs and hotel rooms before eventually leading to a cold-blooded killing on Easter Sunday 1955, when Ellis walked into the Magdala pub in Hampstead, north London, and shot Blakely dead. Blakely was buried at Holy Trinity, but the rebuilding of Holloway Prison some years later meant a new site had to be found for Ellis' body. Her son wanted his mother reunited with Blakely in Penn churchyard, but controversially, Muspratt refused permission for her burial. Ellis was eventually laid to rest at St Mary's Church in Old Amersham, the closest she was to get to her former lover.[10] At the other end of Muspratt's 'reign'—but well after Robert and Emily's time—he allowed the interment of the ashes of the notorious traitor Donald Maclean[11] at Holy Trinity. Maclean had spent much of his childhood at Elm Cottage on Beacon Hill and his ashes were buried by torchlight near the grave of his parents in a secret ceremony in March 1983. Muspratt defended himself by claiming an act of Christian charity, but there was fury in Penn. News of the ceremony reached Fleet street, some said the ashes desecrated the Churchyard and there were moves to get a Commons question tabled to find out who allowed the ashes to be brought from Russia to the UK. Love him or loathe him, life alongside Muspratt was often a 'white knuckle' ride!

Muspratt must have met Robert and Emily soon after they retired to Penn Road—their re-naming of number 67 as '*St Helier*' being an irresistible invitation for him to 'meet and greet' the new couple—and although there is no evidence to suggest that he and Robert were in any sense 'friends', the common circumstances of their lives suggests some empathy existed between the two men. But despite sharing a birthplace, suffering the loss of an elder brother, the powerful thread of chemistry and Muspratt serving in both Malta and Sicily, there were other circumstances that may have led to these men later coming together in Muspratt's maneuverings regarding the Penn-Pennsylvania link, circumstances we explain later.

18.3   Retirement and Benevolence

When Robert and Emily arrived in Penn Road, they would have found Penn village huddled around Holy Trinity. The Vicarage, the Almshouses, the old Parish rooms, the War Memorial Green and The Crown public house formed its nucleus and, with the exception of the Almshouses, this remains so today. Around Holy Trinity and along Church Road, were some of the fine houses of the local important families, together with workers cottages and a Methodists' chapel. Today, many of these are listed, some dating back to the late 17th Century. To many of us, *St Helier* would seem to be custom built for an idyllic retirement, but retirement often means different things to different people. To some it is a curse, to others an opportunity, but it is always a time to reflect on what has passed in our lives. Robert knew that he was more fortunate than many, not only in terms of the circumstances of his birth but in some of the choices he made in life. Equally, he and Emily suffered more than many through the loss of their sons and Emily's arthritic condition which, as the years went by, must have imposed severe limits on what was achievable together. Even so, because of the visit of his friend Ralph Chirnside in March 1976 and the transcript of their conversation (Sect. 18.10), we have some insight into what occupied this couple during retirement. Robert of course had a small workshop—a shed in the garden—where he indulged himself in his life-long love of 'things mechanical'. Family members Richard and John recall visiting 'Bob' and Emily at *St Helier* in the mid 1960s and later. On one visit, Robert proudly showed off his garden workshop to Richard and fiancée Alex, and he was particularly impressed that Alex knew what lathes and milling machines were and how to operate them. 'You've certainly found a good girl there!' was his advice to Richard. But the workshop was not just for show, here Robert did some quite remarkable things. Having cut his 'horological teeth' some years earlier by building a quartz clock, he turned his hand to a 'grandfather' clock which he built from bits and pieces, housed it in a 'contemporaneous' case (according to Richard), and embellished the clock face with the makers name; 'Robert Le Rossignol Beaconsfield'. The clock kept remarkably good time and even Chirnside recalls seeing it on his visit there in March 1976. But there were more practical things for Robert to attend to. He bought a lift for Emily to reach the upstairs. He installed it himself and built a false floor to make sure that no-one could get trapped below it. Robert delighted in demonstrating its effect; as the lift descended, he would stand underneath it. As soon as it touched his head the lift would stop! John and Richard also recall the tale of Emily's 'Parker Knoll' chair. Robert had bought an electric tilt chair for Emily. The design of the original supplied by a specialist company was not to the couple's taste and so he asked if the tilt motor could be fitted to his wife's 'Parker Knoll'. This was deemed too impractical by the company, so 'Bob' did the necessary re-fitting himself in the garden workshop. It gently tilted her favourite chair until she could comfortably rise out of it. Richard and John also recall 'Bob' driving the 'must have' vehicle of the comfortable middle class in those days viz, the classic 1960s Rover 80, and the couple's benevolence to their father and mother who, on having a new house built in North Devon in 1963, received a financial gift from the couple which they used to landscape their new garden.

Robert's workshop was important to him—his wishes were such that its contents were the first thing offered to friends after his passing in 1976. Little is known of Emily's interests during retirement, these would have been dictated by her disability, but what we do know of this couple suggests that the tragic events they had experienced were used positively by them during their retirement, enabling a number of acts of benevolence from which they were able to draw comfort and make some sense of their lives. In this respect the couple followed Hemingway's maxim; not wallowing in personal tragedy but finding the positive, and using it to the good. Chirnside's obituary of Robert 'Robert Le Rossignol, 1884–1976', *Chemistry in Britain*, **13**, 269–271, (1977), recounts a conversation he had shortly before Robert's passing in which Robert said, 'I hope I have been able to make use of some of my rewards to help those less fortunate …'. Clearly a wealthy couple, the financial bedrock of Robert's 'rewards' were undoubtedly his share of Haber's royalties, carefully managed and invested over the years. Through their benevolence over their retirement years, the couple were to give all of their money away to charitable causes without fuss or favour, and with only minimal public recognition. After all, what is the purpose of living if we do not embrace life? And retirement had hardly begun before an opportunity arose for Robert to make a benevolent contribution to the history of fixation, by donating items to the National Science Museum which he had acquired on 02 July 1909, the day he demonstrated the feasibility of what was to become the most important technology of the 20th Century.

18.4   The National Science Museum October 1952—The Ramsay Exhibition

October 1952 saw the centenary of William Ramsay's birth, and the National Science Museum in London organised a 'Ramsay Exhibition' in which exhibits illustrating Ramsay's life were contributed by friends, family, eminent colleagues and a range of scientific institutions. There were hundreds of items in all; photographs, letters, oil paintings, medals and awards, specimens, books, various pieces of apparatus from Ramsay's laboratories together with Travers' original hydrogen liquefier with coils, glass jar and two vacuum vessels.[12]

Throughout his life Robert always felt 'indebted' to UCL, and the exhibition could hardly escape his notice as newspaper articles appeared in the broadsheet press such as *The Times* article; *William Ramsay, 1852–1916, Great Discoverer and Leader of Chemical Research.*, by E. N. da C. Andrade F. R.S. on 02 October 1952, the centenary of Ramsay's birth. And another later in the *Illustrated London News* on 11 October 1952 (p. 593) describing the original hydrogen liquefaction plant installed for Sir William in 1899. We can't be sure if Robert read these articles of course, but what we do know is that on Thursday 16 November 1952 this retired senior member of the GEC Laboratories had arranged to be shown around the exhibition personally by the Keeper of Chemistry and Photography, Mr. Alexander Barclay,[13] an Imperial College (Royal College of Science) educated chemist who had joined the Museum in 1921. In the event, Mr. Barclay was unable to keep the appointment, but the Assistant Keeper, Mr. Frank Greenaway—himself an Oxford trained chemist who joined the Museum in 1949—was able to deputise. During the visit, and in conversation with Mr. Greenaway, Robert revealed that he had been a student at UCL, but left in 1906 to work with

Haber in Karlsruhe on the ammonia problem, and that he still had some of the ammonia produced in 1909 from the prototype fixation plant. The next day, Mr. Greenaway hand-wrote a 'memo' to Mr. Barclay, and it is clear from this, and their subsequent correspondence, that despite a fine chemistry pedigree, neither gentleman had any idea who Robert was, although the Museum's records today describe Robert as 'Haber's British-born assistant [who] was present at the time and acquired a sample of the ammonia made on that historic occasion'. One could be forgiven for thinking that, at the time, the shadow of Haber was still cast long, but neither is there any real evidence here to suggest that Robert had enlightened the young Mr. Greenaway regarding the depth of his involvement with Haber. He seemed to have been left with the impression that Robert was simply someone who worked alongside Haber, and happened to be there when the first liquid ammonia was produced, which in turn all serves to illustrate just how little of Robert's involvement with nitrogen fixation remained in mainstream chemistry at the time, but equally, how reticent Robert seems to have been in promoting himself. Mr. Greenaway wrote;

Mr. Barclay,

On Thursday 16.11.52 I saw your visitor, Mr. Rossignol, and showed him the Ramsay Exhibition. He was very interested and appreciative. Two things arose.

1. *Mr. Rossignol identified himself in the group photograph labeled '1909' in the desk case on the east side of the … case in Geology 1. He said the date must be 1905 or 1906, from his own period at U.C. and the presence of some others in the group. He thought 1906 the more likely.*
2. *Mr. Rossignol worked with Haber at one time, and produced with him the first synthetic ammonia in measurable quantity from a pilot prototype of the plant subsequently used commercially. He still has some of this ammonia in sealed tubes and would be prepared to present it to the Museum. Such an object would have a considerable interest as a relic of a very important development in chemical technology and I should recommend its acceptance.*

The 'sealed tubes' that Robert had kindly offered to donate contained samples of the synthetic liquid ammonia collected by him in the presence of Haber, Mittasch and Kranz on the evening of 02 July 1909. (Chap. 8). Subsequent to Greenaway's memo, Barclay sent a typed letter[14] to Robert (26 November 1952) regarding this generous, and unexpected, offer;

Dear Mr. Rossignol,

I was sorry to miss you when you called at the museum some days ago, I was away sick and could not, unfortunately, keep our appointment. I understand however, that Mt Greenaway met you and showed you round the exhibition. He tells me that you have some sealed tubes of the first synthetic ammonia produced by the Haber process, which you are willing to present to the museum. I much appreciate your kind interest and should like to know a little more about them. Would they bear transport in the post? If not, perhaps you would be up this way again and could bring the specimens with you.

Yours sincerely,

A. BARCLAY

Keeper.

Shortly after the receipt of Barclay's letter, Robert replied with the (hand-written) letter below (30 November 1952) providing a description of the tubes and enclosing a reprint of the paper he and Haber had published in 1913.[12] (The technical production of Ammonia from the elements, *Z. Elektrochem.* **19**, 53–72, (1913). This paper was in German, and although the paper bore Haber's name and his own, the letter hardly provided any further clarification of his role alongside Haber;

Dear Mr. Barclay,

I must apologise for not writing before to thank you for arranging for Mr. Greenaway to show me round the Ramsay Exhibition. I was laid up some days with an unpleasant attack of 'flu'.

The Exhibition interested me very much indeed, I recognised quite a number of the old pieces of apparatus. I was very interested in the photographs of the Chemical and Physical Society. I think I can help to correct the supposed date (1909) of the photo where I am at the extreme right-hand end of one of the rows. This cannot be later than July 1906, because I left at that time, as it also includes photos of two friends of mine, E. W. Millar and C. T. Gimmingham who left about one year before me. I think the correct date is 1905.

Now as to the samples of liquid synthetic ammonia which I mentioned to Mr. Greenaway. I have two glass tubes about 3/8' diameter and 6' long about 2/3 full of liquid ammonia. This ammonia was produced the first time the laboratory scale technical apparatus was run. The date was July 2nd 1909. For details of the apparatus see p.40 of the reprint which I have pleasure in enclosing.

If the Science Museum wishes to have these tubes it would give me great pleasure to present them and I would bring them up personally.

Yours sincerely,

R. Le Rossignol.

With Robert's offer now in writing and 'official', the Museum had to decide what to do and respond accordingly. There was a short, hand written, correspondence between Barclay and the Director Frank Sherwood Taylor (FST) in mid December '52;

Director,

This is interesting, but the tubes will not be much to look at, and in view of their contents, should not, perhaps, be exhibited. However, they are historical and the kind of thing that we might preserve. What do you think?

A.B. 13.12.52

Mr. Barclay,

We should accept these; I think they should be quite safe as long as nothing that would be injured by ammonia is in the case. The volume of ammonia would be only about 6 litres and could do little harm.

FST

Barclay wrote to Robert again to arrange a convenient time to meet him, Robert subsequently bringing his tubes to the Museum on 22 December 1952. The specimens were officially accepted by Sherwood Taylor in a letter dated 30 December 1952, but there was also the inevitable 'housekeeping' to attend to;

Dear Mr. Rossignol,

It is with much pleasure that I am able to accept the first tubes of ammonia liquefied by the Haber process on July 2nd, 1909, which you have generously presented to the National Collections. The historical interest in these specimens will increase as time goes on and I am most grateful to you for depositing them here.

May I ask you to be kind enough to sign and return the
enclosed form for our record purposes?

Yours sincerely,

SHERWOOD TAYLOR,

Director.

Robert completed his donation to the Museum with the letter and form
shown in Plate 18.2 dated 03 January 1953, and notice here that he too uses
the ubiquitous term 'Haber' process. But what prompted Robert to donate
the samples to the Museum in the first place? Could it have been some kind
of final attempt at recognition, after all, what he achieved at Karlsruhe
eclipsed anything of Ramsay's—at least in terms of its impact on humanity—
or could he have simply wanted to associate UCL with this major event in
chemical history? More likely it was a simple act of benevolence. Robert had
hung on to these important historical items for almost half a century, but
what was to become of them? There were no children nor grandchildren to
pass them on to as 'heirlooms', and a Science Museum must have seemed
natural for their final resting place, safe, and available in perpetuity for future
generations to view relics from a day that changed the world.

As Robert would have wished, the tubes he donated remain at the
Museum, inventory number 1952-391. The first tube 1952-391/1, has been
displayed/stored inside a spherical pressure tested glass container since 1986.
The second (duplicate) tube has remained largely in storage. At the time the
author contacted the Museum (March 2011), the first tube was displayed in
the Museum's Smith Centre Exhibition. The second was in the Science
Museum's Small to Medium Object Store at Blythe House. Both tubes are
still available to view today,[15] and although in Barclay's words they are 'not
much to look at …', gaze at them with respect, because these are the 'stuff' of
significant world history, the 'stuff' that allowed Germany to sustain a world
war that changed the face of Europe. Even today, in the fields of northern
France, 100 years after the end of the first war, people are still being killed by
unexploded munitions created by nitrogen 'fixation', and such is the chemical
pollution from these explosives that ground waters are still unsafe, and babies
prevented from drinking some tap waters.[16]

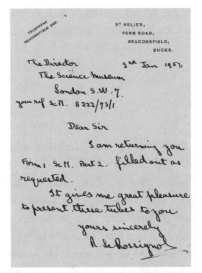

Plate 18.2 Robert's original documentation accompanying his donation to the National Science Museum in January 1953. Courtesy of the Science Museum

18.5   September 1959, the Visit of Johannes Jaenicke

We know little of Robert and Emily's life over the next few years. One can only assume that with progressing age and Emily's condition worsening it was at best quiet, 'comfortable' and filled with the kinds of things that elderly retired couples do, but as for many, painfully punctured by the depressing inevitability of another conflict involving British servicemen and women, this time at Suez in 1956. But towards the end of the decade, Robert was afforded another opportunity to contribute to the history of fixation when 'out of the blue', Johannes Jaenicke, a friend of Haber and a former co-worker on Haber's 'gold from seawater' project, contacted him with a request for his recollections of his time with the 'great man'—Jaenicke having in mind a biography to right what he saw as the serious injustice to Haber's reputation which had persisted over the years. Jaenicke's letter to Robert was 'couched' in quite familiar terms, viz., from one 'Haberian' to another, and his intention was to use his annual holiday to come to the UK to speak to Robert personally about his memories, thereby affording him as little inconvenience as possible. Jaenicke frequently referred to Robert as 'Doktor' Le Rossignol in his correspondence, the first of which was dated 25 August 1959. A rather clumsy translation of his letter (by this author) appears below.

August 25th 1959,

Dear Herr Doktor Le Rossignol,

There is a rumour here in Germany that a biography of Fritz Haber is about to appear. A number of magazines have called for information about Haber but they have unfortunately found only a very weak echo, so that in the light of the devastation caused by the Second World War and its consequences regarding the German archives, there are still wide gaps in the documents.

We have both had the good luck to acquire Haber as a common boss and friend, and I wonder if you were willing to share your memories of him with me. I know that I am not the first to bother you with such curiosity, but I would be delighted if you tried to help me remove the serious injustice which has cast a shadow on his legacy, and restore a dignified and true image of his personality and his work.

I would very much like to save you the trouble of written comments and I would like therefore to ask the question of whether it would suit you if my wife and I, during our vacation in September which we plan to spend in England, if we could meet with you at your convenience.

Please let me know where and when - if at all - you will give me the benefit of a personal meeting with you.

Yours faithfully,

very devotedly,

Johannes Jaenicke.

Robert's modest handwritten reply, *Plate* (18.3), was hardly encouraging for Jaenicke, playing down as it did the extent to which he thought he could help, but nevertheless it was accommodating and he and Emily seemed more than willing to arrange a meeting at *St Helier* at a mutually convenient time;

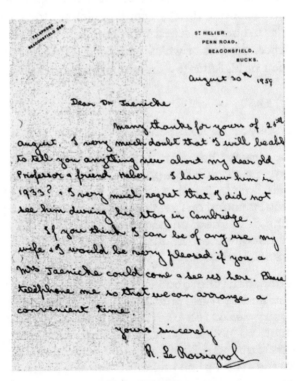

Plate 18.3 Robert's reply to Jaenicke's letter of 30 August 1959. Courtesy of Archiv der Max-Planck-Gesellschaft, Berlin-Dahlem

Jaenicke however leapt at the opportunity to interview one of Haber's 'oldest and most trusted' employees, eager to learn of anything Robert could recall even if it were only his impressions of Haber's powerful personality, and in his reply of 02 September (again translated by this author) he arranged to call Robert from London when he and his wife arrived there for their vacation.

September 1ˢᵗ 1959,

Dear Herr Doktor Le Rossignol,

It is a pleasure for me to have a personal meeting with just yourself and your gracious wife. My wife and I cordially thank you for your kind invitation.

Even if you don't really have any new facts about Haber that you can make known, my wife and I – as part of his former team – will be happy to hear what you thought of his impressive personality. As one of Haber's oldest and most trusted employees, your opinions stand well above others as a competent referee for this period in Haber's life.

In addition, I would like the opportunity to get to know about the other chemists there.

I will call you from London, to explore when our visit would be most suitable.

With the best regards …

Robert and Emily subsequently arranged to receive Dr. and Frau Jaenicke at *St Helier* on 16 September 1959, and Jaenicke's transcript of their lively and extensive conversation is a 'gem'—and quite contrary to Robert's inclination in his letter of 30 August. But despite the transcript being catalogued in the Max Planck Archives, and therefore freely available for the past twenty-five or so years, it has remained biographically dormant. However, it is a missing link in the story of 'fixation'. Here at last is some of the story told by one who was at the heart of the matter, and even though Ralph Chirnside's Obituary of Robert described his friend as 'kindly', never speaking 'ill of anyone', Robert's account of his time with Haber is occasionally acerbic, sometimes sarcastic, but perceptive, frank, often comical and even fifty years after the event still largely fresh in his mind. In the comfort of his own home, with the passing of many of the players in the story of 'fixation' and accompanied by Emily in convivial conversation with the Jaenickes, Robert 'opened up'. In the evening sunshine of his life he 'came out of Haber's shadow', told his side of the story and expressed opinions and observations that would otherwise have been lost to us. That day

Jaenicke reminded Robert that over the years his name had become famous in Germany—'every chemist knows the Le Rossignol valve' he said—and although circumstance had convinced Robert that his name would 'not be remembered for big things …', he was obviously aware that Jaenicke's opinion of him as one of Haber's 'oldest and most trusted' employees would eventually lead to his reflections finding their way into the future biography. Robert therefore put the record straight, properly accounting for his work at Karlsruhe and clarifying once and for all the roles of himself, Haber and Kirchenbauer in the creation of the technology of 'fixation'. Jaenicke's transcript is altogether different from Robert's 1928 paper (Chap. 8) which was written in celebration of Haber's 60th birthday; that was a personal homage, but as an account of the principal characters and the fine detail of 'fixation' it was a disappointment. The transcript however describes the division of work at Karlsruhe, who did what, the origins of some of the technical innovations, and Robert's opinions of friends and colleagues; Much of this detail has already been included where relevant in the book, because Jaenicke's attempt at a biography of course overwhelmed him.[17] But through his transcript we can now accept that at Karlsruhe—and in Robert's words that September day;

> … Haber did the theoretical side, and I did the engineering side … I … did the practical work … I had nothing to do with the business side … [and] … Haber did the whole of the patent matter …

Robert Le Rossignol was therefore responsible for both the fundamental experimentation, and then the engineering, that eventually led to a pilot technology for high pressure 'fixation'. Engineering that Patrick Coffey, (*Cathedrals of Science*, Oxford University Press) describes as—'the most demanding technical chemical project ever undertaken in an academic setting at the time'. He built the laboratory scale apparatus, he worked out the finer detail just as Haber expected of his co-workers and he made the machine work using his skill in mechanics, together with techniques borrowed from Travers' hydrogen liquefier (Chap. 4). No wonder then that Chirnside's obituary of Robert records that 'he always felt indebted to UCL'. Haber of course was a reluctant experimentalist, the time-line of his work (Chap. 2) shows him clearly moving from the 'lab' to the 'log', and the sophisticated engineering described in Chap. 8 would have been beyond his capability. *His* contribution was confined to the theoretical calculations and the management of the business of the BASF contracts and the patents—none of which helped build the machine.[18] Yes, during that time Haber would have fussed around Le Rossignol like a mother hen, suggesting, advising and encouraging as was his way, but he would not have had the practical skills to progress any kind of machine himself even though of course he helped specify exactly what it should do, and according to Robert, Kirchenbauer—although a skilful engineer with whom 'he got on well'—had to be 'advised … what things should be done'.[19]

Over the years Haber became one of the most powerful men in Germany, but at the time his reputation was more modest, so let's puncture the Haber reputation here for a moment. From May 1908 until the demonstration of the final version of the machine at Karlsruhe on 02 July 1909, it appears that Haber was really just a facilitator. It was Haber who provided a salary and the environment for Robert's work; it was Haber who procured samples of 'exotic' materials as candidate catalysts; it was Haber who communicated events to the BASF and orchestrated developments; it was Haber who enthused and kept spirits high when problems inevitably arose and it was Haber who provided probably the greatest incentive of all for Robert to succeed viz, the private contract promising him 40% of his royalties from the BASF, (Chap. 7). But clearly, Robert built the machine and his deep involvement in both the formulation of the 'Haber' process and the creation of the technology has never received the recognition it deserves.[20] Jaenicke's transcript should now inform us, help change this view, and allow us at last to confidently understand how the 'fixation' of nitrogen was achieved. Haber's position in the history of fixation is therefore less assured now than the 'engineering' of his Nobel Prize would otherwise have us believe (Chap. 15).

But along with the recollections of his own work, Robert also accommodated Jaenicke's wishes contained in his letter of 01 September and expressed his opinions regarding other colleagues at Karlsruhe; although Jaenicke's descriptions here are often rambling, clumsy and vague. Besides the observations mentioned earlier in the book, Robert for example, found Engler 'very nice', he says he was also held in high regard by Haber. Bunte of course was an expert in patent matters and helped Robert obtain his *Gebrausmuster* (Chap. 8), but Haber looked upon Bunte 'to get a little old' probably with an eye on eventually taking his position. Bunte also provided a flat for Robert in the early days at Karlsruhe, and Frau Bunte was also very friendly with him. Robert describes Askenasy as having a very good [all-round] knowledge. He had married for a second time, and he and his wife were kind; a daughter from his first marriage use to play with their children and during 'Hitlerism' they bravely helped hide many Jews—an observation which suggests Robert was aware of friends in Germany long after he left. Coates however is singled out for particular attention. 'He owes his life to me!' Robert recalls when Coates and Kirchenbauer worked on what Jaenicke vaguely calls the 'oxide case'[21] which they were testing at 300 atmospheres. 'I was horrified at the [dangerous] way they tested it .. they simply put it out of the window ... Coates and Kirchenbauer used to have a jolly good talking [to]!' But he also recorded the undercurrent of anti-Semitism in Germany against some. Robert's friend Paul Krassa for example married a Russian General's daughter, who was very indignant that she married a Jew. The Habers and the Le Rossignols were at their wedding in Dahlem, Robert noting that she cried the whole time and she was very nervous and sick. After that the couple lost contact with their parents and later left for Chile making their life there. As for the relationship between Haber and Nernst, Robert was careful ... 'I don't like to say too much ... Nernst was not a pleasant gentleman ... [at Hamburg] he told Haber all his work was wrong ...'.

After leaving Robert and Emily, the Jaenickes moved on to Swansea to meet J. E. Coates, still surviving after a career as Professor of Chemistry at University College there! Coates of course presented the Haber Memorial Lecture to the Chemical Society a few years after Haber's death in 1934, and Jaenicke was keen to be allowed access this material along with Coates' recollections, but did he mischievously make Coates aware of Robert's comments? Who knows? However, Robert subsequently received a final enthusiastic letter from Jaenicke;

October 13th 1959,

Dear Herr Doktor Le Rossignol

After an enjoyable trip through southern England and Wales my wife and I wish to return [to Germany] without delay. We are grateful to you and your gracious wife for the hospitality and the information you gave us in your home. We have perceived it as a particular advantage, to have had a personal acquaintance with one of the closest employees of Haber in both a social and academic climate. Through your recollections I feel in many respects more confident and richer than previously and I now have a clearer picture of Haber allowing me to counter some of the more extravagant claims regarding his life.

From our notes, my wife will be able to reconstruct the conversations with yourself and your wife … allowing us to make your relationship with Haber come alive again. But in the meantime, I am still interested in is the smallest things you can remember and, in this respect, it would suffice if you could record such reminiscences in the form of keywords for me so that at a later meeting, we could discuss these orally *in extenso*. [*at full length*]

Professor Coates in Swansea has also given us most valuable information and made available all of the material which he collected for his memorial lecture. We have therefore not only had a very pleasant holiday in England but thanks to everyone's benevolent under-standing of my biography we take home very much more than the memory of a pleasant holiday.

With the best regards, …

After receiving this letter Robert could have rightly assumed that his story would one day be told, and even at the age of seventy-five he could still have reasonably expected that Jaenicke's biography might be published in his lifetime. He may well have looked forward to that, but of course Jaenicke never published. Seventeen years after the meeting at *St Helier* Robert passed away, still largely unrecognised except by the few. The twists of fate that had dogged the proper recognition of his contribution to 'fixation' during his life continued; overwhelmed by the task in the late 1980s, Jaenicke presented all his work to the MPG then 'walked away'; based on his efforts, biographies of Haber began to appear; in all of these the eyes of history have never bothered to look into Haber's shadow, and although some have referred to Robert's letter of 30 August 1959,[22] Jaenicke's transcript—the last *publicly* available contribution Robert was to make to our understanding of fixation—has been overlooked. But surely, not now. Because together with Chirnside's transcript of his conversation with Robert in 1976 it provides the evidence for what we suspected all along. That *he* was the engineer that created the technology of fixation and that his relationship with Haber was closer and far more durable than biographers had previously been able or willing to recognise.

18.6   Elsie Edith Le Sueur and the Beginning of Remarkable Financial Benevolence

The visit of the Jaenickes in September 1959 must have stirred many memories for Robert. After all, fifty years had passed since his work at Karlsruhe. But more memories—those of the family—must have been recalled when a few months later his sister, Elsie Edith, died in Harrow. Elsie had married Arthur Le Sueur in 1917 and they had lived for some time at Westbourne Terrace, Jersey. However, on Arthur's death Elsie moved to Harrow and lived in a substantial house at 53 Gayton Road which ran at right angles to St Johns Road, and just 'around the corner' from the junction of the two roads was Robert at number seven. Indeed, by this time the remaining Le Rossignol siblings had all de-camped from Jersey to the UK. Herbert Sorel living at Balliol House, Putney. Elsie Edith died on Monday 16 May 1960 aged 79[23] and her remains were cremated at Breakspear Crematorium, Ruislip, on the following Friday. We know little of Robert's relationship with his sister but one can only assume it was fond as she chose to live close to Robert and Emily in Harrow in her later years, although at the time of her death the couple had already moved to Penn Road.

But after Elsie's death, with age progressing, (75 at the time) and the meeting with the Jaenickes all reminding him of his own mortality, Robert resumed the benevolence that had began with the 'Haber Foundation' at Karlsruhe[24] in 1917, which continued with his and Emily's contribution to the Red Cross and St John Fund in 1940, (Chap. 17) and with his donation the Science Museum in 1952, a benevolence in keeping both with his humanity and with that of his father Augustin. (Chap. 10) The first of these new gifts occurred less than a year after Elsie's passing when Robert completed a Deed of Charitable Trust with the University of London. (16 March 1961) Robert, described only as 'an old student of the College' in the literature provided for me,[25] gave the University the enormous sum of £50,000 to establish an Endowment Fund for 'the advancement of education generally, and in any manner and in particular at or for the students and graduates of University College in the University of London in the fields of Chemistry, Biochemistry, Chemical Engineering, Physics and Biophysics in such manner as the College Committee may from time to time think fit'. Robert's wish was that two thirds of the income from the 'Robert Le Rossignol Fund' was to be used for the departments of Chemistry, Biochemistry and Chemical Engineering and the remainder for Physics and Biophysics—although this was not an absolute direction. Clearly, through his donation Robert wished to help those less fortunate than himself, particularly the young whose path in life brought them to study the sciences, a path which could be smoothed by some financial assistance. Clause 10 of the Deed also contained the condition that 'no capital or income subject to the trusts and powers hereof shall be at any time paid or applied for the benefit of the grantor [Robert] or any wife to whom he be married', making the donation to the University absolute. Now, writing these lines more than fifty years after Robert's donation, it is difficult to grasp just how significant this sum of money was at the time, but in 1961 the average wage in the UK was between £560—£633[26]—although the author recalls some teachers at his Grammar School earning around £1000 pa—and so it was somewhere between fifty and one hundred times the contemporary average wage. Today (2014) that wage is ~£25,000, so 'do the maths', and you'll see that by any standard it was an extraordinary gift. Robert's wish was that the money be invested in 'equities' and the income generated used to support the Fund, although the University was given full discretion in such matters. The original investment schedule however was not particularly 'pc' investing as it did in, the Birmingham Small Arms Company, the British American Tobacco Company and the Distillers Company, but alongside these were the (nowadays more acceptable), British Oxygen Company, the City of London Real Property Company, the House of Fraser, Imperial Chemical Industries (ICI), the International Nickel Company of Canada, and Unigate Ltd.

Over the years there have been many demands made on the Fund—after all it was spread to support five expensive departments—but it still exists after fifty-five years and although these days it has been much degraded, it still buys the 'odd bit of equipment' for which the university is grateful. With this gift, Robert used what he described to Chirnside as 'some of his rewards' to begin to repay what he always saw as his 'debt to UCL', but there was more significant benefaction to come. This time some of it undoubtedly orchestrated by the Rev. Oscar Muspratt.

18.7   The Penn—Pennsylvania Fellowship

Muspratt was active in all matters regarding Penn and Holy Trinity. He undoubtedly achieved a great deal of good during his tenancy there, but equally such dynamic characters sometimes get carried away by the enthusiasm for their 'pet projects', and that enthusiasm is not always shared by others. And so, it was with Muspratt's determination to revive the contentious issue of the Penn-Pennsylvania link which had last surfaced in 1913;

he was absolutely committed to prove that Penn the Quaker was from the same family as the Penns of Penn Parish. Over the years he was to research and write extensively regarding what he thought was evidence for the link.[5] This in itself was really quite benign, but things started to get rather more heated when he later began to publicly involve the whole village in his plans, thereby raising the profile of what others would have preferred to remain a 'sleepy hollow'—because quite frankly, many villagers simply did not accept the link, saw his efforts as an exercise in self aggrandizement and many were concerned about the implications of 'world-wide' fame for the tiny village if that was what he wanted to achieve. Nietsche's words '… he who enters my home honours me; he who passes by pleases me …', suited most parishioners, and they were an articulate lot the Vicar's parishioners, many were journalists for newspapers or magazines—one for the New Yorker.

Their concerns were well founded, for Muspratt had previous 'form'. For example, he had always celebrated every milestone in the history of Pennsylvania in Penn church. These were often 'show cased'[5] affairs with radio and TV coverage to the 'States', with messages to and from Philadelphia and with representatives of the U.S. military or the U.S. Embassy present. Whenever the Lambeth Conference brought the Bishops of Pennsylvania or Delaware to England they were invited to Penn, and in memory of the first visit by Bishop Oliver James Hart in 1948, a window was dedicated in the Lady Chapel of Holy Trinity. In 1962 Muspratt was invited to Philadelphia as an official guest by the Bishop of Pennsylvania. He met the Governor of the State, opened a session of the Senate with prayers, made nine radio broadcasts and gave two TV interviews.[27]

Much of this it seems was tolerated by the villagers, but less than a year after his trip to Philadelphia, on 06 May 1963, Muspratt orchestrated the foundation of a charitable Trust Deed known as the Penn-Pennsylvania Fellowship in order to emphasise—as he saw it—the historical link between the Parish of Penn, and Pennsylvania. He described this Trust as 'of special and far-reaching interest which, whilst capable of dealing with any charitable purpose generally, would in particular preserve and perpetuate the historical character and associations of the ancient village of Penn'.[28] Now Muspratt always demanded the best in everything he did, and although clearly a colourful fellow he managed to attract the support of two most eminent and respected parishioners to act alongside himself as Trustees, thereby giving the Trust undeniable credibility viz, Wilfred Quixano Henriques, Vice Chairman of Clark Chapman and Co., civil engineers, and Ernest Whitley-Jones,[29] General Manager of Lloyds Bank. A year later on 12 May 1964, a Deed of Gift by Viscount Curzon and others, conveyed ownership of a plot of land behind the old Almshouses—and the Almshouses themselves—to the Trustees thereby establishing the purpose of the Trust, viz, to provide a group of buildings opposite the church to be known as the 'Penn Centre', comprising affordable accommodation ('flatlets') for the elderly, a community hall for the village and car parking facilities, all of which was to prepare Penn for the future. Once again, Muspratt's general vision for the Trust seems to have been largely tolerated. But where would the money come from and what would this new 'Centre' in a tiny ancient village look like?

It was always an intention of the Trust—as its name implies—to raise money in Pennsylvania, but that never happened. A public appeal was another option. There were also many wealthy parishioners in the Penn area and Muspratt was a persuasive character, never 'backward in coming forward' when progressing his agenda. But who would be prepared to 'throw in their lot' with this 'maverick priest'[30] and kick-start the proceedings? Well, in August 1964, the Le Rossignols 'stepped up to the mark' and provided the Trust with a 'magnificent gift'[28] of approximately £40,000 in cash and securities, in two equal parts of £20,000 each, by Robert and Emily. But what prompted this quiet couple to commit to such a potentially contentious arrangement? Confidence in the Trustees must have played a part. as did the Curzon gift, and we have already made the case for some empathy between Robert and Muspratt because of the common circumstances of their lives (Sect. 18.2). But there were other circumstances that may have drawn the couple to conditionally support Muspratt in his endeavours. One of these concerns Oscar's brother Eric, who was born in November 1899 in Essex and who travelled extensively throughout his early years, firstly with his parents and siblings in the home-made gypsy-style caravan that took the family to Jersey, then with his father on some 'hobo-esque' adventures through America, Canada and Australia. Eric led an itinerant life[31] moving through a succession of jobs and travelling around the world. By the age of 26 he had worked as a sailor, been a soldier in the First World War, a pineapple farmer in Australia of course, an odd-jobber in the United States, and a plantation manager in the Solomon Islands. Eric described his travels and adventures in a number of autobiographical books. His final autobiography *Fire of Youth*, was published the year before his death in 1949 in Concord, New South Wales from an overdose of a sleeping draught in just his fiftieth year. It was suspected at the time that he took his own life. Now who knows what passes between a parishioner and his priest. Did these men ever confide in each other regarding the circumstances of their loved-ones passing? If they did, there was never any indication to others outside their relationship, and Robert we know was notoriously reluctant to discuss his sons. But if we are looking for some explanation as to why such a conservative couple were to join with a flamboyant character like Muspratt there may be some understanding here. Even so, whatever their reasons, on 15 August 1964 an *Agreement*[32] was signed between Robert, Emily and the Trustees and the money passed to the Penn-Pennsylvania Fellowship. But unlike Robert's gift to UCL, the couples' support for the Trust was conditional, for they too had an agenda, one which was meant to be as humanitarian and equitable as possible, and it is the wording of the *Agreement* that—common circumstance accepted—provides the final clue to the reason the Le Rossignol's joined with Muspratt's Trust.

According to the *Agreement*, the whole of the 'Le Rossignol Fund' was to be used for the purpose of providing and endowing 'flatlets' on the plot of land in Penn owned by the Trustees, and for building self-contained accommodation for 'elderly persons of limited means'. These flatlets were to be made available to all suitable applicants regardless of their religious denomination or present residence. The Trustees were to allow the Distressed Gentlefolks' Aid Association[33] (the DGAA, today 'Elizabeth Finn Care'), to nominate and care for up to five occupants of the flatlets provided that the Trustees were of the opinion that their nominees would fit in with existing tenants, and the Association was to be consulted regarding any nominations of the Trustees. If the Trustees failed to implement the agreement within two years by placing a contract to build, the Fund, and any income accrued, was to be handed over to the Association to be used for their general purposes. The Le Rossignols also wanted the accommodation to be as modern and up-to-date as possible[28] and, according to a later 'flyer' sent out to anyone interested in renting the flatlets, they were to stand in memory of their sons, both of whom were 'killed in action in the Second World War whilst serving in the Royal Air Force'. The involvement of the DGAA was seen by the Le Rossignols as a critical component in the *Agreement*, allowing the Association to care for residents at the proposed flatlets, and an examination of the historical records of the DGAA[33] held by Elizabeth Finn Care explains why Robert and Emily insisted on their involvement. Indeed, with regard to the DGAA, Robert and Emily's 'men were already on the road'.

The Report of the Chairman of the Council of the DGAA for the year ending 31 March 1961 records … 'a magnificent gift of £50,000 donated by a generous supporter and his wife who live in Buckinghamshire and who have been interested in our work for a number of years …' In the Report, the Council planned to use their gift to extend or replace, 'Merlewood', their existing nursing home in Virginia Water, Surrey. Merlewood—built by leather merchant Thomas Lawrence in 1856, purchased by the DGAA in 1948 and standing in 23 acres of private grounds—was the oldest of the seven homes run by the DGAA at the time, and was regarded as something of a 'Cinderella'. But with the Le Rossignol gift the Report records that 'the Fairy Godmother has now arrived!' In the event, a decision was made to extend the home and increase its capacity from twenty-one residents to thirty-six. The supervising Architect for the build was E. Forster BA FRBIA. The Annual Report of the DGAA for 1963 includes a sketch of the new extension and the Foundation Stone was laid by Robert and Emily on 10 July 1963. By November 1964, shortly after the signing of the *Agreement* at Penn, the extension was officially opened and all thirty-six beds occupied (Plate 18.4).

Here then we can at last make sense of the Le Rossignol's engagement with Muspratt's Penn-Pennsylvania Fellowship. Certainly, common circumstance of life may have played a part, certainly too Muspratt may have been per-suasive in 'drumming up' support for his cause. But clearly Robert and Emily already had a substantial 'track record' in such benefaction and equally they saw Muspratt as an opportunity to progress *their* agenda, already begun at Merlewood. A philanthropic symbiosis if ever there was one. Robert and Emily's gift to the DGAA, the building of the extension and its official opening in November 1964 does not seem to have attracted any great interest from the national Press at the time. But conversely—and undoubtedly driven by Muspratt's 'showmanship'—05 September 1964 saw *The Times* publish the short article, 'Trust to be link with Pennsylvania', which described the general aims of the new Penn-Pennsylvania Trust, the Curzon and Le Rossignol gifts, the collaborating bodies, and naming Muspratt's cousin, the eminent Sir Hugh Casson—whose works were generally characterised by elegance, lightness of touch and sensitivity to surroundings—as consulting architect. The Le Rossignol conditions clearly obliged the Trustees to provide accommodation for the most deserving on a humanitarian rather than parochial basis. Nothing in the *Agreement* aligned itself with Muspratt's Penn-Pennsylvania link neither did it predispose any particular capacity for the flatlets. But the two-year clause meant that the Trustees and the con-sulting architect had to move quickly and plan this aspect of their Penn Centre if they wished to use the Le Rossignol gift. One cannot help feeling too that the agreement was meant to 'rein in' any tendency towards flam-boyance. A year later, on Friday 12 November 1965, after seeking the best legal advice from Lincoln's Inn regarding the Almshouses, Sir Hugh Casson presented and explained the Trust's proposals to a large gathering of parishioners. The old Almshouses were to be demolished, along with the old school already used as a Church Hall. A new hexagonal Hall was to be built, together with a 3-story block of 22 small flats and apartments with parking for up to 60 cars and there was also a recommendation for a relief road to take

Plate 18.4 Merlewood House as it was in the early 1960s and the Architect's sketch for the new Le Rossignol extension which was added to the right of the House in the photo. Although eminently suited to its purpose, by today's standards, the building seems to be a hideous 'carbuncle' grafted onto a classical building. But 1963 was an era of modernism; of Mary Quant, 'Twiggy' the Beatles and the austerity of the '50s was swept away to make room for all things new. Whatever we think of it, the extension allowed the DGAA to extend its care and it was only replaced in the late 1990s by a new single-story build. Photographs courtesy of Elizabeth Finn Care/Turn2Us

traffic away from the village.[18] However, unlike Merlewood whose modern extension was set in private grounds, the proposed new Penn Centre occupied the heart of an ancient village … and there was uproar!

Casson's plans seriously overstepped the villagers' tolerance. Putting aside the dubious historical basis upon which the plan was based, the general feeling was that it was far too ambitious a project, too expensive, too grandiose and far too disruptive.[5] In short, and despite the best efforts of the Le Rossignols, it bore all the hallmarks of the flamboyant Muspratt. One prominent opponent of the scheme described it as 'a cross between an enormous bird cage and [Casson's] elephant house at the [London] zoo, absolutely out of scale and totally out of keeping with the pretty village, the pub and the Church'.[5] The meeting with Sir Hugh was hostile, he was shouted down and the plan was abandoned. Suspicious of what might lay in the future, the *Penn and Tylers Green Society* was set up, headed by leading parishioners. Lord Howe was its President. Ironically, a Vice-President was a descendent of William Penn the Quaker and members were John Betjeman the future Poet Laureate, the actor Rupert Davies (of 'Maigret' fame), Oliver Millar the future surveyor of the Queen's pictures, and authors Alison Uttley and Elizabeth Taylor.[5] A year later, the dispute at Penn was still rumbling on and it even reached the USA when a 'Schadenfreude-like' article appeared in the *Toledo Blade* newspaper (August 1966) in Pennsylvania's neighbouring State of Ohio;

Brotherly Love Soured by William Penn Issue

English Village not anxious for fame debates project to memorialize Quaker

Residents, it reported, had written to Governor Scranton of Pennsylvania earlier in the year; 'the majority of the residents in the village do not wish to see this building proceed, as it will undoubtedly spoil the beauty of the village and there is apparently no local demand for accommodation …' laying the blame for the whole affair firmly at the feet of Penn's 'determined vicar', but no blame ever seemed to be directed towards the Le Rossignols whose motives were seen as entirely genuine. In the event, Muspratt's Penn Centre never materialized. The only part of the plan that was implemented was the demolition of the Almshouses—declared unfit for human habitation and not worthy of repair—and replaced by eight small but well-designed apartments known as 'Penn Mead'[34] flatlets, funded entirely by Robert and Emily's gift. Undoubtedly modern, but modest in scale, elegant in design and standing back from the road in a dip behind the Crown public house, the use of the local warm red brick gave it a classical appearance which blended in well with the old-world character and charm of the ancient village. And so it is today (Plate 18.5).

Plate 18.5 (Top) The old Almshouses prior to demolition with the newly constructed flatlets behind. (Below) The flatlets as they appeared in a 'flyer' sent to those who requested information regarding Penn Mead. Circa 1967. Photographs courtesy of Penn Trust

Robert and Emily's request that the flatlets should be 'modern and up-to-date'—maybe a 'throw back' to the rigours of Ruhleben—was entirely fulfilled by this later Casson design. Each unit of accommodation had a living room and a bedroom, or a living room/bedroom. There was also a kitchen, bathroom and hall. Modern materials were used throughout and there was full central heating, constant hot water, ample storage space and electricity points. The bathrooms were tiled with paneled baths and heated towel rails, and the kitchens boasted sinks with double draining boards, electric cookers and fridge/freezers. In the communal laundry room, there was an automatic washing machine, spin dryer and tumble dryer. A television point in each flatlet was connected to a communal aerial, there was sound proofing and a sophisticated security and door entry system for the added safety of the residents. And all of this in 1967 when the author of these lines lived in a house which often had ice on the inside of the bedroom windows in winter, an outside toilet, and coal fires!

18.8   The Opening Ceremony Penn Mead Flatlets

Robert and Emily's flatlets, given and endowed in memory of their sons and managed by the Distressed Gentlefolk's' Aid Association on behalf of the Penn-Pennsylvania Fellowship, were opened by Sir Henry Floyd, Lord Lieutenant of Buckinghamshire, in a special ceremony in Penn on Saturday 14 October 1967. Sir Henry's speech—preceded by a short service in the Church—summed up the purpose of the building; 'This is not an old peoples home in the ordinary sense of that term, it is a place where old people can find security, comfort and companionship'[35] He continued to describe how the new development blended into the existing character of Penn … 'I came over the other day to look at the building and I was immediately impressed by its appearance. New buildings today so often look like glass matchboxes up on end … [but] … as you come around the bend in the road you can see that this piece of development is all that is best in Old England. I think it's a very fine building.'[35] One of the Trustees, Wilfred Henriques, thanked Sir Henry and appealed to the villagers to lay the discontent of the past few years to rest; 'We earnestly hope that all of you, whether you disagreed with us at the start, will personally try to make this the success it ought to be.'[35] After the speeches, Robert Le Rossignol, standing quietly in the background throughout, presented the keys of the apartments to Sir Henry who ceremoniously unlocked the doors. The apartments were then open for inspection by the villagers accompanied by Mr. Robert Goodell, cultural Affairs officer from the American Embassy in London, invited to represent Penn's 'ancient connection with America' (Plate 18.6).

Over the years, the flatlets—expertly governed by the Fellowship's dedicated Trustees—have proved to be a resounding success. But they were built as part of a pretence viz., the need to 'future-proof' Penn, a pretence

Plate 18.6 The opening ceremony at Penn Mead flatlets. On the extreme left is the Rev. Muspratt, Earl Howe is centre delivering his speech with Sir Henry Floyd to his left. On the extreme right is Robert Le Rossignol, listening politely and holding a plate on which lay the keys to the building. (Photograph courtesy of the Bucks Free Press, Friday 10 October, 1967)

engineered largely to serve the ambitions of one man. The anticipated changes never really happened and the village is as peaceful today as it always has been. For Robert and Emily, the links with Pennsylvania and the need for a Penn Centre were probably never considerations regarding their gift. Theirs was an humanitarian act providing a precious resource where vulnerable people could feel safe and comfortable in their final years. Although they went along with Muspratt, the conditions they imposed on the Trust show that they were probably 'singing from a different hymn sheet'. Today, the links with William Penn have been forgotten, regarded by most as entirely bogus. What remains stands not in homage to the Rev. Muspratt, nor the Quaker, nor Pennsylvania, but as a testament to what decent people can do, and if Robert and Emily 'cheated' a little bit in dedicating Penn Mead to the memory of their sons who 'both died in the war', who could possibly blame them? For all the Trustees past and present, the Le Rossignols wishes in this matter stand immutable and long may Penn Mead survive to continue its work, to provide comfort and security, to remember John and Peter and to condemn the utter futility of war.

18.9   And Now … the End Is Near

When the Merlewood extension was built and Penn Mead opened for 'business' in 1967, Robert and Emily were already in their early eighties. By this time, Emily was entirely crippled by her arthritis, both wheelchair and house-bound. For Robert too the years had taken their toll and although mentally alert, remarkably forward-looking and mobile—a few villagers today can recall seeing him 'about' at the time—a hearing aid was by now the 'order of the day' and Richard and John recall a pronounced shaking of his hand, especially when pouring tea. But the couple were to live long enough yet to derive pleasure from their gifts and to see peace break out in Penn after the opening of Penn Mead, which has now served the community for almost 50 years. But, as for all of us who pass this way just the once, the inevitable came, and for Emily, it was on 26 October 1975. Robert placed a notice in *The Times* on the 28th;

> LE ROSSIGNOL—On October 26th 1975, peacefully at St Helier, Penn Road, Beaconsfield, two weeks after the 65th anniversary of our wedding day. Emily, aged 90, beloved wife of Robert Le Rossignol. Cremation at Amersham, on Friday 31st October at 1.30. No flowers or letters please.

Emily's passing was the result of decades crippled by arthritis, her immobility eventually causing heart failure and thrombosis. In this work we have been unable to present much detail of Emily's life. Like Robert perhaps, she spent her life 'in the shadows'. No photographs remain, those who knew of her at Penn remember her deformed by her condition and naturally reclusive. The photographs we have suggest that Peter Walter resembled Robert when he was young, but John Augustin appears dark and slim and maybe resembled Emily more. However, anonymity is probably precisely what Emily would have wished for. Robert described Emily as his 'beloved wife', a wife he cared for throughout her long and painful illness, and his notice in *The Times* conveys the emotion he felt at his loss. But he clearly wanted a quiet funeral, with no 'fuss', and this too was probably Emily's wish. After her cremation, on 03 November, her ashes were scattered in a dispersal area within the crematorium grounds.[36] The following month, the Penn Parish News Letter contained a brief notice of her passing.

'Life in the shadows' may have been Emily's wish, and during her lifetime that wish was respected. But we can gain some measure of this lady through her Will[37] which shows that she dispersed her estate widely amongst friends, family—in the UK and Germany—employees, and charities. Indeed, a much more vibrant picture of Emily emerges than we could ever have otherwise imagined, a lady with a wide circle of friends and a sound humanitarian foundation to her life. Emily also loved her jewelry. Specific legacies included, diamond brooches, gold rings with diamonds, emeralds, and sapphires, and a gold watch chain, all given to a wide range of friends living locally and further afield. One particularly touching gift, and the first mentioned, refers to her 'Air Force Brooch', presumably passed to a particularly close friend. Other legacies included linen and a treasured oak leaf pattern silver tea set. Pecuniary legacies too were made, especially to those who had been in her employ such as her resident maid, the daily help and the gardener. Emily probably regarded Robert's nephew, Clement Sorel, as a 'son' and he was to receive *St Helier* and all its contents other than those mentioned in specific legacies. Finally, Emily gave the residue of her estate equally to five UK charities demonstrating a wide appreciation of the needs of others viz, 'Oxfam', 'Shelter' (the National Campaign for the Homeless), The National Society for the Prevention of Cruelty to Children, Help the Aged and finally The Missions to Seamen. As a measure of her benefaction, Emily left a net estate valued at £183,265.[38]

18.10   The Visit of Dr. Ralph Chirnside, 29 March 1976

After Emily's death Robert maintained their household staff, including a gardener, a full-time cook and housekeeper. When visiting Robert, Richard especially remembers him ringing for service from the dining table with a small silver bell—undoubtedly 'Upstairs Downstairs'—but Robert was of course 'a man of his time' and such attention was quite normal for him. Shortly after Emily's passing, Robert had to face another loss, that of his elder brother 'Bertie' on the 19 January 1976 (Plate 18.7). Bertie's first wife Mai had died in December 1955 and a year later he married again, one Gladys Vaux, at St Johns Church Putney, Robert acting as witness to the union. Robert and Bertie, it seems were quite close, but with Bertie's passing Robert became the last of his line. Even so, when Ralph Chirnside arrived at *St Helier* shortly after Bertie's death, he still found Robert as optimistic and as 'full of life' as ever.

Chirnside told him that one of the aspects of Robert's disposition that never failed to impress him over all the years they had worked together was his forward-looking approach to life. Many other men his senior, he added,

Plate 18.7   Robert's brother Herbert, outside his home in London. Photograph courtesy of the Le Rossignol family

became more reactionary with age, but not Robert. Indeed, Robert—still an avid reader of *New Scientist* for example—thought that so much was better now; for the young, for morality, for social conditions etc., and that 'the best was yet to be'. But Chirnside's visit that day was not simply a social call on an old friend—but a visit on behalf of the Royal Institute of Chemistry (RIC), Chirnside being a Committee member. The Institute was to celebrate its Centenary the following year (1977) and there was a feeling throughout the Committee that it would be an appropriate occasion to refer to the contribution(s) that had been made by some of its oldest and most distinguished members. Chirnside informed the Committee that he would invite Robert's agreement and co-operation with this endeavour in an attempt to put on record the part Robert played in the production of synthetic ammonia. Chirnside had been given an 'impressive list' of members who were several years older than Robert, but none that he could recognise whose achievements were a match. Even so, he continued, 'very few British chemists today seem to know this'. Chirnside went on to recount his personal experience of contemporary references to Robert's work with Haber. About the time of the Haber Centenary (1968) there was a reference to the 'mellifluously named joint inventor of the Haber process' in an Editorial of *Chemistry and Industry* he tells us, and later, in the 9th Dunn Memorial Lecture in Newcastle, Lord

Wynne-Jones referred to some of the reactions of 'Haber and Le Rossignol', but otherwise he found that Robert's name had seldom appeared, but on the occasion of the Centenary of the RIC he wanted this interview to help put 'the record straight' and Robert receive his proper recognition.

Chirnside tape-recorded his interview with Robert that day and subsequently produced a transcript which passed firstly to Robert, then to Clement, and finally to his next of kin, Richard and John on Clement's passing in November 2014. From the brothers it was kindly loaned to the author of these lines. The original tape-recording however seems to have been lost. Chirnside's transcript begins with a pre-amble covering all that he knows of Robert, things that require clarification, and recollections of their time together as colleagues at the GEC. This pre-amble forms the structure of the subsequent interview which covers Robert's young life in Jersey, his time with Ramsay and Smiles at UCL, his work with Haber, his opinion(s) of the man, the influence of ammonia synthesis on the Great War and Haber's role—dispelling (correctly) the myth that Germany held back on war until the Haber-Bosch process could assure them a supply of explosives, Robert adding 'I doubt if the military machine was aware that ammonia could be converted into nitric acid ...' There was also some insight into Robert's financial dealings with Haber and some discussion of the role Robert and Chirnside played in developing the high-power transmitting valves for the GEC. Finally, Chirnside asks Robert of his opinions on the energy crisis and modern life. At the time, some such as Sir George Porter felt that the cheap production of hydrogen was the answer, Robert disagreed; *we need to solve the 'fusion' process'* was his reply, and the high energy requirements of Haber-Bosch too should be addressed by mimicking biological fixation (Sect. 18.12). As for Robert's reading of the *New Scientist*, Chirnside declared that he put many young chemists to shame, nowadays feeling it unfair to ask them to know about anything that fell outside their immediate daily work! Finally, Chirnside asked Robert 'How do you find the world today?', to which Robert replied;

> I have only one immediate comment. I think it is tragic and wrong that a handful of men can prevent a much larger number of people from going about their work. I would never subscribe to any organisation that sought to order me to withdraw my effort.

And, with an observation that recalled the memo, 'WASTING TIME', (Chap. 17) which criticised the 'slackers' at GEC during the war who, driven by the Trade Union 'Gods', had not 'learned the way to work hard', but also reflected the rampant industrial strife of the 1970s, Robert took his final bow.

Like the Jaenicke interview, we have already incorporated much of what Robert had to say to Chirnside at the appropriate point in the book. This interview followed seventeen years after that with the Jaenickes but nothing from the latter had yet entered the public domain. Generally, in-keeping with what he said in 1959—although there are some contradictions between the two interviews—the transcript has now become Robert's final word on his life and his role in the ammonia synthesis.

18.11   And Finally

Robert was to survive Emily for a further nine months, passing away himself at Wycombe General Hospital on the morning of 26 June in the long and fearsomely hot summer of 1976. A contributory factor was congestive heart failure. The death certificate describes one of the 'Fathers of Fixation' as simply, 'Scientist (retired)'. And so it was that, 'quiet as a domino', Robert Le Rossignol, a kindly private man, passed 'gentle into that good night'. With his executors named as Lloyds Bank, Reading, a notice appeared in *The Times* on 29 June 1976.

> **LE ROSSIGNOL**—On June 26th, after a short illness. Robert Le Rossignol, B.Sc, F.R.I.C., aged 92, of Beaconsfield and Jersey. Cremation at Amersham Crematorium.

With no record of a church service, his cremation took place on a hot sunny day on the 06 July 1976 and his ashes were scattered on 08 July in the same dispersal area as Emily's.[38] No memorials were ever erected to the couple there. Robert's funeral was attended by family members Clement, Richard, and Alfred, Richard recalling that the group traveled from central London by tube that day because Richard's car was being repaired after failing its 'MOT'. At the funeral, the official limousine was occupied by Robert's household staff, the whole being a fairly small affair attended at most by about 20 people. The service was conducted by an elderly and quietly spoken minister; no great eulogies nor oratories were offered and after the service, some, but not the minister, returned to *St Helier* for light refreshment. From Richard's description, it seems probable that the minister that day was not Oscar Muspratt and neither do we know if the modest 'congregation' included Dr. Chirnside, but most likely it did.

The brief notice in *The Times* announced the end of a quite remarkable life, but otherwise his passing was marked only modestly, firstly by a 'mention' in the Penn Parish News Letter of 03 September 1976;

BURIAL

'Lo, I am with you always'

June 4th   Arnold John Romer.

July 6th   Robert Le Rossignol.

Aug. 5th   Linda Mollie Jones.

then later by the sincere Obituary in *Chemistry in Britain* by his friend Ralph Chirnside. In the News Letter, the Rev. Muspratt wrote of the celebration of 04 July at Holy Trinity, of leading a televised procession into Westminster Abbey for a Service of Thanksgiving,[39] of his continuing efforts regarding his treasured Penn-Pennsylvania link, of the death of the wife of the Bishop of Oxford, the illness of his own wife Margaret, and the death of the Church Treasurer Arnold Romer. But other than the 'brief mention', Robert's passing was not elaborated upon. In his Will[37] drawn up just a month before his death, Robert honoured many of Emily's remaining wishes regarding her specific legacies which had not yet been discharged. Substantial pecuniary legacies were also made to family and friends (amounting to £16,000) along with any resident maid, daily help or gardener in his or his wife's employ at the time. The Council of Penn Parish Church received £200. For many of the pecuniary legacies, children were to stand in place of their parent if the parent had predeceased him, and as agreed with Emily, Robert gave *St Helier* to his nephew Clement. But the first gifts mentioned in his Will were to two friends who were to have any of the contents of his workshop they may care to select, and to one of these, Dr. Raymond Sims—a former Trustee of Penn Mead—he gave an additional gift, viz, the grandfather clock that stood on his stairs at *St Helier*; and this was given;

> *... in appreciation of his help and advice which he has so often and without stint given to me ...*

Robert's workshop was clearly important to him, but the scientist had a softer side, throughout his life collecting paintings by the celebrated Jersey artist 'Le Capelain', and his third gift was to the Societé Jersiaise of St Helier, Jersey, who received his whole collection.

Robert's Will was published by his executors in *The Times* on 07 December 1976. It was the first mentioned that day and read;

Latest Wills

Residue to establish Research fund.

Mr. Robert Le Rossignol of Beaconsfield left £141,333 net. After various bequests, he left the residue to University College London to establish a research fund to be known in memory of his wife.

Robert's last wish was that the residue of his estate was to be given to University College Hospital Medical School to establish a Research Fund to be known as the *Emily Le Rossignol Fellowship Fund.* An Award Committee structure was suggested in the Will which, 'as far as possible' he would like realised. The Committee was to use the income generated from the Fund to encourage research at a Post-Doctoral level into some aspect of the 'arthritic condition' and almost forty years after Robert's passing this Fund still exists.

This final act of benefaction in Robert's life inevitably brings us back to Chirnside's transcript, because for of all the material in the twelve pages of typewritten script, there is one illuminating sentence. On a number of occasions in this work we have quoted from Chirnside's Obituary of Robert, especially the phrase 'I hope I have been able to make use of some of my rewards to help those less fortunate'. But that is not what is recorded in the transcript. Rather it was 'I hope I have been able to make proper and prudent use of my rewards, and I have tried to help some others less fortunate …'. With the passing of their sons and with neither daughters-in-law nor grandchildren, Robert and Emily were generous with their monies within the family. But this sentence suggests that they felt they were just the custodians of their wealth, and that that it should be 'consumed', not selfishly, but for the good of the weak and vulnerable thereby mitigating the brutal circumstances by which it had been obtained. And today, anyone learning of Robert and Emily's benefaction over the years couldn't possibly disagree that their distribution of these rewards was indeed 'prudent, proper' and a credit to the humanity of the couple.

Shortly after Robert's funeral, in a letter written to Clement dated 31 July 1976 by Gordon Hope-Morley, Chairman of the Council and Hon. Treasurer of the DGAA, the Council of Management expressed their deepest sympathy to the family and their grateful thanks for the gift to Merlewood and the 'Le Rossignol Fund' at Penn, both of which would be 'a continuing commemoration' of the life of Clement's Uncle and Aunt. Prophetically, the letter went on to say that the gifts were 'proving absolutely invaluable ... particularly now ... that there is such a heavy demand on geriatric beds in the National Health wards and the conditions in the hospitals are so far from re-assuring.' Clement was to eventually sell *St Helier*, but not before he had 'cleared' the house and attended to a number of communications immediately after Robert's death. In a letter dated 09 August 1976, Clement wrote to Dr. Chirnside returning a copy of the transcript of Robert's conversation with him that he had found in the house—but retaining a copy for himself. Responding to an enquiry from the Societé Jersiaise appealing for information regarding Robert, Clement later wrote to the Coordinator of the Biographical Dictionary of Jersey in a letter dated 26 August 1976 explaining the existence of the transcript and later sending a copy of the Obituary upon which it was based. Clement was careful to preserve anything of importance regarding Robert's life; papers, letters and the like. In the event however, Robert seemed to retain very little. An old leather brief case contained a pre-print of Robert's 1936 paper on large transmitting valves (Chap. 16), there was the single page with notes on the ice machines (Chap. 15), Robert's 'filofax' with notes covering the 1930–40s, and the odd letter.

As the years passed by, the family were careful to attend to Robert's memory as they had long felt he had not been given the credit he deserved regarding the discovery of 'fixation'. Sometime after 1983—probably in 1984 —Clement acquired an A3 poster produced by the RIC entitled; *Achievements in Chemistry. Haber, Le Rossignol and Ammonia.* Probably produced for schools, the poster had photographs of both Haber and Robert (the passport photo in *Plate* 1.1) and described the *Haber* process, the nitrogen cycle, world food shortage, Agrochemicals and the RIC. As late as 1994 Clement wrote once more to the Societé Jersiaise responding to a request from them for information on Robert for their newsletter, being at the time the 110th anniversary of Robert's birth. He sent them Chirnside's Obituary, and he also recommended they acquire a copy of Clayton and Algar's history of the GEC Laboratories (Chap. 15). But the years were to take their toll on Clement. In later life he became unwell, somewhat reclusive, frequently hospitalised and latterly unable to deal with any communication. The 'baton' however was taken up by others. Around the time when the first biographies of Haber appeared, Clement's cousin Richard was contacted by a German lady on behalf of the BBC Science unit who were planning a co-production documentary on great scientists of the 20th Century—which was to include Haber. There were requests for photographs of Robert showing that from a German perspective at least he was not entirely forgotten, but in the event nothing materialised. And even after Jaenicke and Chirnside's best efforts, the duty of the family in responding to all requests for information, the appearance of the Haber biographies, a number of BBC radio programs (Chap. 13) and various articles describing Haber's life, little has appeared which has added to the public recognition of Robert Le Rossignol. And that which has appeared is often misinformed. On Clement's death in November 2014, the material collected by him passed to his next of kin. And there it remains, like a latter-day sleeping King Arthur, awaiting the call of his country. Well … perhaps.

18.12   A Lasting Legacy But a Life Forgotten and a Story Abandoned

Robert Le Rossignol's passing hardly caused a ripple of recognition amongst the British chemical community at the time, yet his ground-breaking experimentation and engineering at Karlsruhe helped change the world. Haber's theoretical contributions to the discovery of high-pressure fixation— arguably one of few levels where he was involved 'hands on'—were tested on Robert's work which in turn provided the template for a giant of industrial chemistry and the evidence to support a Third Law of Thermodynamics. Eventually, Robert's work also played a part in four fellow scientists achieving a Nobel Prize. In another life, his contribution(s) to thermionic technology laid the foundations of today's modern world of communication, and a 'spin-off' was the timely development and progression of 'radar', probably the most important technology of the second world war. Yet, his whole life seems to have been something of a paradox, because throughout all of this he has remained anonymous, Today, the author of these lines knows of just one memorial to Robert. Follow in his footsteps if you wish but you'll find nothing of Robert at 17 David Place, 7 St John's Road, *St Helier*, Caesarea Place and the Merlewood extension no longer exist, the latter being replaced to make way for a new single-story wing in the 1990s. He has gone, taking his secrets with him, leaving only a few fading memories amongst those who knew him—and a legacy.

But for us chemists, what a legacy.[40] Those interested in what fixation has achieved in the 100 or so years after the discovery of the technology need look no further than Vaclav Smil's book, *'Enriching the Earth'* (Chap. 2). And amazingly, hardly any of the chemistry involved has changed since Bosch and Mittasch scaled up Le Rossignol's pioneering engineering at Karlsruhe to develop the huge plants at Oppau and Leuna—adding just one fundamental new component, the first *heterogeneous* industrial catalyst. Today, industrial nitrogen fixation—a 'sledgehammer' mimic of one of the most gentle and fundamental reactions of nature—makes roughly the same amount of ammonia as the sum total of all the world's biological nitrogen fixation. The industry now consumes 1–2% of the world's energy output; two percent of the world's natural gas is used to make hydrogen for the process[41]; the world's largest plant in Saudi Arabia produces 3300 tonnes of ammonia day^{-1}; world-wide ~ 140 million tonnes of synthetic ammonia are produced each year and Haber-Bosch is still the only chemical process that uses nitrogen as a feedstock. Over the years, skilful engineering has made many improvements and with an efficiency of around 70%, the modern Haber-Bosch is almost at its theoretical optimum. R&D now concentrates on increasing the capacity and longevity of ammonia plants and improving the catalyst to operate at lower temperatures and pressures thereby simplifying the engineering.

But there was once a more fundamental hope, because even the best optimized ammonia plants consume vast amounts of energy by forcing nitrogen and hydrogen to combine at searing temperatures and bone-crushing pressure. Just a few weeks before he died, Robert and his friend Ralph Chirnside discussed the television film *The chemical dream* which demonstrated contemporary advances in enzyme chemistry. The implication here being that a catalyst might one day be found that emulates nitrogenase, the enzyme found in nitrogen-fixing microbes such as Rhizobia. These microbes associate themselves with bacteria in the roots of certain crops such as the legumes and gently 'unpick' di-nitrogen one bond at a time to form ammonia —*and* at ambient temperature and pressure. But progress in this direction has been difficult. Even for Mittasch's much simpler heterogeneous catalyst perfected in 1910, it took until the 1970–80s for chemists to understand its behaviour viz, 'pushing' electrons into di-nitrogen's antibonding orbitals to weaken the molecule's powerful triple bond. But the natural process has proved difficult to emulate and neither is biological fixation a paragon of efficiency; each molecule of nitrogen requires at least 16 molecules of ATP to make ammonia and the process 'wastes' a quarter of the electrons to make hydrogen gas as a by-product, so researchers today agree that the biggest challenge to Haber-Bosch will come from plant biology—i.e., genetic engineering—which will engineer cereal crops to associate with nitrogen-fixing bacteria in the soil in the same way as the legumes, thereby dramatically reducing the need for artificial fertiliser. Until that happens, it will be 'business as usual' for Haber-Bosch, and with today's plants lasting for 30 years or more, the industry is unlikely to make major changes to the iconic process in the coming decades. And by-the-way, if you fancy an ammonia plant in your garden, you can now easily order one on the internet. Just 'Google' … *buy an ammonia plant.*

With such a legacy it seems extraordinary that in 1959 Robert was moved prophetically—but correctly—to confide in Jaenicke, 'My name will not be remembered for big things … 'and wherever appropriate in this book, we have tried to suggest reasons why this was so, finding that it was largely due to a litany of confounding circumstance, orchestrated by commercial secrecy, a catastrophic world war coupled to a pre-disposition towards Victorian modesty.[42] One could also be excused for thinking that Robert fell into that all too familiar group in science—of partners in discovery being ignored by those claiming the honours.[43] But Robert's contribution was never ignored by Haber—indeed the opposite was true—but even so, no-one has ever bothered to ask why Robert Le Rossignol was so appreciated by Haber. And therein lies the final reason for his anonymity. Haber. A forgotten man after his death in 1934, Jaenicke's 30 years of research deposited at the MPG laid down a treasure trove for biographers and authors have been beguiled by this biographical 'honey-pot'. By overlooking Jaenicke's interview with Robert in 1959 *his* life was largely an empty page. But for this author, the key to fully understanding the story of 'fixation' was Haber's dislike of laboratory practice and to those of us who are similarly inclined, we have had to ask a question of the experimentation and engineering … well, if not Haber, who? All the clues of course lead to just one man and in this book, we have tried to tell the story of his life. And even though Robert may have left us and taken most of his secrets, even today there remain a few who remember him—and indeed Emily—and through their memories we get a last chance to look into the shadows on the canvas of Haber's life. And so finally, join me on a journey to Penn to find out a little more of the man who helped increase the world's population from 1 million at the turn of the last century to over 7 million today. But what would Robert have made of it? One third of us are now obese.

Notes

1. Ronald Blythe (Foreword) and Denis Judd (Editor), *The Private Diaries of Alison Uttley 1932–1971: Author of Little Grey Rabbit and Sam Pig*. (May 2009). ISBN - 10 1844680401, ISBN-13 978-1844680405. This quip is also widely quoted on 'the web'.

2. The term 'Brythonic' was derived by Celticist John Rhys, from the Welsh word '*Brython*', meaning an indigenous Briton as opposed to an Anglo-Saxon or Gael. It refers to a collection of languages, predominately old Welsh, but including Cornish, Breton, Manx and Cumbrian which were widely spoken

in Britain south of the Firth of Forth during the Iron age and the Roman occupation.

3. From; https://www.chilternsaonb.org/uploads/files/AboutTheChilterns/People_ and_History/PDFs/HolyTrinityChurchToBeaconHill.pdf.

4. The 'Quaker's vault' at Holy Trinity was constructed by Thomas Penn, one of the Quaker's sons. Only a church whose patron and vicar fully accepted the family's claim to kinship with the Penns of Penn would have allowed him the privilege of such a major and disruptive undertaking. But see also Green, *op. cit.* (note 5).

5. I am grateful to Miles Green for this information from his book *William Penn and Quaker links with Penn Parish*. published by the Penn and Tylers Green Residents Society, (2009). A catalogue record for this book is available from the British Library. ISBN 9780955579820.

6. Bruce Grant, *Arthur and Eric: An Anglo-Australian Story from the Journal of Arthur Hickman,* Heinemann, (1977). ISBN-10 0855610417 ISBN-13 978-0855610418.

7. Much of what follows here is due to three articles chronicling Muspratt's life written in the *Bucks Free Press* between 26 August 1988 and 09 September 1988, kindly passed to me by Mr. Christopher White, a close friend of the Muspratt.

8. A tradition crafted largely by Oscar's cousin Sir Max Muspratt, who—by the time Robert met Oscar—had been MP for Cheshire, Lord Mayor of Liverpool, lead engineer of the Mersey Tunnel Project, President of the Federation of British Industries and a founder member of the British industrial giant Imperial Chemical Industries (ICI). The chemistry thread indeed ran deep and profound, and Muspratt was able to attend university because of the assistance of Sir Max who expected him to return to the UK and follow in his footsteps. (Taken from an address '*Whither Penn?',* made by Muspratt on 13 November 1965 at the public launching of the Penn-Pennsylvania Trust in Holy Trinity. Courtesy of the Archives of the Trust.

9. Service No.178168 from the *London Gazette,* p. 2093, 11 April 1941.

10. From, 'Is this the time to put this sorry story to rest?', *Keighley News,* at http://www.keighleynews.co.uk/news/441799.print/.

11. Reported in the local newspaper the *Bucks Examiner,* Friday 25 March 1983.

12. Letters and documents regarding the Exhibition and Robert's engagement with it were kindly supplied by John Herrick, Rory Cook and Shani Davis of the National Science Museum through a 'freedom of information' request made by the author by email, (April 2011).

13. Staff at the time are described in, P. Morris, *The Image of Chemistry Presented by the Science Museum, London, in the 20th Century: An International Perspective,* freely available at http://www.hyle.org/journal/issues/12-2/ morris.pdf.

14. The letter bore a stamped, rather than written, signature.
15. Contacting the Museum; Smith Centre, National Science Museum, Exhibition Rd, London SW7 2DD, and National Science Museum, Blythe House, 23 Blythe Road, London, W14 0XQ.
16. From a BBC News report, 21 March 2014, in northern France.
17. Probably a classic example of 'paralysis by analysis'.
18. All the theoretical aspects had been 'put to bed' by the time '*Bestimmung*' had been published in 1908 (Chap. 6). As for the machine, the components necessary and their behaviours were well understood when construction began. There was no more theoretical work to be done - just engineering. Almost *always* the most difficult part.
19. Of those who have recounted the story of 'fixation', Patrick Coffey, (*Cathedrals of Science*. Oxford University Press, (2008)), is the only one who has recognised Le Rossignol as the *principal* project engineer at Karlsruhe, although he provides no evidence to support his view. But he incorrectly involves Kirchenbauer with the design of the high-pressure conical valves. There is of course no mention of Kirchenbauer in Roberts patent (Chap. 8), neither is there any evidence to suggest he received monies from Robert's Gebrausmuster.
20. Robert's painstaking work for the Hamburg (Chap. 5) and '*Bestimmung*' (Chap. 6) papers accurately established the temperature/pressure characteristic of the ammonia equilibrium and formed the experimental basis of what the world came to know as the '*Haber* Process'. But there is a distinction to be made between the *process* and the *technology*. The *process* is simply a concept, a specification, which tells us broadly *what* has to be achieved; ideally this specification accounts for every possible way of implementing the process and for 'fixation' it was described in the 'high pressure' and 'circulation' patents submitted jointly by both men. These suggested a way of making synthetic ammonia viz, `circulate a stoichiometric mixture of nitrogen and hydrogen at high pressure in a closed system, pass it over a catalyst, remove the ammonia formed and refresh and recycle the unused gases.` The 'energy balanced' machine built by Robert Le Rossignol at Karlsruhe was an efficient implementation of this process showing *how* it could be achieved, and it represented the creation of a model of one form of the *technology*.
21. Could it be uranium oxide? Maybe another element say iron. Or it could have been that they were aware of the difficulty with osmium—Jaenicke's description is too vague to pinpoint the oxide in question.
22. See for example Dietrich Stoltzenberg, *Fritz Haber. Chemist, Noble Laureate, German, Jew.* Chemical Heritage Press, Philadelphia, Pennsylvania, (2004), p. 72.
23. *The Times,* 'Deaths', Wednesday 18 May 1960.

24. Subsequently called the *Vereinigten Studienstiftung der Universität Karlsruhe,* (the 'United Study Foundation of the University of Karlsruhe'). See Chap. 7.

25. Kindly supplied to me by Professors Andrea Sella and Alwyn Davies, at UCL by email, (May 2009).

26. Gregory Clark, *What Were the British Earnings and Prices Then? (New Series),* Measuring Worth, (2014). https://measuringworth.com/ukearncpi/.

27. Over the years, two other visits followed. In 1969 he and the Earl and Countess Howe were invited as guests of the Mayor of Philadelphia, and in 1982 Muspratt was in the Mayor's official party on the QE2 following William Penn's journey across the Atlantic and up the Delaware river. On each visit there was a programme of events; meetings, interviews, lectures and dinner parties. He addressed the State legislature, visited the Senate and the UN and made the most of every opportunity to advance his argument that 'proved' Penn was the Quaker's ancestral home. Green, *op. cit.* (note 5).

28. From a letter in the Archives of the Trust written by Muspratt to the Charity Commission on 01 September 1965. Kindly supplied to me by former Chairman Christopher White.

29. Incidentally, Whitley-Jones, was involved in 'Operation Mincemeat' during the second world war. Wishing to invade Italy via Sicily, the Allies needed to convince the Germans that they would in fact do so via Sardinia and Greece. To achieve this, a body was floated ashore in southern Spain, carrying papers which identified him as one 'Major Martin' of the Royal Marines. By providing a letter for a financial demand (£79.19 s.2d) for the corpse Whitley-Jones helped establish his identity. He wrote the letter himself and typed it up on bank notepaper in his own office at Lloyds and personally signed it. The corpse used was that of an unfortunate Welsh tramp, Glyndwr Michael, from Aberbargoed in South Wales who died in London after swallowing rat poison. From, Denis Smyth, *Deathly Deception: The Real Story of Operation Mincemeat,* OUP, Oxford, (September 2011). ISBN-10 019960598X, ISBN-13 978-0199605989.

30. A description of Muspratt used by one of the parishioners on a visit I made to Penn in 2012. Chap. 19.

31. Sourced and adapted from https://www.austlit.edu.au/austlit/page/A42760 and http://apenguinaweek.blogspot.co.uk/2011/07/penguin-no-258-wild-oats-by-eric.html.

32. The *'Agreement'* from the Trusts archives was supplied to me by the former Chairman, Christopher White.

33. A national British charity begun in 1897 by Elizabeth Finn and her daughter to provide help and assistance to relieve the problems of old age, illness, social isolation and disability. Today it is known as Elizabeth Finn Care, the term 'gentlefolk' being dropped as it was deemed too 'Victorian', suggesting members of a particular 'class', the charity wishing to be seen as 'classless'. Today, the flatlets are managed by the Trustees on behalf of the local

authority but Elizabeth Finn Care continues to run a number of care homes for the elderly. I am grateful to Petra Gomersall and Geoffrey Roper, Legacy Officer and Compliance Project Analyst respectively at Elizabeth Finn Care, for taking the time to find the records of the Le Rossignol gift for me. (by letter and email March 2015). www.elizabethfinncare.org.uk. (Turn2us).

34. Presumably named after the trial of William Penn and William Mead in 1610 for preaching to an unlawful assembly. See Green, *op. cit.* (note 5).

35. From an article in the *Bucks Free Press*, 20 October 1967; *Security for Old Folk from £40,000 Gift. Penn Mead Housing Scheme is Launched.*

36. I am grateful to the Staff at the Chilterns Crematorium, Amersham, for providing this information from their records by email (07 March 2012). Emily's ashes were scattered in dispersal area 2.

37. Wills are of course public documents. All the Wills were obtained from, UK Documents, 29 Bunning Way, Islington, London, N7 9UP. (http://ukdocuments.com).

38. Published in *The Times,* 'Latest Wills', Tuesday 30 December 1975.

39. For the life of Dame Sybil Thorndike, apparently, a 'cousin of a cousin' of his. The televised procession however was 'inconveniently displaced' from the national News by the Israeli raid on Entebbe … he added.

40. Much of what appears in the next few paragraphs is from Mark Peplow, 'A fixation with nitrogen', *Chemistry World,* **10**, 5, 48–51, (May 2013).

41. By 'steam reforming' methane to make carbon monoxide and hydrogen.

42. Besides his 1928 paper (Chap. 8) and his conversations with Jaenicke and Chirnside, on only two other occasions has this author been able to find some evidence where Robert has freely and independently promoted his involvement with the discovery of fixation, and even then unconvincingly, viz, his donation of the tubes of ammonia to the Science Museum and one other occasion recalled in the final chapter.

43. Probably one of the most notorious is the popular re-naming of the '*Calvin Cycle*' as the *Calvin-Benson-Bassham* Cycle. Melvin Calvin, Nobel Prizewinner in 1961, deliberately, consistently and controversially ignored the contribution of his colleagues Andrew Benson and James Bassham in its discovery. The cycle traces the path of carbon assimilation in plants from atmospheric CO_2 via photosynthesis to plant products (sugars).

19

February 2012

*A personal pilgrimage to Penn to meet with some
who still remembered the Le Rossignols.*

Time passes. Listen ... Time passes.

‘First voice’, from Dylan Thomas’ ‘Under Milkwood’, 1954.

19.1 To Penn

Ralph Chirnside[1] described Robert as a *‘kindly man, with a simple uncomplicated philosophy of life …’*. Unsurprisingly then, and with seemingly scant
regard for his own historical importance, Robert left little of himself behind,
and without close family able to account intimately for the man, a trip to
Penn was in order to speak to the very few people who still remembered
Robert and Emily before these memories too were lost. Having spent most of
the previous four years researching Robert's background and early academic
work, the time was right for this change of direction. A further motivation for
the visit was the receipt of a copy of Robert's Will which recorded a gift of
£200 to the Council of Penn Parish Church.[2] I was soon able to reach Mr.
Miles Green—Clerk of the Council at the time and a significant local historian—and through him I learnt of the existence of Penn Mead Flatlets.
Subsequently I was able to speak to Mr. Christopher White, former
Chairman of the Penn-Pennsylvania Fellowship, together with a family friend
of his, both of whom remembered Robert when they were much younger.

© Springer Nature Switzerland AG 2020
D. Sheppard, *Robert Le Rossignol*, Springer Biographies,
https://doi.org/10.1007/978-3-030-29714-5_19

Over a number of telephone conversations, it became clear that there were now very few residents left that had any memories of Robert and Emily but significantly, one of Robert's beneficiaries viz, Dr. Raymond Sims—a former Trustee of Penn Mead and a neighbour of Robert—still lived on Penn Road. A few more telephone calls and together we were able to arrange a meeting for the 13 February 2012.

The morning broke bright and clear with a deep blue cloudless winter's sky. Crisp snow lay on much of the frozen ground the whole of the journey to the M4 junction at Slough. Leaving the motorway, the road took me through the town towards Beaconsfield, past a multitude of car sales rooms, burger bars, pubs, offices, storage depots, garages, bruised buildings, busy roads and befuddling forests of advertising hoardings, traffic lights and road signs. Ploughing through the weekend's drift of 'take-away tumbleweed', one recalled Betjeman's lines from his poem *Slough*,[3] and even after all this time, one could still see his point. Driving further north the environment became more suburban; newsagents, hairdressing salons, grocers, mini-markets and 'tidy' gardens replaced the hard and raw industry of Slough, then suddenly, still on the A355 to Beaconsfield, I passed through the trees and green fields into Bucks proper. Busy little Beaconsfield passed in a blink and soon I was on the 'home run' along the still leafy and affluent Penn Road to what was once Robert and Emily's *St. Helier.*

But 67 Penn Road had changed … it had become a building site. I had been pre-warned about the re-development, begun only a few weeks before I arrived, but the extent to which it had already proceeded was disappointing. The old house that Robert and Emily would have recognised was gone, razed to the ground, and on its footprint a spectacular suburban behemoth was rising, retailing at £3.25 m with six bedrooms, seven shower/bathrooms, a drawing room, sitting room/library, dining room, study, family room, kitchen/breakfast room, utility, gym, cinema, garage and gardens of ½ acre. Its modernity and opulence stood in stark contrast to what I knew must have previously been here, testament to the enduring desirability of the area and—of course—to 'progress' (Plate 19.1).

Stopping briefly in the gateway of no 67 to gaze around and to take in what the Le Rossignols must have known for 25 yrs or so, I moved a mile along the road to Penn village, arriving late morning and parking up opposite Holy Trinity where the old Almshouses once stood. Here I met with Christopher—long time local resident and my host for the day—together with Miles, 'on duty' that morning in the old school buildings which had survived the original Casson plan.

Plate 19.1   What was once the site of Robert and Emily's 'St Helier' transforming into 'Clarendon House'. Photograph courtesy of Mr. Christopher White

19.2   Holy Trinity and Penn Mead Flatlets

Over lunch in *The Crown* public house with Christopher, I was able to explain my interest in Robert and Emily and what had happened to their sons —to whose memory of course the Trust's flatlets were originally dedicated. Christopher in turn told me that he remembered Robert but hardly knew him personally, that the Trustees were completely unaware of his scientific importance. From the outset, Christopher has always been enthusiastic to engage with the Le Rossignol story. He has a deep love of the village and its history and he has been a fount of knowledge, indeed this author's 'eyes and ears' in all 'matters Penn'. Through Christopher I have been able to learn of so many aspects of life in Penn and so it was on this first meeting when he provided me with a collection of material from the Trust's Archives that he felt might help me understand the historical circumstances of the Trust and the Le Rossignol involvement. And through Christopher over our lunch that day, I was first told about the local legend that was the Rev. Oscar Muspratt. But pub lunches, like all good things, soon come to an end, and expertly led by Christopher it was time to explore Penn.

Leaving *The Crown*, we drove to the vicarage to pick up the keys to Holy Trinity church from the Rev. Mike Bisset. At the time, it was clear that I had much to learn, and the significance of what I was about to see was frankly lost to me, only becoming more meaningful as I later unpicked the story of the Le Rossignols in Penn. But here in this church, which had stood for almost 1000 years, were the echoes of controversial and quarrelsome vicars, of contentious interments, of dubious family connections, of the lives of those of high birth together with the baptisms, weddings and burials of ordinary mortals long forgotten. The church too was the place where the Rev. Muspratt practiced his art of extravagant celebrations for more than forty years and looking around, somewhere amongst these pews, on occasions, Robert Le Rossignol would have sat, seeing precisely what I saw that day. Christopher was particularly keen to show me the graveyard—in which two plots were reserved for himself and his wife(!)—and using the graveyard 'map' inside the church together with knowledge of Robert's wish to be cremated,[4] we were able to discount Holy Trinity as the resting place for his ashes. But at the time exactly 'where' these were, remained a mystery.

From the church we crossed the road to the flatlets, and walking down the same path that Robert did on that opening day forty-five years ago, they still looked as fresh, 'smart' and as 'right for Penn' as the day they were built. Reaching the main entrance Christopher navigated the security system—a feature of the original design—and we passed into the main entrance hall with its long viewing gallery looking out over the Buckinghamshire hills (Plate 19.2).

Unexpectedly, I found the hallway dominated by a portrait of the Rev. Muspratt in full ecclesiastical 'garb', instantly imposing—for *me* at least—a dimension of Anglican faith to the place and which I later discovered was maybe not in keeping with the non-denominational spirit of the Le Rossignol gift.[5] Learning too of the flamboyance of the Reverend, I later came to regard the portrait as another exercise in self aggrandizement, but nothing could be further from the truth as Christopher once again enlightened me. After the death of his wife Margaret, local resident Vera Allen kept house for him at the Vicarage. They later married and in a curious twist, the couple moved into Penn Mead flatlets together, later moving again to Morgannwg House in Brecon (mid Wales) to be closer to Vera's family. Oscar died there on 08 March 2000 being interred at Holy Trinity alongside Margaret, but later (in 2002), Vera suggested that a portrait of Oscar be hung at Penn Mead in recognition of his inspiration and as a founder. The portrait was therefore Vera's choice, and for me a classic example of *'give a dog a bad name…'*!

Plate 19.2 The tranquil residents' garden at Penn Mead flatlets today. Photograph courtesy of Mr. Christopher White

My surprise regarding the portrait also probably stemmed from my expectation that there would have been some prominent recognition of the Le Rossignols, and turning to the wall just to the right of the portrait I indeed found what I was expecting, but this time a far more modest affair. A simple grey stone plaque, where the gift of '*Mr. and Mrs. Le Rossignol*' is recorded, but almost lost amongst the names of those who opened the building and the Trustees at the time. It wasn't possible to look inside the flats themselves, these were all occupied and of course private, but it was clear from this brief visit that Robert and Emily had left a precious legacy here. And just imagine how fortunate one would consider oneself, spending your remaining days here in Penn, especially if coming from one of those areas up and down the country where others felt it best if '*friendly bombs*' should still fall (Plate 19.3).

19.3   The Memories of Dr. Sims

Leaving the flatlets, I traveled back down the B474 to an afternoon meeting with Dr Raymond Sims. I had previously made a number of telephone calls to Dr Sims to arrange a meeting with him, and he kindly agreed to 'fit-me-in'

Plate 19.3 The unmissable portrait of the Rev. Muspratt that dominates the entrance hall at Penn Mead. Robert it seems, was not only overshadowed by Haber. Photograph by the author, courtesy of Mr. Christopher White

while I was 'up' at Penn. His connection with the Le Rossignols was much stronger than others I met that day. For example, he had been a neighbour for at least ten years before Robert died, he was a Trustee[6] of the Fellowship when the flatlets opened in 1967 (Plate 19.4), he was a scientist—a mathematician and a physicist—and he was also a beneficiary in Robert's Will. Arriving at Dr Sims' home at the agreed time, I was taken to a first-floor sitting room to meet with him. In his nineties, but alert and most welcoming, he began 'Well … what do you want to know?' which I took as my 'cue' to provide him with what I knew—and didn't know—of Robert's life at the time. Having discovered that Dr Sims had a scientific background, I had previously sent him a copy of the English version of the 1913 Haber-Le Rossignol paper[7] and this formed the basis of my discussion of the discovery of 'fixation'. To emphasise the importance of Haber, I brought a number of biographies[8] with me, and kneeling on the floor besides Dr. Sims[9] I took these from my back-pack and laid them out for him to pore over. After looking at the photographs for a while, he was quite unequivocal about Haber 'Unpleasant looking man! …'. Fritz was clearly off to a bad start, but my first

Plate 19.4 In contrast, the stone plaque that records Robert and Emily's gift. Photograph by the author, courtesy of Mr. Christopher White

question was quite obvious … 'How much did Robert tell you of his work with Haber in Germany?'.

It appears that the two men had discussed the German episode but not in any great depth, and only on the odd occasion. Dr. Sims recalled that Robert told him that he had worked alongside a 'senior German' at the time, but he revealed little regarding his role in building the high-pressure machine. Except, that finding a suitable catalyst proved a particularly difficult problem. Indeed, following on from Robert's discussions with Jaenicke in 1959,[10] where—contrary to his 1928 paper[11]—he declared that he really had 'no problems' building the machine, it now seems clear that the catalyst—not the engineering—was the major stumbling block for 'fixation', and understandably so because neither Haber nor Le Rossignol were experts in such matters and both were fixated on pure substances as the only contenders for the role. Robert's opinion that the catalyst was the major problem becomes more credible when we realise that he was recalling events for Dr. Sims from some fifty years earlier, but it was this problem alone that he chose to share. But here it seems we reach the end of the line regarding the detail of the creation of the technology, for Dr. Sims could recall Robert telling him nothing more on the matter.

I therefore moved to ask Dr. Sims about Robert and Emily. He said that he 'got on very well with Robert', but only knew him well in the last two years of his life, adding, 'a lovely man, I liked him a lot …'. He described Robert as 'devoted' to Emily who was confined to her wheelchair, physically 'distorted' because of her severe arthritis, 'totally uncommunicative', never involved in anything, and a recluse.[12] Dr Sims said that he wasn't on 'first name' familiar terms with Emily, and never heard her speak and so he didn't know if her English was good, or bad, or whether she spoke it at all!

I asked if he had any photographs of himself with Robert, but there were none, and so I moved to ask what he knew of their sons. He told me that Robert never mentioned them nor discussed their lives, although of course as a Trustee he knew that they were 'both killed in the war'. However, I was interested to see if there was any resemblance between John Augustin and Emily, and so I showed him the RAF group photograph of 51 OTU (Plate 17.9) and pointed John out. He looked at the photo for some time, but concluded that because she was so deformed, and of course by the time he met her quite old too, it was difficult to say, although there was some possible likeness. I then explained as sensitivity as I could, what had really happened to Peter. He was very surprised as nothing regarding Peter had ever become general—or even private—knowledge in Penn during the couple's time there. I was also able to explain how the news of Peter's death was reported widely at the time and I gave him a copy of the story published in the Daily Mirror (Plate 16.4). He spent some time reading the article and then, clearly moved and slowly shaking his head, he said softly 'why do they do it?'. We agreed that this must have been such a difficult time for Robert and Emily and I wondered if they might have drawn some comfort from their religious beliefs, and so I asked if Robert was a devout man? 'No more than me! ...' was his firm, immediate reply, and knowing from others in Penn that Robert was 'not an active parishioner', and that the Agreement (Chap. 18) between the Le Rossignols and the Fellowship was made 'regardless of [the tenants'] religious denomination', we have to conclude that humanitarian considerations probably played a much more important role in the couples' life. As for Clement Sorel, Robert's nephew, Dr Sims knew nothing at all.

But the question of faith—or the lack of it—drew me quite naturally to ask what he knew of their relationship with the Rev. Muspratt. Having just left Christopher, a close family friend of the Muspratt's, and at the time quite unaware of Oscar's 'Marmite' nature, his reply took me aback. Nasty man! ... 'he scowled, his opinion being that Muspratt was cunning and disingenuous, regarding the Le Rossignols simply as 'a source of funding'. But I suspect that Robert and Emily were canny enough to see through all of that, and they in their turn used Muspratt's enthusiasm to further *their* agenda. But Muspratt is not here to defend himself so we have to leave all of this at the level of petty bickering.

Moving away from contentious issues, I asked about Robert's workshop. He told me that towards the end of his life he was not particularly active there, that the workshop itself was 'modest' and after Robert's death he did not take advantage of the opportunity bequeathed him in his Will (Chap. 18). I told him that Chirnside and family members describes him making a 'clock' in his workshop, one which kept remarkably good time.[1] Did he know anything of that? He said that he must have been referring to the 'grandmother' clock Robert made, and that 'it was made for UCL', but he had no idea what had happened to it. And so, I asked, 'the clock that you inherited from Robert—the one that stood on the stairs in *St Helier*—was a different clock?' Indeed, it was he said—a grandfather clock of course. But to this day he had no idea why Robert gave it to him, and even when I read the Will to him, he was surprised that Robert had regarded his 'advice' so highly. But Dr Sims still had the clock, and though he couldn't enlighten me at all as to its provenance, I was fortunate enough to be allowed to take a photograph of it in the dining room. But there remains some confusion here, because UCL had no recollection of a 'grandmother clock', whilst as children, John and Richard, recall being impressed with the clock Robert had hand made from spare bits and pieces (Chap. 18), but they said that the clock in Dr Sims possession bears no resemblance to the one Robert built. The whereabouts of Robert's hand-built clock therefore remains a mystery (Plate 19.5).

Standing and listening to the heartbeat of Robert's clock, one was indeed aware that 'time passes …' and the need to collect as many of Dr Sims' memories as possible. Returning to the sitting room I finally asked about Robert and Emily's deaths. He remembered little of Emily, but recalled that Robert was [*naturally*] lonely after she died. Emily's death of course was followed a few months later by Robert's elder brother Bertie and he said Robert realised that he had become the last of his line. At Christmas-time in 1975, Dr Sims and his wife Joyce invited Robert to lunch and he proved a

Plate 19.5   Robert Le Rossignol's grandfather clock from the stairs at 'St Helier', still keeping perfect time. Photograph by the author, courtesy of Dr Raymond Sims

most amiable guest, so when Robert passed away just six months later Dr Sims was surprised. Robert's illness was quite short, Dr Sims didn't even know that he had been taken to Wycombe hospital, and when he 'popped over' to *St. Helier* the morning of Robert's death, a nurse or carer[13] there told him that he had died a just few hours earlier. Dr Sims had no recollection of the funeral, or where it was held, he didn't go, and as far as he was aware neither did anyone else.

But there was just one more question—an opinion really—and so I asked … 'Dr Sims, from what I know of Robert, I'm not really sure he would have approved of what I'm doing'. Sitting bolt upright in his chair and facing me square on, he said of his friend, 'Well … if he didn't, he would have told you! …'. But then, slowly settling back, smiling wryly and winking, he added … 'but nicely …'. On that note I left, and three years later I completed my story of Robert Le Rossignol, his life with Haber, his triumphs and his tragedies, and his forgotten benevolence.

And Robert, in my life, I've only ever had a few humble dreams, writing this book was one of them, and by fulfilling my dream I only hope that I have told your story honestly, that I have in some small way helped to give you the recognition you should have had in your lifetime, and that along the way, I haven't hurt anyone.

And after all, isn't that the best kind of dream?

Notes

1. R. C. Chirnside, 'Robert Le Rossignol, 1884–1976', *Chemistry in Britain*, **13**, 269–271, (1977).
2. Wills are of course public documents. All the Wills in this work were obtained from, UK Documents, 29 Bunning Way, Islington, London, N7 9UP. (http://ukdocuments.com).
3. '*Slough*', is a poem from Betjeman's 1937 collection, *Continual Dew*. In it he bemoans the hundreds of factories, the 'trading estates' and the dumped surplus war materials that sprang up there just after the second world war. He exhorts; '*Come, friendly bombs, and fall on Slough! It isn't fit for humans now, there isn't grass to graze a cow. Swarm over, Death! …*' an observation not really targeted at Slough in particular, but more at the march of 'modernity' into our lives.
4. Expressed in his Will.
5. Although as Christopher pointed out to me, the 'non-denominational requirement' strictly applied to potential tenants and was not intended to cover Trustees or Management. The fact that Muspratt was an Anglican had

no bearing on the Trustees decision. Had he been any other religion, Vera's choice of portrait would have been approved.

6. Replacing Sir Kenneth Peppiatt on 21 May 1966, previously Chief Cashier of the Bank of England (1934–1949). He in turn had replaced an original Trustee, Ernest Whitley Jones. During his time as Trustee, Dr Sims was Director General of the National Coal Board (NCB).

7. *The Production of Synthetic Ammonia*, Journal of Industrial and Engineering Chemistry, **5**, 328-321, (1913).

8. Those by Charles, Stoltzenberg and Sollozi-Janze. See notes, Chap. 8.

9. As someone raised in a South Wales pit village in the 1950s and living through the miners strike of 1984, I'm not sure how some of the 'old guys' would have taken to my kneeling before the Director General of the National Coal Board!

10. Jaenicke's and Robert's letters from this period and the transcript of their conversation are held at the Max Planck Gesellschaft, Archiv der MPG, Va. Abt., Rep. 0005, Fritz Haber. Haber Sammlung von Joh. Jaenicke. Nr. 253 (Jaenicke's letters) and Nr. 1496 (conversation transcript).

11. R. Le Rossignol, 'Zur Geschichte der Herstellung des synthetischen Ammoniaks', (The history of the manufacture of synthetic ammonia), *Naturwissenschaften*, **16**, 50, 1070–1071, (December 1928).

12. But of course, given the right conditions) we know that Emily could be courteous and charming (Chap. 18).

13. Death Certificates show that Emily's death was reported by Robert. However, Robert's death was reported by one, Ellen Elizabeth Harman. The same Ms Harman was also left £1000 in Emily's Will, and so she was probably the nurse or carer Dr Sims met that morning.

Appendix A: Heat Capacity and Enthalpy

The Influence of Heat Capacity When Determining the Variation of the Equilibrium Position with Temperature.

Le Chatelier's intervention in the formal world of the mathematics of thermodynamics may be surprising to readers who, familiar with the standard textbook treatment of his Principle, use Le Chatelier as an example of a qualitative or empirical treatment of equilibrium. However, a more thorough look at his life and career reveals a man who consistently integrated theory with practice. Although the following argument is not directly attributable to Le Chatelier, it encompasses his understanding and his concerns.

The heat capacity of a substance is a measure of the heat energy required to increase the temperature of *a unit quantity* of that substance by a *unit degree*. In chemistry we refer to *molar* heat capacities which represent the amount of heat we have to supply to one mole of the substance to raise its temperature by one degree Kelvin. The symbol used to represent molar heat capacity is C. Molar heat capacity can be measured at constant volume C_v (where no *work* is done) or at constant pressure C_p (where the heat absorbed is used to raise the temperature *and* perform work). Measuring the heat capacity at constant volume can be prohibitively difficult for liquids and solids. Even small temperature changes typically require large pressures to maintain a liquid or solid at constant volume implying the containing vessel must be nearly rigid or at least very strong. Instead it is easier to measure the heat capacity at constant pressure (allowing the material to expand or contract as it wishes). For chemists therefore the more usual measurement is C_p.

© Springer Nature Switzerland AG 2020
D. Sheppard, *Robert Le Rossignol*, Springer Biographies,
https://doi.org/10.1007/978-3-030-29714-5

As an example, the molar heat capacity[1] (C_p) of liquid water at 25 °C and atmospheric pressure is 75.3 J K^{-1} mol^{-1}. Under these conditions therefore we have to supply 75.3 J of heat energy to raise the temperature of 18 g of water at 25 °C and normal atmospheric pressure by one degree K. For some increase in temperature ΔT, a substance with a molar heat capacity of C_p Joules K^{-1} mol^{-1} will absorb Q joules of energy according to the equation;

$$Q = n\, C_p\, \Delta T$$

where n is the number of moles involved. On cooling through the same temperature range, Q Joules of heat are lost.

Heat capacities at constant pressure may also vary with temperature. Returning to the water example, although the molar heat capacity C_p for (liquid) water at 25 °C and one atmosphere pressure is 75.3 J K^{-1} mol^{-1}, at 100 °C (steam) it is 37.47 J K^{-1} mol^{-1} and at -10 °C (ice) it is 38.09 K^{-1} mol^{-1}. Why *these* particular values are so different is not important to us but as chemists we have to realise that the heat change accompanying any process, chemical or physical, generally varies with temperature. As a rule, when the temperature rises, additional degrees of freedom may become available to substances. Besides translation, vibrations and rotations may be excited and these too are capable of storing heat energy so that in general, the heat capacity of a substance varies with temperature usually following an equation of the type;

$$C_p = a + bT + cT^2 + dT^3 + \cdots \tag{A.1}$$

$a, b, c, d \ldots$ being constants for the given substance. The water example concerns a physical change but during Haber's time, thermodynamics was sufficiently well advanced to develop the following argument for chemical reactions.

Consider a simple chemical reaction represented by the equation A → B. Suppose we wish to start with reactant at temperature T_1 and proceed to product at T_2. Two routes are possible

Route 1: Allow the reaction to occur at T_1 with a heat of reaction equal to ΔH_1. The temperature of the product is now raised from T_1 to T_2 i.e. ΔT. If we have just one mole of A ($n = 1$) then the total heat change is;

$$\Delta H_1 + (C_p)_B \Delta T$$

where $(C_p)_B$ is the molar heat capacity of the product.

---

[1]Here we use Joules but in Haber's day of course this would have been calories deg^{-1} mol^{-1}.

Route 2: The temperature of the reactant is raised from T_1 to T_2 i.e. ΔT. Allow the reaction to occur at T_2 with a heat of reaction equal to ΔH_2. If we have just one mole of A then the total heat change is;

$$\Delta H_2 + (C_p)_A \Delta T$$

where $(C_p)_A$ is the molar heat capacity of the reactant. The two routes have the same initial state and the same final state. Therefore the heat changes involved must be equal. So,

$$\Delta H_2 + (C_p)_A \Delta T = \Delta H_1 + (C_p)_B \Delta T$$

$$(\Delta H_2 - \Delta H_1)/\Delta T = (C_p)_B - (C_p)_A = \Delta C_p$$

and where ΔC_p clearly represents the difference in the heat capacities between the product and the reactant. If the temperature difference is quite small we may write this equation as;

$$(\partial(\Delta H)/\partial T)_p = \Delta C_p$$

where ΔH represents the heat of reaction at constant pressure. This relationship was first developed by G. R. Kirchhoff and relates the variation of the heat of reaction with temperature at constant pressure to the change in heat capacity accompanying the process.

Integration of the Kirchhoff equation between temperatures 0 and T gives;

$$\Delta H - \Delta H_0 = \int_0^T \Delta C p dT$$

or

$$\Delta H = \Delta H_0 + \int_0^T \Delta C p dT \tag{A.2}$$

ΔH being the change in heat content at temperature T and ΔH_0—an integration constant—is equal to some hypothetical value of the change at absolute zero. To evaluate this integral, we have to express ΔC_p as a function of temperature and given Eq. A.1 we have the means to do this. To show how the integrated form of the Kirchhoff equation allowed Haber's generation to calculate ΔH at any required temperature we examine the obvious example of the ammonia reaction.

$$N_{2(g)} + 3H_{2(g)} = 2NH_{3(g)}$$

where the relevant molar heat capacities[2] of the gaseous reactants and products expressed as a function of temperature according to A.1 are;

$$N_2 \quad C_p = 6.5 + 10^{-3}T \text{ cal deg}^{-1}$$
$$H_2 \quad C_p = 6.5 + 9 \times 10^{-4}T \text{ cal deg}^{-1}$$
$$NH_3 \, C_p = 8.04 + 7 \times 10^{-4}T + 5.1 \times 10^{-6}T^2 \text{ cal deg}^{-1}$$

Using these values, and given that for the formation of *two* moles of ammonia,

$$\Delta C_p = 2(C_p)_{NH3} - \left[(C_p)_{N2} + 3(C_p)_{H2} \right]$$

we have,

$$\Delta C_p = -9.92 - 2.3 \times 10^{-3}T + 10.2 \times 10^{-6}T^2$$

$$\therefore \int_0^T \Delta CpdT = -9.92T - 1.15 \times 10^{-3}T^2 + 3.4 \times 10^{-6}T^3$$

Inserting this result into the integrated form of the Kirchhoff equation in A.2 gives,

$$\Delta H = \Delta H_0 - 9.92T - 1.15 \times 10^{-3}T^2 + 3.4 \times 10^{-6}T^3$$

If ΔH is known at any one temperature T, then it is possible to calculate the integration constant ΔH_0 (again for the formation of *two* moles of ammonia) which in this case turns out to be $-19,000$ cal (or $-9,500$ cal mol^{-1}), so that for the ammonia reaction we have,

$$\Delta H = -19,000 - 9.92T - 1.15 \times 10^{-3}T^2 + 3.4 \times 10^{-6}T^3 \text{cal}$$

enabling ΔH at *any* required temperature to be calculated from simple heat measurements. Using this equation at a temperature of 298 K gives an answer of $-21,968$ cal or $-10,984$ cal mol^{-1}. Converting to Joules we have

---

[2]These figures/approach are taken from Samuel Glasstone's, *Textbook of Physical Chemistry*, 2nd Ed, Macmillan, 1966. Now in calories $\text{deg}^{-1} \text{ mole}^{-1}$ just to be historically sympathetic.

$\Delta H = -46.13$ kJ mole^{-1}, remarkably close to the currently accepted value[3] for the standard heat of formation of ammonia of -46.2 kJ mol^{-1}.

Heat capacity therefore had a powerful influence on enthalpy change but through the van't Hoff equation,

$$(\partial \log_e K / \partial T)_p = \Delta H^\theta / RT$$

and its subsequent integration between temperatures T_1 and T_2,

$$\log_e (K_{p2}/K_{p1}) = -\Delta H/R(1/T_2 - 1/T_1)$$

heat capacity change could also be related to equilibrium constants thereby fully vindicating Le Chatelier's concerns.[4]

---

[3]*Chemistry Data Book*, J. G. Stark, H .G. Wallace, M. L. McGlashan, 2nd Edition in SI, John Murray Publishers, 1982. ISBN 0-7195-3951-X.
[4]Where here ΔH now represents the *average* enthalpy change over the two temperatures.

Appendix B: The Third Law in Action; Thermodynamics of the *n*-Butane to *iso*-Butane Isomerism

The Use of Heat Data to Determine the Equilibrium Constant for the Isomeric Conversion from n–Butane to Iso–Butane.

The thermodynamic data and modus operandi used here are due to Everdell, note 47, Chap. 2. There are minor differences from the modern values but not enough to spoil the argument. Once again calories are used to maintain historical perspective. As always, 1 cal = 4.184 J and R becomes 1.987 cal deg^{-1} mol^{-1}.

ΔH^{θ} for the process,

$$CH_3 \cdot CH_2 \cdot CH_2 \cdot CH_{3(g)} \rightarrow CH_3 \cdot CH(CH_3) \cdot CH_{3(g)}$$

can be determined from the standard heats of combustion of *n*-butane and *iso*-butane and gives the result ΔH^{θ} = −1630 cal. To determine ΔS^{θ} for the process however, we need to know how the molar heat capacities of *n*-butane and *iso*-butane vary with temperature, where this is not known how they can be estimated, and all the relevant molar latent heats. From Appendix A we know that,

$$C_p = a + bT + cT^2 + \cdots$$

where the constants *a, b, c, ..* are determined by experiment. The quantity of heat required to raise the temperature (of one mole of substance) from temperature T to $T + dT$ at constant pressure therefore equals $C_p dT$. The increase in entropy of the system during a *reversible* rise in temperature from T_1 to T_2 is therefore is given by the expression;

© Springer Nature Switzerland AG 2020
D. Sheppard, *Robert Le Rossignol*, Springer Biographies,
https://doi.org/10.1007/978-3-030-29714-5

$$\Delta S = \int_{T1}^{T2} (Cp/T)dT$$

Because the heat capacity varies with temperature, substitution for C_p gives,

$$\Delta S = \int_{T1}^{T2} \left(\frac{a}{T} + b + cT + \cdots\right) dT$$

which integrates to give us,

$$\Delta S = a\ln(T_2/T_1) + b(T_2 - T_1) + c/2(T_2^2 - T_1^2) + \cdots$$

so that ΔS between two temperatures T_2 and T_1 can be calculated provided a, b, c, ... are known. The entropy changes during fusion, evaporation or polymorphic change are simply the latent heats divided by the temperature at which the change takes place. The remaining problem to overcome is to estimate the entropy change between absolute zero and the lowest temperature at which the heat capacity has been measured. The Debye theory of specific heats suggested that the molar thermal capacities of crystalline solids approach zero as the temperature falls to absolute zero, and over a temperature range close to absolute zero they are proportional to the cube of the absolute temperature. Therefore, close to absolute zero we have;

$$C_p = \alpha T^3$$

If we assume that this equation holds from 0-degree K to the lowest temperature at which C_p has been determined (T say), then the value of α can be calculated by making use of the value of C_p at the lowest known temperature. The entropy change over the range 0 to T is then given by,

$$\Delta S = \int_0^T (Cp/T)dT = \int_0^T \alpha T^2 dT = \alpha\left[1/3(T^3)\right]_0^T$$

Frequently, the value calculated by the Debye approximation is small in comparison to the other terms in the summation of the entropies, so that even an appreciable error leads to little effect on ΔS.

Now, returning to the specific example of the transformation of n-butane to iso-butane at one atmosphere, the lowest temperature at which the thermal capacity of solid *n*-butane has been measured is 10 K. The change in entropy from zero to 10 K is therefore calculated with the Debye equation. At 107.55 K crystalline *n*-butane undergoes a change of structure for which the molar latent heat of transformation is 494 cal. This second crystalline modification fuses at 134.89 K, the molar latent heat of fusion being 1113.7 cal. The liquid boils at 272.66 K and its corresponding molar latent heat is 5351 cal. The table below summarises the calculated entropy change (s) from 0 to 298.15 K (Table B.1).

Table B.1 Entropy changes for *n*-butane from 0 to 298.15 K

Temperature, K	Entropy calculated from	Entropy change
0–10	Debye extrapolation	0.15
10–107.55	$\int_{T1}^{T2}(Cp/T)dT$	14.534
107.55	Polymorphic change 494/107.55	4.593
107.55–134.89	$\int_{T1}^{T2}(Cp/T)dT$	4.520
134.89	Fusion 1113.7/134.89	8.255
134.89–272.66	$\int_{T1}^{T2}(Cp/T)dT$	20.203
272.66	Evaporation 5351/272.66	19.62
272.66–298.15	$\int_{T1}^{T2}(Cp/T)dT$	1.95

The *absolute* entropy of *n*-butane at 298.15 K is therefore given by the summation of the values in column 3,

$$S^{298.15}n-\text{butane}_{abs} = S^{298.15}n-\text{butane} - S^{o}n-\text{butane}$$
$$= (74 - 0) = 74\,\text{cal deg}^{-1}\,\text{mol}^{-1}.$$

The lowest temperature at which the molar thermal capacity of solid *iso*-butane at one atmosphere has been measured is 12.53 K. Only one crystalline modification is known, this melts at 113.74 K and the molar latent heat of fusion is 1085.4 cal. The liquid boils at 261.44 K and the molar latent heat of evaporation is 5089.6 cal. The table below summarises the calculated entropy change(s) from 0 to 298.15 K (Table B.2).

The absolute entropy of *iso*-butane at 298.15 K is therefore given by the summation of the values in column 3;

Table B.2 Entropy changes for *iso*-butane from 0 to 298.15 K

Temperature, K	Entropy calculated from	Entropy change
0–12.53	Debye extrapolation	0.247
12.53–113.74	$\int_{T1}^{T2}(Cp/T)dT$	16.115
113.74	Fusion 1085.4/113.74	9.543
113.74–261.44	$\int_{T1}^{T2}(Cp/T)dT$	22.030
261.44	Evaporation 5089.6/261.44	19.468
261.44–298.15	$\int_{T1}^{T2}(Cp/T)dT$	2.910

$$S^{298.15}iso-\text{butane}_{abs} = S^{298.15}iso-\text{butane} - S^0 iso-\text{butane}$$
$$= 70.4 - 0 = 70.4\,\text{cal deg}^{-1}\,\text{mol}^{-1}$$

The entropy change for the isomeric transformation of one mole of *n*-butane to *iso*-butane at one atmosphere and 298.15 K is therefore given by

$$\Delta S^{\theta} = S^{298.15}iso-\text{butane}_{abs} - S^{298.15}n-\text{butane}_{abs}$$
$$= 70.4 - 74.0 = -3.6\,\text{cal deg}^{-1}\text{mol}^{-1}.$$

Another way to look at this is to consider the *reversible* path from *n*-butane to *iso*-butane *via* absolute zero. Firstly, cool one mole of *n*-butane at atmospheric pressure from 298.15 K to absolute zero. The entropy change is -74.0 cal deg^{-1} mol^{-1}. Perform the transformation from *n*-butane to *iso*-butane at absolute zero. If both reactant and product are perfect crystalline solids then $\Delta S^0 = 0$. Now warm the *iso*-butane to 298.15 K, the entropy change being 70.4 cal deg^{-1} mol^{-1}. ΔS^{θ} is the sum of these heat changes, i.e.

$$\Delta S^{\theta} = -74.0 + 0 + 70.4 = -3.6\,\text{cal deg}^{-1}\text{mol}^{-1}.$$

Given that;

$$\Delta G^{\theta} = \Delta H^{\theta} - T\Delta S^{\theta}$$

substitution allows us to calculate the standard free energy change as,

$$\Delta G^{\theta} = -1630 - (298.15 \times (-3.60)) = -556.66\,\text{cal mol}^{-1}$$

and since $-\Delta G^{\ominus} = RT\ln K_p$, we conclude that $K_p = 2.56$. Tiresome as this note may have been to some readers, determining the equilibrium constant in this fashion is much simpler than through experiment!

Appendix C: The Nernst Approximation Formula

C.1 The Nernst Heat Theorem Applied to Gases.

[5]The 'Heat Theorem' strictly only applied to reactions between perfectly crystalline *solids* at absolute zero. Nernst however, found a way to extend its conclusions to gaseous systems—albeit in an approximate form. To avoid mathematical complexity, we consider the simple gaseous reaction below,

$$A_{(g)} \leftrightharpoons B_{(g)}$$

together with the related processes of sublimation and the solid phase reaction viz.,

$$A_{(s)} \leftrightharpoons A_{(g)}$$
$$B_{(s)} \leftrightharpoons B_{(g)} \text{ and}$$
$$A_{(s)} \rightarrow B_{(s)}$$

By looking at the relationships between the thermodynamics of these processes, Nernst was able to apply generalisations from his 'Heat Theorem' to gas reactions.

He looked firstly at the sublimation process in terms of the vapour pressure p^0 of a pure solid, the variation of which with temperature could be represented by the Clapeyron-Clausius equation,

---

[5]Based on a description by Samuel Glasstone in his, *Textbook of Physical Chemistry*, 2nd Ed, Macmillan, 1966, pp 860–864.

© Springer Nature Switzerland AG 2020
D. Sheppard, *Robert Le Rossignol*, Springer Biographies,
https://doi.org/10.1007/978-3-030-29714-5

$$d \log_e p^0 / dT = L_s / RT^2 \qquad \text{(C.1)}$$

where L_s is the molar heat of sublimation, the vapour being assumed ideal. The Kirchhoff equation (Appendix A) for the process can be written approximately as,

$$dL_s / dT = \Delta C'_p$$

where $\Delta C'_p$ is equal to the heat capacity of the vapour *minus* that of the solid at constant pressure. Integration of the Kirchhoff equation gives,

$$L_s = L_{s(0)} + \int_0^T \Delta C' p \, dT \qquad \text{(C.2)}$$

Substituting into Eq. C.1 and integrating gives,

$$\log_e p^0 = -L_{s(0)} / RT + \int dT / RT^2 \int_0^T \Delta C' p \, dT + i \qquad \text{(C.3)}$$

the integration constant i being called the '**true chemical constant**' of the given substance. The second term on the right-hand side of the equation is a product and can therefore be integrated 'by parts', so that,

$$\int dT / RT^2 \int_0^T \Delta C' p \, dT = -\left(\int_0^T \Delta C' p \, dT \right) / RT + \left(\int_0^T \Delta C' p / RT \right) dT$$

From Eq. C.2 we see that in this expression,
$\int_0^T \Delta C' p \, dT = L_s - L_{s(0)}$, so that eventually Eq. C.3 becomes,

$$\log_e p^0 = -L_s / RT + \left(\int_0^T \Delta C' p / RT \right) dT + i \qquad \text{(C.4)}$$

which describes the variation in vapour pressure of a pure solid with temperature, and for two substances A and B, there must also exist *two* 'true chemical constants', i_A and i_B so we can generalise further therefore to suggest that *every* substance has a 'true chemical constant'.

Next, Nernst considered the simple homogeneous gaseous chemical reaction;

$$A_{(g)} \leftrightharpoons B_{(g)}$$

Here, the equilibrium constant K_p is equal to p_B/p_A where the p terms are the equilibrium partial pressures. The van't Hoff equation for this process is therefore given by,

$$d \log_e K_p / dT = \Delta H^G / RT^2$$

where ΔH^G is now the change in heat content for the *gas* reaction and from Kirchhoff's equation,

$$d\Delta H^G / dT = \Delta C_p^G = C_{p(B)}^G - C_{p(A)}^G$$

the C_p^G terms being the heat capacities of the gases or vapours involved. By following the same procedure as before and by analogy with Eq. C.4, we can derive,

$$\log_e K_p = -\Delta H^G / RT + \left(\int_0^T \Delta C^G p / RT \right) dT + i_R \qquad (C.5)$$

where the integration constant i_R now applies to the gas reaction, and Nernst was intrigued to see how this constant was related (if at all) to the 'true chemical constants' of the substances involved i.e., i_A and i_B.

To this end he finally determined the free energy change for the process $A_{(s)} \rightarrow B_{(s)}$. This change may be evaluated at some arbitrary temperature T by means of the van't Hoff device for which Haber had earlier coined the term 'equilibrium box' (Chap. 2). Nernst began with the expression for the free energy change in a gaseous system where one mole of an *ideal* gas at a pressure p is transferred *isothermally* and *reversibly* to one mole of the gas at pressure p', viz.,

$$\Delta G = RT \log_e(p'/p)$$

Assume also that we have three *large* boxes all at the same arbitrary temperature T in which the equilibria $A_{(s)} \leftrightharpoons A_{(g)}$, $B_{(s)} \leftrightharpoons B_{(g)}$ and $A_{(g)} \leftrightharpoons B_{(g)}$ are all established—the last in the *absence* of solid. We assume that the boxes are so large, that the transfer of relatively small amounts of material from one to another results in no appreciable change in the partial pressures of the substances taking part in the equilibria. We further assume that the first box has a wall permeable to A, the second box has a wall permeable to B and the third box has *two* permeable walls, one permeable to A the other to B. Since all three boxes contain processes at equilibrium, ΔG is in each case equal to zero, for no net work can be done by a system at equilibrium. The partial pressures of the gaseous component are p_A^0 in box one, p_B^0 in box two then p_A and p_B in box three. We now transfer one mole of gas A at p_A^0 from box one to box three *via* the permeable walls. The final pressure of the gas is p_A and all equilibria are maintained. The free energy change for this process is therefore,

$$\Delta G = RT \log_e (p_A/p_A^0)$$

Now transfer one mole of gas B at p_B from box three to box two *via* the permeable walls. The final pressure of the gas is p_B^0 and all equilibria are maintained. The free energy change for this process is therefore,

$$\Delta G = RT \log_e (p_B^0/p_B)$$

The overall effect of these transfers is to convert one mole of A into one mole of B i.e., $A_{(s)} \rightarrow B_{(s)}$ and since the free energy change is a *state* change independent of pathway, the overall free energy change of the process is equal to the sum of the changes along the path, therefore we can write,

$$
\begin{aligned}
\Delta G^S &= RT \log_e (p_A/p_A^0) + RT \log_e (p_B^0/p_B) \text{ or} \\
&= RT \left(\log_e (p_B^0/p_A^0) - \log_e (p_B/p_A) \right) \text{ or finally} \qquad \text{(C.6)} \\
&= RT \left(\log_e (p_B^0/p_A^0) - \log_e K_p \right)
\end{aligned}
$$

where K_p is the equilibrium constant for the gaseous reaction and ΔG^S now represents the free energy change of the reaction between the solids.

Assuming the vapours to behave as ideal gases, the vapour pressures of the solids i.e. p_A^0 and p_B^0 are obtained from Eq. C.4, whereas that for K_p is given by Eq. C.5, and if we insert these into Eq. C.6 we are able to develop the following form for ΔG^S, viz.,

$$
\begin{aligned}
\Delta G^S = &\left(L_{S(A)} - L_{S(B)} + \Delta H^G \right) \\
&+ RT \int_0^T \left(\left(\left(\Delta C'_{p(B)} \right) - \Delta C'_{p(A)} - \Delta C_P^G \right)/RT \right) dT \qquad \text{(C.7)} \\
&+ RT(i_B - i_A - i_R)
\end{aligned}
$$

Now, from Hess' Law we can see that the first term on the right hand side of Eq. C.7 is simply ΔH^S, the heat of reaction[6] *for* $A_{(s)} \rightarrow B_{(s)}$. Similarly, the numerator of the integrand is the difference in the heat capacities of A and B in the *solid* state i.e.—ΔC_p^S, so using these observations and by taking the constant $1/R$ outside the integral in the second term of the right-hand side of C.7, we may write,

$$\Delta G^S = \Delta H^S - T \int\limits_0^T \left(\Delta C_p^S/T\right) dT + RT(i_B - i_A - i_R) \tag{C.8}$$

Now, the term $\int_0^T \left(\Delta C_p^S/T\right) dT$ clearly represents the change in entropy for the reaction from 0 to T, which means;

$$\Delta S^S - \Delta S_O^S = \int\limits_0^T \left(\Delta C_p^S/T\right) dT \text{ or}$$

$$\Delta S^S = \int\limits_0^T \left(\Delta C_p^S/T\right) dT + \Delta S_O^S$$

where ΔS_O^S—essentially an integration constant—represents the entropy change of the reaction at absolute zero. Because of his 'Heat Theorem' however, Nernst knew that $\Delta S_O^S = 0$, so that *eqn* C.8 becomes,

$$\Delta G^S = \Delta H^S - T\Delta S^S + RT(i_B - i_A - i_R) \tag{C.9}$$

Now, because $\Delta G = \Delta H - T\Delta S$ *always* applies, the final term in Eq. C.9 must also be zero, which in turn can only be so if $(i_B - i_A - i_R) = 0$, that is $(i_B - i_A) = i_R$. Therefore, the integration constant in Eq. C.5 for the homogeneous *gaseous* reaction is simply the difference in the algebraic sums of the 'true chemical constants' of the products and reactants. Nernst defined this term as $\sum vi$, so that Eq. C.5 may finally be expressed as,

$$\log_e K_p = -\Delta H^G/RT + \left(\int\limits_0^T \Delta C_p^G/RT\right) dT + \sum vi \tag{C.10}$$

---

[6]The heat of reaction $A_{(s)} \rightarrow B_{(s)}$ is equal to the sum of the heats of reaction of the processes $A_{(s)} \rightarrow A_{(g)} + A_{(g)} \rightarrow B_{(g)} + B_{(g)} \rightarrow B_{(s)}$.

Now, the 'true chemical constants' involved may be determined from vapour pressure and heat capacity measurements by means of Eq. C.4, and so it became possible to evaluate K_p (hence ΔG) for a gas reaction from thermal and vapour pressure data alone. Although this result has been derived for a very simple reaction, it holds in more complex cases. Indeed, the conclusions hold even when the reaction is heterogeneous, $\sum vi$ only involving terms for those reactants and products involving *gases*.

C.2 The Nernst Approximation Formula

The importance of Nernst's 'Heat Theorem' when applied to gaseous systems was that he was able to use it to deduce the integration constant(s) involved. However, the application of eqn C.10 is obviously difficult since it requires data on heat capacity and vapour pressure down to the absolute zero, Nernst therefore developed a more applicable, but more approximate form for the equation. The Clapeyron-Clausius equation for the sublimation of a pure solid can be expressed in the form; $dp^0/dT = L_s/T(v^G - v^S)$ where v^G and v^S represent the volumes of the gaseous and solid states respectively, L_s is the heat of sublimation and p^0 is the vapour pressure of the pure solid. Using the *empirical* formula;

$$p^0(v^G - v^S) = RT(1 - p^0/p_C)$$

where p_c is the critical pressure[7] of the gas, substitution for (v^G-v^S) in the Clapeyron-Clausius equation gives;

$$(dp^0/dT)(1/p^0)(1 - p^0/p_C) = L_s/RT^2$$

which in turn can be written as,[8]

$$(d\log_e p^0/dT)(1 - p^0/p_C) = L_s/RT^2$$

If we replace L_s by *another* empirical expression, viz.,

$$L_s = (L_{s(0)} + aT - bT^2)(1 - p^0/p_C)$$

---

[7]The maximum temperature at which a gas can be liquefied, i.e., the temperature above which liquid *cannot* exist, is the 'critical temperature'. The pressure required to cause liquefaction at this temperature is the 'critical pressure'.

[8]$d\log_e x/dx = 1/x$, hence $d\log_e x = dx/x$, if $x = p^0$ then $d\log_e p^0 = dp^0/p^0$ so that $(dp^0/dT)(1/p^0) = d\log_e p^0/dT$.

where $L_{s(0)}$ is the heat of sublimation at absolute zero and a and b are constants for the substance, and then integrate we get,

$$\log_e p^0 = -\left(L_{s(0)}/RT\right) + (a/R)\ln T - (b/R)T + i \qquad (C.11)$$

where i should be at least approximately equal to the 'true chemical constant' for the substance involved. Indeed, Eq. C.11 is an approximate form of Eq. C.4. From measurements of the heat capacities of gases, Nernst concluded that the value 3.5 may be used for the constant a for *all* substances, so that a/R is approximately 1.75 and converting to common logarithms Eq. C.11 finally becomes;

$$\log_{10} p^0 = -L_{s(0)}/4.57RT + 1.75\log_{10} T - (b/4.57)T + I \qquad (C.12)$$

which in turn represents a more applicable form of Eq. C.4.

Now, we might reasonably expect the constant I to be equal to $i/2.303$, but because of the approximations involved it becomes a *new* empirical constant which was then called the '**conventional chemical constant**' for the substance. Using Eq. C.12, and other empirical relationships discovered to exist between I and physical properties such as boiling point, critical pressure and vapour pressure near the critical point, Nernst was able to determine values for I for various gases based on pressures in atmospheres. Some of these (from Glasstone) are shown below;

Hydrogen 1.6	Chlorine 3.1	Nitric Oxide 3.5
Methane 2.5	Carbon dioxide 3.2	Carbon monoxide 3.5
Nitrogen 2.6	Ammonia 3.3	Bromine 3.5
Oxygen 2.8	Hydrogen bromide 3.3	Water 3.6
Hydrogen chloride 3.0	Hydrogen iodide 3.4	Iodine 3.9

and apart from hydrogen they all fall between 2.5 and 4.0.

There is a clear parallel between Eqs. C.4 and C.5 in that both have a similar form. Because Nernst was able to reduce C.4 to the more applicable C.12, by analogy he reasoned that similar approximations should apply to C.5. Following the method used above, Nernst developed an expression for K_p for a gas reaction analogous to that for vapour pressure. Hence,

$$\log_{10} K_p = -\Delta H_0^G/4.57T + \sum v\,1.75\log_{10} T - (\beta/4.57)T + \sum vI \qquad (C.13)$$

where the variation of ΔH^G with temperature is given by;

$$\Delta H^G = \Delta H_0^G + \sum v\, 3.5T + \beta T^2$$

and where $\sum v$—the change in the number of molecules in the reaction—refers to *gaseous* reactants only, whilst $\sum vI$ similarly represents the change in conventional chemical constants between *gaseous* products and reactants.

Equation C.13 is admittedly approximate, but Nernst proposed to simplify it further by neglecting terms involving β *and replacing H_0^G* by ΔH, the value at ordinary temperatures, thereby finally giving the Nernst *approximation formula* for gas reactions, viz.,

$$\log_{10} K_p = -\Delta H/4.57T + \sum v\, 1.75 \log_{10} T + \sum vI \qquad (C.14)$$

Now in spite of its approximate nature, Eq. C.14 gave reasonably good results in certain cases, especially when $\sum v = 0$—probably because of the partial cancellation of errors. As an example of its use consider the 'water gas' reaction,

$$CO_2(g) + H_2(g) = CO(g) + H_2O(g)$$

for which $\sum v = 0$, $\sum v\ I = (3.5 + 3.6) - (3.2 + 1.6) = 2.3$ and $\Delta H = +10{,}000$ cal. Hence,

$$\log_{10} K_p = -10{,}000/4.57T + 2.3$$

At 1073 K, $\log_{10} K_p$ is 0.26 so that K_p is 1.82, in 'fair' agreement with the 'modern' value of 0.92.

Appendix D: The Haber(-Le Rossignol) Equations 1907–1914

The Haber—(Le Rossignol) Equations Between 1907 and 1914.

The equations used by Haber-Le Rossignol in their 1907 (note 16, Chap. 2) and 1908 (note 1, Chap. 6) papers for the reaction;

$$\frac{1}{2}N_{2(g)} + \frac{3}{2}H_{2(g)} = NH_{3(g)}$$

were based on an expression for the 'reaction energy' A, viz.,

$$A = \left(Q_0 - \sigma_p T \log_e T - \sigma' T^2\right) - RT \log_e K_p + \text{const}.T$$

which had already been developed by Haber as early as 1905 in Lecture Three of his book, '*The Thermodynamics of Technical Gas Reactions*'. This expression was subsequently applied to various gaseous equilibria—including ammonia —in Lecture 5, the Appendix to which of course was later translated by Le Rossignol. Haber's expression was developed in terms of what we would understand today as the 'work function', together with an appreciation of the importance of heat capacity. Indeed, in his book, Haber pointed out that;

> Le Chatelier, to be sure, showed long ago .. the importance of specific heats, but this work was not accorded the general notice which it deserved. The result is that our knowledge of the subject is limited and the number of cases at our disposal is scanty.

© Springer Nature Switzerland AG 2020
D. Sheppard, *Robert Le Rossignol*, Springer Biographies,
https://doi.org/10.1007/978-3-030-29714-5

The work function is a measure of the maximum (reversible) work of *all* kinds available in a given isothermal process. The symbol A, is taken from the German word for work, '*arbeit*'. In Haber's book this term was referred to as the 'reaction energy'. An enthalpy term was referred to as the 'reaction heat' and given the symbol Q, so that $Q_{p(T)}$ represents the 'reaction heat' at a given pressure and temperature. Generally, it is unnecessary to give some physical significance to any of the thermodynamic properties of a system, such as S and H. They are mathematical quantities important simply because they are state functions and because they can be related to one another by mathematical expressions. However, students often find their significance easier to grasp if they are given some kind of physical form, such as entropy in terms of the order or disorder of the system. This kind of metaphor guided Haber's thinking when trying to derive an equation for the total amount of work of all kinds available from a chemical reaction, 'energy' of course, simply being the capacity to *perform* work. In natural science Haber argued, we generally measure forces by opposing them with other quantifiable forces. Mechanical forces can be measured by opposing them with a balance and weights, electrical forces by opposite electrical force, but it is not possible to measure the 'driving force' of a chemical reaction with opposing forces of a chemical nature. In such a case, we have to relate A to other thermodynamic properties. Haber explained that when a change of state occurs in an isothermal system, measurable heat is absorbed or evolved. For vapourisation, some energy becomes 'latent', for condensation the same energy appears out of its 'latent condition'. Haber then spoke of the 'latent heat' of a chemical reaction progressing at constant temperature and volume, meaning the difference between the maximum amount of work we can get A, and the corresponding decrease in the total energy, U. decrease of total energy under const vol for gas reacts is equal to the heat evolved. And for const pres enthalpy. Defining the amount of heat 'used up', 'hidden' or 'becoming latent' as $-q$, Haber derived his fundamental equation;

$$A - U = -q$$

'Latent' of course refers to *entropy*, and so with $S = q/T$, he introduced temperature T into the equation quite simply;

$$A = U - TS \tag{1}$$

and where the reaction energy is now related to the other thermodynamic properties, S, U and T. Today, A is known as the 'work function'. Haber's main contribution to the thermodynamics of gas reactions was his consideration of heat capacity *and* entropy. Over small temperature ranges or low temperatures he argued, heat capacity and entropy changes little, but for large temperature ranges, the influence of both factors have to be considered.

Haber then proceeded by substituting, term by term, in Eq. 1. In gaseous reactions, the reaction heat, $Q_{p(T)}$, can be substituted for the *decrease* of total energy.

$$A = Q_T - TS \qquad (2)$$

At the time, Kirchoff's law (Appendix A) permitted the determination of reaction heat in terms of the expression; $Q_{p(T)} = Q_0 + \sigma'_p T + \sigma'' T^2 + \ldots$, where Q_0 represents the reaction heat at absolute zero and the remaining terms relate to differences in heat capacities, so that ignoring terms involving T^3, substitution into Eq. 2 gave;

$$A = \left(Q_0 + \sigma'_p T + \sigma'' T^2 \right) - TS$$

and where we begin to 'glimpse' the final form of the equation. This final form was developed by substituting for the entropy term which involved a consideration of the change of entropy from that of the reactants to that of the gaseous mixture. By 1914[9] a more detailed equation emerged which Haber also quoted in his Nobel acceptance speech of 1920 (note 46, Chap. 15), but all three equations (for 1907, 1908, and 1914) were simply variations of the same expression re-worked in light of the availability of more accurate heat capacity data. The work function is probably less familiar to physical chemists these days but we can develop all these equations by using a more 'modern' terminology.

The van't Hoff equation was introduced in Appendix A viz.,

$$(\partial \log_e K / \partial T)_p = \Delta H^\theta / RT^2$$

In many cases, and for ideal gases, the heat of reaction does not vary much with the partial pressures of the reactants/products and so ΔH^θ can be replaced by ΔH, the heat of reaction at any arbitrary pressure. Therefore, we can write;

$$(\partial \log_e K / \partial T)_p = \Delta H / RT^2$$

For a small temperature range, we can further assume that ΔH remains constant so that the equation can be integrated to give;

---

[9]F. Haber, *Z. Elektrochem.*, 20, 603, (1914).

$$\log_e K_p = -(\Delta H/RT) + I$$

where I, the constant of integration, can be determined if both ΔH and K_p are known at one temperature within the range to which the approximation applies. But if, like Haber, we wish to integrate the equation over a temperature range[10] so great that the variation in ΔH must be taken into account, then Kirchhoff's Law must be used to construct an expression for ΔH as a function of temperature. Equation A.2 in Appendix A shows that Kirchhoff may be expressed;

$$\Delta H = \Delta H_0 + \int_0^T \Delta Cp\, dT$$

ΔH being the change in heat content at temperature T, and ΔH_0—an integration constant—is equal to some hypothetical value of the change at absolute zero. Substitution for ΔH in van't Hoff's equation gives;

$$d\log_e K_p/dT = \Delta H_0/RT^2 + 1/RT^2 \int_0^T \Delta CpdT \qquad (D.1)$$

Now, from Eq. A.1, we know that the variation of the heat capacity with temperature at constant pressure for an ideal gas takes the form,

$$C_p = a + bT + cT^2 + \cdots$$

and the same form of expression applies to the change in heat capacity (ΔC_p) between product(s) and reactant(s) for a chemical reaction, so that the definite integral in Eq. D.1 can be expressed as,

$$\int_0^T \Delta CpdT = \int_0^T (a + bT + cT^2 + \cdots)dT$$

This evaluates to give;

$$\int_0^T \Delta CpdT = aT + (b/2)T^2 + (c/3)T^3 + \cdots$$

Substitution of the result into Eq. D.1 gives,

---

[10] Haber and Le Rossignol examined the ammonia reaction over the temperature range 700–1000 °C. By the time of the 1914 paper the range had been extended to 200–1000 °C.

$$dlog_e K_p/dT = \Delta H_0/RT^2 + 1/RT^2(aT + (b/2)T^2 + (c/3)T^3 + \cdots) \text{ or}$$
$$dlog_e K_p/dT = \Delta H_0/RT^2 + (a/RT + b/2R + cT/3R + \cdots)$$

Taking the constant $1/R$ outside the right-hand expression and forming the indefinite integral we get;

$$\int d\log_e K_p/dT = 1/R \int \Delta H_0/T^2 + a/T + b/2 + (cT)/3 + \cdots$$

Hence, on integration,

$$\log_e K_p = 1/R\left(-\Delta H_0/T + a\log_e T + bT/2 + cT^2/6 + \ldots I\right)$$

which expands to give Eq. D.2 viz.,

$$\log_e K_p = -\Delta H_0/RT + (a/R)\log_e T + (b/2R)T + (c/6R)T^2 + \cdots + I/R \quad (D.2)$$

where I, the constant of integration, is a value which can be deduced by the experimental determination of K_p at temperatures within the range to which the expression applies.[11]

Continuing now with Eq. D.2, and converting to common logarithms ($\log_e K_p = 2.303\log_{10}K_p$) then simplifying, we get the expression;

$$\log_{10} K_p = -\Delta H_0/2.303RT + (a/R)\log_{10} T + (b/4.606R)T$$
$$+ (c/13.818R)T^2 + I/2.303R \quad (D.3)$$

Using D.1 and D.3 all the equations used by Haber-Le Rossignol in their publications can be deduced.

<u>The 1907 paper</u>; F. Haber and R. Le Rossignol: '*Über das Ammoniak-Gleichgewicht*'

We can easily express Eq. D.2 in the form used by Haber here. Multiplying both sides by RT and ignoring terms involving c, (i.e., being *less* accurate in terms of the effect of heat capacity) we get;

$$RT\log_e K_p = -\Delta H_0 + aT\log_e T + (b/2)T^2 + I.T \text{ or}$$
$$0 = \left(-\Delta H_0 + aT\log_e T + (b/2)T^2\right) - RT\log_e K_p + I.T$$

---

[11]See for example page 204, 'Thermodynamics of Technical Gas Reactions' where Haber determines I for the ammonia reaction by comparing experimental and calculated results.

which in the terminology of the day, and at equilibrium where $A = 0$, is entirely equivalent to Haber's 'reaction energy' equation;

$$A = \left(Q_0 - \sigma_p T \log_e T - \sigma'T^2\right) - RT \log_e K_p + \text{const.T}$$

The term in parenthesis therefore is simply the 'heat of reaction' at absolute zero and some constant pressure P (i.e., the enthalpy change), adjusted to the temperature of the experiment by including the variation in the heat capacity between reactant(s) and product(s). Haber conventionally assigned the symbol $Q_{p(T)}$ to this term resulting in the expression;

$$0 = Q_{p(T)} - RT \log_e K_p + \text{const. } T$$

so that given $Q_{p(T)}$ and the relevant heat capacity terms, Q_0 $(-\Delta H_0)$ could be determined. Converting to common logarithms $(\log_e K_p = 2.303\log_{10} K_p)$, substituting for $Q_{p(T)}$ (using the accepted experimentally determined value of 12,000 cal *evolved* mol^{-1} of ammonia formed), then rearranging in terms of $\log_{10} K_p$ and simplifying gave;

$$\log_{10} K_p = \frac{12,000}{4.57T} + \frac{\text{const}}{4.57}$$

With the constant determined experimentally at -26.93, the equation became;

$$\log_{10} K_p = \frac{12,000}{4.57T} - 5.8927$$

The 1908 paper; F. Haber and R. Le Rossignol: '*Bestimmung des Ammoniak-Gleichgewicht unter Druck*'

In '*Bestimmung …*' Haber used the heat capacity data from Lecture 5 of his book viz.,

NH$_3$ $C_p = 6.1 + 0.004T + 7.82 \times 10^{-6}T^2 \text{ cal deg}^{-1}$

Permanent gases $C_p = 13.28 + 0.0006T \text{ cal deg}^{-1}$

where the term '*permanent gases*' refers to the heat capacity of a stoichiometric nitrogen: hydrogen mixture according to $\left(\frac{1}{2}C_pN_2 + \frac{3}{2}C_pH_2\right)$. So given that;

$$\Delta C_p = C_pNH_3 - \left(\frac{1}{2}C_pN_2 + \frac{3}{2}C_pH_2\right)$$

substitution for the various heat capacities gives;

$$\Delta C_{\mathrm{p}} = -7.18 + 0.0034T + 7.82 \times 10^{-6}T^2 \mathrm{cal\ deg}^{-1}$$

Since in general $\Delta C_{\mathrm{p}} = a + bT + cT^2 + \ldots$, we can extract the following values for the ammonia reaction;

$$a = -7.18, b = 3.4 \times 10^{-3} \text{ and } c = 7.82 \times 10^{-6}$$

With an average 'heat of reaction' $Q_{\mathrm{p(T)}}$ of $-12,000$ cal mol^{-1} within the temperature range 700–1000 °C, Haber used his 'reaction energy' equation to deduce Q_0 ($-\Delta H_0$) = 10,100 cal mol^{-1}. Using Eq. D.3 with $\Delta H_0 = -10,100$ cal mol^{-1}, $R = 1.987$ cal deg^{-1} mol^{-1}, $I = 21.98$ and a, b, c as above, we arrive at Haber's final equation viz.,

$$\log_{10} Kp = 2207/T - 3.613 \log_{10} T + 3.7 \times 10^{-4}T + 2.9 \times 10^{-7}T^2 + 4.8$$

the very small difference between this expression and that in '*Bestimmung…*' probably being due to the values of the various conversion constants in use at the time.

The 1914 paper; F. Haber, *Untersuchungen Über Ammoniak: Sieben Mitteilungen. I. Allgemeine Übersicht des Stoffes und der Ergebnisse.* Z. Elektrochem., 20, (597–604) 603, (1914).[12]

By 1914 more accurate heat data for reactants and 'permanent gases' had become available allowing Haber to determine the ammonia equilibrium over the temperature range 200–1000 °C. The figures shown below are actually taken from Samuel Glasstone's, *Textbook of Physical Chemistry*, 2nd Ed, Macmillan, 1966 and although more 'modern' they approximate very well to those used by Haber;

$$N_2 \quad C_p = 6.5 + 10^{-3}T \text{ cal deg}^{-1}$$
$$H_2 \quad C_p = 6.5 + 9 \times 10^{-4}T \text{ cal deg}^{-1}$$
$$NH_3 \quad C_p = 8.04 + 7 \times 10^{-4}T + 5.1 \times 10^{-6}T^2 \text{ cal deg}^{-1}$$

From this data and his equation for the 'reaction energy' Haber was now able to determine Q_0 ($-\Delta H_0$) = 9,500 cal mol^{-1} and $I = 9.55$ respectively. So given that;

---

[12]'Investigations on Ammonia: Seven Communications. I. A General survey of the subject and of the results.'

$$\Delta C_p = C_p NH_3 - \left(\frac{1}{2} C_p N_2 + \frac{3}{2} C_p H_2 \right)$$

substitution and reduction gave;

$$\Delta C_p = -4.96 - 1.15 \times 10^{-3} T + 5.1 \times 10^{-6} T^2 \text{cal deg}^{-1}$$

Once again $\Delta C_p = a + bT + cT^2 + \cdots$, so in light of the more accurate heat data we can now extract another set of values for the ammonia reaction viz.,

$$a = -4.96, b = -1.15 \times 10^{-3} \text{ and } c = 5.1 \times 10^{-6}$$

Using Eq. D.3 with $\Delta H_0 = -9,500$ cal mol^{-1}, $R = 1.987$ cal deg^{-1} mol^{-1}, $I = 9.55$ and a, b, c as above, we arrive at the final equation used by Haber to calculate equilibrium constants between 200 and 1000 °C viz.,

$$\log_{10} Kp = 9,500/4.571T - (4.96/1.987) \log_{10} T$$
$$- \left(0.575 \times 10^{-3}/4.571 \right) T + \left(0.85 \times 10^{-6}/4.571 \right) T^2 + 2.1$$

The small difference between this equation and that published by Haber in 1914 being due to us using rather more 'modern' values for the variables and the difference in the various conversion constants. But Haber's equation remains remarkably accurate even by more modern standards. This equation was also quoted in Coates' tribute (note 1, Chap. 2) and Haber's Nobel Acceptance speech (note 46, Chap. 2).

Appendix E: Thermionic Valves

A Simple Explanation of the Function of a Thermionic Valve.

In general terms, the simplest 'valve', the 'triode', consists of a negative *cathode* (filament), a '*grid*' through which electrons can pass, and a positive *anode* or 'plate', all housed inside an insulating glass envelope under very high vacuum. The grid is positioned between the anode and the cathode, and the anode plate—often cylindrical—encloses both. A simple diagram clarifies the basic architecture (Plate E.1).

'Primary' electrons are forced out of the cathode by heating it electrically, just like the filament in an incandescent lamp. The electron 'cloud' that forms around the cathode is attracted towards the anode because of the high voltage maintained between the two. The flow of electrons from cathode to anode (i.e., the electric current passing thru' the device) can be controlled by an electrostatic field applied to the grid(s). Thus, a change of potential at the grid will vary the flow of electrons from cathode to anode. The valve can be constructed so that a small variation of potential on the grid can cause a large variation of current between cathode and anode—this being the basis of any signal amplification by the valve. Equally, by altering the potential at the grid, electrons can be prevented from reaching the anode during the reverse cycle

© Springer Nature Switzerland AG 2020
D. Sheppard, *Robert Le Rossignol*, Springer Biographies,
https://doi.org/10.1007/978-3-030-29714-5

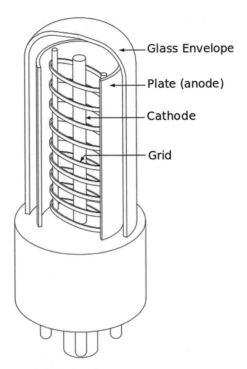

Plate E.1 Structure of a triode vacuum tube. Additional grids give rise to 'tetrode', 'pentode' ... valves etc. Diagram courtesy of https://www.commons.wikimedia.org, the free media repository. Licensed under the Creative Commons Share-Alike 2.5 Generic License

of an alternating signal providing a pulsed rectifier, and if the grid is used to allow, or prevent, electrons ever reaching the anode, we have a switch or binary storage device.

Appendix F: Robert Le Rossignol's Patents 1909–1956

Robert Le Rossignol's Patents.

This table does not include the original circulation (13.10.1908) and high pressure (14.09.1909) base patents (see Chap. 8). The original documents for all the patents can be downloaded free of charge from http://depatisnet.dpma.de/.

No.	Publication number	Application date ▲	Publication date	Applicant/Owner	Title
1	CA000000521347A		31.01.1956	M O Valve Co Ltd	[EN] Gas filled thermionic valves [FR] Soupapes Thermioniques Remplies de Gaz
2	CA000000215217A		17.01.1922	Gasgluhlicht AG Deutsche Patra Patent Treuhand	[EN] Process of Exhausting Electrical Glow Lamps [FR] Procede A Faire LE Vide Dans Lampes Electriques
3	CA000000199094A		06.04.1920	Rossignol Robert Le	[EN] Process and Apparatus for Exhausting Electrical Glow Lamps [FR]

(continued)

© Springer Nature Switzerland AG 2020
D. Sheppard, *Robert Le Rossignol*, Springer Biographies,
https://doi.org/10.1007/978-3-030-29714-5

(continued)

No.	Publication number	Application date ▲	Publication date	Applicant/Owner	Title
4	GB000190815065A	16.07.1908	01.04.1909	Rossignol Robert Le, DE	Procede ET Appareil A Faire Le Vide Dans Ampoules Electriques [En] Conical Screw Down Valve with an Angle of Inclination between 85° and 90°
5	US000001202995A	13.08.1909	31.10.1916	Basf AG, DE	[EN] Production of Ammonia
6	US000000971501A	13.08.1909	27.09.1910	Basf AG, DE	[En] Production of Ammonia
7	CA000000133527A	14.08.1909	06.06.1911	Badische Anilin and Soda Fabrik, DE Le Rossignol Robert, DE	[En] Ammonia Production [FR] Production D'Ammoniaque
8	CA000000131566A	14.08.1909	07.03.1911	Basf AG, DE Heber F, DE Le Rossignol Robert, DE	[En] Process of Producing Ammonia [FR] Procede de Production D'Ammoniaque
9	US000000999025A	18.05.1910	25.07.1911	Basf AG, DE	[En] Process of Making Ammonia
10	CA000000135875A	25.05.1910	03.10.1911	Basf AG, DE Haber Fritz, DE Le Rossignol Robert, DE	[En] Ammonia Production [FR] Production D'Ammoniac
11	US000001006206A	01.07.1910	17.10.1911	Basf AG, DE	[En] Production of Ammonia
12	CA000000135876A	04.07.1910	03.10.1911	Basf AG, DE Haber Fritz, DE Le Rossignol Robert, DE	[En] Ammonia Production [FR] Production D'Ammoniac
13	US000001298569A	30.01.1917	25.03.1919	Gen Electric, US	[En] Apparatus For Exhausting Incandescent Lamps
14	GB000000148008A	07.05.1919	29.07.1920	Robert Le Rossignol	[EN] Improvements relating to airships
15	US000001430118A	09.07.1920	26.09.1922	Patra Patent Treuhand	[EN] Process and apparatus for sealing lamps

(continued)

(continued)

No.	Publication number	Application date ▲	Publication date	Applicant/Owner	Title
16	GB000000239736A	30.10.1924	17.09.1925	Albert Charles Bartlett Gen Electric Co Ltd Herbert William Benjamin Gardi Robert Le Rossignol	[EN] Improvements in electric discharge tubes
17	GB000000257999A	09.06.1925	09.09.1926	Albert Charles Bartlett Brian Stephen Gossling Gen Electric Co Ltd Robert Le Rossignol	[EN] Improvements in thermionic valves
18	GB000000260716A	10.09.1925	11.11.1926	Gen Electric Co Ltd Robert Le Rossignol	[EN] Improvements in or relating to the construction of electric discharge tubes
19	US000001632870A	17.11.1926	21.06.1927	Gen Electric Co Ltd	[EN] Thermionic valve
20	GB000000393379A	26.02.1932	08.06.1933	Gen Electric Co Ltd Robert Le Rossignol	[EN] An improved method for modulating high frequency transmitters
21	GB000000421848A	25.10.1933	01.01.1935	Gen Electric Co Ltd Robert Le Rossignol	[EN] Improvements in or relating to electric discharge devices
22	GB000000431582A	15.01.1934	11.07.1935	Gen Electric Co Ltd Robert Le Rossignol	[EN] Improvements in or relating to electric arc converters
23	GB000000461212A	05.12.1935	12.02.1937	M O Valve Co Ltd Robert Le Rossignol	[EN] Improvements in or relating to thermionic valves
24	GB000000466830A	31.01.1936	07.06.1937	Gen Electric Co Ltd Robert Le Rossignol	[EN] Improvements in switches for converting electric currents

(continued)

(continued)

No.	Publication number	Application date ▲	Publication date	Applicant/Owner	Title
25	GB000000469481A	11.02.1936	27.07.1937	Gen Electric Co Ltd Robert Le Rossignol	[EN] Improvements in or relating to switches for converting electric currents
26	GB000000464164A	26.02.1936	13.04.1937	M O Valve Co Ltd Robert Le Rossignol	[EN] Improvements in or relating to electric discharge devices
27	GB000000476758A	10.07.1936	15.12.1937	MI O Valve Company Ltd Robert Le Rossignol	[EN] Improvements in or relating to thermionic valves and other electric discharge devices
28	GB000000474933A	15.07.1936	10.11.1937	M O Valve Co Ltd Robert Le Rossignol Stephen Michael Duke	[EN] Improvements in or relating to electric discharge devices with indirectly heated cathodes
29	GB000000470212A	17.07.1936	11.08.1937	M O Valve Co Ltd Robert Le Rossignol	[EN] Improvements in or relating to electric discharge devices with external cooled anodes
30	GB000000490385A	15.02.1937	15.08.1938	M O Valve Co Ltd Robert Le Rossignol	[EN] Improvements in demountable electric discharge devices
31	GB000000496710A	07.10.1937	05.12.1938	MI O Valve Company Ltd Robert Le Rossignol Stephen Michael Duke	[EN] Improvements in thermionic valves

(continued)

(continued)

No.	Publication number	Application date ▲	Publication date	Applicant/Owner	Title
32	GB000000497115A	23.12.1937	13.12.1938	M O Valve Co Ltd Robert Le Rossignol	[EN] Improvements in large thermionic valves
33	GB000000514651A	12.05.1938	14.11.1939	Donald Arthur Boyland M O Valve Co Ltd Robert Le Rossignol	[EN] Improvements in air-cooled thermionic valves
34	GB000000519490A	24.10.1938	28.03.1940	Gen Electric Co Ltd Robert Le Rossignol	[EN] Improvements in thermionic amplifiers
35	GB000000530296A	21.06.1939	09.12.1940	M O Valve Co Ltd Robert Le Rossignol	[EN] Improvements in the cooling of .external metal parts of electric discharge devices
36	GB000000589959A	20.10.1943	04.07.1947	M O Valve Co Ltd Robert Le Rossignol	[EN] Improvements in gas-tight joints between metals and ceramics
37	GB000000916150A	23.06.1947	23.01.1963	Gen Electric Co Ltd Robert Le Rossignol	[EN] Improvements in glands for shafts rotatable at high peripheral speeds
38	GB000000663304A	20.04.1949	19.12.1951	M O Valve Co Ltd Robert Le Rossignol	[EN] Improvements in or relating to gas-filled thermionic valves
39	GB000000666175A	06.05.1949	06.02.1952	M O Valve Co Ltd Robert Le Rossignol	[EN] Improvements in or relating to directly heated cathodes for thermionic valves

Printed in the United States
by Baker & Taylor Publisher Services